VOLUME NINETY FOUR

ADVANCES IN BOTANICAL RESEARCH

Regulation of Nitrogen-Fixing Symbioses in Legumes

ADVANCES IN BOTANICAL RESEARCH

Series Editor

Jean-Pierre Jacquot
Professor, Membre de l'Institut Universitaire de France, Unité Mixte de Recherche INRA, UHP 1136 "Interaction Arbres Microorganismes", Université de Lorraine, Faculté des Sciences, Vandoeuvre, France

VOLUME NINETY FOUR

Advances in
BOTANICAL RESEARCH
Regulation of Nitrogen-Fixing Symbioses in Legumes

Edited by

PIERRE FRENDO
INRA, CNRS, Université Côte d'Azur, ISA, France

FLORIAN FRUGIER
Institute of Plant Sciences Paris Saclay (IPS2), CNRS, INRA, Univ Paris Diderot,Univ Paris-Sud, Univ d'Evry, Université Paris-Saclay, Gif-sur-Yvette, France

CATHERINE MASSON-BOIVIN
LIPM, Université de Toulouse, INRA, CNRS, Castanet-Tolosan, France

ELSEVIER

ACADEMIC PRESS
An imprint of Elsevier

Academic Press is an imprint of Elsevier
125 London Wall, London EC2Y 5AS, United Kingdom
525 B Street, Suite 1650, San Diego, CA 92101, United States
50 Hampshire Street, 5th Floor, Cambridge, MA 02139, United States
The Boulevard, Langford Lane, Kidlington, Oxford OX5 1GB, United Kingdom

First edition 2020

Notices
Knowledge and best practice in this field are constantly changing. As new research and experience broaden our understanding, changes in research methods, professional practices, or medical treatment may become necessary.

Practitioners and researchers must always rely on their own experience and knowledge in evaluating and using any information, methods, compounds, or experiments described herein. In using such information or methods they should be mindful of their own safety and the safety of others, including parties for whom they have a professional responsibility.

To the fullest extent of the law, neither the Publisher nor the authors, contributors, or editors, assume any liability for any injury and/or damage to persons or property as a matter of products liability, negligence or otherwise, or from any use or operation of any methods, products, instructions, or ideas contained in the material herein.

ISBN: 978-0-08-102798-1
ISSN: 0065-2296

For information on all Academic Press publications visit our website at https://www.elsevier.com/books-and-journals

Publisher: Zoe Kruze
Acquisitions Editor: Sam Mahfoudh
Editorial Project Manager: Peter Llewellyn
Production Project Manager: Divya Krishna Kumar
Cover Designer: Greg Harris

Typeset by TNQ Technologies

Contents

Contributors

Manuel Becana
Departamento de Nutrición Vegetal, Estación Experimental de Aula Dei, Consejo Superior de Investigaciones Científicas (CSIC), Zaragoza, Spain

Anke Becker
Center for Synthetic Microbiology (SYNMIKRO) and Faculty of Biology, Philipps-Universität Marburg, Marburg, Germany

Stephane Boivin
Laboratoire des Symbioses Tropicales et Méditerranéennes, INRA, Montpellier, France

Eric Boncompagni
INRA, CNRS, Université Côte d'Azur, ISA, France

Renaud Brouquisse
INRA, CNRS, Université Côte d'Azur, ISA, France

Claude Bruand
Laboratoire des Interactions Plantes-Microorganismes (LIPM), Université de Toulouse, INRA, CNRS, INSA, Castanet-Tolosan, France

Delphine Capela
LIPM, Université de Toulouse, INRA, CNRS, Castanet-Tolosan, France

Delia Casas-Pastor
Center for Synthetic Microbiology (SYNMIKRO) and Faculty of Biology, Philipps-Universität Marburg, Marburg, Germany

Frédéric Debellé
LIPM, Université de Toulouse, INRA, CNRS, Castanet-Tolosan, France

George C. diCenzo
Department of Biology, Queen's University, Kingston, ON, Canada

Turlough M. Finan
Department of Biology, McMaster University, Hamilton, ON, Canada

Pierre Frendo
INRA, CNRS, Université Côte d'Azur, ISA, France

Georg Fritz
Center for Synthetic Microbiology (SYNMIKRO) and Faculty of Biology, Philipps-Universität Marburg, Marburg, Germany

Florian Frugier
Institute of Plant Sciences Paris Saclay (IPS2), CNRS, INRA, Univ Paris Diderot, Univ Paris-Sud, Univ d'Evry, Université Paris-Saclay, Gif-sur-Yvette, France

Barney A. Geddes
Department of Biology, McMaster University, Hamilton, ON, Canada

Véronique Gruber
Institute of Plant Sciences Paris Saclay (IPS2), CNRS, INRA, Univ Paris Diderot, Univ Paris-Sud, Univ d'Evry, Université Paris-Saclay, Gif-sur-Yvette, France

Christine Hervé
LIPM, Université de Toulouse, INRA, CNRS, Castanet-Tolosan, France

Théophile Kazmierczak
Institute of Plant Sciences Paris Saclay (IPS2), CNRS, INRA, Univ Paris Diderot, Univ Paris-Sud, Univ d'Evry, Université Paris-Saclay, Gif-sur-Yvette, France; BIOtransfer, Montreuil, France

Jason Kearsley
Department of Biology, McMaster University, Hamilton, ON, Canada

Marc Lepetit
Laboratoire des Symbioses Tropicales et Méditerranéennes, INRA, Montpellier, France

Juan Liang
Genetics, Faculty of Biology, Ludwig Maximilians University Munich, Germany

Shengbin Liu
Institute of Plant Sciences Paris Saclay (IPS2), CNRS, INRA, Univ Paris Diderot, Univ Paris-Sud, Univ d'Evry, Université Paris-Saclay, Gif-sur-Yvette, France

Kévin Magne
Institute of Plant Sciences Paris Saclay (IPS2), CNRS, INRA, Univ Paris Diderot, Univ Paris-Sud, Univ d'Evry, Université Paris-Saclay, Gif-sur-Yvette, France

Macarena Marín
Genetics, Faculty of Biology, Ludwig Maximilians University Munich, Germany

Manuel A. Matamoros
Departamento de Nutrición Vegetal, Estación Experimental de Aula Dei, Consejo Superior de Investigaciones Científicas (CSIC), Zaragoza, Spain

Ulrike Mathesius
Division of Plant Science, Research School of Biology, Australian National University, Australia

Mame Diarra Mbengue
LIPM, Université de Toulouse, INRA, CNRS, Castanet-Tolosan, France

Doreen Meier
Center for Synthetic Microbiology (SYNMIKRO) and Faculty of Biology, Philipps-Universität Marburg, Marburg, Germany

Eliane Meilhoc
Laboratoire des Interactions Plantes-Microorganismes (LIPM), Université de Toulouse, INRA, CNRS, INSA, Castanet-Tolosan, France

Peter Mergaert
Institute for Integrative Biology of the Cell, UMR9198, CEA, CNRS, Université Paris-Saclay, Gif-sur-Yvette, France

Richard Morton
Department of Biology, McMaster University, Hamilton, ON, Canada

Pascal Ratet
Institute of Plant Sciences Paris Saclay (IPS2), CNRS, INRA, Univ Paris Diderot, Univ Paris-Sud, Univ d'Evry, Université Paris-Saclay, Gif-sur-Yvette, France

Mingxing Tang
LIPM, Université de Toulouse, INRA, CNRS, Castanet-Tolosan, France

Rafael E. Venado
Genetics, Faculty of Biology, Ludwig Maximilians University Munich, Germany

Li Yang
INRA, CNRS, Université Côte d'Azur, ISA, France

Preface

Nitrogen (N) is one of the major macronutrients needed for plant growth. It is an essential element of all amino acids and a component of nucleic acids. N is also a constituent of chlorophyll, which allows plants to use solar radiation as a source of energy for converting inorganic carbon into organic molecules. However, whereas N is the most abundant element in the atmosphere, it is frequently the most limiting plant nutrient in soils. In contrast to eukaryotes that are not able to reduce atmospheric dinitrogen (N_2) in an N form that can be used by plants, many prokaryotes are able to convert N_2 into ammonia (NH_3). These prokaryotes include soil bacteria that fix N in free-living conditions and associate with plants, such as *Azotobacter* or *Azospirillum*, and bacteria that fix N in more tight endosymbiotic conditions, such as rhizobia and *Frankia* with legumes and actinorhizal plants respectively. In these mutualistic associations, bacteria deliver N sources to the plant whilst plants provide bacteria with photosynthates and a specialized ecological niche that favors their growth.

This volume entitled "Regulation of Nitrogen-Fixing Symbioses in Legumes", written by expert scientists in their different research fields, consists of 11 chapters presenting recent scientific advances related to biological N fixation in legumes. It addresses various aspects of current research, ranging from the diversity of symbioses to mechanisms that regulate plant-bacteria interactions across the different symbiotic stages, from establishment to senescence.

The rhizobium-legume symbiosis is a highly complex symbiosis implementing two main and overlapping processes: the *de novo* formation of a new organ, the nodule, and the infection of roots and of neo-formed nodules. These two processes rely on a continuous molecular dialog between the two partners. In the first chapter, Mame D. Mbengue and colleagues focus on the central symbiotic pathway, leading to nodulation and infection, the Nod factor signaling pathway, which involves the bacterial production of lipochitooligosaccharidic molecules and their perception and transduction in plants, and discuss its evolutionary origin. In chapter two, Ulrike Mathesius shows how, following Nod factor perception, symbiosis-recruited plant hormonal pathways coordinate rhizobial infection and nodule organogenesis. This chapter additionally covers how systemic pathways acting at the whole plant level regulate nodulation. Manuel Matamoros and Manuel

Becana depict in chapter three the crucial role of the redox state in the regulation of symbiosis, which occurs through the production of reactive oxygen and N species (ROS/RNS), and how antioxidant defense mechanisms are modulated in both partners along the different stages of the symbiosis.

Rhizobial bacteria can take different ways to enter the roots of legume, and nodules can differentiate into various shapes and types. In chapter four, Raphael E. Venado and colleagues describe the plant epidermal and cortical genetic reprogramming that sustains the different root entry modes leading to intracellular accommodation. In chapter five, Shengbin Liu and colleagues review the different nodule structures, their evolution, and the regulatory mechanisms underlying their differentiation and identity maintenance.

Following the infection stage, internalized bacteria differentiate into N-fixing bacteroids. This stage is however transient and, ultimately, bacteroids degenerate and nodules senesce. In chapter six, Peter Mergaert describes the coordinated differentiation of host plant cells and bacterial endosymbionts. The physiological and transcriptomic modifications occurring during nodule senescence and the contribution of ROS/RNS and hormones in developmentally or environmentally-induced senescence are reviewed by Théophile Kazmierczak and colleagues in chapter seven.

Rhizobia, the N fixing bacterial partners of legumes, are phylogenetically and genetically diverse bacteria belonging to hundreds of species distributed among alpha- and beta-proteobacteria. In chapter eight, Barney A. Geddes and colleagues review the general and specific features of rhizobial genomes, with a focus on key symbiotic nodulation and N fixation genes. In chapter nine, Mixing Tang and Delphine Capela interpret this rhizobial diversity as resulting from a two step-evolutionary scenario, *i.e.* horizontal transfer of key symbiotic genes in various taxonomic branches followed by the recruitment of local functions and regulatory circuitries to optimize symbiosis. As an example of how alpha-rhizobia have optimized regulatory circuitries to cope with complex soil and plant environments, Doreen Meier and colleagues describe the role of sigma factors in the regulation of bacterial gene expression in chapter ten.

The symbiotic success of a rhizobial strain in controlled conditions and in fields relies on many factors, including bacterial and plant traits. In the last chapter, Stéphane Boivin and Marc Lepetit review the different features required for a competitive and efficient N fixing symbiosis in an agronomical environment, which involves diverse rhizobial communities.

Implications for improving field inoculation strategies to boost legume crop yields in an agro-ecological context are discussed.

We hope that this volume will be useful to graduate students and researchers in a broad range of fields including plant physiology, cell biology, genomics, and evolutionary biology. We thank all the authors for their contributions to this volume.

Pierre Frendo, Florian Frugier and Catherine Masson-Boivin

CHAPTER ONE

Nod factor signaling in symbiotic nodulation

Mame Diarra Mbengue, Christine Hervé and Frédéric Debellé*
LIPM, Université de Toulouse, INRA, CNRS, Castanet-Tolosan, France
*Corresponding author: E-mail: frederic.debelle@inra.fr

Contents

Abstract

In the rhizobial legume symbiosis, an exchange of signals leads to the specific infection of the host plant by its rhizobial symbiont and the development of specialized organs, the nodules, in which rhizobia fix nitrogen. In response to flavonoids secreted by legume roots, rhizobia notably produce lipochitooligosaccharidic molecules called Nod factors (NFs). Following extensive research, we now understand better the mechanisms of perception and transduction of the NF signal by the legume host. NF signaling results in a wide transcriptional reprogramming that leads to coordinated multistep processes of infection and nodule organogenesis, involving many plant and bacterial molecular components. Recent data suggest that the NF biosynthesis pathway evolved from non rhizobial bacteria and that NF perception/

Advances in Botanical Research, Volume 94
ISSN 0065-2296
https://doi.org/10.1016/bs.abr.2019.10.002

signaling in legumes adapted components preexisting in plant signaling pathways controlling in particular the establishment of the more ancient arbuscular mycorrhizal symbiosis.

Introduction

Nitrogen fixing symbioses between rhizobial bacteria and their legume hosts are characterized by the formation on the roots of the host plant of new organs, the nodules, in which bacteria differentiated in bacteroids fix atmospheric dinitrogen. These associations are often specific, a given rhizobium being able to induce nodule formation on a limited number of legume hosts (Dénarié, Debellé, & Rosenberg, 1992). The various rhizobium-legume symbioses exhibit many variations in the mode of infection (through root hairs or intercellular), the type of nodules formed (determinate or indeterminate), or their functioning (Hadri, Spaink, Bisseling, & Brewin, 1998). The symbiotic relationship results from a multi-step process in which the tightly associated partners exchange various signals, making difficult the analysis of the molecular mechanisms involved. However, since the rhizobium-legume symbiosis is not obligatory, non-lethal mutations causing defects in the nodulation process can be obtained in both symbiotic partners, making genetics a powerful approach to dissect the molecular mechanisms controlling nodulation.

1. Identification of Nod factors, rhizobial lipochitooligosaccharidic signals with a major role in nodulation and host range control

1.1 Identification of *nod* genes

The development of molecular genetics of Gram-bacteria in the 1980s was instrumental in our understanding of the molecular mechanisms controlling the nodulation process. The first studies were carried out in model rhizobia such as *Rhizobium leguminosarum* and *Sinorhizobium meliloti*, and made use in particular of the availability of broad host range cloning vectors adapted to rhizobia, and the introduction of methods for random and site-directed mutagenesis by transposons. Early work had shown that genetic determinants of nodulation and host range were located on plasmids (Johnston et al., 1978) and rhizobial mutants affected in nodulation were isolated. This led to the first cloning of a locus controlling nodulation genes by complementation of a *nod*-mutation (Long, Buikema, & Ausubel, 1982). Further work involving site-directed mutagenesis, physical mapping and sequencing allowed the identification of a series of *nod* genes in various rhizobial species (see reviews Dénarié et al., 1992; Schlaman, Phillips, & Kondorosi, 1998; Downie, 1998). Some of these genes (*nodDABCIJ*) were found in all rhizobium species and were called common *nod* genes. Other *nod* genes, called host specific *nod* genes, were found only in some rhizobial strains. Mutations in the *nodABC* genes usually had a strong effect on rhizobium infection and nodulation, while the effect of mutations in other *nod* genes varied according to the nature of the gene.

1.2 *Nod* gene expression

The use of reporter genes made possible the analysis of *nod* gene expression. The *nodD* gene was found constitutively expressed in rhizobial growth medium while the expression of the other *nod* genes required the addition to the medium of root exudates together with a functional *nodD* gene (Long, 1989). NodD is a transcriptional activator of the LysR family. It can bind to a specific sequence upstream the *nod* gene operons, called the *nod* box, and activate the expression of other *nod* genes in the presence of compounds secreted by roots (Schlaman et al., 1998). The *nod* gene inducing compounds were isolated from exudates of various legume plants. They are generally flavonoids, secondary metabolites biosynthesized by the phenylpropanoid pathway, a metabolic pathway which expanded in legumes leading to a great diversity of molecules. In spite of repeated attempts, no crystal structure of a NodD protein with and without ligands has been obtained and there is little biochemical evidence of flavonoid binding to NodD (Peck, Fisher, Bliss, & Long, 2013). However direct binding of NodD to flavonoids is supported by the existence of *nodD* mutants with altered specificity with regard to flavonoids (Peck et al., 2013; Schlaman et al., 1998). The detailed mechanisms of *nod* gene activation are also poorly understood: NodD could bind to the *nod* box as a dimer or higher order oligomer, and NodD binding to DNA could be stimulated by flavonoids. The way NodD interacts with RNA polymerase to activate *nod* gene expression is not known (Schlaman et al., 1998). This first step of the molecular dialog between the legume host and rhizobium contributes to host range specificity: the NodD protein of a rhizobial species activates *nod* gene expression in the presence of the set of flavonoids produced by compatible host species. Different NodD proteins can have different inducer specificities and a *nodD* mutation in a rhizobial species thus sometimes cannot be complemented by the *nodD* gene of another rhizobium species (Spaink, Wijffelman, Pees, Okker, & Lugtenberg, 1987). In addition, some rhizobium species such as *S. meliloti* carry several copies of *nodD* that encode proteins with different inducer specificities, or that can activate *nod* gene expression in the absence of flavonoids like NodD3 (Dénarié et al., 1992). Some flavonoids can also act as anti-inducers, inhibiting the activation of *nod* gene expression (Schlaman et al., 1998). Interestingly, different sets of flavonoids can be secreted at different stages of the symbiotic process. In *Medicago truncatula*, luteolin, the first flavonoid identified as a *nod* gene inducer is mostly produced by germinating seeds. In root exudates, another compound with a higher *nod* gene inducing activity than luteolin was identified, methoxychalcone. Synthesis of this compound is controlled by the enzyme chalcone-O-methyl transferase and genes encoding this protein were found induced in root hairs in the presence of rhizobia (Liu & Murray, 2016), suggesting a role for methoxychalcone in the infection process. Recent results on the *Mesorhizobium loti-Lotus japonicus* symbiosis indicate that the two *nodD* genes in *M. loti* might act at different stages in the symbiotic process. NodD2 is mainly involved in induction of *nod* genes in the rhizosphere and within nodules, while *nodD1* acts primarily within root hair infection threads (Kelly et al., 2018). The nature of the Lotus flavonoid compounds involved in *nod* gene activation at these different stages of the symbiotic process is still unknown. However a *Lotus* gene encoding a chalcone isomerase involved in isoflavonoid biosynthesis is induced in the presence of rhizobia or NFs and specifically expressed within root hairs. Mutant analysis indicated that it might be responsible for the production of an inducer specifically recognized by NodD1 in infected root hairs. In addition to induction by flavonoids, *Rhizobium tropici* CIAT899 *nod* gene expression can be induced by salt or osmotic stress (Del Cerro et al., 2019). *R. tropici* CIAT899 is highly resistant to various abiotic stresses, and it contains five *nodD* genes. Among these, *nodD1* is required for *nod* gene induction by flavonoids, while induction by salt or osmotic stress requires *nodD2* (Del Cerro et al., 2019). This might be a strategy to ensure nodulation under abiotic stress conditions.

Other mechanisms contribute to the regulation of *nod* gene expression, which vary from species to species (Schlaman et al., 1998). In *S. meliloti*, another regulator of the LysR family, *syrM*, participates together with *nodD3* in *nod* gene induction. In *Bradyrhizobium diazoefficiens*, a two component regulator *nodVW* controls *nod* gene induction by flavonoids, in parallel to the control exerted by *nodD1*. In this species, quorum sensing also plays a role in *nod* gene regulation (Calatrava-Morales, McIntosh, & Soto, 2018; Jitacksorn & Sadowsky, 2008). Negative regulators of *nod* gene expression have also been characterized, such as *nolR* in *S. meliloti*, and *nolA* in *B. diazoefficiens* (Schlaman et al., 1998). *nod* gene expression is thus tightly regulated in response to various developmental and environmental conditions by specific mechanisms in different species, which also coordinate *nod* gene expression with that of other symbiosis-related processes (see below).

1.3 *Nod* genes and Nod factors

The biochemical function of the proteins encoded by *nod* genes remained elusive for several years until a link between the genetic determinants of nodulation and the production of secreted molecules was established: it was shown that sterile supernatants of rhizobial cultures grown in the presence of flavonoids could induce root or root hair deformations on homologous hosts and that this activity depended on the common *nodABC* genes (Van Brussel et al., 1986; Faucher et al., 1988). Using an alfalfa root hair deformation assay, active extracellular molecules called Nod factors (NFs) were extracted from cultures of *S. meliloti* overexpressing the *nod* genes. *S. meliloti* NFs consist of a backbone of four to five N-acetyl glucosamine residues, N-acylated at the non-reducing end by various unsaturated C16 fatty acids, O-acetylated at the non-reducing end and sulfated at the reducing end (Fig. 1) (Dénarié, Debellé, & Promé, 1996; Lerouge et al., 1990). NFs have since been purified from other rhizobial species. In most cases, similar molecules have been obtained: mono N-acylated chitooligosaccharides of generally four or five N-acetylglucosamine residues, carrying different substitutions, mainly on the two terminal glucosamine residues (Fig. 1) (see reviews by Dénarié et al., 1996; Perret, Staehelin, & Broughton, 2000; D'Haeze & Holsters, 2002). Rhizobial strains generally produce a mixture of NFs with slightly different structures, whose composition can vary with environmental conditions. No original NF structures have been described in recent years, suggesting that all common NF substitutions have been identified. However, purification and characterization of new LCOs from rhizobial cultures remain tricky particularly when they are produced in small amounts. Identification of new *nod* genes by large-scale genome or metagenome sequencing might help discover new NF molecules.

1.4 Biosynthesis of Nod factors

Since the expression of *nod* genes is required for NF production, *nod* genes were likely to encode the enzymes necessary for NF biosynthesis, with the

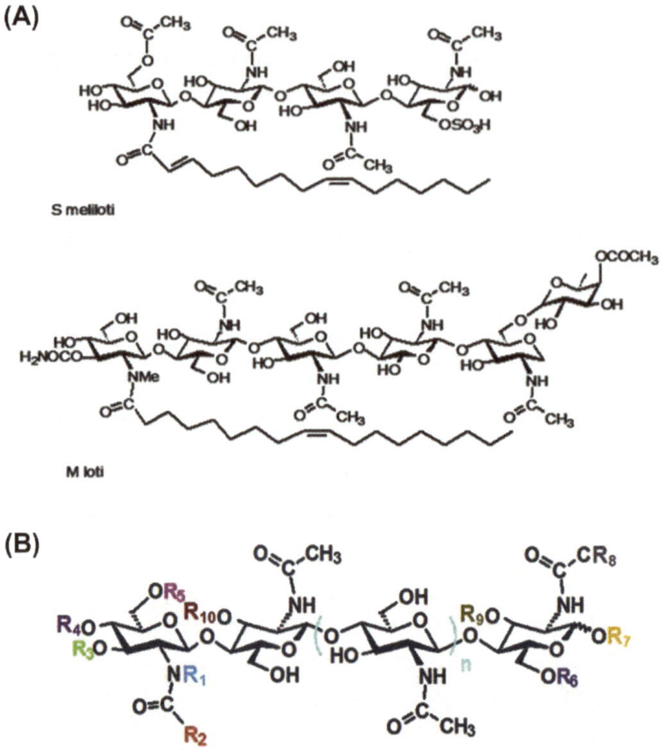

Fig. 1 ***Structure of Nod factors produced by various rhizobia.*** The structure of *S. meliloti* and *M. loti* NFs is represented in (A). In (B) is shown the structure of the chitooligosaccharidic backbone common to all NFs (n = 2 or 3 in most cases) and the positions of the various substitutions (R1 to R9) that have been described in NFs produced by different rhizobial strains. For a detailed list of substitutions and genes controlling them, see Dénarié et al. (1996); Perret et al. (2000); and D'Haeze & Holsters (2002).

common *nodABC* genes directing biosynthesis of the lipooligosaccharidic core common to all NFs, and the host specific *nod* genes controlling the rhizobium species specific substitutions of NFs. The role of each gene was assessed by different approaches: analysis of the structure of NFs produced by *nod* mutants, similarity of Nod proteins to proteins of known function, and direct enzymatic assays (Dénarié et al., 1996; Perret et al., 2000; D'Haeze & Holsters, 2002). For example, *nodH* mutants of *S. meliloti* produced non sulfated NF suggesting a role for *nodH* in NF sulfation; the *NodH* sequence is similar to that of known sulfotransferases; and purified NodH exhibits LCO sulfotransferase activity *in vitro*. The common NodC, NodB and NodA proteins were shown to carry out the synthesis of the chitooligosacharride backbone, its N-deacetylation at the non-reducing end, and its acylation, respectively. Most of the enzymatic steps leading to the biosynthesis of NFs have been characterized although uncertainties remain about the order in which they are carried out (Broughton, Jabbouri, & Perret, 2000; Poinsot et al., 2016). Little is known about the mechanisms of NF secretion. The *nodIJ* genes, which encode the ATPase and permease of an ABC-type transport system, are found in all NF producing rhizobia, and play a role in NF secretion (Downie, 1998). However other genes are likely involved as well, since *nodIJ* mutants still secrete NFs and have a weak symbiotic phenotype (Downie, 1998).

1.5 Nod factor degradation

NF concentration can be regulated by modulating NF biosynthesis but also NF degradation. Indeed, NFs can be cleaved and inactivated by host enzymes such as Nod Factor hydrolase 1 in *M. truncatula* (MtNFH1) or chitinase 5 in *L. japonicus* (LjCHIT5) (Cai et al., 2018; Malolepszy et al., 2018). Contrary to related chitinases, MtNFH1 hydrolyzes neither chitin nor chitin fragments and seems to have a high cleavage preference for *S. meliloti* LCOs. Upon inoculation by *S. meliloti*, MtNFH1 was found to accumulate in the infection chamber at the curled tip of root hairs and mutants affected in *MtNFH1* exhibited delayed root hair infection, abnormal nodule branching and hypertrophy (Cai et al., 2018). In contrast, in *Ljchit5* mutants, the progression of infection threads in the cortex and of intracellular infections were inhibited, leading to non-functional nodules (Malolepszy et al., 2018). This suggests that in both systems, fine tuning of NF levels is required for proper infection and nodulation of host plants.

1.6 Biological activity of Nod factors and importance for host range specificity

Purified NFs can induce at very low concentrations many symbiotic-like responses specifically on the roots of legume hosts (Hadri & Bisseling, 1998; D'Haeze & Holsters, 2002). These include modifications of root hair growth, rearrangement of cytoplasm in outer cortical cells leading to the formation of preinfection threads, and initiation of cell division in the inner root cortex at the origin of nodule primordium formation. At the cellular level, NFs induce depolarization of the epidermis cytoplasmic membrane, production of reactive oxygen species, and different ion fluxes. Among these, NF induced-calcium spiking in and around the cell nucleus has been extensively used to monitor plant response to LCOs (Ehrhardt,

Wais, & Long, 1996; Kelner, Leitão, Chabaud, Charpentier, & de Carvalho-Niebel, 2018). NFs also induce tissue-specific responses at the transcriptional level originally revealed by the identification of early nodulin (*ENOD*) genes (Hadri & Bisseling, 1998), and then by whole genome transcriptome analyses (Breakspear et al., 2014; Damiani, Drain et al., 2016; Damiani, Pauly, Puppo, Brouquisse, & Boscari, 2016; Jardinaud et al., 2016). LCOs also induce lateral root formation, and might participate in the formation of biofilms (Fujishige et al., 2008; Oláh, Brière, Bécard, Dénarié, & Gough, 2005).

NFs are additionally important determinants of host range. Analysis of NF structure in various rhizobial species generally showed a correlation between the type of NF produced and host range. For example, two symbionts of soybean belonging to different rhizobial genera, *Sinorhizobium fredii* and *B. diazoefficiens*, produce similar NFs. In addition, NF substitutions, which are encoded by host specific nodulation genes, are important for their biological activity. Mutations in these genes or their transfer to another rhizobial strain can result in changes of the host range. Depending on plants, the requirements for NF structure are not always stringent. For example, *Phaseolus* can be nodulated by rhizobia producing NFs with different structures. In addition, NFs are not the only determinants of host range: beside the nature of the plant flavonoid signal, rhizobial surface components or secreted protein effectors interacting with their plant receptor also play a role in partner choice (Wang, Liu, & Zhu, 2018).

Through the regulatory activity of flavonoids and NodD proteins, NF biosynthesis is connected to other biological processes that may play a role in the establishment of the symbiotic interaction. For example, the *S. meliloti nodD3* and *syrM* genes were found to induce the synthesis of exopolysaccharides, rhizobial surface components that play an important role in the infection process. More recently, genomic and transcriptomic analyses performed in different rhizobial species showed that a number of operons regulated by *nodD* and flavonoids through binding to *nod* boxes were not involved in NF biosynthesis. In particular, in the *Sinorhizobium* NGR234 and HH103 strains and in *B. diazoefficiens,* flavonoids induce the expression of the type III secretion system machinery and the synthesis of effector proteins that are injected in the host plant in a process similar to that observed with many plant pathogens (Jimenez-Guerrero et al., 2017). These effectors can have positive or negative effects on the symbiotic process and contribute to host range specificity (Deakin & Broughton, 2009; Miwa & Okazaki, 2017; Nelson & Sadowsky, 2015). In *S. fredii* HH103 and *R. tropici* 899, genes coding for the biosynthesis of the phytohormone auxin are induced by NodD and flavonoids (Jimenez-Guerrero et al., 2017). Interestingly, auxin and NFs were recently found to have a synergistic effect on lateral root development (Herrbach et al., 2017). Contrary to NF biosynthetic genes that are found in most rhizobia, these NodD/flavonoid responsive genes are found only in some rhizobial species, indicating lineage specific adaptations to NF induced nodulation and infection by rhizobia. In addition, plant flavonoids can also induce responses in rhizobia in a *nodD* independent way, which are likely involved in the rhizospheric life of rhizobia.

2. Nod factor perception and signaling in the host plant

Purified NFs from a rhizobium species were shown to induce, at very low concentrations and in a NF structure-dependent way, symbiotic responses specifically on roots of the legume host. This suggested that these molecules could be perceived by high-affinity receptors, and that a signaling pathway then transduced the NF signal to start the nodulation and rhizobial infection programs. In order to identify components of this signaling pathway, legume mutants affected in the root-nodule symbiosis (RNS) were therefore tested for their responses to purified NFs. Several of these mutants were impaired in their responses to NFs, indicating that the mutations likely affected NF signaling (Catoira et al., 2000). The genes involved in the responses to NFs in legume hosts were then ordered in a putative linear symbiotic signaling pathway. Upstream of the pathway were genes essential for all responses to NF, likely involved in NF perception, such as Nod Factor Perception (*NFP*) in *M. truncatula or* Nod factor Receptor 5 (*NFR5*) in *L. japonicus*. The calcium spiking response to NFs allowed further ordering of downstream signaling genes: genes like *MtNFP (LjNFR5)*, Does not Make Infections 1 (*MtDMI1) (LjCASTOR/POLLUX)*, and *MtDMI2* or SYMBIOSIS RECEPTOR-LIKE KINASE in Lotus *japonicus (LjSYMRK)* were required for calcium spiking while *MtDMI3 or* Calcium/CalModulin-dependent protein Kinase in *L. japonicus (LjCCaMK)* was not. In contrast to the above described genes, Nodulation Signaling Pathway 1 and 2 (*NSP*1 and *NSP2)* seemed dispensable for root hair deformation by NFs, suggesting that they acted downstream of *DMI3. L. japonicus* gain of function mutants able to nodulate in the absence of rhizobia also helped to decipher the NF signaling pathway and new actors of the pathway were additionally identified by looking for interactors of known components. Some of the genes involved, among them *DMI 1, 2,3* were shown to also control the establishment of the symbiosis with endomycorrhizal fungi and thus to define a common symbiotic signaling pathway (CSSP). Genes with similar functions were found in *M. truncatula, L. japonicus* and *Pisum sativum,* indicating conservation of the NF signaling pathway in different legumes (Oldroyd, 2013; Oldroyd, Murray, Poole, & Downie, 2011). Map-based cloning of these genes mostly in *M. truncatula* and *L. japonicus* confirmed that they were likely to control a signal perception/transduction pathway, and biochemical and cellular approaches shed light on the molecular mechanisms involved, although many aspects remain poorly understood despite extensive studies. The signaling pathway can be divided in three main steps: perception of microbial signals, transduction of the signals and transcriptional activation of responses leading to the accommodation of rhizobia in the host plant (Fig. 2).

2.1 Perception

In *L. japonicus* and *M. truncatula,* two putative NF receptors, LjNFR5 and MtNFP, were identified as lysin-motif (LysM) receptor-like kinases, with three LysM domains (known to bind N-acetyl glucosamine containing oligosaccharides) constituting the extracellular part of the proteins. The intracellular kinase domain of these proteins is inactive, suggesting that an interaction with other receptors with active kinase domains is required to initiate a signal transduction. Indeed, LysM-RLK with active kinase domains having crucial roles in infection and nodulation were identified in both *L. japonicus* (Nod factor Receptor 1, LjNFR1) and *M. truncatula* (LysM domain-containing receptor-like kinase 3, MtLYK3). Biochemical approaches showed that LjNFR1 and LjNFR5 interact in a heterologous system (Madsen et al., 2011). In *M. truncatula,* NFP and LYK3 localized in the same cell layers in the apical part of the nodule (Moling et al., 2014), where they probably also form a heterodimeric complex (Pietraszewska-Bogiel et al., 2013). However they could not be detected in the root epidermis, probably due to their low levels. NF perception slightly differs between *L. japonicus* and *M. truncatula*: both *NFR1* and *LYK3* are required for infection and nodulation, but contrary to *LjNFR1* mutants, *MtLYK3* mutants still exhibit some responses to NFs (Arrighi et al., 2006; Limpens et al., 2003; Madsen et al., 2003; Radutoiu et al., 2003; Smit et al., 2007). This might be due to redundancy of LYK3 with other *M. truncatula* LysM-RLKs. Indeed *M. truncatula* and *L. japonicus* carry genes closely related to *LjNFR1/MtLYK3*. One of these, *LjNFRe*, which has a spatio-temporal

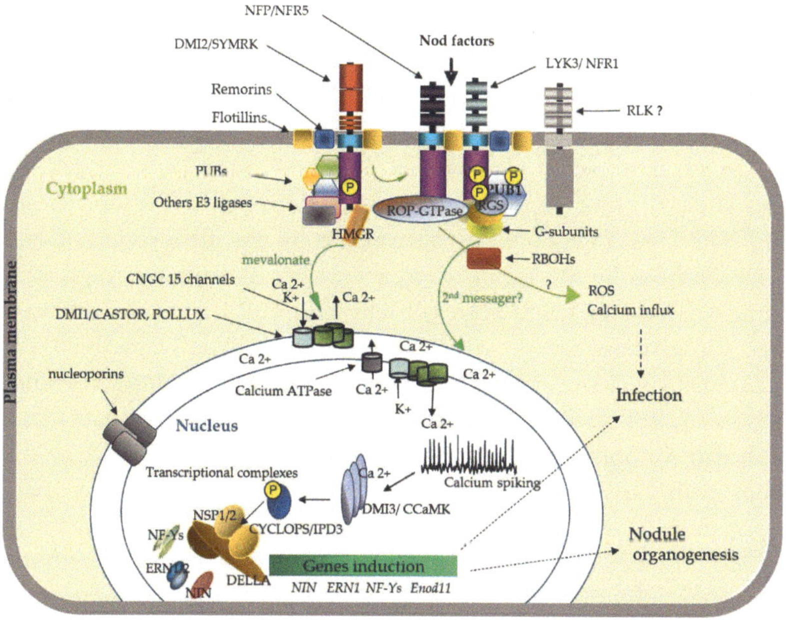

Fig. 2 ***A model for Nod Factor signaling in* Medicago truncatula *and* Lotus japonicus *symbiotic nodulation.*** Nod factor (NF) perception is mediated by a set of receptor-like kinases containing a LysM extracellular domain, including NFP/NFR5 and LYK3/NFR1. Other proteins, and among those other LysM-RLKs, are likely contributing to NF perception. During rhizobial infection, ROP-GTPases interact with LysM-RLKs and activate RBOHs, which produce reactive oxygen species and play a positive role in rhizobial infection and nodulation. The DMI2/SYMRK receptor is probably associated to the LysM perception complex within structured micro-domains in the plasma membrane, organized by the scaffolding proteins flotillins and remorins. DMI2/SYMRK could act as a co-receptor leading to the activation of intracellular signaling pathways. DMI2/SYMRK interacts with many partners including several E3 ubiquitin ligase (PUBs, SIE3, SINA4 and LIN/CERBERUS) that could participate in its turnover. The 3-hydroxy-3-methylglutaryl-CoA-reductase (HMGR) enzyme also interacts with DMI2, and could lead to the production of mevalonate as a secondary messenger to transmit the NF signal from the plasma membrane to the cell nucleus. An additional mechanism that is proposed to activate nuclear calcium spiking in *L. japonicus* involves heterotrimeric G-protein complexes that interact with and are phosphorylated by NFR1. This activation would lead to the production of an unknown secondary messenger promoting the calcium spiking. These secondary messengers would activate DMI1/POLLUX and CASTOR cation channels, which interact with three calcium channels from the CNGC15 family. The Ca2+ influx through CNGC15 channels is countered by a DMI1-dependent K+ efflux into the nuclear envelope, resulting in nuclear membrane repolarization to initiate calcium spiking in response to NFs. A calcium ATPase (MCA8) maintains the nuclear calcium concentration, and three nucleoporins identified in *L. japonicus* are also necessary for calcium spiking. In the nucleus, DMI3/CCaMK decodes calcium oscillations, and phosphorylates IPD3/CYCLOPS, a transcriptional activator. The phosphorylated form of IPD3/CYCLOPS would interact with various transcription factors (DELLA, NSP1/2) within a large transcriptional complex, and induce a transcriptional cascade leading to nodule organogenesis and/or rhizobial infection, which involves additional transcription factors (NIN, NF-Ys, and ERN1/2).

expression pattern and signaling properties distinct to those of LjNFR1, was recently shown to contribute also to NF signaling (Murakami et al., 2018).

Transformation of *LjNFR5* and *LjNFR1* into *M. truncatula* led to nodulation of the transformed plants by *M. loti*, the *L. japonicus* symbiont, indicating a role of these genes in host range determination (Radutoiu et al., 2007). The host range might be determined by specific recognition of the NF substitutions by the legume host receptors. Accordingly, high affinity binding of *M. loti* LCOs to both full-length LjNFR1 and LjNFR5 was reported (Broghammer et al., 2012). In addition, NFRe is also able to bind NFs *in vitro* (Murakami et al., 2018). In contrast, in *M. truncatula*, no binding of LCOs to NFP or LYK3 was demonstrated, while high affinity binding was shown for another LysM-RLK with an inactive kinase, MtLYR3, a protein with no evidence of symbiotic role based on mutant analyses (Fliegmann et al., 2013). However, no effect of the rhizobia specific NF substitutions on receptor-ligand affinity was detected in any of these binding experiments. By domain swapping and amino acid substitution, the NFR5/NFP LysM2 domains, and in particular a leucine residue in these domains, were shown as important for NF recognition (Bensmihen, de Billy, & Gough, 2011; Radutoiu et al., 2007). Crystal structures of the receptors in the presence of their ligands are however necessary to decipher the basis of specific LCO recognition. Since biochemical analysis of membrane proteins such as LysM-RLKs is difficult, a currently used approach consists in the characterization of the structure and NF binding properties of the extracellular domain of symbiotic LysM-RLK. However, NF binding *in vivo* could be different from that observed with the extracellular domain of NF receptors: NF receptors are indeed likely part of protein complexes that are located in plasma membrane nanodomains, which probably change along the symbiotic process (Ott, 2017). Interestingly, a *L. japonicus* LysM-RLK with a symbiotic role is involved in the response to modified exopolysaccharides produced by *M. loti*, raising the question of the range of ligands recognized by LysM-RLKs (Kawaharada et al., 2015).

2.2 Transduction

MtDMI2 (*LjSYMRK*) is positioned downstream of the LysM-RLK receptors: in contrast to *nfp* mutants, *dmi2* mutants still exhibit some responses to rhizobia and NFs, such as the early calcium influx (Endre et al., 2002). Over expression of full length *LjSYMRK* or the kinase domain of *MtDMI2* led to nodule induction in the absence of rhizobia, and this effect depended on the *DMI3* (*LjCCaMK*) and *DMI1* (*LjCastor LjPollux*) genes. This indicates that DMI2 acts upstream of DMI1 and DMI3 (Ried, Antolín-Llovera, & Parniske, 2014; Saha, Dutta, Bhattacharya, & DasGupta, 2014).

In *dmi2/symrk* mutants, initiation of nodule organogenesis is prevented, and neither infection pockets nor infection threads are observed indicating a role of *DMI2/SYMRK* in the early stages of both symbiotic programs (Catoira et al., 2000; Endre et al., 2002; Stracke et al., 2002). In *L. japonicus*, the *symrk-14* mutation (affecting the extracellular domain) uncouples epidermal and cortical symbiotic responses (Kosuta et al., 2011). These phenotypes highlight a major role for *DMI2/SYMRK* in both early events at the root epidermis, where the initial infection occurs, and also in the cortical signaling pathway required for the initiation of nodule primordia and cortical infections. However, how DMI2/SYMRK acts during signal transduction to initiate these symbiotic responses remains unclear.

Many plant developmental programs are controlled by RLKs, which act in protein complexes to perceive an extracellular ligand leading to the activation of intracellular signaling pathways. MtDMI2 (LjSYMRK) could thus act as a coreceptor of LysM-RLK proteins to transduce the NF signal (Osakabe, Yamaguchi-Shinozaki, Shinozaki, & Tran, 2013). DMI2/SYMRK is a receptor-like kinase with an extracellular domain containing three leucine-rich repeats (LRRs) and 2 malectin-like domains (MLDs). MLDs are known to bind oligosaccharide ligands but none have been identified for DMI2/SYMRK. In contrast, LjSYMRK was found by coimmunopurification to bind the NFR5 and NFR1 receptors (Antolín-Llovera, Ried, & Parniske, 2014; Ried et al., 2014). This fits with the report that in *M. truncatula*, the Symbiotic Remorin 1 (MtSYMREM1) interacts with both DMI2 and the symbiotic receptors NFP and LYK3, probably as a scaffolding protein, to regulate spatially the signaling complex along the symbiotic interaction (Lefebvre et al., 2010; Liang et al., 2018).

With the aim to identify targets of symbiotic kinases, quantitative proteomic and phosphoproteomic analyses were performed in *M. truncatula*, which measured rapid changes in the phosphorylation status of about 13 000 sites in 7739 proteins upon NF perception (Grimsrud et al., 2010; Rose, Volkening et al., 2012, Rose, Grimsurd et al., 2012; Valdés-López et al., 2018). Soybean phosphoproteomic analysis of root hair cells identified more than 250 phosphopeptides corresponding to 240 phosphoproteins significantly regulated in response to inoculation with *Bradyrhizobium japonicum* (Nguyen et al., 2012). This approach led to the identification in *M. truncatula* of a new regulator, the Phosphorylated Protein 1 (EPP1) required for nodulation initiation (Valdés-López et al., 2018). However further work is required to check if any of the proteins characterized by these approaches are direct targets of DMI2 or symbiotic LysM-RLKs.

Several evidences indicate that the amount of the DMI2 (SYMRK) protein is important for proper signaling and is thus tightly regulated (Antolín-Llovera et al., 2014; Pan, Stonoha-Arther, & Wang, 2018). Indeed, RLK signaling not only depends on the perception of the ligand but also needs a precise amount of active membrane-associated RLK. Several studies underline the importance of a fine-tuned regulation of DMI2/SYMRK by its extra cellular domain (Antolín-Llovera et al., 2014; Li et al., 2018; Pan et al., 2018). MLD integrity is required for the DMI2 protein to function properly during infection and early nodule development (Pan et al., 2018). Indeed, MLD stabilizes the full-length DMI2/SYMRK protein (Antolín-Llovera et al., 2014; Pan et al., 2018). On the other side, the conserved GDPC motif connecting LRR and MLD domains is required for MLD cleavage of the SYMRK protein in response to rhizobium (Antolín-Llovera et al., 2014).

Ubiquitination is a major mechanism associated with plasma membrane proteins to regulate the turnover of some receptors (Tian & Xie, 2013). This process occurs through the sequential activation of a cascade of reactions catalyzed by three classes of enzymes referred as E1, E2 and E3 (reviewed in Vierstra, 2009). Symbiotic RLKs were found to interact with different E3 ubiquitin ligases, among which three Plant U-box (PUB) ARMADILLO E3 ubiquitin ligases. MtPUB1 was identified as a negative regulator of nodulation and as a direct interactor of the intracellular kinase domain of the symbiotic RLKs LYK3 (Mbengue et al., 2010) and DMI2 (Vernié et al., 2016). Analyses of a *pub1* mutant affected in the U-box domain revealed that the ubiquitin ligase activity of PUB1 is essential for its negative regulatory function, nevertheless the symbiotic receptors DMI2 and LYK3 do not appear to be the targets of the PUB1 ubiquitination activity. Instead, another PUB ARMADILLO E3 ligase, MtPUB2, mediates MtDMI2 ubiquitination and the associated 26S proteasome-dependent degradation *in vitro* and *in planta* (Liu et al., 2018). MtPUB2 is a positive regulator of nodulation and by a cascade of phosphorylation and ubiquitination regulations, MtDMI2-MtPUB2 forms a negative feedback loop which could participate in the control of nodulation homeostasis (Liu et al., 2018). A third different PUB E3-ubiquitin ligase, LjPUB13, specifically ubiquitinates the kinase domain of NFR5 and has a positive regulatory role during the initial stages of nodulation (Tsikou et al., 2018).

In addition to PUB E3-ubiquitin ligases, the *L. japonicus* SYMRK-INTERACTING E3 ubiquitin ligase (SIE3) was shown to bind and to ubiquitinate SYMRK to positively regulate nodulation (Yuan et al., 2012). The SEVEN IN ABSENTIA 4 (LjSINA4), another type of E3 ubiquitin ligase, negatively affects the rhizobial infection process and interacts with and destabilizes SYMRK, but ubiquitination of SYMRK was not detected (Den Herder et al., 2008). Finally, Lumpy INfections (*LINs)* and *CERBERUS*, in *M. truncatula* and *L. japonicus* respectively, encode E3 ligases containing U-box, armadillo (ARM) and WD-40 repeats, and control rhizobial infection inside root hairs (Kiss et al., 2009; Yano et al., 2009).

PUB, SIE3, LIN/CERBERUS and SINA4 proteins belong to three structurally very different families of E3 ligases (Fig. 3) and may thus have different modes of action (Hervé, Lefebvre, & Cullimore, 2011). And, to further complicate these regulatory processes, a deubiquitinating enzyme, ASSOCIATED MOLECULE WITH THE SH3 DOMAIN OF STAM 1 (AMSH 1) was recently identified as a positive regulator of nodulation (Małolepszy et al., 2015). Currently, it is therefore difficult to get from the literature a clear picture of the role of ubiquitination in nodulation. One of the main issues remains the identification of the E3 ubiquitin ligases targets.

Additional components of the symbiosis signaling pathway downstream of the receptor complex have been identified by looking for interactors of components of this complex. For example, in a yeast two-hybrid screen, MtDMI2 was found to interact with 3-hydroxy-3-methylglutaryl CoA

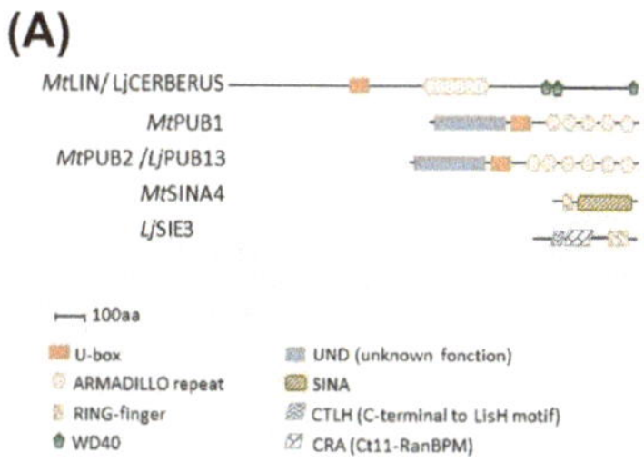

(B)

E3 ubiquitin ligase	Receptor	Target	Role in nodulation
*Mt*PUB1	*Mt*LYK3 *Mt*DMI2	?	Negative regulator
*Mt*PUB2	*Mt*DMI2	V	Positive regulator ; proteasome-mediated degradation of target ; nodulation homeostasis
*Lj*PUB13	*Lj*NFR5	V	Positive regulator ; initial stages of nodulation
*Lj*SIE3	*Lj*SYMRK	V	Positive regulator
*Lj*SINA4	*Lj*SYMRK	?	Negative regulator of infection
*Mt*LIN	?	?	Infection in root hairs
*Lj*CERBERUS	?	?	Infection in root hairs, growth, differentiation of nodules

Fig. 3 ***E3 ubiquitin ligases involved in*** **Medicago truncatula** ***and*** **Lotus japonicus** ***symbiotic nodulation.*** (A) Schematic representation of E3 ubiquitin ligases (adapted from Hervé et al., 2011). (B) Summary of different types of E3 ubiquitin ligases and their role in nodulation. The column « receptor » indicates the symbiotic receptor like-kinases which interact with each E3 ubiquitin ligase, and in the next column, « V » indicates that the ubiquitination of these RLKs by the corresponding E3 ubiquitin ligase was validated.

Reductase 1 (HMGR1), an enzyme involved in the biosynthesis of mevalonate, a precursor of isoprenoid secondary metabolites (Kevei et al., 2007). RNAi silencing of HMGR1 reduced nodulation and responses to NFs, such as calcium spiking and *ENOD11* marker gene expression (Venkateshwaran et al., 2015), indicating a role of HMGR1 in NF signaling. In addition, exogenous application of mevalonate induced calcium spiking in *M. truncatula* root epidermal cells, and this effect depended on DMI1 but not DMI2, indicating that the mevalonate signal acts downstream of DMI2 but upstream of DMI1. Finally, the application of mevalonate to kidney cells expressing the nuclear cation channel DMI1 (see below) induced calcium spiking, suggesting an interaction between mevalonate and DMI1 (Venkateshwaran et al., 2015). Thus, mevalonate or products of its metabolism could act as secondary messengers to transmit the NF signal from the plasma membrane to the cell nucleus.

More recently in *L. japonicus*, a receptor-like cytoplasmic kinase interacting with LjNFR5 was identified by a coimmunoprecipitation-based proteomic approach (Wong et al., 2019). This protein named NFR5-interacting cytoplasmic kinase 4 (NiCK4), was able to bind and phosphorylate LjNFR5 and LjNFR1 *in vitro*. The corresponding gene has an expression profile similar to that of LjNFR5 and its mutation resulted in decreased nodulation. Interestingly, this protein was found to relocate to the nucleus upon NF treatment in an *NFR1/NFR5* dependent manner indicating that it could contribute directly to the transmission of the NF signal from the plasma membrane, where receptors are located, to the nucleus where transcriptional activation leading to nodulation occurs (Wong et al., 2019).

Another mechanism could contribute to the transmission of the NF signal from the plasma membrane to the nucleus. Pharmacological approaches indicated a role in NF signaling of heterotrimeric G-protein complexes comprising Gα, Gβ, and Gγ subunits and the Regulator of G-protein Signaling (RGS) protein. Mastoparan 7 (Mas7), a G-protein agonist was shown to induce early nodulin gene expression and calcium spiking in *M. truncatula* roots in the absence of rhizobia (Pingret, Journet, & Barker, 1998; Sun, Miwa, Downie, & Oldroyd, 2007). This required DMI3 but not NFP, DMI1 or DMI2. More recently in soybean, RGS, Gβ, and Gγ were shown to positively regulate nodule formation, while Gα appears to be a negative regulator (Choudhury & Pandey, 2013, 2015). In addition, RGS interacts with and is phosphorylated by the LysM-RLK NFR1. It is thus proposed that active, phosphorylated RGS maintains Gα proteins in their trimeric inactive conformation, and initiates the signaling that leads to calcium spiking. However the role of putative second messengers and other signaling components acting downstream the G-protein complex remains unknown.

Three nucleoporins were also identified in *L. japonicus* as necessary for various steps of the root nodule symbiosis including calcium spiking (Groth et al., 2010; Kanamori et al., 2006; Saito et al., 2007). Nucleoporins are components of the nuclear pore complexes that control the exchange of macromolecules across the nuclear membranes and play a role in many different physiological processes. However, the symbiotic Lotus mutants affected in nucleoporins have only slight growth or developmental phenotypes and it is not clear how the mutations inhibit calcium spiking. In addition, mutations in genes encoding such proteins were so far not identified in screens of *M. truncatula* symbiotic mutants, suggesting different mechanisms and/or levels of redundancy between the two model legumes.

It took more than 10 years to identify the nuclear-localized ion channels and calcium pump required to generate the nucleo-cytoplasmic calcium spiking observed in response to NFs (Charpentier, 2018). DMI1 in *M. truncatula* and CASTOR and POLLUX in *L. japonicus* were first characterized as putative cation channels likely to control ion transport upstream of calcium spiking (Ané et al., 2004; Charpentier et al., 2008; Venkateshwaran et al., 2012). In addition, Cyclic Nucleotide-Gated Channels (CNGC15a, CNGC15b, and CNGC15c) were identified as calcium channels located on the nuclear membrane, which interact with DMI1 and mediate nuclear calcium oscillations in response to symbiotic LCOs (Charpentier et al., 2016). A model proposed that the influx of calcium through CNGC15 channels is countered by a DMI1-dependent potassium efflux into the nuclear envelope, resulting in nuclear membrane repolarization to enable a sustained calcium spiking (Charpentier et al., 2016). This model may however need revision due to the recent study of Kim et al. (2019) indicating that CASTOR/DMI1 is more likely a calcium channel. Finally, the nuclear calcium concentration is maintained by the action of the M. truncatula calcium ATPase MCA8 (Capoen et al., 2011). Overall, through these ion fluxes, the NF signal perceived at the plasma membrane is transmitted into the nucleus to generate calcium oscillations that are decoded by DMI3 and associated proteins to initiate transcriptional regulations (see below; Charpentier, 2018).

In addition, Morieri and collaborators proposed a model in which NFs, during the rhizobial infection process, would activate two pathways, one involving calcium spiking, as described above, and another one acting independently of calcium spiking (Morieri et al., 2013). This second pathway includes small Rho-Of-Plants (ROP) GTPases, which bind to and can be activated by the NFP and LYK3 components of the symbiotic receptor complex (Ke et al., 2012). During rhizobial infection, ROP-GTPases activate NADPH oxidases (Kiirika et al., 2012), also named Respiratory Burst Oxidase Homologs (RBOHs), which produce reactive oxygen species and play a positive role in rhizobial infection and nodulation (Marino et al., 2011).

2.3 Transcriptional activation

MtDMI3 (*LjCCaMK*) is a central regulator of NF-dependent early signaling. *DMI3* mutants are affected in infection and nodulation, but not in calcium spiking, indicating that this gene acts downstream of calcium spiking in NF signaling. DMI3 was identified as a calcium and calmodulin dependent protein kinase (CCaMK) (Lévy et al., 2004; Mitra et al., 2004), and was shown to have a nuclear localization (Smit et al., 2005). This confirmed the importance of calcium signaling in nodulation and made DMI3 a good candidate to decode oscillations of calcium concentration in the nucleus. Autoactive forms of DMI3 can induce spontaneous nodulation in the absence of rhizobia (Gleason et al., 2006; Takeda, Maekawa, & Hayashi, 2012; Tirichine et al., 2006) and can rescue calcium-spiking deficient mutants. This underlines the central role of DMI3 and suggests that the primary function of NF signaling components acting upstream of the calcium spiking (described above) is the activation of DMI3. Contrary to more common calcium-dependent protein kinases (CDPKs), DMI3 can bind calcium both directly through three EF-hand domains, and indirectly through calmodulin calcium-binding domains. Although the structure of the DMI3 protein is not known, genetic and modeling analyses allowed to propose a putative mechanism for the activation of DMI3 by calcium (Miller et al., 2013). At basal calcium concentrations, binding of calcium to the EF-hands inhibits the activity of DMI3 by inducing an autophosphorylation of the protein, which promotes bond formation between the kinase and the calmodulin-binding domains. When the calcium concentration increases, calcium/calmodulin binds to DMI3, which blocks its autophosphorylation and promotes substrate phosphorylation.

In order to understand how DMI3 controls the expression of genes leading to nodulation, it was important to identify target substrates. Interacting

Protein of DMI3 (IPD3) was identified as a DMI3 interactor in a yeast two hybrid screen (Messinese et al., 2007). IPD3 localizes to the nucleus and can be phosphorylated by DMI3. Mutants in *IPD3* and its *L. japonicus* ortholog *CYCLOPS* are impaired in rhizobial colonization but exhibit less severe phenotypes than *dmi3* mutants, suggesting a partial functional redundancy. CYCLOPS was shown to act as a transcriptional activator that binds DNA in a phosphorylation-dependent manner (Singh, Katzer, Lambert, Cerri, & Parniske, 2014). Two phosphorylated serine residues of CYCLOPS are necessary for its activity and phosphomimetic versions of CYCLOPS are able to trigger root nodule organogenesis in the absence of rhizobia and CCaMK (DMI3). Upon phosphorylation by CCaMK, CYCLOPS can bind to the promoter of the Nodule Inception (NIN) transcription factor, and induces a transcriptional cascade leading to root nodule development. IPD3/CYCLOPS acts within a larger transcriptional complex that includes several GRAS-domain containing transcription factors such as Nodulation Signaling Pathway 1 and 2 (NSP1/2) and DELLA proteins that are required for rhizobial infection (Floss, Levy, Lévesque-Tremblay, Pumplin, & Harrison, 2013; Fonouni-Farde et al., 2016; Jin et al., 2016; Pimprikar et al., 2016). DELLA proteins are required for gibberellic acid (GA) signaling, and then likely participate in the coordination of early symbiotic signaling with plant development. The NIN transcription factor acts as a master coordinator of responses leading to nodule organogenesis and infection, together with the Ethylene Response Factor Required for Nodulation 1 (ERN1) and NF-YA transcription factors (Liu, Rutten et al., 2019, Liu, Breakspear, 2019; Schauser, Roussis, Stiller, & Stougaard, 1999). A number of studies shed light on the hierarchical relationships between these transcriptional regulators, mainly by monitoring symbiotic gene expression in various lines overexpressing or mutated for these TFs (Cerri et al., 2012; Fonouni-Farde et al., 2016; Laloum et al., 2014; Liu, Rutten et al., 2019, Liu, Breakspear, 2019; Soyano, Kouchi, Hirota, & Hayashi, 2013). However the interpretation of these analyses is difficult due to partial redundancy, feedback regulations and subtle spatio-temporal expression patterns of these genes.

Beside GA signaling, other hormone signaling pathways, involving notably cytokinins and auxins, interact with NF signaling to ensure the coordination between symbiotic responses induced in the root epidermis, pericycle and cortex tissues, which are associated respectively with infection by rhizobia and cell divisions leading to nodule primordium formation (Boivin, Fonouni-Farde, & Frugier, 2016; Gamas, Brault, Jardinaud, &

Frugier, 2017, Buhian & Bensmihen, 2018; see also Chapter 12: Hormonal interactions in the regulation of the nitrogen-fixing legume Rhizobium symbiosis, by Mathesius, 2019).

3. Origin and evolution of Nod factor signaling

3.1 Origin and evolution of rhizobial nod genes

For many years the common *nod* genes were considered as indispensable to establish RNS on legumes as well as on *Parasponia*, the only known non-legume plant able to nodulate with rhizobia. The presence of these genes was the main criterium to identify a bacterial strain as a rhizobium. However in 2007, it was shown that photosynthetic *Bradyrhizobium* strains that efficiently fix nitrogen on some *Aeschynomene* plant species lack *nodABC* genes (Giraud et al., 2007). The mechanism inducing nodule formation in this biological system remains not known, as in spite of an extensive mutagenesis, no strict non-nodulating bacterial mutant was obtained (Bonaldi et al., 2010). In addition, there was no evidence of the production by photosynthetic *Bradyrhizobium* of a diffusible bacterial signal able to induce symbiotic responses. More recently, a *nodC* mutant of *Bradyrhizobium elkanii*, unable to produce NFs, was reported to induce nodule formation on soybean, although with a lower efficiency than the WT strain (Okazaki et al., 2016; Okazaki, Kaneko, Sato, & Saeki, 2013). This nodule induction pathway depended on an intact type 3 secretion system suggesting that this mutant could bypass NF perception/signaling likely through the action of effectors secreted through this secretion machinery.

Soil Actinobacteria of the genus *Frankia* can induce nodulation and establish symbiotic nitrogen fixing symbioses with a diverse group of plants in the orders Cucurbitales, Fagales and Rosales, which belong to the same clade of plants able to establish nitrogen fixing symbioses as the Legume family-containing Fabales order (Soltis et al., 1995). Many studies aimed at identifying putative NF-like molecules produced by *Frankia* and involved in actinorhizal plant nodulation. These efforts are hampered by the difficulty to genetically manipulate *Frankia* strains, and even to grow them in the absence of the host plant. Sequencing of the genome of a number of *Frankia* strains able to nodulate *Alnus* or *Casuarina* plants (strains that belong to *Frankia* cluster I) showed that close homologs of *nodABC* genes were absent in these *Frankia* strains, indicating that other mechanisms than LCO signaling are likely involved in nodulation (Normand et al., 2007). Indeed, molecules secreted by *Frankia* and potentially involved in early symbiotic signaling have been described, based on biological tests such as root hair deformation, the induction of the *Casuarina NIN* symbiotic gene, or the induction of calcium spiking (Chabaud et al., 2016; Cissoko et al., 2018). These molecules have biochemical properties different from that of LCOs (more hydrophilic and resistant to chitinases), but their exact nature remains to be determined. Interestingly, canonical *nodABC* genes were more recently identified in genomes of uncultured *Frankia* strains Dg1 and Dg2 (strains that belong to *Frankia* cluster II) whose sequence was obtained by metagenome analysis of samples isolated from nodules of an actinorhizal plant, *Datisca glomerata* (Nguyen et al., 2016; Persson et al., 2015). The Dg1 *nodABC* genes are expressed in nodules and the *nodC* gene could partially complement a mutation of the *R. leguminosarum nodC* gene, and rescue the nodulation on its legume host, *P. sativum*. The Dg1 and Dg2 genomes also contain distant homologs of *nodIJ*, and the Dg2 genome contains two copies of the *nodH* sulfotransferase gene. However, since these *Frankia* strains cannot be cultured, the production of LCOs could not be validated. More recently, the genome of a *Frankia* isolate from cluster III was sequenced and found to contain the *nodABCIJH* genes (Ktari et al., 2017). This strain can be cultivated, so that it should be possible to test its LCO production capacity. However, since none of the *Frankia* strains can be genetically manipulated, the importance of *nodABC* and LCO production in the nodulation of actinorhizal plants will remain difficult to establish. In addition, no *nodD*-like gene was identified in the *nodABC* containing *Frankia* genomes, raising the question of how *nodABC* gene expression is regulated.

The identification of canonical *nodABC* genes in *Frankia* strains suggests a role for LCO signaling in the nodulation of some actinorhizal plants. It also sheds light on a possible origin and evolution of LCO signaling in bacteria. The phylogenetic analysis of *nodA* is particularly informative. For a long time, this gene encoding an acyl transferase was considered as specific to rhizobia, as it could not be identified in non-rhizobial members of the alpha- and beta-Proteobacteria, even in the form of distantly related paralogs. In contrast, more recent studies revealed that NodA homologs can be detected across the whole Actinobacteria phylum, including non-nodulating genera, and exhibit a higher diversity than the rhizobial NodA proteins (Persson et al., 2015). In a NodA phylogenetic tree, the *Frankia* proteins cluster with the rhizobial proteins and are basal to the rhizobial NodA orthologs. This suggests that the *nodA* gene responsible for the acyl transfer step in

LCO biosynthesis originated in Actinobacteria, and was later transmitted to Proteobacteria by lateral transfer. Similarly, the *nodBC* genes are likely to have originated in Actinobacteria before being transferred to Proteobacteria, although the analysis of phylogenetic trees is more complex since these genes encoding a deacetylase and a glycosyl transferase, respectively, are members of multigene families. The presence of *nodABC* genes in Actinobacteria that are not known to nodulate raises the question of a non-symbiotic role of these genes. In contrast to *nodABC,* the *nodIJ* genes that are also common to all rhizobia are likely to have a different origin. Although in *Frankia* genes similar to *nodIJ* are found in the vicinity of *nodBC* genes, their similarity with rhizobial *nodIJ* genes is low and they probably originate from a different lineage. Thus, the *nodIJ* genes likely evolved independently in Actinobacteria and Proteobacteria (Aoki, Ito, & Iwasaki, 2013). With respect to NodH, the actinobacterial sequences form a sister group of rhizobial NodH sequences, suggesting that a gene transfer occurred between rhizobia and *Frankia,* or that both groups acquired the gene from a common ancestor.

Phylogenetic analyses also indicate that the *Frankia* cluster II, in which the *nodABC* were first identified, is in basal position relative to other *Frankia* groups. This suggests that the ancestor of symbiotic *Frankia* contained the canonical *nod* genes, which were transferred to Proteobacteria and have been lost in some strains of *Frankia* clusters I and III. Within the Proteobacteria, the *nodABC* genes spread to various branches of the Rhizobiales (alpha-Proteobacteria) and Burkholderiales (beta-Proteobacteria) orders by horizontal gene transfer. The high frequency of *nod* gene horizontal transfer can be detected through comparisons of phylogenetic trees of *nod* genes with those of 16S rRNA or other non-symbiotic genes (Andrews et al., 2018). The frequent incongruence detected between housekeeping and symbiosis gene trees fits with the older observation that *nod* genes and genes involved in nitrogen fixation are located on plasmids or chromosome symbiotic islands that are transferable from strain to strain. Accordingly, a transfer of *nod* genes has been observed in a few cases (Sullivan, Patrick, Lowther, Scott, & Ronson, 1995; Nandasena, O'Hara, Tiwari, & Howieson, 2006; Nandasena, O'Hara, Tiwari, Sezmiş, & Howieson, 2007), and interestingly, such transfer of *nod* genes was recently found to be induced by plant flavonoids similar to those inducing *nod* gene expression (Ling et al., 2016). This induction of gene transfer was however not dependent on NodD but on a transcriptional regulator of the same family.

Horizontal transfer of symbiotic genes seems frequent within species or among related species (Andrews et al., 2018; Remigi, Zhu, Young, & Masson-Boivin, 2016; Rogel, Ormeno-Orrillo, & Martinez-Romero, 2011). It is often detected when analyzing rhizobia isolated from nodules of a given legume species: for example, rhizobia isolated from pea nodules generally belong to several different species related to *R. leguminosarum* and contain very similar *nod* genes acquired by horizontal transfer between these species, which are designated as the symbiovar *viciae*. The *nod* genes in different species of the symbiovar *viciae* can be more similar than those found in *R. leguminosarum* strains which nodulate other hosts like *Trifolium* or *Phaseolus*. Transfer of the *nod* genes from one symbiovar to another one in the same species can result in a change of host range. Transfer of *nod* genes between genera seems less frequent but does occur: the *Sesbania* symbiont *Agrobacterium* IRBG74 has *nod* genes very similar to those of *Sinorhizobium* strains nodulating the same legume, indicating transfer of *nod* genes between the *Sinorhizobium* and *Agrobacterium* genera. In contrast, the *Bradyrhizobium* and *Sinorhizobium* strains nodulating soybean carry divergent *nod* genes. There is also evidence of lateral transfer of *nod* genes from alpha-Proteobacteria to beta-Proteobacteria: the *nod* genes of *Burkholderia* species nodulating South African legumes seem related to those of *Mesorhizobium* nodulating the same legumes (Andrews et al., 2018).

A recent large scale analysis of Rhizobiales bacteria found in soil and root-associated microbiomes reinforced these evolutionary hypotheses (Garrido-Oter et al., 2018). Rhizobiales were found in a high relative abundance in soil, root, and rhizosphere samples indicating a good adaptation to the root environment. Nearly 1000 members of this order were isolated and had their genome sequenced. A minority of them (less than 0.3 % of bacteria isolated from roots) had *nod* genes and thus the potential to nodulate. Analysis of all available Rhizobiales genome sequences allowed a large scale phylogenetic reconstruction of ancestral characters. This confirmed that the most common ancestor of Rhizobiales likely did not have canonical *nod* genes, and that the ability to nodulate was acquired multiple times after the divergence of the major groups of Rhizobiales, probably through multiple horizontal gene transfers after acquisition from a non-Rhizobiales bacterium. By testing representative members of the different Rhizobiales groups isolated from roots, it was shown that the majority had a root growth promoting activity, indicating a plant commensal lifestyle that likely predated and perhaps favored the acquisition of *nod* genes and their biological success. Similar processes might have occurred in the Burkholderiales order of the beta-Proteobacteria which contains nodulating rhizobia as well as plant growth promoting bacteria. Although there are a few unconfirmed reports of the presence of nodulating rhizobia outside these two groups (Martínez-Hidalgo & Hirsch, 2017), further analysis of the nodule and rhizophere microbiome of various legumes with genome and metagenome sequencing is required to get a more detailed understanding. This would also help to assess the spread of *nod* genes to other groups of bacteria abundant in the soil and in the rhizosphere, which might have been limited by a lesser pre-adaptation of these taxa to symbiosis, the difficulty of genetic transfer across large phylogenetic distances, and/or various environmental factors.

The analysis of the Rhizobiales genomes also showed that some clades were richer in nodulating strains (*Mesorhizobium*, *Bradyrhizobium*) than other ones (*Agrobacterium*), and that after an initial acquisition of *nod* genes in a lineage, some strains in that lineage could lose the ability to nodulate. Comparison of nodulating with non-nodulating bacteria did not provide evidence of whole-genome level signatures of adaptation (Garrido-Oter et al., 2018). However, nodulation and infection of legumes by rhizobia require many genes in addition to the *nodABC* genes, and these have been likely recruited from the genome of the *nod* gene acquiring strain. These genes can be ubiquitous or lineage-specific genes. For example, the *S. meliloti* genes involved in the synthesis of succinoglycan exopolysaccharides are required for plant infection, and are found in *Sinorhizobium* species but not in other rhizobial lineages (Tian et al., 2012). Similarly lineage-specific genes involved in the translocation of effectors through type 3 or type 4 secretion systems also play a role in modulating infection and host range. Accommodation of acquired symbiotic functions likely required adaptations in these genes together with the coordination between the expression of plasmid-located symbiotic genes and chromosome encoded genes necessary for successful symbiosis (Jiao et al., 2018; Remigi et al., 2016). Such modifications in the recipient genome after the acquisition of *nod* genes have been detected in experimental evolution experiments carried out on rhizobia (Capela et al., 2017; Remigi, Masson-Boivin, & Rocha, 2019).

3.2 A single origin for nodulation in plants?

Root nodule nitrogen fixing symbioses occur in several orders of Angiosperms that all also include non-nodulating plants. In addition, a polyphyletic group of bacteria in Proteobacteria and Actinobacteria induces a symbiotic process which exhibits a great diversity in the modes of infection, types, and functioning of nodules. It was therefore important to assess whether common mechanisms controlled these different symbiotic processes, and in particular to establish the role of the NF signaling pathway. In *Parasponia*, a member of the *Cannabaceae* family nodulated by rhizobia, NFs were shown to induce calcium spiking, and an autoactive form of CCaMK was able to induce nodule formation on *Parasponia* roots in the absence of rhizobia (Granqvist et al., 2015; Op den Camp et al., 2011). This indicated that components of the NF signaling pathway play a role in rhizobial nodulation on hosts outside of the *Fabaceae* family.

In *Aeschynomene evenia,* a legume nodulated by photosynthetic *Bradyrhizobium* strains lacking the *nodABC* genes required for NF biosynthesis, RNAi inactivation of the closest homologs of *DMI2* and *DMI3* impaired nodule development, and expression of a deregulated form of *AeDMI3* led to nodule formation in the absence of rhizobia (Fabre et al., 2015). This showed that at least some components of the NF signaling pathway are required for this NF-independent nodulation. In contrast, orthologs of the *NFP* and *LYK3* genes were not consistently found in the *Aeschynomene* hosts of the *nod*-independent *Bradyrhizobium*, suggesting that they are likely to be dispensable for nodulation in these species. Screening and characterization of plant symbiotic mutants should help to understand how *A. evenia* recognizes its rhizobial symbiont (Arrighi et al., 2012).

In the case of the *Frankia*-actinorhizal plant symbiosis, genes in the NF signaling pathway, orthologous of *DMI2, DMI3*, and *NIN*, are required for nodule induction in *Casuarina* (Gherbi et al., 2008; Svistoonoff et al., 2013). The putative role of LysM-RLKs in actinorhizal plant symbioses with *Frankia* containing or not *nod* genes however remains to be assessed.

There is therefore strong evidence that, although each type of symbiosis presents specificities, common mechanisms are involved in the establishment of RNS in these different plant families. In addition, all nodulating plants belong to a monophyletic group of Angiosperms including the orders Fabales, Fagales, Cucurbitales, and Rosales, the so-called nitrogen-fixing clade (Soltis et al., 1995). This suggests a single origin of nodulation in an ancestor of this clade. To explain that members of this clade are not able to nodulate, one has to assume that, once acquired, the ability to nodulate was lost in some lineages. Indeed, the comparison of genomes from nodulating and non-nodulating plants in the nitrogen-fixing clade revealed that non-nodulating plants had lost genes such as *NIN* or *NFP/NFR5* that are required for RNS (Griesmann et al., 2018; van Veltzen et al., 2018; van Velzen, Doyle, & Geurts, 2018). A single origin of nodulation would also imply that there are genes or regulatory elements specific to the nitrogen fixation clade and present in all nodulating species. Finding these would help to understand the origin of nodulation.

Early in the analysis of legume genetic determinants of nodulation, it was observed that legume mutants affected in the RNS were also defective for the establishment of the arbuscular endomycorrhizal symbiosis (AMS) (Duc, Trouvelot, Gianinazzi-Pearson, & Gianinazzi, 1989), and later on several components of the NF signaling pathway were found to also control the AMS, constituting the so-called common symbiotic signaling pathway (see above). Since the AMS is much more ancient than the RNS (450 million years vs. 80 million years), the hypothesis was that the RNS had coopted key components of the AMS (Parniske, 2008). Mutants in genes controlling NF perception (MtNFP/LYK3 and LjNFR5/NFR1) were not originally described as altered in AMS, raising doubts about the role of chitin-like molecules in AMS establishment. However Myc-LCOs, with a structure similar to that of NFs (Maillet et al., 2011), as well as short-chain chitin oligomers (Myc-COs), were identified in AM fungi spore exudates and found to induce calcium spiking in roots of *M. truncatula* as well as other legume and non-legumes plants (Genre et al., 2013; Sun et al., 2015). The implication of these chitin-like molecules reinforces that LysM-RLK proteins are involved in their perception and thus in mycorrhization. Indeed, silencing of NFP/NFR5 homologs in tomato led to a strong reduction in mycorrhization (Buendia, Wang, Girardin, & Lefebvre, 2016). In legumes, inactivation of the *MtLYK3* gene decreases mycorrhization (Zhang et al., 2015). In addition, two LysM-RLKs, homologous to the Arabidopsis chitin receptors AtCERK1 and AtLYK5, MtLYK9 and MtLYR4, have been identified as candidate CO receptors (Bozsoki et al., 2017). However, the precise role of LCOs and COs in the AM symbiosis remains poorly understood. No gene similar to the canonical *nodABC* genes has been yet identified in AM fungi sequenced genomes. The biosynthetic pathways leading to LCO and CO production in AMF are therefore not known, although a number of chitin synthases involved in fungal cell wall synthesis could be involved. Recent work indicates that LCOs and LCO signaling pathways also play a role in the establishment of some ectomycorrhizal symbioses (Cope et al., 2019). Moreover, COs and LCOs seem to be produced by a large range of fungi, symbiotic or not (Plett, 2018). A diverse range of organisms carry genes likely to control the synthesis of COs, which could thus have a wide role as signaling molecules in various biological processes beyond symbiosis (Dénarié et al., 1996). In particular, recognition of chitin-like molecules plays an important role in plant-pathogen interactions. Understanding how plants distinguish CO and LCO producing rhizobia, AMF and pathogenic fungi to induce appropriate responses is a challenge for future research (Zipfel & Oldroyd, 2017).

References

Andrews, M., De Meyer, S., James, E., Stępkowski, T., Hodge, S., Simon, M., et al. (2018). Horizontal transfer of symbiosis genes within and between rhizobial genera: Occurrence and importance. *Genes, 9*, 321. https://doi.org/10.3390/genes9070321.

Ané, J. M., Kiss, G. B., Riely, B. K., Penmetsa, R. V., Oldroyd, G. E., Ayax, C., et al. (2004). Medicago truncatula DMI1 required for bacterial and fungal symbioses in legumes. *Science, 303*, 1364–1367. https://doi.org/10.1126/science.1092986.

Antolín-Llovera, M., Ried, M. K., & Parniske, M. (2014). Cleavage of the SYMBIOSIS RECEPTOR-LIKE KINASE ectodomain promotes complex formation with Nod factor receptor 5. *Current Biology, 24*, 422–427. https://doi.org/10.1016/j.cub.2013.12.053.

Aoki, S., Ito, M., & Iwasaki, W. (2013). From β- to α-proteobacteria: The origin and evolution of rhizobial nodulation genes nodIJ. *Molecular Biology and Evolution, 30*, 2494–2508. https://doi.org/10.1093/molbev/mst153.

Arrighi, J. F., Barre, A., Amor, B. B., Bersoult, A., Soriano, L. C., Mirabella, R., et al. (2006). The Medicago truncatula lysine motif-receptor-like kinase gene family includes NFP and new nodule-expressed genes. *Plant Physiology, 142*, 265–279. https://doi.org/10.1104/pp.106.084657.

Arrighi, J. F., Cartieaux, F., Brown, S. C., Rodier-Goud, M., Boursot, M., Fardoux, J., et al. (2012). Aeschynomene evenia, a model plant for studying the molecular genetics of the Nod-independent rhizobium-legume symbiosis. *Molecular Plant-microbe Interactions, 25*, 851–861. https://doi.org/10.1094/MPMI-02-12-0045-TA.

Bensmihen, S., de Billy, F., & Gough, C. (2011). Contribution of NFP LysM domains to the recognition of Nod factors during the Medicago truncatula/Sinorhizobium meliloti symbiosis. *PLoS One, 6*, e26114. https://doi.org/10.1371/journal.pone.0026114.

Boivin, S., Fonouni-Farde, C., & Frugier, F. (2016). How auxin and cytokinin phytohormones modulate root microbe interactions. *Frontiers of Plant Science, 7*, 1240. https://doi.org/10.3389/fpls.2016.01240.

Bonaldi, K., Gourion, B., Fardoux, J., Hannibal, L., Cartieaux, F., Boursot, M., et al. (2010). Large-scale transposon mutagenesis of photosynthetic Bradyrhizobium sp. strain ORS278 reveals new genetic loci putatively important for nod-independent symbiosis with Aeschynomene indica. *Molecular Plant-Microbe Interactions, 23*, 760–770. https://doi.org/10.1094/MPMI-23-6-0760.

Bozsoki, Z., Cheng, J., Feng, F., Gysel, K., Vinther, M., Andersen, K. R., et al. (2017). Receptor-mediated chitin perception in legume roots is functionally separable from Nod factor perception. *Proceedings of the National Academy of Sciences, 114*, E8118–E8127. https://doi.org/10.1073/pnas.1706795114.

Breakspear, A., Liu, C., Roy, S., Stacey, N., Rogers, C., Trick, M., et al. (2014). The root hair "infectome" of Medicago truncatula uncovers changes in cell cycle genes and reveals a requirement for auxin signaling in rhizobial infection. *The Plant Cell, 26*, 4680–4701. https://doi.org/10.1105/tpc.114.133496.

Broghammer, A., Krusell, L., Blaise, M., Sauer, J., Sullivan, J. T., Maolanon, N., et al. (2012). Legume receptors perceive the rhizobial lipochitin oligosaccharide signal molecules by direct binding. *Proceedings of the National Academy of Sciences, 109*, 13859–13864. https://doi.org/10.1073/pnas.1205171109.

Broughton, W. J., Jabbouri, S., & Perret, X. (2000). Keys to symbiotic harmony. *Journal of Bacteriology, 182*, 5641–5652. https://doi.org/10.1128/JB.182.20.5641-5652.2000.

Buendia, L., Wang, T., Girardin, A., & Lefebvre, B. (2016). The LysM receptor-like kinase SlLYK10 regulates the arbuscular mycorrhizal symbiosis in tomato. *New Phytologist, 210*, 184–195. https://doi.org/10.1111/nph.13753.

Buhian, W. P., & Bensmihen, S. (2018). Mini-Review: Nod factor regulation of phytohormone signaling and homeostasis during rhizobia-legume symbiosis. *Frontiers of Plant Science, 9*. https://doi.org/10.3389/fpls.2018.01247.

Cai, J., Zhang, L. Y., Liu, W., Tian, Y., Xiong, J. S., Wang, Y. H., et al. (2018). Role of the Nod factor hydrolase MtNFH1 in regulating Nod factor levels during rhizobial infection and in mature nodules of Medicago truncatula. *The Plant Cell, 30*, 397–414. https://doi.org/10.1105/tpc.17.00420.

Calatrava-Morales, N., McIntosh, M., & Soto, M. (2018). Regulation mediated by N-Acyl Homoserine Lactone quorum sensing signals in the Rhizobium-Legume symbiosis. *Genes, 9*, 263. https://doi.org/10.3390/genes9050263.

Capela, D., Marchetti, M., Clérissi, C., Perrier, A., Guetta, D., Gris, C., et al. (2017). Recruitment of a lineage-specific virulence regulatory pathway promotes intracellular infection by a plant pathogen experimentally evolved into a legume symbiont. *Molecular Biology and Evolution, 34*, 2503–2521. https://doi.org/10.1093/molbev/msx165.

Capoen, W., Sun, J., Wysham, D., Otegui, M. S., Venkateshwaran, M., Hirsch, S., et al. (2011). Nuclear membranes control symbiotic calcium signaling of legumes. *Proceedings of the National Academy of Sciences, 108*, 14348–14353. https://doi.org/10.1073/pnas.1107912108.

Catoira, R., Galera, C., de Billy, F., Penmetsa, R. V., Journet, E. P., Maillet, F., et al. (2000). Four genes of Medicago truncatula controlling components of a Nod factor transduction pathway. *The Plant Cell, 12*, 1647–1665. https://doi.org/10.1105/tpc.12.9.1647.

Cerri, M. R., Frances, L., Laloum, T., Auriac, M. C., Niebel, A., Oldroyd, G. E., et al. (2012). Medicago truncatula ERN transcription factors: Regulatory interplay with NSP1/NSP2 GRAS factors and expression dynamics throughout rhizobial infection. *Plant Physiology, 160*, 2155–2172. https://doi.org/10.1104/pp.112.203190.

Chabaud, M., Gherbi, H., Pirolles, E., Vaissayre, V., Fournier, J., Moukouanga, D., et al. (2016). Chitinase-resistant hydrophilic symbiotic factors secreted by Frankia activate both Ca^{2+} spiking and NIN gene expression in the actinorhizal plant *Casuarina glauca*. *New Phytologist, 209*, 86–93. https://doi.org/10.1111/nph.13732.

Charpentier, M. (2018). Calcium signals in the plant nucleus: Origin and function. *Journal of Experimental Botany, 69*, 4165–4173. https://doi.org/10.1093/jxb/ery160.

Charpentier, M., Bredemeier, R., Wanner, G., Takeda, N., Schleiff, E., & Parniske, M. (2008). Lotus japonicus CASTOR and POLLUX are ion channels essential for perinuclear calcium spiking in legume root endosymbiosis. *The Plant Cell, 20*, 3467–3479. https://doi.org/10.1105/tpc.108.063255.

Charpentier, M., Sun, J., Martins, T. V., Radhakrishnan, G. V., Findlay, K., Soumpourou, E., et al. (2016). Nuclear-localized cyclic nucleotide–gated channels mediate symbiotic calcium oscillations. *Science, 352*, 1102–1105. https://doi.org/10.1126/science.aae0109.

Choudhury, S. R., & Pandey, S. (2013). Specific subunits of heterotrimeric G proteins play important roles during nodulation in soybean. *Plant Physiology, 162*, 522–533. https://doi.org/10.1104/pp.113.215400.

Choudhury, S. R., & Pandey, S. (2015). Phosphorylation-dependent regulation of G-protein cycle during nodule formation in soybean. *The Plant Cell, 27*, 3260–3276. https://doi.org/10.1105/tpc.15.00517.

Cissoko, M., Hocher, V., Gherbi, H., Gully, D., Carré-Mlouka, A., Sane, S., et al. (2018). Actinorhizal signaling molecules: Frankia root hair deforming factor shares properties with NIN inducing factor. *Frontiers of Plant Science, 9*. https://doi.org/10.3389/fpls.2018.01494.

Cope, K. R., Bascaules, A., Irving, T. B., Venkateshwaran, M., Maeda, J., Garcia, K., et al. (2019). The ectomycorrhizal fungus laccaria bicolor produces lipochitooligosaccharides and uses the common symbiosis pathway to colonize populus roots. *The Plant Cell*. https://doi.org/10.1105/tpc.18.00676.

Damiani, I., Drain, A., Guichard, M., Balzergue, S., Boscari, A., Boyer, J. C., et al. (2016). Nod factor effects on root hair-specific transcriptome of Medicago truncatula: Focus on plasma membrane transport systems and reactive oxygen species networks. *Frontiers of Plant Science, 7*, 794. https://doi.org/10.3389/fpls.2016.00794.

Damiani, I., Pauly, N., Puppo, A., Brouquisse, R., & Boscari, A. (2016). Reactive oxygen species and nitric oxide control early steps of the legume–rhizobium symbiotic interaction. *Frontiers of Plant Science, 7*, 454. https://doi.org/10.3389/fpls.2016.00454.

Deakin, W. J., & Broughton, W. J. (2009). Symbiotic use of pathogenic strategies: Rhizobial protein secretion systems. *Nature Reviews Microbiology, 7*, 312. https://doi.org/10.1038/nrmicro2091.

Del Cerro, P., Megías, M., López-Baena, F. J., Gil-Serrano, A., Pérez-Montaño, F., & Ollero, F. J. (2019). Osmotic stress activates nif and fix genes and induces the Rhizobium tropici CIAT 899 Nod factor production via NodD2 by up-regulation of the nodA2 operon and the nodA3 gene. *PLoS One, 14*, e0213298. https://doi.org/10.1371/journal.pone.0213298.

Den Herder, G., De Keyser, A., De Rycke, R., Rombauts, S., Van de Velde, W., Clemente, M. R., et al. (2008). Seven in absentia proteins affect plant growth and nodulation in Medicago truncatula. *Plant Physiology, 148*, 369–382. https://doi.org/10.1104/pp.108.119453.

Dénarié, J., Debellé, F., & Promé, J. C. (1996). Rhizobium lipo-chitooligosaccharide nodulation factors: Signaling molecules mediating recognition and morphogenesis. *Annual Review of Biochemistry, 65*, 503–535. https://doi.org/10.1146/annurev.bi.65.070196.002443.

Dénarié, J., Debellé, F., & Rosenberg, C. (1992). Signaling and host range variation in nodulation. *Annual Review of Microbiology, 46*, 497–531. https://10.1146/annurev.mi.46.100192.002433.

Downie, J. A. (1998). Functions of rhizobial nodulation genes. In H. P. Spaink, A. Kondorosi, & P. J. Hooykaas (Eds.), *The rhizobiaceae: Molecular biology of model plant-associated bacteria* (pp. 387–402). Dordrecht: Springer. https://doi.org/10.1007/978-94-011-5060 6_20.

Duc, G., Trouvelot, A., Gianinazzi-Pearson, V., & Gianinazzi, S. (1989). First report of non-mycorrhizal plant mutants (Myc−) obtained in pea (Pisum sativum L.) and fababean (Vicia faba L.). *Plant Science, 60*, 215–222. https://doi.org/10.1016/0168-9452(89)90169-6.

D'Haeze, W., & Holsters, M. (2002). Nod factor structures, responses, and perception during initiation of nodule development. *Glycobiology, 12*, 79R–105R. https://doi.org/10.1093/glycob/12.6.79R.

Ehrhardt, D. W., Wais, R., & Long, S. R. (1996). Calcium spiking in plant root hairs responding to Rhizobium nodulation signals. *Cell, 85*, 673–681. https://doi.org/10.1016/S0092-8674(00)81234-9.

Endre, G., Kereszt, A., Kevei, Z., Mihacea, S., Kaló, P., & Kiss, G. B. (2002). A receptor kinase gene regulating symbiotic nodule development. *Nature, 417*, 962. https://doi.org/10.1038/nature00842.

Fabre, S., Gully, D., Poitout, A., Patrel, D., Arrighi, J. F., Giraud, E., et al. (2015). Nod factor-independent nodulation in Aeschynomene evenia required the common plant-microbe symbiotic toolkit. *Plant Physiology, 169*, 2654–2664. https://doi.org/10.1104/pp.15.01134.

Faucher, C., Maillet, F., Vasse, J., Rosenberg, C., Van Brussel, A. A., Truchet, G., et al. (1988). Rhizobium meliloti host range nodH gene determines production of an alfalfa-specific extracellular signal. *Journal of Bacteriology, 170*, 5489–5499. https://doi.org/10.1128/jb.170.12.5489-5499.1988.

Fliegmann, J., Canova, S., Lachaud, C., Uhlenbroich, S., Gasciolli, V., Pichereaux, C., et al. (2013). Lipo-chitooligosaccharidic symbiotic signals are recognized by LysM receptor-like kinase LYR3 in the legume Medicago truncatula. *ACS Chemical Biology, 8*, 1900–1906. https://doi.org/10.1021/cb400369u.

Floss, D. S., Levy, J. G., Lévesque-Tremblay, V., Pumplin, N., & Harrison, M. J. (2013). DELLA proteins regulate arbuscule formation in arbuscular mycorrhizal symbiosis. *Proceedings of the National Academy of Sciences, 110*, E5025–E5034. https://doi.org/10.1073/pnas.1308973110.

Fonouni-Farde, C., Tan, S., Baudin, M., Brault, M., Wen, J., Mysore, K. S., et al. (2016). DELLA-mediated gibberellin signalling regulates Nod factor signalling and rhizobial infection. *Nature Communications, 7*, 12636. https://doi.org/10.1038/ncomms12636.

Fujishige, N. A., Lum, M. R., De Hoff, P. L., Whitelegge, J. P., Faull, K. F., & Hirsch, A. M. (2008). Rhizobium common nod genes are required for biofilm formation. *Molecular Microbiology, 67*, 504–515. https://doi.org/10.1111/j.1365-2958.2007.06064.x.

Gamas, P., Brault, M., Jardinaud, M. F., & Frugier, F. (2017). Cytokinins in symbiotic nodulation: When, where, what for? *Trends in Plant Science, 22*, 792–802. https://doi.org/10.1016/j.tplants.2017.06.012.

Garrido-Oter, R., Nakano, R. T., Dombrowski, N., Ma, K. W., Team, T. A., McHardy, A. C., et al. (2018). Modular traits of the rhizobiales root microbiota and their evolutionary relationship with symbiotic rhizobia. *Cell Host & Microbe, 24*, 155–167. https://doi.org/10.1016/j.chom.2018.06.006.

Genre, A., Chabaud, M., Balzergue, C., Puech-Pagès, V., Novero, M., Rey, T., et al. (2013). Short-chain chitin oligomers from arbuscular mycorrhizal fungi trigger nuclear Ca^{2+} spiking in Medicago truncatula roots and their production is enhanced by strigolactone. *New Phytologist, 198*, 190–202. https://doi.org/10.1111/nph.12146.

Gherbi, H., Markmann, K., Svistoonoff, S., Estevan, J., Autran, D., Giczey, G., et al. (2008). SymRK defines a common genetic basis for plant root endosymbioses with arbuscular mycorrhiza fungi, rhizobia, and Frankiabacteria. *Proceedings of the National Academy of Sciences, 105*, 4928–4932. https://doi.org/10.1073/pnas.0710618105.

Giraud, E., Moulin, L., Vallenet, D., Barbe, V., Cytryn, E., Avarre, J. C., et al. (2007). Legumes symbioses: Absence of nod genes in photosynthetic bradyrhizobia. *Science, 316*, 1307–1312. https://doi.org/10.1126/science.1139548.

Gleason, C., Chaudhuri, S., Yang, T., Muñoz, A., Poovaiah, B. W., & Oldroyd, G. E. (2006). Nodulation independent of rhizobia induced by a calcium-activated kinase lacking autoinhibition. *Nature, 441*, 1149. https://doi.org/10.1038/nature04812.

Granqvist, E., Sun, J., den Camp, R. O., Pujic, P., Hill, L., Normand, P., et al. (2015). Bacterial-induced calcium oscillations are common to nitrogen-fixing associations of nodulating legumes and nonlegumes. *New Phytologist, 207*, 551–558. https://doi.org/10.1111/nph.13464.

Griesmann, M., Chang, Y., Liu, X., Song, Y., Haberer, G., Crook, M. B., et al. (2018). Phylogenomics reveals multiple losses of nitrogen-fixing root nodule symbiosis. *Science, 361*, eaat1743. https://doi.org/10.1126/science.aat1743.

Grimsrud, P. A., den Os, D., Wenger, C. D., Swaney, D. L., Schwartz, D., Sussman, M. R., et al. (2010). Large-scale phosphoprotein analysis in Medicago truncatula roots provides insight into in vivo kinase activity in legumes. *Plant Physiology, 152*, 19–28. https://doi.org/10.1104/pp.109.149625.

Groth, M., Takeda, N., Perry, J., Uchida, H., Dräxl, S., Brachmann, A., et al. (2010). NENA, a Lotus japonicus homolog of Sec13, is required for rhizodermal infection by arbuscular mycorrhiza fungi and rhizobia but dispensable for cortical endosymbiotic development. *The Plant Cell, 22*, 2509–2526. https://doi.org/10.1105/tpc.109.069807.

Hadri, A. E., & Bisseling, T. (1998). Responses of the plant to Nod factors. In H. P. Spaink, A. Kondorosi, & P. J. J. Hooykaas (Eds.), *The rhizobiaceae: Molecular biology of model plant-associated bacteria* (pp. 403–416). Dordrecht, The Netherlands: Springer Science+Business Media. https://doi.org/10.1007/978-94-011-5060-6_21.

Hadri, A. E., Spaink, H. P., Bisseling, T., & Brewin, N. J. (1998). Diversity of root nodulation and rhizobial infection processes. In H. P. Spaink, A. Kondorosi, & P. J. J. Hooykaas (Eds.), *The rhizobiaceae: Molecular biology of model plant-associated bacteria* (pp. 347–360). Dordrecht, The Netherlands: Springer Science+Business Media. https://doi.org/10.1007/978-94-011-5060-6_18.

Herrbach, V., Chirinos, X., Rengel, D., Agbevenou, K., Vincent, R., Pateyron, S., et al. (2017). Nod factors potentiate auxin signaling for transcriptional regulation and lateral root formation in Medicago truncatula. *Journal of Experimental Botany, 68*, 569–583. https://doi.org/10.1093/jxb/erw474.

Hervé, C., Lefebvre, B., & Cullimore, J. (2011). How many E3 ubiquitin ligase are involved in the regulation of nodulation? *Plant Signaling & Behavior, 6*, 660–664. https://doi.org/10.4161/psb.6.5.15024.

Jardinaud, M. F., Boivin, S., Rodde, N., Catrice, O., Kisiala, A., Lepage, A., et al. (2016). A laser dissection-RNAseq analysis highlights the activation of cytokinin pathways by Nod factors in the Medicago truncatula root epidermis. *Plant Physiology, 171*, 2256–2276. https://doi.org/10.1104/pp.16.00711.

Jiao, J., Ni, M., Zhang, B., Zhang, Z., Young, J. P. W., Chan, T. F., et al. (2018). Coordinated regulation of core and accessory genes in the multipartite genome of Sinorhizobium fredii. *PLoS Genetics, 14*, e1007428. https://doi.org/10.1371/journal.pgen.1007428.

Jimenez-Guerrero, I., Acosta-Jurado, S., del Cerro, P., Navarro-Gómez, P., López-Baena, F., Ollero, F., et al. (2017). Transcriptomic studies of the effect of nod gene-inducing molecules in rhizobia: Different weapons, one purpose. *Genes, 9*, 1. https://doi.org/10.3390/genes9010001.

Jin, Y., Liu, H., Luo, D., Yu, N., Dong, W., Wang, C., et al. (2016). DELLA proteins are common components of symbiotic rhizobial and mycorrhizal signalling pathways. *Nature Communications, 7*, 12433. https://doi.org/10.1038/ncomms12433.

Jitacksorn, S., & Sadowsky, M. J. (2008). Nodulation gene regulation and quorum sensing control density-dependent suppression and restriction of nodulation in the Bradyrhizobium japonicum-soybean symbiosis. *Applied and Environmental Microbiology, 74*, 3749–3756. https://doi.org/10.1128/AEM.02939-07.

Johnston, A. W. B., Beynon, J. L., Buchanan-Wollaston, A. V., Setchell, S. M., Hirsch, P. R., & Beringer, J. E. (1978). High frequency transfer of nodulating ability between strains and species of Rhizobium. *Nature, 276*, 634–636. https://doi.org/10.1038/276634a0.

Kanamori, N., Madsen, L. H., Radutoiu, S., Frantescu, M., Quistgaard, E. M., Miwa, H., et al. (2006). A nucleoporin is required for induction of Ca2+ spiking in legume nodule development and essential for rhizobial and fungal symbiosis. *Proceedings of the National Academy of Sciences of the United States of America, 103*, 359–364. https://doi:10.1073/pnas.0508883103.

Kawaharada, Y., Kelly, S., Nielsen, M. W., Hjuler, C. T., Gysel, K., Muszyński, A., et al. (2015). Receptor-mediated exopolysaccharide perception controls bacterial infection. *Nature, 523*, 308. https://doi.org/10.1038/nature14611.

Ke, D., Fang, Q., Chen, C., Zhu, H., Chen, T., Chang, X., et al. (2012). The small GTPase ROP6 interacts with NFR5 and is involved in nodule formation in Lotus japonicus. *Plant Physiology, 159,* 131–143. https://doi.org/10.1104/pp.112.197269.

Kelly, S., Sullivan, J. T., Kawaharada, Y., Radutoiu, S., Ronson, C. W., & Stougaard, J. (2018). Regulation of Nod factor biosynthesis by alternative NodD proteins at distinct stages of symbiosis provides additional compatibility scrutiny. *Environmental Microbiology, 20,* 97–110. https://doi.org/10.1111/1462-2920.14006.

Kelner, A., Leitão, N., Chabaud, M., Charpentier, M., & de Carvalho-Niebel, F. (2018). Dual color sensors for simultaneous analysis of calcium signal dynamics in the nuclear and cytoplasmic compartments of plant cells. *Frontiers of Plant Science, 9,* 245. https://doi.org/10.3389/fpls.2018.00245.

Kevei, Z., Lougnon, G., Mergaert, P., Horváth, G. V., Kereszt, A., Jayaraman, D., et al. (2007). 3-Hydroxy-3-methylglutaryl coenzyme A reductase 1 interacts with NORK and is crucial for nodulation in Medicago truncatula. *The Plant Cell, 19,* 3974–3989. https://doi.org/10.1105/tpc.107.053975.

Kiirika, L. M., Bergmann, H. F., Schikowsky, C., Wimmer, D., Korte, J., Schmitz, U., et al. (2012). Silencing of the Rac1 GTPase MtROP9 in Medicago truncatula stimulates early mycorrhizal and oomycete root colonizations but negatively affects rhizobial infection. *Plant Physiology, 159,* 501–516. https://doi.org/10.1104/pp.112.193706.

Kim, S., Zeng, W., Bernard, S., Liao, J., Venkateshwaran, M., Ané, J. M., et al. (2019). Ca(2+)-regulated Ca(2+) channels with an RCK gating ring control plant symbiotic associations. *Nature Communications, 10,* 3703. https://doi.org/10.1038/s41467-019-11698-5.

Kiss, E., Oláh, B., Kaló, P., Morales, M., Heckmann, A. B., Borbola, A., et al. (2009). LIN, a novel type of U-box/WD40 protein, controls early infection by rhizobia in legumes. *Plant Physiology, 151,* 1239–1249. https://doi.org/10.1104/pp.109.143933.

Kosuta, S., Held, M., Hossain, M. S., Morieri, G., MacGillivary, A., Johansen, C., et al. (2011). Lotus japonicus symRK-14 uncouples the cortical and epidermal symbiotic program. *The Plant Journal, 67,* 929–940. https://doi.org/10.1111/j.1365-313X.2011.04645.x.

Ktari, A., Nouioui, I., Furnholm, T., Swanson, E., Ghodhbane-Gtari, F., Tisa, L. S., et al. (2017). Permanent draft genome sequence of Frankia sp. NRRL B-16219 reveals the presence of canonical nod genes, which are highly homologous to those detected in Candidatus Frankia Dg1 genome. *Standards in Genomic Sciences, 12,* 51. https://doi.org/10.1186/s40793-017-0261-3.

Laloum, T., Baudin, M., Frances, L., Lepage, A., Billault-Penneteau, B., Cerri, M. R., et al. (2014). Two CCAAT-box-binding transcription factors redundantly regulate early steps of the legume-rhizobia endosymbiosis. *The Plant Journal, 79,* 757–768. https://doi.org/10.1111/tpj.12587.

Lefebvre, B., Timmers, T., Mbengue, M., Moreau, S., Hervé, C., Tóth, K., et al. (2010). A remorin protein interacts with symbiotic receptors and regulates bacterial infection. *Proceedings of the National Academy of Sciences, 107,* 2343–2348. https://doi.org/10.1073/pnas.0913320107.

Lerouge, P., Roche, P., Faucher, C., Maillet, F., Truchet, G., Promé, J. C., et al. (1990). Symbiotic host-specificity of Rhizobium meliloti is determined by a sulphated and acylated glucosamine oligosaccharide signal. *Nature, 344,* 781. https://doi.org/10.1038/344781a0.

Lévy, J., Bres, C., Geurts, R., Chalhoub, B., Kulikova, O., Duc, G., et al. (2004). A putative Ca^{2+} and calmodulin-dependent protein kinase required for bacterial and fungal symbioses. *Science, 303,* 1361–1364. https://doi.org/10.1126/science.1093038.

Li, H., Chen, M., Duan, L., Zhang, T., Cao, Y., & Zhang, Z. (2018). Domain swap approach reveals the critical roles of different domains of SymRK in root nodule symbiosis in Lotus japonicus. *Frontiers of Plant Science, 9*. https://doi.org/10.3389/fpls.2018.00697.

Liang, P., Stratil, T. F., Popp, C., Marin, M., Folgmann, J., Mysore, K. S., et al. (2018). Symbiotic root infections in Medicago truncatula require remorin-mediated receptor stabilization in membrane nanodomains. *Proceedings of the National Academy of Sciences, 115*, 5289–5294. https://doi.org/10.1073/pnas.1721868115.

Limpens, E., Franken, C., Smit, P., Willemse, J., Bisseling, T., & Geurts, R. (2003). LysM domain receptor kinases regulating rhizobial Nod factor-induced infection. *Science, 302*, 630–633. https://doi.org/10.1126/science.1090074.

Ling, J., Wang, H., Wu, P., Li, T., Tang, Y., Naseer, N., et al. (2016). Plant nodulation inducers enhance horizontal gene transfer of Azorhizobium caulinodans symbiosis island. *Proceedings of the National Academy of Sciences, 113*, 13875–13880. https://doi.org/10.1073/pnas.1615121113.

Liu, C. W., Breakspear, A., Guan, D., Cerri, M. R., Jackson, K., Jiang, S., et al. (2019b). NIN acts as a network hub controlling a growth module required for rhizobial infection. *Plant Physiology, 179*, 1704–1722. https://doi.org/10.1104/pp.18.01572.

Liu, C. W., & Murray, J. (2016). The role of flavonoids in nodulation host-range specificity: An update. *Plants, 5*, 33. https://doi.org/10.3390/plants5030033.

Liu, J., Deng, J., Zhu, F., Li, Y., Lu, Z., Qin, P., et al. (2018). The MtDMI2-MtPUB2 negative feedback loop plays a role in nodulation homeostasis. *Plant Physiology, 176*, 3003–3026. https://doi.org/10.1104/pp.17.01587.

Liu, J., Rutten, L., Limpens, E., van der Molen, T., van Velzen, R., Chen, R., et al. (2019a). A remote cis-regulatory region is required for NIN expression in the pericycle to initiate nodule primordium formation in Medicago truncatula. *The Plant Cell, 31*, 68–83. https://doi.org/10.1105/tpc.18.00478.

Long, S. R. (1989). Rhizobium genetics. *Annual Review of Genetics, 23*, 483–506. https://doi.org/10.1146/annurev.ge.23.120189.002411.

Long, S. R., Buikema, W. J., & Ausubel, F. M. (1982). Cloning of Rhizobium meliloti nodulation genes by direct complementation of Nod– mutants. *Nature, 298*, 485. https://doi.org/10.1038/298485a0.

Madsen, E. B., Antolín-Llovera, M., Grossmann, C., Ye, J., Vieweg, S., Broghammer, A., et al. (2011). Autophosphorylation is essential for the in vivo function of the Lotus japonicus Nod factor receptor 1 and receptor-mediated signalling in cooperation with Nod factor receptor 5. *The Plant Journal, 65*, 404–417. https://doi.org/10.1111/j.1365-313X.2010.04431.x.

Madsen, E. B., Madsen, L. H., Radutoiu, S., Olbryt, M., Rakwalska, M., Szczyglowski, K., et al. (2003). A receptor kinase gene of the LysM type is involved in legume perception of rhizobial signals. *Nature, 425*, 637. https://doi.org/10.1038/nature02045.

Maillet, F., Poinsot, V., André, O., Puech-Pagès, V., Haouy, A., Gueunier, M., et al. (2011). Fungal lipochitooligosaccharide symbiotic signals in arbuscular mycorrhiza. *Nature, 469*, 58. https://doi.org/10.1038/nature09622.

Małolepszy, A., Urbański, D. F., James, E. K., Sandal, N., Isono, E., Stougaard, J., et al. (2015). The deubiquitinating enzyme AMSH 1 is required for rhizobial infection and nodule organogenesis in Lotus japonicus. *The Plant Journal, 83*, 719–731. https://doi.org/10.1111/tpj.12922.

Malolepszy, A., Kelly, S., Sørensen, K. K., James, E. K., Kalisch, C., Bozsoki, Z., et al. (2018). A plant chitinase controls cortical infection thread progression and nitrogen-fixing symbiosis. *Elife, 7*, e38874. https://doi.org/10.7554/eLife.38874.001.

Marino, D., Andrio, E., Danchin, E. G., Oger, E., Gucciardo, S., Lambert, A., et al. (2011). A Medicago truncatula NADPH oxidase is involved in symbiotic nodule functioning. *New Phytologist, 189*, 580–592. https://doi.org/10.1111/j.1469-8137.2010.03509.x.

Martínez-Hidalgo, P., & Hirsch, A. M. (2017). The nodule microbiome: N2-fixing rhizobia do not live alone. *Phytobiomes, 1*, 70–82. https://doi.org/10.1094/PBIOMES-12-16-0019-RVW.

Mathesius, U. (2019). Hormonal interactions in the regulation of the nitrogen-fixing legume Rhizobium symbiosis. In P. Frendo, C. Masson, & F. Frugier (Eds.), *Advances in Botanical Research* (vol. 94). pp. 41–66.

Mbengue, M., Camut, S., de Carvalho-Niebel, F., Deslandes, L., Froidure, S., Klaus-Heisen, D., et al. (2010). The Medicago truncatula E3 ubiquitin ligase PUB1 interacts with the LYK3 symbiotic receptor and negatively regulates infection and nodulation. *The Plant Cell, 22*, 3474–3488. https://doi.org/10.1105/tpc.110.075861.

Messinese, E., Mun, J. H., Yeun, L. H., Jayaraman, D., Rougé, P., Barre, A., et al. (2007). A novel nuclear protein interacts with the symbiotic DMI3 calcium-and calmodulin-dependent protein kinase of Medicago truncatula. *Molecular Plant-microbe Interactions, 20*, 912–921. https://doi.org/10.1094/MPMI-20-8-0912.

Miller, J. B., Pratap, A., Miyahara, A., Zhou, L., Bornemann, S., Morris, R. J., et al. (2013). Calcium/Calmodulin-dependent protein kinase is negatively and positively regulated by calcium, providing a mechanism for decoding calcium responses during symbiosis signaling. *The Plant Cell, 25*, 5053–5066. https://doi.org/10.1105/tpc.113.116921.

Mitra, R. M., Gleason, C. A., Edwards, A., Hadfield, J., Downie, J. A., Oldroyd, G. E., et al. (2004). A Ca^{2+}/calmodulin-dependent protein kinase required for symbiotic nodule development: Gene identification by transcript-based cloning. *Proceedings of the National Academy of Sciences, 101*, 4701–4705. https://doi.org/10.1073/pnas.0400595101.

Miwa, H., & Okazaki, S. (2017). How effectors promote beneficial interactions. *Current Opinion in Plant Biology, 38*, 148–154. https://doi.org/10.1016/j.pbi.2017.05.011.

Moling, S., Pietraszewska-Bogiel, A., Postma, M., Fedorova, E., Hink, M. A., Limpens, E., et al. (2014). Nod factor receptors form heteromeric complexes and are essential for intracellular infection in Medicago nodules. *The Plant Cell, 26*, 4188–4199. https://doi.org/10.1105/tpc.114.129502.

Morieri, G., Martinez, E. A., Jarynowski, A., Driguez, H., Morris, R., Oldroyd, G. E., et al. (2013). Host-specific Nod-factors associated with Medicago truncatula nodule infection differentially induce calcium influx and calcium spiking in root hairs. *New Phytologist, 200*, 656–662. https://doi.org/10.1111/nph.12475.

Murakami, E., Cheng, J., Gysel, K., Bozsoki, Z., Kawaharada, Y., Hjuler, C. T., et al. (2018). Epidermal LysM receptor ensures robust symbiotic signalling in Lotus japonicus. *Elife, 7*, e33506. https://doi.org/10.7554/eLife.31826.001.

Nandasena, K. G., O'Hara, G. W., Tiwari, R. P., & Howieson, J. G. (2006). Rapid in situ evolution of nodulating strains for Biserrula pelecinus L. through lateral transfer of a symbiosis island from the original mesorhizobial inoculant. *Applied and Environmental Microbiology, 72*, 7365–7367. https://doi.org/10.1128/AEM.00889-06.

Nandasena, K. G., O'Hara, G. W., Tiwari, R. P., Sezmiş, E., & Howieson, J. G. (2007). In situ lateral transfer of symbiosis islands results in rapid evolution of diverse competitive strains of mesorhizobia suboptimal in symbiotic nitrogen fixation on the pasture legume Biserrula pelecinus L. *Environmental Microbiology, 9*, 2496–2511. https://doi.org/10.1111/j.1462-2920.2007.01368.x.

Nelson, M. S., & Sadowsky, M. J. (2015). Secretion systems and signal exchange between nitrogen-fixing rhizobia and legumes. *Frontiers of Plant Science, 6*, 491. https://doi.org/10.3389/fpls.2015.00491.

Nguyen, T. H. N., Brechenmacher, L., Aldrich, J. T., Clauss, T. R., Gritsenko, M. A., Hixson, K. K., et al. (2012). Quantitative phosphoproteomic analysis of soybean root hairs inoculated with Bradyrhizobium japonicum. *Molecular & Cellular Proteomics, 11*, 1140–1155. https://doi.org/10.1074/mcp.M112.018028.

Nguyen, T. V., Wibberg, D., Battenberg, K., Blom, J., Heuvel, B. V., Berry, A. M., et al. (2016). An assemblage of Frankia Cluster II strains from California contains the canonical nod genes and also the sulfotransferase gene nodH. *BMC Genomics, 17*, 796. https://doi.org/10.1186/s12864-016-3140-1.

Normand, P., Lapierre, P., Tisa, L. S., Gogarten, J. P., Alloisio, N., Bagnarol, E., et al. (2007). Genome characteristics of facultatively symbiotic Frankia sp. strains reflect host range and host plant biogeography. *Genome Research, 17*, 7–15. https://doi.org/10.1101/gr.5798407.

Okazaki, S., Kaneko, T., Sato, S., & Saeki, K. (2013). Hijacking of leguminous nodulation signaling by the rhizobial type III secretion system. *Proceedings of the National Academy of Sciences, 110*, 17131–17136. https://doi.org/10.1073/pnas.1302360110.

Okazaki, S., Tittabutr, P., Teulet, A., Thouin, J., Fardoux, J., Chaintreuil, C., et al. (2016). Rhizobium–legume symbiosis in the absence of nod factors: Two possible scenarios with or without the T3SS. *The ISME Journal, 10*, 64. https://doi.org/10.1038/ismej.2015.103.

Oláh, B., Brière, C., Bécard, G., Dénarié, J., & Gough, C. (2005). Nod factors and a diffusible factor from arbuscular mycorrhizal fungi stimulate lateral root formation in Medicago truncatula via the DMI1/DMI2 signalling pathway. *The Plant Journal, 44*, 195–207. https://doi.org/10.1111/j.1365-313X.2005.02522.x.

Oldroyd, G. E. (2013). Speak, friend, and enter: Signalling systems that promote beneficial symbiotic associations in plants. *Nature Reviews Microbiology, 11*, 252. https://doi.org/10.1038/nrmicro2990.

Oldroyd, G. E., Murray, J. D., Poole, P. S., & Downie, J. A. (2011). The rules of engagement in the legume-rhizobial symbiosis. *Annual Review of Genetics, 45*, 119–144. https://doi.org/10.1146/annurev-genet-110410-132549.

Op den Camp, R., Streng, A., De Mita, S., Cao, Q., Polone, E., & Geurts, R. (2011). LysM-type mycorrhizal receptor recruited for rhizobium symbiosis in nonlegume Parasponia. *Science, 331*, 909–912. https://doi.org/10.1126/science.1198181.

Osakabe, Y., Yamaguchi-Shinozaki, K., Shinozaki, K., & Tran, L. S. P. (2013). Sensing the environment: Key roles of membrane-localized kinases in plant perception and response to abiotic stress. *Journal of Experimental Botany, 64*, 445–458. https://doi.org/10.1093/jxb/ers354.

Ott, T. (2017). Membrane nanodomains and microdomains in plant–microbe interactions. *Current Opinion in Plant Biology, 40*, 82–88. https://doi.org/10.1016/j.pbi.2017.08.008.

Pan, H., Stonoha-Arther, C., & Wang, D. (2018). Medicago plants control nodulation by regulating proteolysis of the receptor-like kinase DMI2. *Plant Physiology, 177*, 792–802. https://doi.org/10.1104/pp.17.01542.

Parniske, M. (2008). Arbuscular mycorrhiza: The mother of plant root endosymbioses. *Nature Reviews Microbiology, 6*, 763. https://doi.org/10.1038/nrmicro1987.

Peck, M. C., Fisher, R. F., Bliss, R., & Long, S. R. (2013). Isolation and characterization of mutant Sinorhizobium meliloti NodD1 proteins with altered responses to luteolin. *Journal of Bacteriology, 195*, 3714–3723. https://doi.org/10.1128/JB.00309-13.

Perret, X., Staehelin, C., & Broughton, W. J. (2000). Molecular basis of symbiotic promiscuity. *Microbiology and Molecular Biology Reviews, 64*, 180–201. https://doi.org/10.1128/mmbr.64.1.180-201.2000.

Persson, T., Battenberg, K., Demina, I. V., Vigil-Stenman, T., Heuvel, B. V., Pujic, P., et al. (2015). Candidatus Frankia datiscae Dg1, the actinobacterial microsymbiont of Datisca glomerata, expresses the canonical nod genes nodABC in symbiosis with its host plant. *PLoS One, 10*, e0127630. https://doi.org/10.1371/journal.pone.0127630.

Pietraszewska-Bogiel, A., Lefebvre, B., Koini, M. A., Klaus-Heisen, D., Takken, F. L., Geurts, R., et al. (2013). Interaction of Medicago truncatula lysin motif receptor-like kinases, NFP and LYK3, produced in Nicotiana benthamiana induces defence-like responses. *PLoS One, 8*, e65055. https://doi.org/10.1371/journal.pone.0065055.

Pimprikar, P., Carbonnel, S., Paries, M., Katzer, K., Klingl, V., Bohmer, M. J., et al. (2016). A CCaMK-CYCLOPS-DELLA complex activates transcription of RAM1 to regulate arbuscule branching. *Current Biology, 26*, 987–998. https://doi.org/10.1016/j.cub.2016.01.069.

Pingret, J. L., Journet, E. P., & Barker, D. G. (1998). Rhizobium Nod factor signaling: Evidence for a G protein–mediated transduction mechanism. *The Plant Cell, 10*, 659–671. https://doi.org/10.1105/tpc.10.5.659.

Plett, K. L. (2018). Fresh knowledge for an old relationship: New discoveries in molecular mycorrhizal research. *New Phytologist, 217*, 26–28. https://doi.org/10.1111/nph.14875.

Poinsot, V., Crook, M. B., Erdn, S., Maillet, F., Bascaules, A., & Ané, J. M. (2016). New insights into Nod factor biosynthesis: Analyses of chitooligomers and lipo-chitooligomers of Rhizobium sp. IRBG74 mutants. *Carbohydrate Research, 434*, 83–93. https://doi.org/10.1016/j.carres.2016.08.001.

Radutoiu, S., Madsen, L. H., Madsen, E. B., Felle, H. H., Umehara, Y., Grønlund, M., et al. (2003). Plant recognition of symbiotic bacteria requires two LysM receptor-like kinases. *Nature, 425*, 585. https://doi.org/10.1038/nature02039.

Radutoiu, S., Madsen, L. H., Madsen, E. B., Jurkiewicz, A., Fukai, E., Quistgaard, E. M., et al. (2007). LysM domains mediate lipochitin–oligosaccharide recognition and Nfr genes extend the symbiotic host range. *The EMBO Journal, 26*, 3923–3935. https://doi.org/10.1038/sj.emboj.7601826.

Remigi, P., Masson-Boivin, C., & Rocha, E. P. (2019). Experimental evolution as a tool to investigate natural processes and molecular functions. *Trends in Microbiology*. https://doi.org/10.1016/j.tim.2019.02.003.

Remigi, P., Zhu, J., Young, J. P. W., & Masson-Boivin, C. (2016). Symbiosis within symbiosis: Evolving nitrogen-fixing legume symbionts. *Trends in Microbiology, 24*, 63–75. https://doi.org/10.1016/j.tim.2015.10.007.

Ried, M. K., Antolín-Llovera, M., & Parniske, M. (2014). Spontaneous symbiotic reprogramming of plant roots triggered by receptor-like kinases. *Elife, 3*, e03891. https://doi.org/10.7554/eLife.03891.

Rogel, M. A., Ormeno-Orrillo, E., & Romero, E. M. (2011). Symbiovars in rhizobia reflect bacterial adaptation to legumes. *Systematic & Applied Microbiology, 34*, 96–104. https://doi.org/10.1016/j.syapm.2010.11.015.

Rose, C. M., Venkateshwaran, M., Grimsrud, P. A., Westphall, M. S., Sussman, M. R., Coon, J. J., et al. (2012). Medicago PhosphoProtein database: A repository for Medicago truncatula phosphoprotein data. *Frontiers of Plant Science, 3*, 122. https://doi.org/10.3389/fpls.2012.00122.

Rose, C. M., Venkateshwaran, M., Volkening, J. D., Grimsrud, P. A., Maeda, J., Bailey, D. J., et al. (2012). Rapid phosphoproteomic and transcriptomic changes in the rhizobia-legume symbiosis. *Molecular & Cellular Proteomics, 11*, 724–744. https://doi.org/10.1074/mcp.M112.019208.

Saha, S., Dutta, A., Bhattacharya, A., & DasGupta, M. (2014). Intracellular catalytic domain of symbiosis receptor kinase hyperactivates spontaneous nodulation in absence of rhizobia. *Plant Physiology, 166*, 1699–1708. https://doi.org/10.1104/pp.114.250084.

Saito, K., Yoshikawa, M., Yano, K., Miwa, H., Uchida, H., Asamizu, E., et al. (2007). NUCLEOPORIN85 is required for calcium spiking, fungal and bacterial symbioses, and seed production in Lotus japonicus. *The Plant cell, 19*, 610–624. https://doi.org/10.1105/tpc.106.046938.

Schauser, L., Roussis, A., Stiller, J., & Stougaard, J. (1999). A plant regulator controlling development of symbiotic root nodules. *Nature, 402*, 191. https://doi.org/10.1038/46058.

Schlaman, H. R. M., Phillips, D. A., & Kondorosi, E. (1998). Genetic organization and transcriptional regulation of rhizobial nodulation genes. In H. P. Spaink, A. Kondorosi, & P. J. J. Hooykaas (Eds.), *The rhizobiaceae: Molecular biology of model plant-associated bacteria* (pp. 361–386). Dordrecht: Springer. https://doi.org/10.1007/978-94-011-5060-6_19.

Singh, S., Katzer, K., Lambert, J., Cerri, M., & Parniske, M. (2014). CYCLOPS, a DNA-binding transcriptional activator, orchestrates symbiotic root nodule development. *Cell Host & Microbe, 15*, 139–152. https://doi.org/10.1016/j.chom.2014.01.011.

Smit, P., Limpens, E., Geurts, R., Fedorova, E., Dolgikh, E., Gough, C., et al. (2007). Medicago LYK3, an entry receptor in rhizobial nodulation factor signaling. *Plant Physiology, 145*, 183–191. https://doi.org/10.1104/pp.107.100495.

Smit, P., Raedts, J., Portyanko, V., Debellé, F., Gough, C., Bisseling, T., et al. (2005). NSP1 of the GRAS protein family is essential for rhizobial Nod factor-induced transcription. *Science, 308*, 1789–1791. https://doi.org/10.1126/science.1111025.

Soltis, D. E., Soltis, P. S., Morgan, D. R., Swensen, S. M., Mullin, B. C., Dowd, J. M., et al. (1995). Chloroplast gene sequence data suggest a single origin of the predisposition for symbiotic nitrogen fixation in angiosperms. *Proceedings of the National Academy of Sciences, 92*, 2647–2651. https://doi.org/10.1073/pnas.92.7.2647.

Soyano, T., Kouchi, H., Hirota, A., & Hayashi, M. (2013). Nodule inception directly targets NF-Y subunit genes to regulate essential processes of root nodule development in Lotus japonicus. *PLoS Genetics, 9*(3), e1003352. https://doi.org/10.1371/journal.pgen.1003352.

Spaink, H. P., Wijffelman, C. A., Pees, E., Okker, R. J., & Lugtenberg, B. J. J. (1987). Rhizobium nodulation gene nodD as a determinant of host specificity. *Nature, 328*, 337–340. https://doi.org/10.1038/328337a0.

Stracke, S., Kistner, C., Yoshida, S., Mulder, L., Sato, S., Kaneko, T., et al. (2002). A plant receptor-like kinase required for both bacterial and fungal symbiosis. *Nature, 417*, 959. https://doi.org/10.1038/nature00841.

Sullivan, J. T., Patrick, H. N., Lowther, W. L., Scott, D. B., & Ronson, C. W. (1995). Nodulating strains of Rhizobium loti arise through chromosomal symbiotic gene transfer in the environment. *Proceedings of the National Academy of Sciences, 92*, 8985–8989. https://doi.org/10.1073/pnas.92.19.8985.

Sun, J., Miller, J. B., Granqvist, E., Wiley-Kalil, A., Gobbato, E., Maillet, F., et al. (2015). Activation of symbiosis signaling by arbuscular mycorrhizal fungi in legumes and rice. *The Plant Cell, 27*, 823–838. https://doi.org/10.1105/tpc.114.131326.

Sun, J., Miwa, H., Downie, J. A., & Oldroyd, G. E. (2007). Mastoparan activates calcium spiking analogous to Nod factor-induced responses in Medicago truncatula root hair cells. *Plant Physiology, 144*, 695–702. https://doi.org/10.1104/pp.106.093294.

Svistoonoff, S., Benabdoun, F. M., Nambiar-Veetil, M., Imanishi, L., Vaissayre, V., Cesari, S., et al. (2013). The independent acquisition of plant root nitrogen-fixing symbiosis in Fabids recruited the same genetic pathway for nodule organogenesis. *PLoS One, 8*, e64515. https://doi.org/10.1371/journal.pone.0064515.

Takeda, N., Maekawa, T., & Hayashi, M. (2012). Nuclear-localized and deregulated calcium-and calmodulin-dependent protein kinase activates rhizobial and mycorrhizal responses in Lotus japonicus. *The Plant Cell, 24*, 810–822. https://doi.org/10.1105/tpc.111.091827.

Tian, C. F., Zhou, Y. J., Zhang, Y. M., Li, Q. Q., Zhang, Y. Z., Li, D. F., et al. (2012). Comparative genomics of rhizobia nodulating soybean suggests extensive recruitment of lineage-specific genes in adaptations. *Proceedings of the National Academy of Sciences, 109*, 8629–8634. https://doi.org/10.1073/pnas.1120436109.

Tian, M., & Xie, Q. (2013). Non-26S proteasome proteolytic role of ubiquitin in plant endocytosis and endosomal trafficking (F). *Journal of Integrative Plant Biology, 55*, 54–63. https://doi.org/10.1111/jipb.12007.

Tirichine, L., Imaizumi-Anraku, H., Yoshida, S., Murakami, Y., Madsen, L. H., Miwa, H., et al. (2006). Deregulation of a Ca^{2+}/calmodulin-dependent kinase leads to spontaneous nodule development. *Nature, 441*, 1153. https://doi.org/10.1038/nature04862.

Tsikou, D., Ramirez, E. E., Psarrakou, I. S., Wong, J. E., Jensen, D. B., Isono, E., et al. (2018). A Lotus japonicus E3 ligase interacts with the Nod Factor Receptor 5 and positively regulates nodulation. *BMC Plant Biology, 18*, 217. https://doi.org/10.1186/s12870-018-1425-z.

Valdés-López, O., Jayaraman, D., Maeda, J., Delaux, P. M., Venkateshwaran, M., Isidra-Arellano, M. C., et al. (2018). A novel positive regulator of the early stages of root nodule symbiosis identified by phosphoproteomics. *Plant and Cell Physiology, 60*, 575–586. https://doi.org/10.1093/pcp/pcy228.

Van Brussel, A. A., Zaat, S. A., Cremers, H. C., Wijffelman, C. A., Pees, E., Tak, T. E. U. N., et al. (1986). Role of plant root exudate and Sym plasmid-localized nodulation genes in the synthesis by Rhizobium leguminosarum of Tsr factor, which causes thick and short roots on common vetch. *Journal of Bacteriology, 165*, 517–522. https://doi.org/10.1128/jb.165.2.517-522.1986.

van Velzen, R., Doyle, J. J., & Geurts, R. (2018). A resurrected scenario: Single gain and massive loss of nitrogen-fixing nodulation. *Trends in Plant Science, 24*, 49–57. https://doi.org/10.1016/j.tplants.2018.10.005.

van Velzen, R., Holmer, R., Bu, F., Rutten, L., van Zeijl, A., Liu, W., et al. (2018). Comparative genomics of the nonlegume Parasponia reveals insights into evolution of nitrogen-fixing rhizobium symbioses. *Proceedings of the National Academy of Sciences, 115*, E4700–E4709. https://doi.org/10.1073/pnas.1721395115.

Venkateshwaran, M., Cosme, A., Han, L., Banba, M., Satyshur, K. A., Schleiff, E., et al. (2012). The recent evolution of a symbiotic ion channel in the legume family altered ion conductance and improved functionality in calcium signaling. *The Plant Cell, 24*, 2528–2545. https://doi.org/10.1105/tpc.112.098475.

Venkateshwaran, M., Jayaraman, D., Chabaud, M., Genre, A., Balloon, A. J., Maeda, J., et al. (2015). A role for the mevalonate pathway in early plant symbiotic signaling. *Proceedings of the National Academy of Sciences, 112*, 9781–9786. https://doi.org/10.1073/pnas.1413762112.

Vernié, T., Camut, S., Camps, C., Rembliere, C., de Carvalho-Niebel, F., Mbengue, M., et al. (2016). PUB1 interacts with the receptor kinase DMI2 and negatively regulates rhizobial and arbuscular mycorrhizal symbioses through its ubiquitination activity in Medicago truncatula. *Plant Physiology, 170*, 2312–2324. https://doi.org/10.1104/pp.15.01694.

Vierstra, R. D. (2009). The ubiquitin–26S proteasome system at the nexus of plant biology. *Nature Reviews Molecular Cell Biology, 10*, 385–397. https://doi.org/10.1038/nrm2688.

Wang, Q., Liu, J., & Zhu, H. (2018). Genetic and molecular mechanisms underlying symbiotic specificity in legume-Rhizobium interactions. *Frontiers of Plant Science, 9*, 313. https://doi.org/10.3389/fpls.2018.00313.

Wong, J. E. M. M., Nadzieja, M., Madsen, L. H., Bücherl, C. A., Dam, S., Sandal, N. N., et al. (2019). A Lotus japonicus cytoplasmic kinase connects Nod factor perception by the NFR5 LysM receptor to nodulation. *Proceedings of the National Academy of Sciences, 116*, 14339–14348. https://doi.org/10.1073/pnas.1815425116.

Yano, K., Shibata, S., Chen, W. L., Sato, S., Kaneko, T., Jurkiewicz, A., et al. (2009). CERBERUS, a novel U-box protein containing WD-40 repeats, is required for formation of the infection thread and nodule development in the legume–Rhizobium symbiosis. *The Plant Journal, 60*, 168–180. https://doi.org/10.1111/j.1365-313X.2009.03943.x.

Yuan, S., Zhu, H., Gou, H., Fu, W., Liu, L., Chen, T., et al. (2012). A ubiquitin ligase of symbiosis receptor kinase involved in nodule organogenesis. *Plant Physiology, 160,* 106–117. https://doi.org/10.1104/pp.112.199000.
Zhang, X., Dong, W., Sun, J., Feng, F., Deng, Y., He, Z., et al. (2015). The receptor kinase CERK 1 has dual functions in symbiosis and immunity signalling. *The Plant Journal, 81,* 258–267. https://doi.org/10.1111/tpj.12723.
Zipfel, C., & Oldroyd, G. E. (2017). Plant signalling in symbiosis and immunity. *Nature, 543,* 328–336. https://doi.org/10.1038/nature22009.

CHAPTER TWO

Hormonal interactions in the regulation of the nitrogen-fixing legume-*Rhizobium* symbiosis

Ulrike Mathesius*
Division of Plant Science, Research School of Biology, Australian National University, Australia
*Corresponding author: E-mail: ulrike.mathesius@anu.edu.au

Contents

Abstract

Plant hormones have long been associated with the ability of legumes to form nitrogen-fixing symbioses. During the nodulation program, two events are required: The formation of a new root organ, the nodule, and the successful infection of rhizobia into the root, and later the nodule. Cell-type specific hormone synthesis, perception and downstream responses are required in the epidermis, the root cortex and in vascular tissues to coordinate infection and nodule development. Recent progress in this area has demonstrated the crucial role of plant hormones in feedback regulation of nodule development and infection. Hormonal communication between shoot and root has also been linked to the systemic autoregulation of nodule numbers. Future studies will need to focus on how legumes regulate nodule development separately from other organogenesis processes like lateral root formation, even though the same hormones are involved in both. Similarly, balancing nodulation with restraints of abiotic and biotic stresses will require integration of hormone studies in plants facing several

Advances in Botanical Research, Volume 94
ISSN 0065-2296
https://doi.org/10.1016/bs.abr.2019.09.010

simultaneous challenges. While most studies have focused on the effects of single hormones, it will also be necessary to develop new tools for single cell hormone analyses and concurrent analysis of multiple hormones.

1. Introduction

The interaction between rhizobia and their legume hosts results in the formation of nitrogen-fixing root nodules. The symbiosis starts with the recognition of plant flavonoids, which activates Nod factor synthesis by rhizobia. Nod factor recognition by the host controls activation/repression of defense responses, the formation of an infection thread (or crack-entry infection in some legumes) and initiation of nodule development (Oldroyd, 2013). This involves signaling between the infected root hair and the inner layers of the root where nodule development starts. Typically, in temperate legumes forming indeterminate nodules, like *Medicago truncatula* or pea, nodule development starts in the pericycle, endodermis and inner cortex (Xiao et al. 2014), while in legumes forming determinate nodules, like *Lotus japonicus* and soybean, cell divisions start in the outer/middle cortex and the pericycle (Hirsch, 1992). Because nitrogen fixation and nodule formation require continuous input of carbon, the symbiosis is autoregulated to optimize nodule numbers. This involves a systemic signaling circuit between root and shoot (autoregulation of nodulation, AON), in which rhizobia induce signaling peptides that move to the shoot, where they are perceived by a receptor kinase. This triggers the movement of a shoot-to-root signal preventing the formation of further nodules (reviewed by Mortier, Holsters & Goormachtig, 2012b; Reid, Ferguson, Hayashi, Lin, & Gresshoff, 2011). At later stages of the symbiosis, in ineffective symbioses or under stressful conditions, the plant induces nodule senescence and loss of nitrogen fixation (Van de Velde et al., 2006).

While rhizobia synthesize some plant hormones, there is so far no strong evidence that these are essential for nodule initiation or infection and this topic is not covered in this review. However, hormones synthesized by rhizobia might play a role in enhancing the efficiency of the symbiosis. In contrast, spatiotemporal control of hormone synthesis, transport and perception by the host were shown to be essential for successful infection and nodule development. This review focusses on the role of the phytohormones abscisic acid, auxin, cytokinin, ethylene, gibberellic acid, jasmonic acid and strigolactones. Due to space constraints, not all studies could be mentioned here. For a broader review on the role of phytohormones and related signals in legume-rhizobia interactions, including other hormones and signals and the influence of hormone synthesis by rhizobia, please refer to Ferguson and Mathesius (2014), Bensmihen (2015), and Liu, Zhang, Yang, Yu, and Wang (2018).

2. Abscisic acid

Abscisic acid (ABA) is well known for its roles in drought responses and seed ripening, but also plays roles in the root development and biotic interactions. ABA has emerged as an important regulator of both rhizobial infection and nodule development. In several legumes, external addition of ABA in the low micro-molar range negatively influences nodule numbers and infections (e.g. Ding et al., 2008; Liang, Mitchell, & Harris, 2007; Suzuki et al., 2004), while an ABA inhibitor was shown to increase nodule numbers in *L. japonicus* (Suzuki et al., 2004).

The negative effect of ABA on nodulation is supported by several studies of ABA-deficient mutants or transgenic plants: Expression of a dominant negative allele of Arabidopsis *abi1-1* (*abscisic acid insensitive1-1*) led to hyper-nodulation in *M. truncatula*, with nodules being normally infected and nitrogen-fixing (Ding et al., 2008). In contrast, the *sta1* (*sensitivity to aba1*) mutant, showed reduced ABA sensitivity during nodulation and effective Nod factor signaling (Ding et al., 2008). A *L. japonicus* mutant (*enf1*, *enhanced nitrogen fixation 1*) with reduced ABA content similarly showed enhanced infection, nodulation and nitrogen fixation (Tominaga et al., 2009). The ABA-insensitive *L. japonicus Beyma* mutant nodulates normally but nodules are smaller (Biswas, Chan, & Gresshoff, 2009). The Medicago *latd/nip1* (*lateral root organ defective/numerous infections and polyphenolics*), which shows altered ABA sensitivity, forms small white nodules with defects in meristem activity, infection thread formation, bacterial release and nitrogen fixation (Bright, Liang, Mitchell, & Harris, 2005; Liang Mitchell & Harris, 2007; Veereshlingham et al., 2004). *LATD/NIP1* is expressed in nodule and root meristems, and its expression is inhibited by ABA (Yendrek et al., 2010). The *latd/nip1* mutant is also defective in lateral root development, but while ABA can rescue the defective lateral root meristems it does not restore the nodulation defect (Liang Mitchell & Harris, 2007). NIP1 encodes a high-affinity nitrate transporter, although the *latd/nip1* mutant shows normal nitrate inhibition of nodulation (Bagchi et al., 2012; Yendrek et al., 2010). It will be interesting to find out whether NIP1 also transports other substrates, possibly hormones, which could account for the mutant's ABA phenotypes (Harris & Dickstein, 2010).

ABA acts by inhibiting Nod factor signaling, including calcium spiking that is rapidly activated downstream of Nod factor perception and induction of several early nodulation genes (Ding et al., 2008). ABA likely acts also downstream of cytokinin signaling (see below) because ABA application inhibits cytokinin-induced nodulin induction and the formation of spontaneous nodules in the Lotus *snf2* mutant, which has a constitutive cytokinin response (Ding et al., 2008).

Interestingly, the same range of ABA concentrations that inhibits nodulation increases lateral root numbers in *M. truncatula* (Ding et al., 2008; Liang & Harris, 2005), while reducing lateral root formation in non-legumes. This suggests that ABA root responses might be distinct in legumes and it will be interesting to find if this is associated with the ability to nodulate (Liang & Harris, 2005).

3. Auxin

The plant hormone auxin represents a group of related hormones, of which indole-3-acetic acid (IAA) is the most studied and likely the most abundant one. However, other forms of auxin are present in plants and may play specific functions, although these remain largely unexplored. Auxin is generally known for its role in the control of cell division, differentiation, and elongation; these functions are essential for organ initiation and differentiation throughout the plant (Benková et al., 2003) and this is similar for nodule development (Mathesius, 2008).

Auxin accumulation in dividing organ primordia is found throughout plant development (Benková et al., 2003). This is mirrored by increased auxin responses in nodule primordia detected through the expression of auxin responsive reporters (*GH3* or *DR5*) in the pericycle, endodermis and inner cortex during indeterminate nodule development (e.g. Breakspear et al., 2014; Mathesius et al., 1998; Ng et al., 2015; van Noorden et al., 2007). In *M. truncatula*, other auxin responses, for example expression of *SAUR* (*SMALL AUXIN UP*) genes is also induced at the infection site (Breakspear et al., 2014). This auxin response is accompanied by an increased auxin (IAA) content at the rhizobial infection site within 24 h (Ng et al., 2015). In legumes forming determinate nodules, increased auxin responses are similarly found in dividing outer/middle cortex cells (e.g. Ng & Mathesius, 2018; Suzaki et al., 2012; Takanashi, Sugiyama, & Yazaki, 2011; Turner et al., 2013). In mature nodules, auxin maxima typically disappear from the infected nodule zone but remain in the vascular tissue and in the nodule meristem at least in indeterminate nodules (Breakspear et al., 2014; Fisher, Gaillard, Fellbaum, Subramanian, & Smith, 2018; Franssen et al., 2015; Guan et al., 2013; Mathesius et al., 1998; Ng & Mathesius, 2018; Suzaki et al., 2012; Takanashi et al., 2011; Turner et al., 2013).

Auxin is likely to play several parallel roles in infection, cell cycle control, vascular tissue development and differentiation, indicated by the large overlap in auxin and rhizobia-induced responses in roots (van Noorden et al., 2007; Herrbach et al., 2017). One of auxin's functions is in the cell cycle control in nodule meristems through regulation of the auxin-induced cyclin, CycA2 (Roudier et al., 2003). In turn, silencing of the cell cycle regulator *CDC16* reduced auxin sensitivity and led to increased nodulation in *M. truncatula* (Kuppusamy et al., 2009). A role for auxin in infection was suggested because of an increased auxin response in successfully infected root hairs; this auxin response might be necessary for the control of defense responses (Breakspear et al., 2014). This is supported by a defective infection phenotype of the auxin response mutant, *arf16a* (*auxin response factor 16a*) of *M. truncatula* (Breakspear et al., 2014). In addition, Nadzieja, Kelly, Stougaard, and Reid (2018) demonstrated a Nod factor-dependent rapid accumulation of auxin in infected root hairs in *L. japonicus*, and that this was due to increased auxin synthesis through LjTAR1 (TRYPTOPHAN AMINOTRANSFERASE-RELATED), whose action was shown to be necessary for infection. This was accompanied by induction of genes controlling auxin response (*LjARF16*, homologous to *MtARF16a*) and auxin conjugation (*LjGH3; GRETCHEN HAGEN3*). Auxin is likely to interact with cytokinin (see below) during infection as auxin application increased cytokinin synthesis, and cytokinin application or constitutive cytokinin response activation in the *Ljlhk1* (*lotus histidine kinase 1*) cytokinin receptor mutant increased auxin responses in root hairs (Nadzieja et al., 2018).

Many elegant studies in Arabidopsis have shown that the localization of auxin in specific cell types is largely a result of directed auxin transport through members of the PIN (PIN-formed), ABC/PGP (ATP-BINDING CASSETTE/P-GLYCOPROTEIN) and AUX1/LAX (AUXIN RESISTANT/LIKE AUX1) auxin transporters (Petrášek & Friml, 2009). Mathematical modeling of auxin maxima forming in nodule primordia has suggested that inhibition of auxin export is the most likely mechanism for auxin localization, with additional roles for local auxin synthesis, inhibition or breakdown and altered auxin responses (Deinum, Geurts, Bisseling, & Mulder, 2012; Deinum, Kohlen, & Geurts, 2016). There is evidence for all of these possible mechanisms, although with differences in indeterminate and determinate nodules (Kohlen, Ng, Deinum, & Mathesius, 2017; Ng & Mathesius, 2018). Evidence for a role of auxin transport control in nodule initiation more directly comes from measurements of acropetal auxin transport, which is inhibited temporarily preceding nodule primordium formation and auxin accumulation in *M. truncatula, Trifolium repens* and *Vicia sativa* (Boot, van Brussell, Tak, Spaink & Kijne, 1999; Mathesius et al., 1998; Wasson, Pellerone, & Mathesius, 2006). However, this auxin transport inhibition does not occur in *L. japonicus*, instead, rhizobia increase acropetal auxin transport in this legume forming determinate nodules (Ng & Mathesius, 2018). Interestingly, auxin transport inhibitors can induce pseudonodules in legumes for indeterminate, but not determinate nodules, supporting the idea that the requirement for auxin transport is different between legumes forming different nodule types (Ng & Mathesius, 2018).

Currently it is unknown which auxin transporters mediate auxin transport inhibition in legumes. Several *PIN* and *LAX* genes are expressed during early nodule development (de Billy, Grosjean, May, Bennett & Cullimore., 2001; Huo, Schnabel, Hughes, & Frugoli, 2006) and in mature nodules (Roux et al., 2014; Sańko-Sawczenko, Łotocka, & Czarnocka, 2016). Silencing of individual *MtPIN* genes has been shown to reduce nodule numbers (Huo et al., 2006). Nod factor application downregulates expression of *MtPIN9*, which might reflect the reduced auxin transport observed (Plet et al., 2011), but this would require genetic support to be established. The auxin importer *Mt*LAX2 is expressed during the early and late stages of nodule development and *Mtlax2* mutants exhibit reduced nodule numbers (Roy et al., 2017). The importance of auxin efflux and influx control is supported by evidence that application of both auxin import inhibitors (Roy et al., 2017) or auxin export inhibitors (van Noorden, Ross, Reid, Rolfe, & Mathesius, 2006) reduces nodule numbers.

In addition to changes in *PIN* or *LAX* gene expression, inhibition of their activity could contribute to altered auxin transport during nodulation. Flavonoids can inhibit polar auxin transport by affecting *PIN* gene expression, intracellular cycling, and activity of ABC-type transporters (Peer & Murphy, 2007). Silencing the flavonoid pathway in *M. truncatula* was shown to prevent nodulation as well as the inhibition of auxin transport by rhizobia (Wasson et al., 2006). Flavonols are the most likely candidates with roles as auxin transport inhibitors in *M. truncatula* (Zhang, Subramanian, Stacey, & Yu, 2009). Flavonols and other flavonoids are induced in the root within the time that auxin transport changes occur (Ng et al., 2015). In addition, application of certain flavonoids to roots of the non-nodulating *Mtcre1* (*cytokinin receptor1*) mutant, which is defective in cytokinin signaling, restored a lack of auxin transport control, auxin responses and nodule initiation in this mutant (Ng et al., 2015). This, and the observation that auxin responses are induced in the constitutive cytokinin signaling mutant *lhk1* in *L. japonicus*, places auxin changes downstream of cytokinin signaling during indeterminate and determinate nodule formation (see below; Ng et al., 2015; Plet et al., 2011; Suzaki et al., 2012).

Additional mechanisms that could explain auxin accumulation in nodule primordia include the induction of local auxin synthesis or release, or reduced auxin breakdown. The auxin synthesis gene, *GmYUC2a* (YUCCA2a) is induced during nodulation in soybean and its overexpression causes defects in nodule development, but also root hair and root development (Wang et al., 2019). Similarly, the gene encoding the auxin biosynthesis enzyme LjTAR is expressed in developing *L. japonicus* nodule primordia (Suzaki et al. 2012). Increased expression of auxin conjugate hydrolase genes suggests that the release of auxin from conjugated forms could contribute to increasing auxin concentration during nodulation in *M. truncatula* (Campanella, Smith, Leibu, Wexler, & Ludwig-Müller, 2008). In soybean, three *GH3* genes, encoding auxin conjugases, are expressed during nodule development, and their silencing causes specific effects on nodule numbers and maturation (Damodaran, Westphall, Kisely, Jez, & Subramanian, 2017). In addition, auxin retention through reduced breakdown of auxin in nodule primordia could be mediated by flavonoids accumulating in dividing cortical cells that can alter the activity of IAA oxidases (Mathesius, 2001).

Altered auxin sensitivity could also account for some of the observed auxin responses during nodulation. Indeed, silencing of genes encoding the auxin receptors TIR1 and AFB3 in soybean led to reduced infection events and nodule numbers, while *TIR1A* and *C* overexpression increased nodule numbers (Cai et al., 2017). Throughout the plant, auxin responses can be altered by microRNAs that target auxin receptors or auxin response genes (Weijers & Wagner, 2016). The *TIR1/AFB3* gene family is targeted by miRNA393 and overexpression of miRNA393 in soybean reduced TIR1/AFB3 expression and reduced nodule numbers (Cai et al., 2017). Similarly, reduced nodule numbers were found when overexpressing the miRNA393 in *M. truncatula* (Mao, Turner, Yu, & Subramanian, 2013). Similarly, overexpression of the microRNA160a, targeting auxin response factor (ARF) genes of the ARF10/16/17 family, reduced nodule numbers (Bustos-Sanmamed et al., 2013). Overexpression of miRNA390 also resulted in inhibition of rhizobial infection and reduced nodule numbers, while inactivation of miRNA390, leading to the inactivation of the auxin response factors ARF2, 3 and 4, increased nodule numbers (Hobecker et al., 2017). Finally, in soybean, the miRNA167 targets the expression of *ARF8a* and *b* and the repression of these ARF genes increased nodulation (Wang et al., 2015). These studies overall suggest that a reduction in specific auxin responses promotes nodulation, notably through the action of miRNAs, but the exact spatiotemporal control remains unclear. It is also possible that silencing of miRNAs affects more than the currently known targets. For example, in soybean, overexpression of the miRNA160 led to an auxin hypersensitivity as expected, but a cytokinin hyposensitivity was also noted (Turner et al., 2013), pointing to a possible indirect effect of auxin modulation on cytokinin responses. Indeed, nodule numbers in these plants could be rescued by cytokinin addition (Nizampatnam, Schreier, Damodaran, Adhikari, & Subramanian., 2015). The reduced auxin responses suggested to occur in these studies would explain why nodulation is generally inhibited by external application of auxin (van Noorden et al., 2006), in contrast to lateral root formation. The action of miRNAs could also explain the drop in auxin responses in differentiating nodules, but does not easily account for the increased auxin responses in nodule primordia.

In addition to the localized changes in auxin accumulation and transport, long distance inhibition of auxin transport has been linked to the control of nodule numbers via AON in *M. truncatula.* An inhibition of auxin transport from the shoot to the root was observed in inoculated wild type plants, but not in the *M. truncatula* AON mutant *sunn (super numeric nodules),* and the supernodulation phenotype of this mutant could be alleviated by an application of auxin transport inhibitors at the shoot/root junction (van Noorden et al., 2006). This long distance auxin transport control is dependent on carbon and nitrogen supply to the root in an AON/SUNN-dependent manner, indicating that auxin transport could reflect changes in resource partitioning (Jin, Watt, & Mathesius, 2012). However, these observations require studies of AON mutants in other legume species, to determine notably if this response is conserved between legumes forming determinate and indeterminate nodules.

4. Cytokinin

Similar to auxins, cytokinins generally have roles in plant development and cell cycle control. While cytokinins are represented by a group of related adenine-derived hormones, most studies have not differentiated between different cytokinin functions and these forms will therefore be referred to as 'cytokinin'. Cytokinin signaling via specific cytokinin receptors is essential for nodule initiation in legumes forming both determinate and indeterminate nodules, but the role of cytokinin differs between negative actions in rhizobial infection and positive actions in nodule organogenesis (Gamas, Brault, Jardinaud, & Frugier, 2017). For example, Lotus and Medicago mutants defective in cytokinin perception are unable to form nodules but can still be infected by rhizobia, although this differs between *L. japonicus*, which is hyperinfected, and *M. truncatula*, which is not (Gonzalez-Rizzo, Crespi, & Frugier, 2006; Murray et al., 2007; Plet et al., 2011). In contrast, mutants with a gain-of-function in cytokinin signaling can spontaneously, in the absence of rhizobia, activate cortical cell divisions and nodule-like organs (Tirichine et al., 2007). In addition, treatment of legume roots with cytokinin can induce cortical cell divisions and "pseudo-nodules" that show similar nodulation gene responses and requirements as rhizobia-induced nodules (e.g. Heckmann et al., 2011; Mathesius, Charon, Rolfe, Kondorosi, & Crespi, 2000). This ability is conserved in nodulating legumes, but absent in non-nodulating legumes and non-legumes, suggesting that it might be an important step in the evolution of nodule organogenesis (Gauthier-Coles, White, & Mathesius, 2018).

Cytokinin effects involve a *de novo* cytokinin biosynthesis during nodulation. In *M. truncatula*, rhizobia induce the cytokinin synthesis genes *LOG1* and *2* (*LONELY GUY1* and *2*) in dividing cortical cells and nodule meristems, and their silencing reduces nodule numbers (Mortier et al., 2014). Similarly, isopentenyl transferase genes are activated in infected roots, and silencing of *IPT3* in *L. japonicus* inhibited nodulation and caused early nodule senescence (Chen et al., 2013). Further, the *L. japonicus IPT2* and *LOG4* genes were shown to be required for the initial cytokinin accumulation in roots during infection (Reid et al., 2017), and in soybean, the *IPT2* gene was induced in roots after inoculation (Mens, Haaima, Gresshoff, & Ferguson, 2018). Increased concentrations of several cytokinins were measured within 3 h of Nod factor treatment in *M. truncatula*, and this was partially dependent on the positive function of NSP1 and NSP2 as well as on the negative function of ethylene signaling through the ethylene response gene *SKL/EIN2* (SICKLE/ETHYLENE INSENSITIVE 2; van Zeijl et al., 2015b). Cytokinin concentrations are tightly controlled by negative feedback inhibitions, for example through the oxidation by cytokinin oxidases (CKX). Silencing of the *L. japonicus CKX3* gene increased cytokinin concentrations and led to a reduced nodulation, suggesting that cytokinin concentrations require a window of optimum concentration (Murray et al., 2007; Reid, Heckmann, Novak, Kelly, & Stougaard, 2016).

The rhizobium-induced cytokinin synthesis then activates cytokinin receptors, in the inner cortex but also in root hairs (Jardinaud et al., 2016; van Zeijl et al., 2015b). The CRE1 and LHK1 receptors are the main nodulation-specific cytokinin receptors in *M. truncatula* and *L. japonicus*, respectively (Murray et al., 2007; Plet et al., 2011), but show partially redundant functions with other cytokinin receptors that show slightly different expression patterns in roots and nodules (Boivin, Fonouni-Farde, & Frugier, 2016; Held et al., 2014). Cytokinin perception acts upstream of several key transcription factors necessary for nodulation, including *NIN (NODULE INCEPTION)*, *NSP2 (NODULATION SIGNALING PATHWAY2)*, *ERN1 (ETHYLENE RESPONSE FACTOR REQUIRED FOR NODULATION1)*, *NF-YA1 (NUCLEAR FACTOR YA1)*, and *bHLH476* (Ariel et al., 2012; Heckmann et al., 2011; Murray et al., 2007; Plet et al., 2011). In the cortex, CRE1 is required for NIN activation, while NIN was proposed to also control the expression of *CRE1* in a positive feedback loop (Plet et al., 2011; Vernié et al., 2015). Recently, the identification of a remote, upstream cis-regulatory part of the *NIN* promoter in *M. truncatula* suggested that *NIN* activation by cytokinin in the pericycle precedes cortical activation and is essential for nodule formation (Liu et al., 2019).

The epidermal response to Nod factors is associated not only with an increased cytokinin biosynthesis but also with an induction of cytokinin response genes within 4 h, preceding the activation of cytokinin responses in inner cortical cells (Jardinaud et al., 2016), whereas in *L. japonicus*, cortical cytokinin responses, evaluated using the TCSn (Two Component System new) cytokinin responsive promoter, precede epidermal responses (Reid et al., 2017). This is consistent with the hyperinfection phenotype of the *hit1/lhk1* mutant of *L. japonicus* (Murray et al., 2007). The *M. truncatula* CRE1 receptor is also necessary to negatively regulate the Nod factor induction of the *ENOD11* early nodulation gene associated to rhizobial infections (Jardinaud et al., 2016). In turn, ectopic expression of the *AtCKX3* cytokinin oxidase/dehydrogenase gene in *M. truncatula* epidermal cells led to reduced cytokinin concentrations, and also reduced the activation of *ENOD11* by Nod factors (Jardinaud et al., 2016). It is likely that the observed positive and negative roles of cytokinin in the cortex and root hairs highlight the delicate control of specific concentration-dependent and cell-type specific cytokinin responses optimal for infection and nodule development.

Cytokinin signaling is also involved in a feedback inhibition of nodulation as the induction of peptides of the CLE family, which activate AON, is dependent on cytokinin signaling and CRE1/LHK1 (Mortier, De Wever, Vuylstreke, Holsters & Goormachtig, 2012a; Mortier et al., 2010; Tsikou et al., 2018) and, in turn, overexpression of the MtCLE12 (AON-inducing) peptide causes activation of the *MtRR8* (*RESPONSE* REGULATOR8) cytokinin response gene (Saur, Oakes, Djordjevic, & Imin, 2011). A study in *L. japonicus* suggested that an AON-dependent induction of *IPT3* in the shoot was responsible for the translocation of specific cytokinins to the root and that this was part of the AON mechanism to inhibit nodulation (Sasaki et al., 2014). In *M. truncatula, IPT3* was similarly induced in the shoot in an AON-dependent manner, whereas other *IPT* genes were induced both in WT and in an AON (*sunn-3*) mutant (Azarakhsh, Lebedeva, & Ludova, 2018). Whether these transcript changes led to altered cytokinin accumulation was not investigated. This result was however not observed in soybean, where the homolog of *IPT3* was induced by rhizobia in the shoot, but not in an AON-dependent manner (Mens, Haaima, Gresshoff & Ferguson, 2018). Furthermore, feeding of high concentrations of cytokinin into the shoot inhibited nodulation, with possible toxic effects, while low concentrations of cytokinin application promoted nodulation. Therefore, the potential role of cytokinins as a shoot-derived inhibitor in AON requires further studies.

5. Ethylene

The gaseous, diffusible hormone ethylene plays important roles in plants, most notably in biotic and abiotic stress responses, fruit ripening and senescence. Application of ethylene to legume roots inhibits nodulation, while ethylene synthesis inhibitors consistently increase nodule numbers in most legumes, with soybean and *Sesbania rostrata* being notable exceptions (D'Haeze et al., 2003; Guinel, 2015). This suggests that ethylene is generally a negative regulator of nodulation, by affecting both infection and nodule development. This is supported by the hypernodulation and hyperinfection phenotypes of ethylene insensitive legume mutants, including the *EIN2/* SKL mutant of *M. truncatula* (Penmetsa & Cook, 1997; Penmetsa et al., 2008), the *ein2* mutant of pea (Foo, McAdam, Weller, & Reid, 2016), the soybean *ein2-1 ein2-2* double mutant (Miyata, Kawaguchi, & Nakagawa, 2013), and the *L. japonicus ein2a ein2b* double mutant (Reid et al., 2018). The latter *L. japonicus* mutant additionally showed a reduced nitrogen fixation. In *M. truncatula*, the ethylene *ein2/skl* hypernodulation phenotype is determined by the root, and this is different from the shoot-dependent systemic action of AON (Penmetsa, Frugoli, Smith, Long, & Cook, 2003). Ethylene additionally dictates the position of nodule initiation that occurs preferentially opposite to protoxylem poles, likely due to higher ethylene synthesis in front of phloem elements (Heidstra et al., 1997). Consistently, nodule positioning appears in a more random pattern around the stele in the *Mtskl* mutant (Penmetsa & Cook, 1997).

Ethylene is synthesized by the conversion of S-adenosyl methionine to 1-aminocyclopropane-1-carboxylic acid (ACC) by ACC synthases (ACS), which is converted to ethylene by ACC oxidases (ACO). Ethylene synthesis is rapidly induced upon Nod factor treatment or rhizobial infection. Increased ethylene concentrations were detected in *Sinorhizobium meliloti*-infected *Medicago sativa* roots (Ligero, Lluch, & Olivares, 1986) and sensitive ethylene measurements in *L. japonicus* showed an ethylene increase within 6 h of inoculation, with a peak at 24 h, which was dependent on early Nod factor signaling, but independent of NIN, CYCLOPS and ERN1 action (Reid et al., 2018). *ASC* expression was induced in rhizobial infected root hairs of *M. truncatula* (Larrainzar et al., 2015) and within 30 min in response to Nod factors (van Zeijl et al., 2015b). The accumulation of an ACO protein was increased within 3 days after rhizobial infection, but not after 24 h (Prayitno, Imin, Rolfe, & Mathesius, 2006a). The induction of ethylene synthesis is accompanied by changes in the expression of ethylene receptors and response genes during nodulation (Wang et al., 2017).

Ethylene acts by inhibiting calcium spiking signaling and preventing infection thread formation (Oldroyd, Engstrom, & Long, 2001). Ethylene signaling additionnally controls the accumulation of defense-related proteins and this might contribute to the regulation of infection events (Prayitno et al., 2006a). The expression of a large number of rhizobia-induced genes was potentiated in the *Mtskl* mutant, but the induced genes showed different responses at 1 and 6 h after rhizobium inoculation, suggesting stage-specific roles of ethylene (Larrainzar et al., 2015). The observation of upregulated ethylene signaling and synthesis genes during *M. truncatula* nodule senescence also suggests a role for ethylene in inducing senescence (Van de Velde et al., 2006).

Ethylene and cytokinin signaling interact during nodulation, likely to continuously balance ethylene and cytokinin levels in infected root hairs (van Zeijl et al., 2015b). In *M. truncatula*, the induction of *ACS* was dependent on cytokinin signaling. In turn, cytokinin synthesis is negatively regulated by ethylene signaling. Inhibition of spontaneous nodule formation by ethylene in the *lhk1* mutant, which carries an autoactive cytokinin receptor, places ethylene signaling downstream of cytokinin perception (Tirichine et al., 2007; Tirichine, James, Sandal, & Stougaard, 2006). Double mutants of *Mtskl* and *Mtcre1* showed intermediate nodule numbers between those observed in each single mutant, suggesting that both hormone signaling pathways can act independently on nodule development (Plet et al., 2011).

Ethylene signaling was also reported to be required for pseudonodule formation induced by auxin transport inhibitors, placing ethylene action below the auxin transport regulation (Rightmyer & Long, 2011). However, there is also evidence that ethylene acts upstream of auxin because ethylene can influence auxin transport (Prayitno, Rolfe, & Mathesius, 2006b), and may thus have an effect on nodule development and numbers. For example, increased expression of *PIN1* and *PIN2* and auxin accumulation in the *Mtskl* mutant during the early stages of nodulation may participate in increasing nodule numbers (Prayitno et al., 2006b).

6. Gibberellins

Gibberellins (GAs) were first identified as a class of plant hormones with stimulating effects on plant growth. GAs have been reported to have both positive and negative effects on nodulation in different legumes and at different stages of nodulation. For example, GA application was shown to inhibit nodulation and infection in *L. japonicus*, where GA inhibited early Nod factor signaling through NSP2 and NIN (Maekawa et al., 2009). Similarly, exogenous addition of active GA to *M. truncatula* reduced nodule numbers and infection events, while a GA synthesis inhibitor increased nodule numbers in a concentration-dependent manner (Fonouni-Farde et al., 2016). However, GA application in pea did not inhibit nodule formation at physiological concentrations (Ferguson, Reid, & Ross, 2005).

GA signaling is mediated by DELLA proteins that act as repressors of GA responses (Davière & Achard, 2013). In *M. truncatula*, DELLAs localize in the root tip, vascular tissue and nodule meristems (Fonouni-Farde et al., 2016). *della* single, double and triple mutants of *M. truncatula* showed reduced numbers of nodules and infection events and reduced expression of nodulation genes (Floss, Levy, Levesque-Tremblay, Pumplin, & Harrison, 2013; Fonouni-Farde et al., 2016; Jin et al., 2016), whereas ectopic overexpression of *DELLA* in epidermal cells increased infections and expression of the infection marker *ENOD11*, suggesting that GAs act as negative regulators of infection, at least in *M. truncatula* (Fonouni-Farde et al., 2016). DELLA proteins were shown to interact with the transcription factors NSP2 and NF-YA1 to regulate the expression of the ERN1 gene (Fonouni-Farde et al., 2016). A parallel study showed a DELLA interaction with IPD3 (INTERACTING PROTEIN OF DMI3) and CCaMK (CALCIUM CALMODULIN KINASE), and demonstrated that DELLAs were acting upstream of IPD3 phosphorylation induced by rhizobium also to regulate ERN1 (Jin et al., 2016). This study additionally suggested that DELLAs may be able to bridge protein complexes involving IPD3 and NSP2 to regulate downstream nodulation genes.

In soybean and pea, GAs also have reported positive roles in nodulation (Ferguson, Foo, Reid, & Ross, 2011; Ferguson Reid & Ross, 2005; Hayashi, Gresshoff, & Ferguson, 2014). In pea, GA deficient-mutants show reduced nodule numbers with aberrant structures indicating that GA is necessary for nodulation (Ferguson et al., 2011; Ferguson, Reid, & Ross, 2005). It is likely that endogenous concentrations of GA require strict spatiotemporal control to establish optimum concentration 'windows' in different cell types during nodulation. Accordingly, detailed studies in GA synthesis and signaling mutants of pea have confirmed the negative role of GA in the infection process in root hairs, similarly as in *L. japonicus* and *M. truncatula*, but suggested an additional positive role for GA in nodule development in the cortex (McAdam, Reid, & Foo, 2018). It has to be noticed however that GA also strongly affects roots and notably the cortex patterning, which may indirectly impair nodule organogenesis (Fonouni-Farde et al., 2019). The McAdam et al. (2018) study also showed that GA biosynthesis as well as signaling through DELLA proteins is required for nitrogen fixation, even though another study suggested that GA may be more related to an inhibition of nodule senescence (Serova, Tsyganova, Tikhonovich, & Tsyganov, 2019). Accordingly, genes involved in GA degradation are induced during nodule senescence (Van de Velde et al., 2006). The GA deficient *na* mutant of pea is also characterized by increased ethylene levels, and double mutant studies demonstrated that GA partly acts through ethylene signaling to affect nodule development, but acts independently of ethylene on infection (McAdam et al., 2018). In *M. truncatula*, GA and cytokinins negatively regulate each other, and GA inhibits the cytokinin regulation of the *NSP2* and *ERN1* Nod factor signaling genes (Fonouni-Farde et al., 2017, 2018). In addition, a GA-insensitive DELLA protein is able to partially complement the low nodulation of the *cre1* cytokinin receptor mutant (Fonouni-Farde et al., 2017). In pea, double mutants with defects in AON and GA biosynthesis suggest that GA acts independently of AON (Ferguson et al., 2011).

7. Jasmonic acid

Jasmonic acid (JA) is a small lipid that acts as a plant hormone in many biotic and abiotic stress responses. In particular, JA is required for the induction of wound responses and systemic resistance to herbivores but also plays roles in plant development. During nodulation, JA acts during early infection events, where JA inhibits infection, possibly through affecting calcium spiking signaling and nodulin gene expression (*RIP1 [Rhizobium Induced Peroxidase1]*; *ENOD11* and *NIN*) in *M. truncatula* and *L. japonicus* (Nakagawa & Kawaguchi, 2006; Sun et al., 2006). JA's effects on calcium spiking resemble the inhibitory effects of ethylene mentioned above, but JA appears to have a unique function in altering the frequency of calcium spiking (Sun et al., 2006). It is likely that ethylene and JA interact because JA effects are exacerbated in the ethylene-insensitive *skl* mutant (Sun et al., 2006). External addition of JA reduces nodule numbers, however it is not well understood whether this is only a result of altering early signaling during infection or whether JA has additional direct effects on nodule development (Sun et al., 2006). JA may also play a role during nodule senescence because some genes of the JA biosynthesis pathway are induced in senescing nodules (Van de Velde et al., 2006). In contrast to phenotypes induced by external JA application, silencing or overexpressing the JA biosynthesis gene encoding an allene oxide cyclase (AOC), which is expressed in uninfected nodule cells in *M. truncatula,* did not affect nodule numbers (Zdyb et al., 2011).

Application of methyl jasmonate (MeJA) was implicated in AON shoot-to-root signaling because it strongly reduces nodule numbers in wild type plants as well as in the AON mutant *har1* of *L. japonicus* after application to the shoot (Nakagawa & Kawaguchi, 2006). The inhibitory effects were seen from the early stages of rhizobial infection, including on root hair curling and infection. The role of JA in AON is supported by the induction during nodulation of JA biosynthesis genes in soybean leaves in a *GmNARK*-dependent manner (Kinkema & Gresshoff, 2008; Seo et al., 2006). JA's systemic effect resembles the systemic resistance mediated by JA during plant-pathogen infections (van Loon, Bakker, & Pieterse, 1998) and it would be interesting to test the involvement of JA in concomitant infections of plants with pathogens and rhizobia.

8. Strigolactones

Strigolactones (SLs) are multifaceted hormones that were first identified as root exudates stimulating spore germination and hyphal growth of mycorrhizal fungal symbioses and controlling parasitic weed germination, but they also have roles in shoot branching and root development (reviewed by De Cuyper & Goormachtig, 2017).

In general, SLs appear to have a positive role in nodulation, at least at low concentrations. In several legumes, nodule numbers were found to be slightly increased after the application of the synthetic SL, GR24 (Foo & Davies, 2011; Liu et al., 2013; Soto et al., 2010). However, while low concentrations of GR24 increased nodule numbers, higher concentrations inhibited nodulation in *M. truncatula* (De Cuyper et al., 2015). While direct measurements of SLs during nodulation have not been made, the SL biosynthesis gene *DWARF27* is induced by Nod factors within 3 h in *M. truncatula* roots (van Zeijl et al., 2015a), and the SL biosynthesis gene *MtCCD8 (CAROTENOID CLEAVING DEOXYGENASE8)* is upregulated during primordia formation and in infected root hairs (Breakspear et al., 2014). SL synthesis was found to be dependent on the essential nodulation transcription factors NSP1 and NSP2 (Liu et al., 2011) as well as the early Nod factor signaling components LYK3, DMI2 and DMI3, but not NIN (McAdam et al., 2017). Despite their increased synthesis during nodulation, SLs are not strictly necessary for nodulation, as the pea *ccd8* mutant, which contains almost no strigolactones at all, still forms nitrogen-fixing nodules, although significantly less than wild type roots (Foo & Davies, 2011; McAdam et al., 2017). Similarly, the *L. japonicus ccd7* mutant shows very low nodule numbers (Liu et al., 2013). While stimulating nodule numbers, SLs do not seem to act downstream of the AON pathway, as double mutants between the SL-deficient pea mutant *Psccd8* and AON mutants still show supernodulation (Foo, Ferguson, & Reid, 2014). SLs also appear to act independently of ethylene during nodulation (McAdam et al., 2017).

It is likely that the positive role of SLs is due to an enhancement of rhizobial infection, as SL deficient mutants show reduced infection, which can be alleviated with addition of GR24 (McAdam et al., 2017). This was not associated with altered exudation of Nod gene-inducing flavonoids or altered bacterial growth and occurred after the induction of root hair curling. Interestingly, GR24 does not enhance nodulation in lupin, which is infected through crack-entry, rather than infection threads, suggesting that SLs might have a specific role in root hair infection (McAdam et al., 2017).

9. Summary and outlook

Hormone pathways evolved long before the relatively recent evolution of nodulation in legumes. The role of plant hormones in development and in the response to environmental stresses had to be adapted to the regulation of nodulation. While our picture of the hormonal regulation of nodulation is steadily growing (summarized in Fig. 1), we still only have a sketchy understanding of the specific role of hormones in nodulation as opposed to other developmental processes, for example through the involvement of specific receptors or response genes during nodulation, some of which might have arisen from gene duplication and neofunctionalisation (Op den Camp et al., 2011). The CRE1 receptor in Medicago is a specific example of a receptor that appears to have taken on a nodulation primary role. For most of the other plant hormones, it remains unknown which pathway determines their specific action in nodulation. This would be particularly interesting to know for auxin, which requires specific localization and differential action in nodule or lateral root primordia (Bensmihen, 2015). Two indications that the predisposition for nodulation in legumes might be determined by altered hormone responses are that legumes show specific responses to both ABA (Liang & Harris, 2005) and cytokinin (Gauthier-Coles et al., 2018), but which genes account for these differential responses is unknown. For hormones that mediate stress responses, like ethylene, JA or ABA, it will be interesting to find out whether their inhibitory role in nodulation is linked to the necessity to inhibit nodulation under stressful conditions.

It is clear that all of the discussed plant hormones show strict concentration- and cell type-dependent signaling/response actions, although many of them remain to be elucidated in detail and require the generation of more

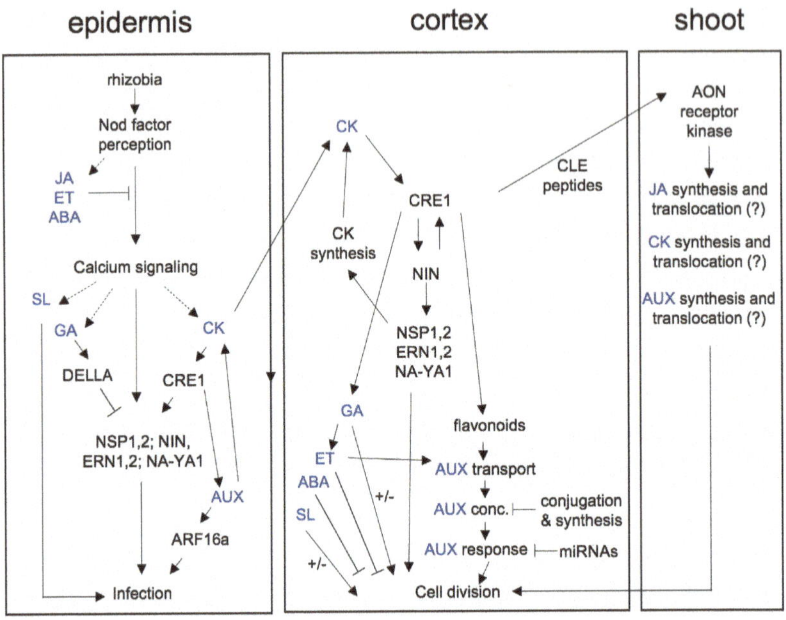

Fig. 1 Model of known hormone interactions during nodulation in legumes. The action of plant hormones (blue) and their potential targets is shown with arrows (positive interactions) and blunt-end arrows (negative interactions) for actions known to occur during infection in the roots hairs, for the activation of cell divisions in the cortex/pericycle, and for autoregulation of nodulation (AON) via the shoot. Not all interactions mentioned in the text are presented here. *ABA*, abscisic acid; *AUX*, auxin; *CK*, cytokinin; *ET*, ethylene; *GA*, gibberellin; *JA*, jasmonic acid; *SL*, strigolactone; *AON*, Autoregulation Of Nodulation; *ARF*, Auxin Response Factor; *CLE*, CLAVATA3 (Clv3)/Endosperm Surrounding Region (ESR)-related; *CRE1*, Cytokinin receptor 1; *ERN1*, Ethylene Response Factor Required For Nodulation 1; *NF-YA1*, NUCLEAR FACTOR YA1; *NIN*, Nodule Inception; *NSP*, Nodulation Signaling Pathway; *TML*, Too Much Love. *The figure was modified from Gamas et al. (2017).*

specific hormone-sensitive reporter genes and the use of tissue-specific promoters. A recent example is the sensitive dual reporter system that has been used for concurrent auxin and cytokinin measurements in soybean (Fisher et al., 2018). The action of plant hormones is complicated by their complex interactions with each other — and these might differ in different cell types, as exemplified in the differential action of cytokinin in the epidermis compared to the cortex (Gamas et al., 2017). Another puzzle remains about the specific roles of various hormone metabolites, for example, the specific roles of the many different chemical forms of auxins and cytokinins has remained largely unexplored.

As legumes in the field are often under the attack of various pathogens or herbivores at the same time as they are infected by symbionts like rhizobia and mycorrhizal fungi, it would further be very interesting to study the coordination of hormone signaling in multi-partite interactions in the future.

Acknowledgments

UM was supported by the Australian Research Council (DP150102002), by the Hermon Slade Foundation (HSF 16-02) and by the Australian Research Council and The Grains Research and Development Corporation (IC170100005).

References

Ariel, F., Brault-Hernandez, M., Laffont, C., Huault, E., Brault, M., Plet, J., et al. (2012). Two direct targets of cytokinin signaling regulate symbiotic nodulation in *Medicago truncatula*. *Plant Cell, 24*, 3838–3852.

Azarakhsh, M., Lebedeva, M. A., & Ludova, L. A. (2018). Identification and expression analysis of *Medicago truncatula Isopentenyl Transferase Genes* (*IPTs*) involved in local and systemic control of nodulation. *Frontiers of Plant Science, 9*, 304.

Bagchi, R., Salehin, M., Adeyemo, O. S., Salazar, C., Shulaev, V., Sherrier, D. J., et al. (2012). Functional assessment of the *Medicago truncatula* NIP/LATD protein demonstrates that it is a high-affinity nitrate transporter. *Plant Physiology, 160*, 906–916.

Benková, E., Michniewicz, M., Sauer, M., Teichmann, T., Seifertová, D., Jürgens, G., et al. (2003). Local, efflux-dependent auxin gradients as a common module for plant organ formation. *Cell, 115*, 591–602.

Bensmihen, S. (2015). Hormonal control of lateral root and nodule development in legumes. *Plants, 4*, 523–547.

de Billy, F., Grosjean, C., May, S., Bennett, M., & Cullimore, J. V. (2001). Expression studies on *AUX1*-like genes in *Medicago truncatula* suggest that auxin is required at two steps in early nodule development. *Molecular Plant-Microbe Interactions, 14*, 267–277.

Biswas, B., Chan, P. K., & Gresshoff, P. M. (2009). A novel ABA insensitive mutant of *Lotus japonicus* with a wilty phenotype displays unaltered nodulation regulation. *Molecular Plant, 2*, 487–499.

Boivin, S., Fonouni-Farde, C., & Frugier, F. (2016). How auxin and cytokinin phytohormones modulate root microbe interactions. *Frontiers of Plant Science, 7*, 1240.

Boot, K. J. M., van Brussel, A. A. N., Tak, T., Spaink, H. P., & Kijne, J. W. (1999). Lipochitin oligosaccharides from *Rhizobium leguminosarum* bv. *viciae* reduce auxin transport capacity in *Vicia sativa* subsp nigra roots. *Molecular Plant-Microbe Interactions, 12*, 839–844.

Breakspear, A., Liu, C., Roy, S., Stacey, N., Rogers, C., Trick, M., et al. (2014). The root hair "infectome" of *Medicago truncatula* uncovers changes in cell cycle genes and reveals a requirement for auxin signaling in rhizobial infection. *Plant Cell, 26*, 4680–4701.

Bright, L., Liang, Y., Mitchell, D. M., & Harris, J. M. (2005). *LATD*, a gene required for both nodule and root development. *Molecular Plant-Microbe Interactions, 18*, 521–532.

Bustos-Sanmamed, P., Mao, G., Deng, Y., Elouet, M., Kahn, G. A., Bazin, J., et al. (2013). Overexpression of miR160 affects root growth and nitrogen-fixing nodule number in *Medicago truncatula*. *Functional Plant Biology, 40*, 1208–1220.

Cai, Z., Wang, Y., Zhu, L., Tian, Y., Chen, L., Sun, Z., et al. (2017). GmTIR1/GmAFB3-based auxin perception regulated by miR393 modulates soybean nodulation. *New Phytologist, 215*, 672–686.

Campanella, J. J., Smith, S. M., Leibu, D., Wexler, S., & Ludwig-Müller, J. (2008). The auxin conjugate hydrolase family of *Medicago truncatula* and their expression during the interaction with two symbionts. *Journal of Plant Growth Regulation, 27*, 26–38.

Chen, Y., Chen, W., Li, X., Jiang, H., Wu, P., Xia, K., et al. (2013). Knockdown of *LjIPT3* influences nodule development in *Lotus japonicus*. *Plant and Cell Physiology, 55*, 183–193.

Damodaran, S., Westfall, C. S., Kisely, B. A., Jez, J. M., & Subramanian, S. (2017). Nodule-enriched GRETCHEN HAGEN 3 enzymes have distinct substrate specificities and are important for proper soybean nodule development. *International Journal of Molecular Science, 18*, 2547.

Davière, J. M., & Achard, P. (2013). Gibberellin signalling in plants. *Development, 140*, 1147–1151.

De Cuyper, C., Fromentin, J., Yocgo, R. E., De Keyser, A., Guillotin, B., Kunert, K., et al. (2015). From lateral root density to nodule number, the strigolactone analogue GR24 shapes the root architecture of *Medicago truncatula*. *Journal of Experimental Botany, 66*, 137–146.

De Cuyper, C., & Goormachtig, S. (2017). Strigolacones in the rhizosphere: Friend or foe? *Molecular Plant-Microbe Interactions, 30*, 683–690.

Deinum, E. E., Geurts, R., Bisseling, T., & Mulder, B. M. (2012). Modeling a cortical auxin aximum for nodulation: Different signatures of potential strategies. *Frontiers of Plant Science, 3*, 1–19.

Deinum, E. E., Kohlen, W., & Geurts, R. (2016). Quantitative modelling of legume root nodule primordium induction by a diffusive signal of epidermal origin that inhibits auxin efflux. *BMC Plant Biology, 16*, 254.

Ding, Y., Kalo, P., Yendrek, C., Sun, J., Liang, Y., Marsh, J. F., et al. (2008). Abscisic acid coordinates Nod factor and cytokinin signaling during the regulation of nodulation in *Medicago truncatula*. *Plant Cell, 20*, 2681–2695.

D'Haeze, W., De Rycke, R., Mathis, R., Goormachtig, S., Pagnotta, S., Verplancke, C., et al. (2003). Reactive oxygen species and ethylene play a positive role in lateral root base nodulation of a semiaquatic legume. *Proceedings of the National Academy of Sciences of the United States of America, 100*, 11789–11794.

Ferguson, B. J., Foo, E., Reid, J. B., & Ross, J. J. (2011). Relationship between gibberellin, ethylene and nodulation in *Pisum sativum*. *New Phytologist, 189*, 829–842.

Ferguson, B. J., & Mathesius, U. (2014). Phytohormone regulation of legume-rhizobia interactions. *Journal of Chemical Ecology, 40*, 770–790.

Ferguson, B. J., Reid, J. B., & Ross, J. J. (2005). Nodulation phenotypes of gibberellin and brassinosteroid mutants of *Pisum sativum*. *Plant Physiology, 138*, 2396–2405.

Fisher, J., Gaillard, P., Fellbaum, C. R., Subramanian, S., & Smith, S. (2018). Quantitative 3D imaging of cell level auxin and cytokinin response ratios in soybean roots and nodules. *Plant, Cell and Environment, 41*, 2080–2092.

Floss, D. S., Levy, J. G., Levesque-Tremblay, V., Pumplin, N., & Harrison, M. J. (2013). DELLA proteins regulate arbuscule formation in arbuscular mycorrhizal symbiosis. *Proceedings of the National Academy of Sciences of the United States of America, 110*, E5025–E5034.

Fonouni-Farde, C., Kisiala, A., Brault, M., Emery, R. J. N., Diet, A., & Frugier, F. (2017). DELLA1-mediated gibberellin signaling regulates cytokinin-dependent symbiotic nodulation. *Plant Physiology, 175*, 1795–1806.

Fonouni-Farde, C., McAdam, E., Nichols, D., Diet, A., Foo, E., & Frugier, F. (2018). Cytokinins and the CRE1 receptor influence endogenous gibberellin levels in *Medicago truncatula*. *Plant Signaling & Behavior, 13*, e1428513.

Fonouni-Farde, C., Miassod, A., Laffont, C., Morin, H., Bendahmane, A., Diet, A., et al. (2019). Gibberellins negatively regulate the development of *Medicago truncatula* root system. *Scientific Reports, 9*, 2335.

Fonouni-Farde, C., Tan, S., Baudin, M., Brault, M., Wen, J., Mysore, K. S., et al. (2016). DELLA-mediated gibberellin signaling regulates Nod factor signaling and rhizobial infection. *Nature Communications, 7*, 12636.

Foo, E., & Davies, N. W. (2011). Strigolactones promote nodulation in pea. *Planta, 234*, 1073–1081.

Foo, E., Ferguson, B. J., & Reid, J. B. (2014). The potential roles of strigolactones and brassinosteroids in the autoregulation of nodulation pathway. *Annals of Botany, 113*, 1037–1045.

Foo, E., McAdam, E. L., Weller, J. L., & Reid, J. B. (2016). Interactions between ethylene, gibberellins, and brassinosteroids in the development of rhizobial and mycorrhizal symbioses of pea. *Journal of Experimental Botany, 67*, 2413–2424.

Franssen, H. J., Xiao, T. T., Kulikova, O., Wan, X., Bisseling, T., Scheres, B., et al. (2015). Root developmental programs shape the *Medicago truncatula* nodule meristem. *Development, 142*, 2941–2950.

Gamas, P., Brault, M., Jardinaud, M.-F., & Frugier, F. (2017). Cytokinins in symbiotic nodulation: When, where and what for? *Trends in Plant Science, 22*, 792–802.

Gauthier-Coles, C., White, R., & Mathesius, U. (2018). Nodulating legumes are distinguished by a sensitivity to cytokinin in the root cortex leading to pseudonodule development. *Frontiers of Plant Science, 9*, 1901.

Gonzalez-Rizzo, S., Crespi, M., & Frugier, F. (2006). The *Medicago truncatula* CRE1 cytokinin receptor regulates lateral root development and early symbiotic interaction with *Sinorhizobium meliloti*. *Plant Cell, 18*, 2680–2693.

Guan, D., Stacey, N., Liu, C., Wen, J., Mysore, K. S., Torrez-Jerez, I., et al. (2013). Rhizobial infection is associated with the development of peripheral vasculature in nodules of *Medicago truncatula*. *Plant Physiology, 162*, 107–115.

Guinel, F. (2015). Ethylene, a hormone at the center-stage of nodulation. *Frontiers of Plant Science, 6*, 1121.

Harris, J. M., & Dickstein, R. (2010). Control of root architecture and nodulation by the LATD/NIP transporter. *Plant Signaling & Behavior, 5*, 1365–1369.

Hayashi, S., Gresshoff, P. M., & Ferguson, B. J. (2014). Mechanistic action of gibberellins in legume nodulation. *Journal of Integrative Plant Biology, 56*, 971–997.

Heckmann, A. B., Sandal, N., Bek, A. S., Madsen, L. H., Jurkiewicz, A., Nielsen, M. W., et al. (2011). Cytokinin induction of root nodule primordia in *Lotus japonicus* is regulated by a mechanism operating in the root cortex. *Molecular Plant-Microbe Interactions, 24*, 1385–1395.

Heidstra, R., Yang, W. C., Yalcin, Y., Peck, S., Emons, A. M., van Kammen, A., et al. (1997). Ethylene provides positional information on cortical cell division but is not involved in Nod factor-induced root hair tip growth in *Rhizobium*-legume interaction. *Development, 124*, 1781–1787.

Held, M., Huo, H., Miri, M., Huyng, C., Ross, L., Hossein, M. S., et al. (2014). *Lotus japonicus* cytokinin receptors work partially redundantly to mediate nodule formation. *Plant Cell, 26*, 678–694.

Herrbach, V., Chirinos, X., Rengel, D., Agbevenou, K., Vincent, R., Pateyron, S., et al. (2017). Nod factors potentiate auxin signaling for transcriptional regulation and lateral root formation in *Medicago truncatula*. *Journal of Experimental Botany, 68*, 569–583.

Hirsch, A. M. (1992). Developmental biology of legume nodulation. *New Phytologist, 122*, 211–237.

Hobecker, K. V., Reynoso, M. A., Bustos-Sanmamed, P., Wen, J., Mysore, K. S., Crespi, M., et al. (2017). The microRNA390/TAS3 pathway mediates symbiotic nodulation and lateral root growth. *Plant Physiology, 174*, 2469–2486.

Huo, X. Y., Schnabel, E., Hughes, K., & Frugoli, J. (2006). RNAi phenotypes and the localization of a protein::GUS fusion imply a role for *Medicago truncatula PIN* genes in nodulation. *Journal of Plant Growth Regulation, 25*, 156–165.

Jardinaud, M. F., Boivin, S., Rodde, N., Catrice, O., Kisiala, A., Legage, A., et al. (2016). A laser dissection-RNAseq analysis highlights the activation of cytokinin pathways by nod factors in the *Medicago truncatula* root epidermis. *Plant Physiology, 171*, 2256–2276.

Jin, Y., Liu, H., Luo, D., Yu, N., Dong, W., Wang, C., et al. (2016). DELLA proteins are common components of symbiotic rhizobial and mycorrhizal signalling pathways. *Nature Communications, 7*, 12433.

Jin, J., Watt, M., & Mathesius, U. (2012). The autoregulation gene *SUNN* mediates changes in root organ formation in response to nitrogen through alteration of shoot-to-root auxin transport. *Plant Physiology, 159*, 489–500.

Kinkema, M., & Gresshoff, P. M. (2008). Investigation of downstream signals of the soybean autoregulation of nodulation receptor kinase GmNARK. *Molecular Plant-Microbe Interactions, 21*, 1337–1348.

Kohlen, W., Ng, J. L. P., Deinum, E. E., & Mathesius, U. (2017). Auxin transport, metabolism and signaling during nodule initiation: Indeterminate and determinate nodules. *Journal of Experimental Botany, 69*, 229–244.

Kuppusamy, K. T., Ivashuta, S., Bucciarelli, B., Vance, C. P., Gantt, J. S., & VandenBosch, K. A. (2009). Knockdown of *CELL DIVISION CYCLE16* reveals an inverse relationship between lateral root and nodule numbers and a link to auxin in *Medicago truncatula. Plant Physiology, 151*, 1155–1166.

Larrainzar, E., Riely, B. K., Kim, S. C., Carrasquilla Garcia, N., Yu, H.-J., Hwang, H. J., et al. (2015). Deep sequencing of the *Medicago truncatula* root transcriptome reveals a massive and early interaction between nod factor and ethylene signals. *Plant Physiology, 169*, 233–265.

Liang, Y., & Harris, J. M. (2005). Response of root branching to abscisic acid is correlated with nodule formation both in legumes and nonlegumes. *American Journal of Botany, 92*, 1675–1683.

Liang, Y., Mitchell, D. M., & Harris, J. M. (2007). Abscisic acid rescues the root meristem defects of the *Medicago truncatula latd* mutant. *Developmental Biology, 304*, 297–307.

Ligero, F., Lluch, C., & Olivares, O. (1986). Evolution of ethylene from roots of *Medicago sativa* plants inoculated with *Rhizobium meliloti. Journal of Plant Physiology, 125*, 361–366.

Liu, W., Kohlen, W., Lillo, A., den Camp, R. O., Ivanov, S., Hartog, M., et al. (2011). Strigolactone biosynthesis in *Medicago truncatula* and rice requires the symbiotic GRAS-type transcription factors NSP1 and NSP2. *Plant Cell, 23*, 3853–3865.

Liu, J., Novero, M., Charnikhova, T., Ferrandino, A., Schubert, A., Ruyter-Spira, C., et al. (2013). Carotenoid cleavage dioxygenase 7 modulates plant growth, reproduction, senescence, and determinate nodulation in the model legume *Lotus japonicus. Journal of Experimental Botany, 64*, 1967–1981.

Liu, J., Rutten, L., Limpens, E., van der Molen, T., van Velzen, R., Chen, R., et al. (2019). A remote *cis*-regulatory region is required for *NIN* expression in the pericycle to initiate nodule primordium formation in *Medicago truncatula. Plant Cell, 31*, 68–83.

Liu, H., Zhang, C., Yang, J., Yu, N., & Wang, E. (2018). Hormone modulation of legume-rhizobial symbiosis. *Journal of Integrative Plant Biology, 60*, 632–648.

van Loon, L. C., Bakker, P. A. H. M., & Pieterse, C. M. J. (1998). Systemic resistance induced by rhizosphere bacteria. *Annual Review of Phytopathology, 36*, 453–483.

Maekawa, T., Maekawa-Yoshikawa, M., Takeda, N., Imaizumi-Anraku, H., Murooka, Y., & Hayashi, M. (2009). Gibberellin controls the nodulation signaling pathway in *Lotus japonicus. The Plant Journal, 58*, 183–194.

Mao, G., Turner, M., Yu, O., & Subramanian, S. (2013). miR393 and miR164 influence indeterminate but not determinate nodule development. *Plant Signaling & Behavior, 8*, e26753.

Mathesius, U. (2001). Flavonoids induced in cells undergoing nodule organogenesis in white clover are regulators of auxin breakdown by peroxidase. *Journal of Experimental Botany, 52*, 419–426.

Mathesius, U. (2008). Auxin: At the root of nodule development? *Functional Plant Biology, 35*, 651–668.

Mathesius, U., Charon, C., Rolfe, B. G., Kondorosi, A., & Crespi, M. (2000). Temporal and spatial order of events during the induction of cortical cell divisions in white clover by *Rhizobium leguminosarum* bv. *trifolii* inoculation or localized cytokinin addition. *Molecular Plant-Microbe Interactions, 13*, 617–628.

Mathesius, U., Schlaman, H. R., Spaink, H. P., Sautter, C., Rolfe, B. G., & Djordjevic, M. A. (1998). Auxin transport inhibition precedes root nodule formation in white clover roots and is regulated by flavonoids and derivatives of chitin oligosaccharides. *The Plant Journal, 14*, 23–34.

McAdam, E. L., Hugill, C., Fort, S., Samain, E., Cottaz, S., Davies, N. W., et al. (2017). Determining the site of action of strigolactones during nodulation. *Plant Physiology, 175*, 529–542.

McAdam, E. L., Reid, J. B., & Foo, E. (2018). Gibberellins promote nodule organogenesis but inhibit the infection stages of nodulation. *Journal of Experimental Botany, 69*, 2117–2130.

Mens, C., Li, D., Haaima, L. E., Gresshoff, P. M., & Ferguson, B. J. (2018). Local and systemic effect of cytokinins on soybean nodulation and regulation of their *Isopentenyl Transferase* (*IPT*) biosynthesis genes following rhizobia inoculation. *Frontiers of Plant Science, 9*, 1150.

Miyata, K., Kawaguchi, M., & Nakagawa, T. (2013). Two distinct EIN2 genes cooperatively regulate ethylene signaling in *Lotus japonicus*. *Plant and Cell Physiology, 54*, 1469–1477.

Mortier, V., De Wever, E., Vuylsteke, M., Holsters, M., & Goormachtig, S. (2012a). Nodule numbers are governed by interaction between CLE peptides and cytokinin signaling. *The Plant Journal, 70*, 367–376.

Mortier, V., Den Herder, G., Whitford, R., Van de Velde, W., Rombauts, S., D'Haeseleer, K., et al. (2010). CLE peptides control *Medicago truncatula* nodulation locally and systemically. *Plant Physiology, 153*, 222–237.

Mortier, V., Hosters, M., & Goormachtig, S. (2012b). Never too many? How legumes control nodule numbers. *Plant, Cell and Environment, 35*, 245–258.

Mortier, V., Wasson, A., Jaworek, P., De Keyser, A., Decroos, M., Holsters, M., et al. (2014). Role of *LONELY GUY* genes in indeterminate nodulation on *Medicago truncatula*. *New Phytologist, 202*, 582–593.

Murray, J. D., Karas, B. J., Sato, S., Tabata, S., Amoyt, L., & Szczyglowski, K. (2007). A cytokinin perception mutant colonized by *Rhizobium* in the absence of nodule organogenesis. *Science, 315*, 101–104.

Nadzieja, M., Kelly, S., Stougaard, J., & Reid, D. (2018). Epidermal auxin biosynthesis facilitates rhizobial infection in *Lotus japonicus*. *The Plant Journal, 95*, 101–111.

Nakagawa, T., & Kawaguchi, M. (2006). Shoot-applied MeJA suppresses root nodulation in *Lotus japonicus*. *Plant and Cell Physiology, 47*, 176–180.

Ng, J. L. P., Hassan, S., Truong, T. T., Hocart, C. H., Laffont, C., Frugier, F., et al. (2015). Flavonoids and auxin transport inhibitors rescue symbiotic nodulation in the *Medicago truncatula* cytokinin perception mutant *cre1*. *Plant Cell, 27*, 2210–2226.

Ng, J. L. P., & Mathesius, U. (2018). Acropetal auxin transport inhibition is involved in indeterminate but not determinate nodule formation. *Frontiers of Plant Science, 9*, 169.

Nizampatnam, N. R., Schreier, S. J., Damodaran, S., Adhikari, S., & Subramanian, S. (2015). microRNA160 dictates stage-specific auxin and cytokinin sensitivities and directs soybean nodule development. *The Plant Journal, 84*, 140–153.

van Noorden, G. E., Kerim, T., Goffard, N., Wiblin, R., Pellerone, F. I., Rolfe, B. G., et al. (2007). Overlap of proteome changes in *Medicago truncatula* in response to auxin and *Sinorhizobium meliloti*. *Plant Physiology, 144*, 1115–1131.

van Noorden, G. E., Ross, J. J., Reid, J. B., Rolfe, B. G., & Mathesius, U. (2006). Defective long distance auxin transport regulation in the *Medicago truncatula super numeric nodulation* mutant. *Plant Physiology, 140*, 1494–1506.

Oldroyd, G. E. D. (2013). Speak, friend, and enter: Signalling systems that promote beneficial symbiotic associations in plants. *Nature Reviews Microbiology, 11*, 252–263.

Oldroyd, G. E., Engstrom, E. M., & Long, S. R. (2001). Ethylene inhibits the Nod factor signal transduction pathway of *Medicago truncatula. Plant Cell, 13*, 1835–1849.

Op den Camp, R. H. M., De Mita, S., Lillo, A., Cao, Q., Limpens, E., Bisseling, T., et al. (2011). A phylogenetic strategy based on a legume-specific whole genome duplication yields symbiotic cytokinin type-a response regulators. *Plant Physiology, 157*, 2013–2022.

Peer, W. A., & Murphy, A. S. (2007). Flavonoids and auxin transport: Modulators or regulators? *Trends in Plant Science, 12*, 556–563.

Penmetsa, R. V., & Cook, R. D. (1997). A legume ethylene-insensitive mutant hyperinfected by its rhizobial symbiont. *Science, 257*, 527–530.

Penmetsa, R. V., Frugoli, J. A., Smith, L. S., Long, S. R., & Cook, D. R. (2003). Dual genetic pathways controlling nodule number in *Medicago truncatula. Plant Physiology, 131*, 998–1008.

Penmetsa, R. V., Uribe, P., Anderson, J., Lichtenzveig, J., Gish, J.-C., Nam, Y. W., et al. (2008). The *Medicago truncatula* ortholog of Arabidopsis *EIN2, sickle*, is a negative regulator of symbiotic and pathogenic microbial associations. *The Plant Journal, 55*, 580–595.

Petrášek, J., & Friml, J. (2009). Auxin transport routes in plant development. *Development, 136*, 2675–2688.

Plet, J., Wasson, A., Ariel, F., Le Signor, C., Baker, D., Mathesius, U., et al. (2011). MtCRE1-dependent cytokinin signaling integrates bacterial and plant cues to coordinate symbiotic nodule organogenesis in *Medicago truncatula. The Plant Journal, 65*, 622–633.

Prayitno, J., Imin, N., Rolfe, B. G., & Mathesius, U. (2006a). Identification of ethylene-mediated protein changes during nodulation in *Medicago truncatula* using proteome analysis. *Journal of Proteome Research, 5*, 3084–3095.

Prayitno, J., Rolfe, B. G., & Mathesius, U. (2006b). The ethylene insensitive *sickle* mutant of *Medicago truncatula* shows altered auxin transport regulation during nodulation. *Plant Physiology, 142*, 168–180.

Reid, D. E., Ferguson, B. J., Hayashi, S., Lin, Y. H., & Gresshoff, P. M. (2011). Molecular mechanisms controlling legume autoregulation of nodulation. *Annals of Botany, 108*, 789–795.

Reid, D. E., Heckmann, A. B., Novak, O., Kelly, S., & Stougaard, J. (2016). CYTOKININ OXIDASE/DEHYDROGENASE3 maintains cytokinin homeostasis during root and nodule development in *Lotus japonicus. Plant Physiology, 170*, 1060–1074.

Reid, D., Liu, H., Kelly, S., Kawaharada, Y., Mun, T., Andersen, S. U., et al. (2018). Dynamics of ethylene production in response to compatible Nod factor. *Plant Physiology, 176*, 1764–1772.

Reid, D., Nadzieja, M., Novak, O., Heckmann, A. B., Sandal, N., & Stougaard, J. (2017). Cytokinin biosynthesis promotes cortical cell responses during nodule development. *Plant Physiology, 175*, 361–375.

Rightmyer, A. P., & Long, S. (2011). Pseudonodule formation by wild type and symbiotic mutant *Medicago truncatula* in response to auxin transport inhibitors. *Molecular Plant-Microbe Interactions, 24*, 1372–1384.

Roudier, F., Fedorova, E., Lebris, M., Lecomte, P., Györgyey, J., Vaubert, D., et al. (2003). The *Medicago* species A2-type cyclin is auxin regulated and involved in meristem formation but dispensable for endoreduplication-associated developmental programs. *Plant Physiology, 131*, 1091–1103.

Roux, B., Rodde, M., Jardinaud, M.-F., Timmers, T., Sauviac, L., Cottret, L., et al. (2014). An integrated analysis of plant and bacterial gene expression in symbiotic root nodules using laser-capture microdissection coupled to RNA sequencing. *The Plant Journal, 77*, 817–837.

Roy, S., Robson, F. C., Lilley, J. L., Liu, C. W., Cheng, X., Wen, J., et al. (2017). MtLAX2, a functional homologue of the auxin importer AtAUX1, is required for nodule organogenesis. *Plant Physiology, 174*, 326–338.

Sasaki, T., Suzaki, T., Soyano, T., Kojima, M., Sakakibara, H., & Kawaguchi, M. (2014). Shoot-derived cytokinins systemically regulate root nodulation. *Nature Communications, 5*, 4983.

Saur, I. M., Oakes, M., Djordjevic, M. A., & Imin, N. (2011). Crosstalk between the nodulation signaling pathway and the autoregulation of nodulation in *Medicago truncatula*. *New Phytologist, 190*, 865–874.

Sańko-Sawczenko, I., Łotocka, B., & Czarnocka, W. (2016). Expression analysis of *PIN* genes in root tips and nodules of *Medicago truncatula*. *International Journal of Molecular Sciences, 17*, 1197.

Seo, H. S., Li, J., Lee, S. Y., Yu, J. W., Kim, K. H., Lee, S. H., et al. (2006). The hypernodulating *nts* mutation induces jasmonate synthetic pathway in soybean leaves. *Molecular Cell, 24*, 185–193.

Serova, T. A., Tsyganova, A. V., Tikhonovich, I. A., & Tsyganov, V. E. (2019). Gibberellins inhibit nodule senescence and stimulate nodule meristem bifurcation in pea (*Pisum sativum* L.). *Frontiers of Plant Science, 10*, 285.

Soto, M. J., Fernández-Aparicio, M., Castellanos-Morales, V., García-Garrido, J. M., Ocampo, J. A., Delgado, M. J., et al. (2010). First indications for the involvement of strigolactones on nodule formation in alfalfa (*Medicago sativa*). *Soil Biology and Biochemistry, 42*, 383–385.

Sun, J., Cardoza, V., Mitchell, D. M., Bright, L., Oldroyd, G., & Harris, J. M. (2006). Crosstalk between jasmonic acid, ethylene and Nod factor signaling allow integration of diverse inputs for regulation of nodulation. *The Plant Journal, 46*, 961–970.

Suzaki, T., Yano, K., Ito, M., Umehara, Y., Suganuma, N., & Kawaguchi, M. (2012). Positive and negative regulation of cortical cell division during root nodule development in *Lotus japonicus* is accompanied by auxin response. *Development, 139*, 3997–4006.

Suzuki, A., Akune, M., Kogiso, M., Imagama, Y., Osuki, K., Uchiumi, T., et al. (2004). Control of nodule number by the phytohormone abscisic acid in the roots of two leguminous species. *Plant and Cell Physiology, 45*, 914–922.

Takanashi, K., Sugiyama, A., & Yazaki, K. (2011). Involvement of auxin distribution in root nodule development of *Lotus japonicus*. *Planta, 234*, 73–81.

Tirichine, L., James, E. K., Sandal, N., & Stougaard, S. (2006). Spontaneous root-nodule formation in the model legume *Lotus japonicus*: A novel class of mutants nodulates in the absence of rhizobia. *Molecular Plant-Microbe Interactions, 19*, 373–382.

Tirichine, L., Sandal, N., Madsen, L. H., Radutoiu, S., Albrektsen, A. S., Sato, S., et al. (2007). A gain-of-function mutation in a cytokinin receptor triggers spontaneous root nodule organogenesis. *Science, 315*, 104–107.

Tominaga, A., Nagata, M., Futsuki, K., Abe, H., Uchiumi, T., Abe, M., et al. (2009). Enhanced nodulation and nitrogen fixation in the abscisic acid low-sensitive mutant *enhanced nitrogen fixation1* of *Lotus japonicus*. *Plant Physiology, 151*, 1965–1976.

Tsikou, D., Yan, Z., Holt, D. B., Abel, N. B., Reid, D. E., Madsen, L. H., et al. (2018). Systemic control of legume susceptibility to rhizobial infection by a mobile microRNA. *Science, 362*, 233–236.

Turner, M., Nizampatnam, N. R., Baron, M., Coppin, S., Damodaran, S., Adhikari, S., et al. (2013). Ectopic expression of miR160 results in auxin hypersensitivity, cytokinin hyposensitivity, and inhibition of symbiotic nodule development in soybean. *Plant Physiology, 162*, 2042–2055.

Van de Velde, W., Perez-Guerra, J. C., De Keyser, A., De Rycke, R., Rombauts, S., Maunoury, N., et al. (2006). Aging in legume symbiosis: A molecular view on nodule senescence in *Medicago truncatula*. *Plant Physiology, 141*, 711–720.

Veereshlingam, H., Haynes, J. G., Sherrier, D. J., Penmetsa, R. V., Cook, D. R., & Dickstein, R. (2004). *nip*, a symbiotic *Medicago truncatula* mutant that forms root nodules with aberrant infection threads and plant defense-like response. *Plant Physiology, 136*, 3692–3702.

Vernié, T., Kim, J., Frances, L., Ding, Y., Sun, J., Guan, D., et al. (2015). The NIN transcription factor coordinates diverse nodulation programs in different tissues of *Medicago truncatula* roots. *Plant Cell, 27*, 3410–3424.

Wang, Y., Li, K., Chen, L., Zou, Y., Liu, H., Tian, Y., et al. (2015). miRNA167-directed regulation of the auxin response factors GmARF8a and GmARF8b is required for soybean nodulation and lateral root development. *Plant Physiology, 168*, 101–116.

Wang, Y., Yang, W., Zuo, Y., Zhu, L., Hastwell, A. H., Chen, L., et al. (2019). GmYUC2a is important for mediating auxin biosynthesis during root development and nodulation in soybean. *Journal of Experimental Botany, 70*(12), 3165–3176. https://doi.org/10.1093/jxb/erz144.

Wang, Y., Yuan, J., Yang, W., Zhu, L., Su, C., Wang, X., et al. (2017). Genome wide identification and expression profiling of ethylene receptor genes during soybean nodulation. *Frontiers of Plant Science, 8*, 859.

Wasson, A., Pellerone, F., & Mathesius, U. (2006). Silencing the flavonoid pathway in *Medicago truncatula* inhibits root nodule formation and prevents auxin transport regulation by rhizobia. *Plant Cell, 18*, 1617–1629.

Weijers, D., & Wagner, D. (2016). Transcriptional responses to the auxin hormone. *Annual Review of Plant Biology, 67*, 539–574.

Xiao, T. T., Schilderink, S., Moling, S., Deinum, E. E., Kondorosi, E., Franssen, H., et al. (2014). Fate map of *Medicago truncatula* root nodules. *Development, 141*, 3517–3528.

Yendrek, C. R., Lee, Y. C., Morris, V., Liang, Y., Pislariu, C. I., Burkart, G., et al. (2010). A putative transporter is essential for integrating nutrient and hormone signaling with lateral root growth and nodule development in *Medicago truncatula. The Plant Journal, 62*, 100–112.

Zdyb, A., Dechenko, K., Heumann, J., Mrosk, C., Grzeganek, P., Göbel, C., et al. (2011). Jasmonate biosynthesis in legume and actinorhizal nodules. *New Phytologist, 189*, 568–579.

van Zeijl, A., Liu, W., Xiao, T. T., Kohlen, W., Yang, W. C., Bisseling, T., et al. (2015a). The strigolactone biosynthesis gene DWARF27 is co-opted in rhizobium symbiosis. *BMC Plant Biology, 15*, 260.

van Zeijl, A., Op den Camp, R. H., Deinum, E. E., Charnikhova, T., Franssen, H., Op den Camp, H. J., et al. (2015b). Rhizobium lipo-chitooligosaccharide signaling triggers accumulation of cytokinins in *Medicago truncatula* roots. *Molecular Plant, 8*, 1213–1226.

Zhang, J., Subramanian, S., Stacey, G., & Yu, O. (2009). Flavones and flavonols play distinct critical roles during nodulation of *Medicago truncatula* by *Sinorhizobium meliloti. The Plant Journal, 57*, 171–183.

CHAPTER THREE

Redox control of the legume-*Rhizobium* symbiosis

Manuel A. Matamoros* and Manuel Becana*
Departamento de Nutrición Vegetal, Estación Experimental de Aula Dei, Consejo Superior de Investigaciones Científicas (CSIC), Zaragoza, Spain
*Corresponding authors: E-mail: m.matamoros@csic.es, becana@eead.csic.es

Contents

Abstract

The onset of the nitrogen-fixing legume-rhizobium symbiosis is a complex process that requires elaborate communication between the two partners. Nodule metabolism is very active and continuously generates reactive oxygen species (ROS) and reactive nitrogen species (RNS). During the evolution of aerobic life, ROS and RNS have been

Advances in Botanical Research, Volume 94
ISSN 0065-2296
https://doi.org/10.1016/bs.abr.2019.09.011

recruited as versatile signaling molecules due to their chemical properties and capacity to interact with enzymes and transcription factors. Numerous studies have demonstrated the presence of nitric oxide, hydrogen peroxide and other ROS and RNS at different stages of symbiosis, from early recognition between plant and bacteria to nodule senescence. Antioxidant metabolites and enzymes and other proteins such as phytoglobins finely regulate ROS and RNS concentrations, thereby allowing the beneficial and critical participation of these molecules as signals in many aspects of nodule physiology. Here we review the contribution of ROS, RNS and antioxidants to the redox control of the onset of symbiosis and subsequent nodule development, paying attention to both developmental (aging) and stress-induced senescence. Because of their importance in regulation and signaling, a substantial part of this review is devoted to post-translational redox modifications involving, among others, residues of methionine, cysteine and tyrosine of nodule proteins.

Abbreviations

Glb	Phytoglobin
Gpx	Glutathione peroxidase
Grx	Glutaredoxin
GSH	Glutathione
hGSH	Homoglutathione
Lb	Leghemoglobin
NTR	NADPH-thioredoxin reductase
Prx	Peroxiredoxin
PTM	Post-translational modification
RNS	Reactive nitrogen species
ROS	Reactive oxygen species
SOD	Superoxide dismutase
SNF	Symbiotic N_2 fixation
Trx	Thioredoxin

1. Introduction

In nitrogen-poor soils many legumes are able to establish N_2-fixing symbioses with bacteria collectively known as rhizobia. Essentially, the mutual recognition of plant and bacteria relies on a complex exchange of molecular signals, involving flavonoids exudated from the roots and nodulation (Nod) factors produced by the rhizobia. In most crop legumes, as well as in model legumes, bacteria infection takes place through root hairs. The root hairs curl and trap the bacteria, which penetrate the epidermis and cortex through tubular structures known as infection threads. Bacteria are released from infection threads into the symbiosomes, organelle-like structures surrounded by a membrane, the symbiosomal or peribacteroid membrane, derived from the root cell plasma membrane. The cytoplasm of infected cells ultimately becomes filled with symbiosomes where bacteria differentiate into bacteroids that reduce (fix) N_2 into ammonia.

Legume nodules provide optimal conditions for the expression of bacterial nitrogenase, whose activity contributes fixed nitrogen to ecosystems and croplands (Oldroyd, 2013). In simple terms, nodules can be classified as indeterminate or determinate according to their growth pattern. Indeterminate nodules, formed by the model legume *Medicago truncatula* and crops such as pea, alfalfa and vetch, contain a persistent meristem and are generally elongated with a longitudinal gradient of age. Indeed, in a typical indeterminate nodule, four zones can be distinguished from the apex (distal) to the base (proximal) region: zone I (meristem), zone II (invasion), zone III (infected) and zone IV (senescent). Determinate nodules, formed by the model legume *Lotus japonicus* and crops such as soybean, common bean and cowpea, lack permanent meristems and are usually spherical. In this case, symbiotic N_2 fixation (SNF) takes place in the central infected zone, which contains also uninfected or interstitial cells and is surrounded by a multi-layered cortex or nodule parenchyma (Fig. 1; Dupont et al., 2012; Minchin, James, & Becana, 2008).

In the last decades, the legume-rhizobium symbiosis has been the focus of intensive research. Most plant and bacterial genes essential for symbiosis have been characterized and the complete genome sequences of many rhizobial and legume species are available (Wang et al., 2017). Transcriptomic studies have greatly expanded our knowledge of developmental and stress-induced nodule senescence (Pérez Guerra et al., 2010). Also, physiological, biochemical and molecular approaches have provided a wealth of information on the antioxidants of nodules (Becana, Matamoros, Ramos, Rubio, & Sainz, 2014; Becana, Matamoros, Udvardi, & Dalton, 2010; Dupont et al., 2012; Puppo et al., 2005) and on the metabolic exchange between the two symbiotic partners (Udvardi & Poole, 2013). Readers are referred to all those articles for useful information complementary to this review.

Reactive oxygen species (ROS), such as superoxide radicals and hydrogen peroxide (H_2O_2), and reactive nitrogen species (RNS), such as nitric oxide (NO), peroxynitrite ($ONOO^-$) and nitrosothiols, are produced at high concentrations in plant cells under severe stress. The excess of ROS and RNS may overwhelm the antioxidant defenses of the plant, causing cellular damage and ultimately death. However, under physiological or mild stressful conditions, the spatio-temporal production and the concentrations of ROS

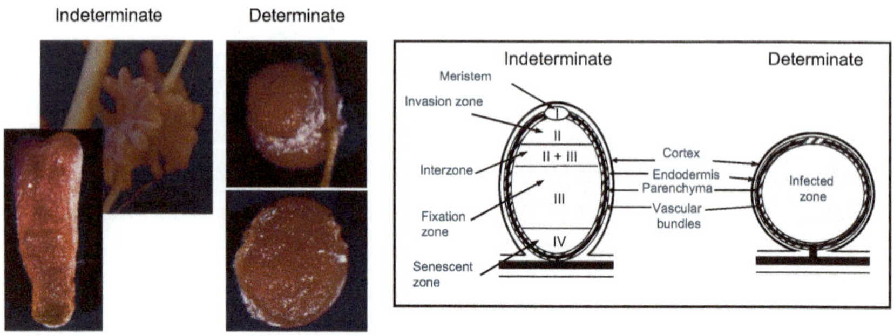

	Indeterminate	Determinate
Host plant	Alfalfa, pea, *Medicago truncatula*	Bean, soybean, *Lotus*
Geographic origin	Template	Tropical and subtropical
Nodule shape	Elongated	Spherical
Initial cell divisions	Inner cortex	Outer cortex
Nodule growth	Cell division. Persistent meristem	Cell expansion
Flavonoids inducing *nod* genes	Isoflavones	Flavones, flavonones
Export of assimilated nitrogen	Amides	Ureides (*Lotus* is an exception)

Fig. 1 Main characteristics of indeterminate and determinate nodules. Both types differ structurally and biochemically. In the photographs, the red color in the fixation zone (III) of the *M. truncatula* indeterminate nodule and in the infected zone of the *L. japonicus* determinate nodule is due to the high concentration of leghemoglobin (Lb). The green-brownish color in the senescent zone (IV) of the indeterminate nodule is indicative of Lb degradation to biliverdin-like pigments.

and RNS are kept under control by a plethora of antioxidant enzymes and metabolites (Becana et al., 2010; Puppo et al., 2005). This tight regulation permits both types of reactive molecules to play essential roles as signals ('oxidative and nitrosative signaling') during plant development and in perception, adaptation and tolerance to stress (Umbreen et al., 2018; Waszczak, Carmody, & Kangasjärvi, 2018). Likewise, it has been shown that superoxide, H_2O_2 and NO are produced in all developmental stages of the legume-rhizobium symbiosis, from rhizobial infection to nodule maturation and senescence (Cárdenas, Martínez, Sánchez, & Quinto, 2008; Jamet, Mandon, Puppo, & Hérouart, 2007; Rubio et al., 2004; Santos, Hérouart, Sigaud, Touati, & Puppo, 2001). Here, we provide an update of findings related to the antioxidant system of legume nodules and then focus on the latest advances on redox-based post-translational modifications (PTMs) of nodule proteins. These PTMs can readily modulate protein activity in response to developmental or environmental cues and thus have a major impact on SNF.

2. Production of reactive oxygen and nitrogen species

2.1 Root infection and nodule primordia

Several studies have reported the rapid and transient generation of ROS in response to rhizobial Nod factors during the early stages of symbiosis (Bueno et al., 2001; Cárdenas et al., 2008; Santos et al., 2001). ROS act downstream of Nod factors in the signaling pathways that lead to the initiation of nodule primordia and are required for an adequate progression of infection threads (Cárdenas et al., 2008; Jamet et al., 2007; Rubio et al., 2004; Santos et al., 2001). Although ROS can be generated from different sources (Becana et al., 2000), there is substantial evidence that points to cell membrane NADPH oxidases (also termed 'respiratory burst oxidase homologues', Rbohs) as the most important ROS-generating system during the first stages of symbiosis (for reviews see Marino, Dunand, Puppo, & Pauly, 2012; Montiel, Arthikala, Cárdenas, & Quinto, 2016). In common bean, down-regulation of *RbohA* and *RbohB* impairs infection thread formation and affects nodule development and function; conversely, *RbohB* overexpression increases the number of infection events and nodule number (Arthikala et al., 2017; Montiel et al., 2016). In plants, small GTP-binding proteins are key regulators of ROS generation by Rbohs. In *M. truncatula*, knockdown plants in the small ρ-type GTPase MtROP9 do not generate ROS in response to *Sinorhizobium meliloti* and show major alterations in the infection process (Kiirika et al., 2012).

NO is a gaseous free radical that performs multiple signaling and regulatory functions in plants (Umbreen et al., 2018). During the early interaction between *L. japonicus* and *Mesorhizobium loti* a transient production of NO is observed after only 4 h of inoculation. Remarkably, NO production is induced by rhizobial outer membrane lipopolysaccharides and occurs only in the presence of compatible bacteria, which suggests a role for NO in plant-bacteria recognition (Murakami et al., 2011; Nagata et al., 2008). Likewise, in the *M. truncatula*-*S. meliloti* symbiosis, NO is detected in infection threads and nodule primordia after two and four days of inoculation, respectively (del Giudice et al., 2011). In *M. truncatula* NO modulates the expression of numerous genes necessary for the onset of symbiosis and plays a crucial role in the repression of the plant's defense responses (Boscari et al., 2013).

2.2 Mature N_2-fixing nodules

SNF by nitrogenase consumes high amounts of ATP and reducing power that are provided by bacteroid respiration. Both the bacteroidal and mitochondrial electron transport chains inevitably generate ROS. Another potentially important source of ROS in nodules is oxidation of heme and [Fe—S] clusters of proteins. Leghemoglobin (Lb) transports and delivers O_2 to the symbiosomes at a low steady concentration (Appleby, 1984) and is essential for SNF (Ott et al., 2005). Mutant nodules lacking Lb generate high concentrations of superoxide, probably as a result of activation of Rboh enzymes (Wang et al., 2019). Autoxidation of oxygenated Lb generates superoxide at significant levels, a process that is favored by the acid pH of senescent nodules (Becana & Klucas, 1992; Puppo, Rigaud, & Job, 1981). The nitrogenase components are irreversibly inactivated by O_2 probably because of the partial reduction of O_2 to ROS by [Fe—S] clusters (Robson & Postgate, 1980). Other abundant bacteroid proteins contributing to ROS production include ferredoxin, the proximal electron donor to nitrogenase, and hydrogenases, membrane-bound enzymes involved in the recycling of H_2 produced by nitrogenase activity (Becana et al., 2000). Rbohs participate also in ROS production in mature nodules. In *M. truncatula*, *RbohA* expression is induced in nodules and its down-regulation decreases SNF and expression of nitrogenase genes (Marino et al., 2012).

Several technical approaches have been used to localize NO production in mature nodules. The fluorescent dye 4,5-diaminofluorescein and a *S. meliloti* strain engineered as NO biosensor proved that NO is produced in zone III of *M. truncatula* nodules (Baudouin, Pieuchot, Engler, Pauly, & Puppo, 2006; Cam et al., 2012). Complementary studies with fluorescent probes and electron paramagnetic resonance spectroscopy of the nitrosyl-ferrous leghemoglobin (LbNO) complexes have shown NO production in a variety of legume nodules and treatment conditions (Calvo-Begueria et al., 2018; Horchani et al., 2011; Mathieu, Moreau, Frendo, Puppo, & Davies, 1998; Meilhoc, Blanquet, Cam, & Bruand, 2013; Sánchez et al., 2010). It should nevertheless be borne in mind that detection of NO by fluorescent dyes is prone to many artifacts, as concluded by Calvo-Begueria et al. (2018). Notably, these authors detected NO in the nodule parenchyma of soybean nodules, where the O_2 diffusion barrier is located (Minchin, James, & Becana, 2008). This observation suggests that NO is engaged in the control of O_2 entry into the infected zone. A few potential sources of NO in mature nodules have been proposed but most evidence supports the contribution of the nitrate reductase activities of both symbiotic partners (Horchani et al., 2011). The presence of an arginine-dependent (NO synthase-like) activity in nodules was also surmised on the basis of studies with inhibitors of animal NO synthases (Horchani et al., 2011; and references therein). However, as occurs with the detection of similar activities in other plant systems, the protein(s) responsible for this reaction remain(s) elusive. The use of several inhibitors of plant nitrate reductase and animal NO synthase activity suggested that in mature nodules the contribution of these enzymes to NO synthesis is secondary with respect to the bacteroid denitrification pathway (Calvo-Begueria et al., 2018).

2.3 Senescent nodules

Legume nodule senescence is a highly organized and regulated process characterized, among other factors, by the degradation of the two symbiotic partners and the switch of the nodule status from carbon sink to general nutrient source (Van de Velde et al., 2006). Extensive studies have been performed on both nodule developmental (aging) and stress-induced senescence (eg. Becana et al., 2010; Pérez Guerra et al., 2010; Puppo et al., 2005; Yuan et al., 2017; and references therein). The developmental and dark-stress induced senescence of *M. truncatula* nodules have been compared in detail at the structural and molecular levels (Pérez Guerra et al., 2010). Features shared by both processes include decreases of nitrogenase expression/activity and Lb content and up-regulation of cysteine proteinase transcripts. However, some interesting structural differences have been noted, such as degradation of symbiosomal membranes during aging but not during dark stress (Pérez Guerra et al., 2010). At the biochemical level, some studies have reported increases of H_2O_2, lipid peroxides and protein carbonyls in the two types of senescing nodules (Alesandrini, Mathis, Van de Sype, Hérouart, & Puppo, 2003; Evans et al., 1999; Loscos, Matamoros, & Becana, 2008). However, other studies found no evidence of oxidative stress associated to nodule aging (Groten et al., 2005). Most probably, alterations in redox homeostasis as a result of an increase in ROS production and a decrease in antioxidants may be perceived by redox sensors that, in turn, modify the signaling pathways that trigger senescence (Becana et al., 2010; Puppo et al., 2005). It is also conceivable that localized oxidative modifications, for example in the mitochondria, can be conveyed to the rest of the cell and regulate the senescence process (Matamoros et al., 2013).

Increasing evidence suggests that NO plays a key role in leaf and nodule senescence. It has been proposed that NO can act as either a positive or a negative regulator depending on the plant organ (Bruand & Meilhoc, 2019). In leaves NO delays senescence and the expression of an NO-scavenging protein (the bacterial flavohemoglobin Hmp, which shows NO dioxygenase activity) induces leaf senescence in Arabidopsis (Mishina, Lamb, & Zeier, 2007). The underlying molecular mechanisms are still poorly defined, but NO-mediated transcriptional regulation or protein PTMs might be involved. On the contrary, NO triggers nodule senescence. Higher NO concentrations have been detected during the later stages of the symbiosis. As occurs in leaves, activation of specific NO-mediated signaling pathways may induce senescence. Alternatively, senescence might result from NO-mediated inactivation of key nodule proteins such as glutamine synthetase (GS) and Lb (see below), or from the inhibition of nitrogenase and the bacterial and mitochondrial respiratory chains (Bruand & Meilhoc, 2019, and references therein).

3. Antioxidant systems

3.1 Ascorbate, glutathione and associated enzymes in nodules

Ascorbate (vitamin C) and glutathione (GSH; γGlu-Cys-Gly) are major antioxidants and redox buffers of plants and are present at millimolar concentrations in legume nodules. Both metabolites act as antioxidants in their own right as ROS scavengers but also indirectly by being substrates of the enzymes of the ascorbate-GSH pathway (Foyer & Noctor, 2011). In this pathway, ascorbate peroxidase catalyzes the reduction of H_2O_2 by ascorbate, which is oxidized to monodehydroascorbate (ascorbate free radical) and dehydroascorbate. Ascorbate is then regenerated by NADH-dependent monodehydroascorbate reductase and GSH-dependent dehydroascorbate reductase. Finally, NADPH-dependent glutathione reductase reduces glutathione disulfide back to GSH. In nodules, the four enzymes are present in the cytosol and mitochondria and probably also in the plastids and peroxisomes (Dalton, Russell, Hanus, Pascoe, & Evans, 1986; Iturbe-Ormaetxe, Matamoros, Rubio, Dalton, & Becana, 2001; Matamoros et al., 2006) and their activities are positively correlated to SNF (Dalton, Langeberg, & Treneman, 1993) (Fig. 2).

Ascorbate is synthesized primarily by the L-galactose pathway in leaf cells (Smirnoff, 2018) and probably also in nodules (Loscos et al., 2008). Aging and stress-induced senescence entails a progressive decline of ascorbate biosynthesis and regeneration capacity in nodules (Loscos et al., 2008).

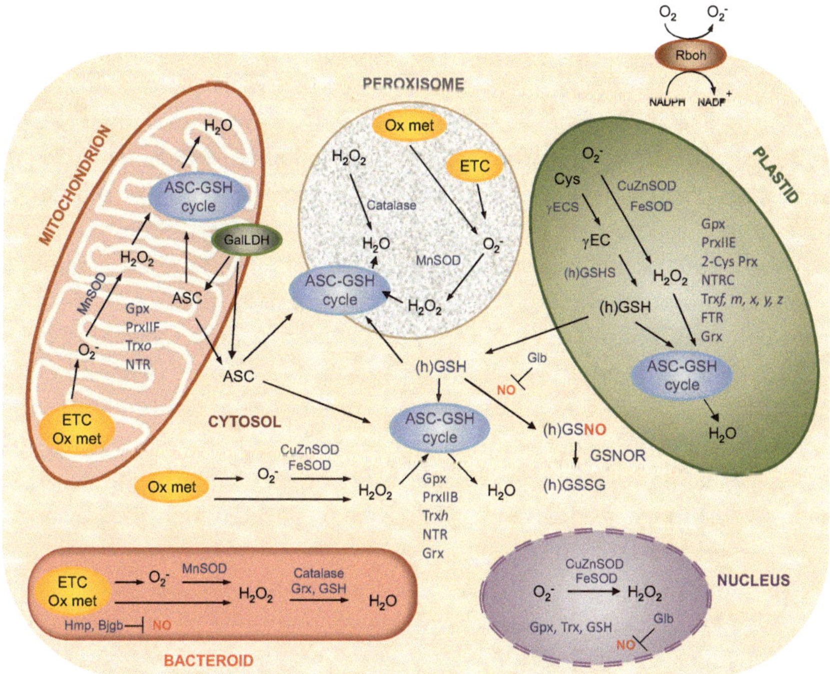

Fig. 2 Schematics showing the subcellular localization of antioxidant enzymes and metabolites in legume nodules. *ASC*, ascorbate; *γEC*, γ-glutamylcysteine; *γECS*, γ-glutamylcysteine synthetase; *ETC*, electron transport chain; *FTR*, ferredoxin-thioredoxin reductase; *GalLDH*, L-galactono-1,4-lactone dehydrogenase; *(h)GSHS*, (homo)glutathione synthetase; *Ox met*, oxidative metabolism.

Increasing the ascorbate content in legumes has therefore biotechnological interest because it could render plants more nutritious and tolerant to environmental stresses. Recent results show that *M. truncatula* plants overexpressing key genes of the L-galactose pathway have higher rates of SNF at the cost of a lower growth (Matamoros, Dalton, & Becana, 2017; Torres-Jerez, Huertas-Ruz, Lara-Dampier, Dalton, & Udvardi, 2017).

In plants GSH is involved in multiple physiological processes including redox homeostasis, regulation of cell cycle, responses to stress and heavy metal detoxification (Noctor et al., 2012). In legumes homoglutathione (hGSH; γGlu-Cys-βAla) may partially or completely replace GSH depending on the plant species and organ (Frendo et al., 1999; Klapheck, 1988; Matamoros, Moran, Iturbe-Ormaetxe, Rubio, & Becana, 1999). This can be illustrated by the thiol distribution in the two model legumes: *M. truncatula* produces GSH in all the plant and hGSH only in the roots and nodules

(Frendo et al., 1999), whereas *L. japonicus* produces almost exclusively hGSH in the roots and leaves and both thiols only in the nodules (Matamoros et al., 2003). The synthesis of GSH and hGSH takes place by two sequential reactions. The first step is catalyzed by γ-glutamylcysteine synthetase and the second one by specific glutathione or homoglutathione synthetases (Frendo et al., 1999; Klapheck, 1988; Matamoros et al., 1999). Interestingly, in the two model legumes the homoglutathione synthetase gene arose from a duplication event and both genes are tandemly arranged in the chromosome. Homoglutathione synthetase has a higher affinity for β-alanine than for glycine (Frendo et al., 2001; Iturbe-Ormaetxe et al., 2002). To date there is not conclusive proof for specific functions of GSH or hGSH in legumes, although the differential transcriptional response of glutathione synthetase and homoglutathione synthetase to NO and hormones suggests that the functions of the two thiols do not completely overlap (Clemente et al., 2012; Innocenti et al., 2007). This issue will need to be settled by using gene-specific knockout mutants.

GSH and hGSH play essential roles in nodulation (Becana et al., 2010; Frendo et al., 2005). In *M. truncatula* the down-regulation of any of the three genes involved in GSH and hGSH synthesis resulted in plants with fewer nodules and lower rates of SNF. Conversely, the overexpression of γ-glutamylcysteine synthetase increased the GSH content and SNF (El Msehli et al., 2011). Bacteroids actively express the genes of the GSH biosynthetic pathway (Roux et al., 2014) and contain most of the GSH (~85%) of the nodules (Fig. 2; Matamoros et al., 2013). The importance of bacterial GSH for the symbiosis was established with the use of mutants. Thus, the *S. meliloti gshA* strain lacking γ-glutamylcysteine synthetase did not nodulate *M. truncatula*, whereas the *gshB* mutants deficient in glutathione synthetase formed nodules with reduced SNF (Cheng, Karunakaran, East, Munoz-Azcarate, & Poole, 2017; Harrison et al., 2005; Muglia, Comai, Spegazzini, Riccillo, & Aguilar, 2008). The enzyme glutathione reductase of *S. meliloti* bacteroids is also critical for symbiosis and nodule redox homeostasis because the corresponding *gor* mutant, which shows a decreased ratio of reduced to oxidized glutathione (GSH/GSSG), has a lower SNF efficiency (Tang et al., 2018).

3.2 Catalase

Catalases are tetrameric heme proteins localized to the peroxisomes that catalyze the decomposition of H_2O_2 to water and O_2. Because of their low affinity for H_2O_2 they may be efficient antioxidants only at high

H_2O_2 concentrations. Unlike Arabidopsis, which has three catalase genes that are differentially regulated during development and in response to environmental factors (Mhamdi et al., 2010), a single catalase gene has been identified in *L. japonicus* (Becana et al., 2014). In white lupin nodules, catalase was immunolocalized in peroxisomes of infected cells and found to decrease during nitrate-induced senescence (Lorenzo, Lucas, Vivo, & de Felipe, 1990). As for the microsymbiont, *S. meliloti* contains three catalase genes encoding two monofunctional (KatA and KatC) and one bifunctional (KatB) catalase-peroxidase enzymes. Interestingly, the bacteroids of the single mutants behave similarly to those of the wild-type strain, but the nodules with bacteroids lacking both KatA and KatC enzymes show dramatic declines in SNF (Jamet, Sigaud, Van de Sype, Puppo, & Hérouart, 2003).

3.3 Superoxide dismutases

Superoxide dismutases (SODs) are metalloenzymes that catalyze the dismutation of superoxide radicals to H_2O_2 and O_2. They can be classified as CuZnSODs, MnSODs and FeSODs based on their metal cofactors. All of them have been identified in the nodule host cells: CuZnSOD and FeSOD in the cytosol, plastids and nuclei, and MnSOD in the mitochondria (Fig. 2). In addition, bacteroids contain a MnSOD with significant homology to the plant protein (Becana et al., 2014; Rubio et al., 2007). The transcripts and proteins of SOD isoforms are differentially localized in determinate and indeterminate nodules, suggesting specific roles for the three classes of SODs during nodule development. Also, co-localization studies suggest a role for CuZnSOD in cell wall growth and in the progression of infection threads by providing the H_2O_2 required for the cross-linking of extensins in the extracellular matrix and in the lumen of infection threads (Rubio et al., 2004, 2007).

3.4 Thiol peroxidases, thioredoxins and glutaredoxins

Peroxiredoxins (Prxs) and glutathione peroxidases (Gpxs) are ubiquitous non-heme enzymes involved in redox homeostasis. Prxs and Gpxs differ in protein sequence and structure but possess similar biochemical properties, functioning as thiol peroxidases in most organisms (Rouhier & Jacquot, 2005). They reduce H_2O_2 and organic peroxides with electrons donated by a catalytic peroxidatic cysteine residue, which results in the oxidation of the thiol group and the formation of a disulfide bond (Dietz, 2011; Passaia & Margis-Pinheiro, 2015). Vertebrate Gpxs contain selenocysteine in their active site and are reduced back to the active form by GSH, whereas plant

Gpxs have catalytic cysteines that are generally reduced by thioredoxins (Trxs) and NADPH-thioredoxin reductase (NTR) in the cytosol and mitochondria, or by Trxs and ferredoxin-thioredoxin reductase in the chloroplasts. Prxs are classified in four groups based on the protein sequence and structure: 1-Cys Prx, PrxQ, PrxII and 2-Cys Prx (Dietz, 2011). Seven Prxs genes were identified in *L. japonicus* and all of them, except 1-Cys Prx, are expressed in nodules. Immunoblot and proteomic analyses detected mitochondrial PrxIIF, cytosolic PrxIIB and plastidic PrxIIE and 2-Cys Prx in the nodules (Fig. 2; Dam et al., 2014; Tovar-Méndez et al., 2011). The genome of *L. japonicus* contains six *Gpx* genes, of which *LjGpx1* and *LjGpx3* are highly expressed in nodules, especially in the infected zone. LjGpx1 localizes to the plastids and nuclei and LjGpx3 to the cytosol and endoplasmic reticulum, and both enzymes seem to prevent lipid peroxidation under stress conditions (Matamoros et al., 2015).

Trxs are classified in seven groups based on protein sequence and localization: Trx*f*, Trx*m*, Trx*x*, Trx*y* and Trx*z* localize in the chloroplasts, Trx*h* in the cytosol and Trx*o* in the mitochondria and nuclei (Martí et al., 2009; Meyer, Buchanan, Vignols, & Reichheld, 2009). The nodules of *L. japonicus* contain functional NTR-Trx systems in the cytosol and mitochondria, and probably a ferredoxin-thioredoxin reductase-Trx system in the plastids. Moreover, NTRC, an enzyme that contains both NTR and Trx domains and may act as a complete NTR-Trx system, was detected in the plastids (Dam et al., 2014; Tovar-Méndez et al., 2011). More recently, another type of Trx, termed Trx*s*, has been characterized in *M. truncatula* (Ribeiro et al., 2017). Two of the four isoforms, Trx*s1* and Trx*s2*, are predominantly expressed in nodules and have an atypical catalytic site, lack classical disulfide reductase activity and have an N-terminal signal peptide for targeting to the secretory pathway. Trx*s* have no orthologues in non-legumes or legumes with determinate nodules like *L. japonicus* or soybean (Alkhalfioui et al., 2008). The Trx*s1* isoform is targeted to the symbiosomes, where it modulates the redox state of nodule-specific cysteine-rich peptides. These plant peptides are involved in the terminal differentiation of *S. meliloti* into bacteroids (Ribeiro et al., 2017).

Glutaredoxins (Grxs) are small and ubiquitous redox regulators functionally related to Trxs which catalyze the reduction of disulfide bonds using GSH as preferred electron donor (Meyer et al., 2009). Grxs also participate in deglutathionylation reactions and in the assembly of [Fe—S] clusters (Moseler et al., 2015). They form a complex family with more than 30 isoforms identified in vascular plants that are subdivided into four classes.

To our knowledge, only a few Grxs (class I and class II) have been identified so far in nodules (Dam et al., 2014; Tovar-Méndez et al., 2011). The class II Grxs detected in *L. japonicus* are homologues to Arabidopsis Grxs S16 and S17. Grx S16 plays a key role in the control of vegetative growth and Grx S17 is involved in development and associates with cytosolic [Fe—S] components (Iñigo et al., 2016; Knuesting et al., 2015; Rey, Becuwe, Tourrette, & Rouhier, 2017). If these proteins perform similar functions in the nodules awaits investigation. Regarding the bacterial partner, the *S. meliloti* genome encodes three Grxs. Mutation of *Smgrx1* had the strongest phenotype, resulting in nodule abortion and absence of bacteroid differentiation; in contrast, *SmGrx2* mutation impaired nodule development but not bacteroid differentiation and *SmGrx3* mutation had no effect on symbiotic performance (Benyamina et al., 2013). Interestingly, the same authors reported that the deficiency of SmGrx2, but not of the other SmGrxs, affects the iron homeostasis of bacteroids, reinforcing the view that each SmGrx isoform plays specific roles during symbiosis.

3.5 *S*-nitrosoglutathione reductase

A major way by which NO signaling is transmitted is through *S*-nitrosylation of target proteins. This PTM may regulate protein function and has been involved in processes such as the response to stress, hormone signaling and development (Astier et al., 2012). The extent of *S*-nitrosylation is in part modulated by the intracellular level of *S*-nitrosoglutathione (Begara-Morales et al., 2018). This is regulated by *S*-nitrosoglutathione reductase, which catalyzes the NADH-dependent reduction of *S*-nitrosoglutathione producing glutathione disulfide and ammonia (Lindermayr, 2017). Expression analysis unveiled the presence of one functional *S*-nitrosoglutathione reductase gene in leaves and roots and two functional genes in nodules (Matamoros et al., 2019). The function of these enzymes in the N_2-fixing symbiosis remains unknown.

3.6 Hemoglobins

Hemoglobins perform multiple functions in all organisms. In addition to Lbs, legumes, like other plants, contain non-symbiotic hemoglobins, now termed phytoglobins (Glbs; Hill, Hargrove, & Arredondo-Peter, 2016), that are expressed in all tissues. Glbs occur at micromolar concentrations and are grouped in three classes based on phylogenetic analyses and biochemical properties (Smagghe et al., 2007, 2009). Class 1 and class 2 Glbs are structurally similar to animal myoglobin and hemoglobin, with a tertiary structure based on a 3-on-3 (3/3) α-helical fold, whereas class 3 Glbs have homology to bacterial truncated hemoglobins and a 2-on-2 (2/2) α-helical sandwich structure (Wittenberg, Bolognesi, Wittenberg, & Guertin, 2002). Class 1 Glbs display very high O_2 affinities and are thus unsuitable for O_2 transport and delivery (Smagghe et al., 2009). They are induced by hypoxia and flooding in several species and confer stress tolerance, at least in part through the ability of the oxyferrous hemoglobins ($Glb^{2+}O_2$) to dioxygenate NO to NO_3^- by their NO dioxygenase activity (Gardner, 2012; Hebelstrup, Shah, & Igamberdiev, 2013; Hill, 2012). Class 2 Glbs have O_2 affinities similar to Lb and their functions are not well defined, although they are involved in plant development and organogenesis (Hebelstrup et al., 2013; Hill, 2012). Class 3 Glbs have unknown functions in plants, although at least one class 3 Glb, THB1, of the unicellular green alga *Chlamydomonas reinhardtii* is able to scavenge NO *in vivo* (Sanz-Luque et al., 2015).

In Arabidopsis each type of Glb is represented by a single gene, whereas *L. japonicus* express two class 1 (LjGlb1-1 and LjGlb1-2), one class 2 (LjGlb2) and two class 3 (LjGlb3-1 and LjGlb3-2) Glbs (Bustos-Sanmamed et al., 2011). The expression of *LjGlb1-1*, *LjGlb2* and *LjGlb3-1* is very high in nodules relative to other plant organs, suggesting that the respective proteins are required for symbiosis. Overexpression of *LjGlb1-1* increases nodulation and nitrogenase activity, reduces NO level in nodules and delays nodule senescence (Fukudome et al., 2019; Shimoda et al., 2009). Studies with knockout or knockdown mutants of *LjGlb1-1* demonstrated its involvement in the infection process and in the regulation of NO during the initial stages of symbiosis (Fukudome et al., 2016). These authors suggested that the duration and amplitude of the NO signal is regulated by Glbs and is crucial for the onset of symbiosis.

The bacterial partner of the symbiosis also contains hemoglobins. The flavohemoglobin Hmp of *S. meliloti* contributes to regulate NO concentration in nodules and is important for the establishment and function of symbiosis (Cam et al., 2012). The typical symbiont of soybean, *Bradyrhizobium japonicum*, contains a single-domain hemoglobin, Bjgb, that may be implicated also in NO detoxification (Sánchez et al., 2011). Besides hemoglobins, bacteroids contain other proteins that may control NO concentration. Respiratory nitric oxide reductases (Nor), which reduce NO to N_2O in the denitrification pathway, are important for NO homeostasis because mutants defective in Nor (*norB* in *S. meliloti* or *norC* in *B. japonicum*) accumulate NO, as detected by fluorescent probes or electron paramagnetic resonance of

LbNO complexes (Calvo-Begueria et al., 2018; Meilhoc et al., 2013; Sánchez et al., 2011).

4. Protein post-translational modifications related to redox signaling

Proteins are exposed to enzymes and redox reactive compounds that can modify their chemical structures (Fig. 3). These PTMs can be reversible (disulfide bonds, methionine sulfoxides, *S*-nitrosylation) or irreversible (carbonylation, glycation). Theoretically, redox modifications can influence protein activity, stability and localization. However, the effect on protein function is difficult to anticipate and the same modification may have contrasting effects in different proteins. Therefore, a case study approach is often necessary.

4.1 Methionine sulfoxidation

Methionine residues can be readily oxidized to a mixture of methionine-*S*-sulfoxide and methionine-*R*-sulfoxide. The oxidation can be reverted by two methionine sulfoxide reductases, MsrA and MsrB, that respectively reduce the *S* and *R* epimers and are ubiquitous in all organisms (Tarrago, Laugier, & Rey, 2009). Methionine oxidation may alter protein structure and function making this modification of regulatory significance in redox signaling. For example, the *in vivo* oxidation of Met-538 in Arabidopsis nitrate reductase inhibits phosphorylation of Ser-534, and some protein kinases can be directly activated by methionine oxidation (Erickson et al., 2008; Hardin, Larue, Oh, Jain, & Huber, 2009). In humans, this modification has also been associated to protein aggregation and degradation during age-related diseases (Stadtman, Moskovitz, & Levine, 2003).

Little is known about methionine oxidation in legume nodules. In bean, the ratio of methionine to methionine sulfoxide did not change for most proteins in aging nodules. One exception is GS, a key enzyme of nodule carbon and nitrogen metabolism. The GS-N1 isoform contains two methionine residues that are more oxidized to sulfoxides in senescing nodules than in young nodules, but the *in vivo* relevance of this modification could not be determined (Matamoros et al., 2013). The rhizobial proteomes contain several proteins with homology to MsrA and MsrB. It is still uncertain, however, whether these enzymes are able to reduce the methionine sulfoxides back to methionines in proteins.

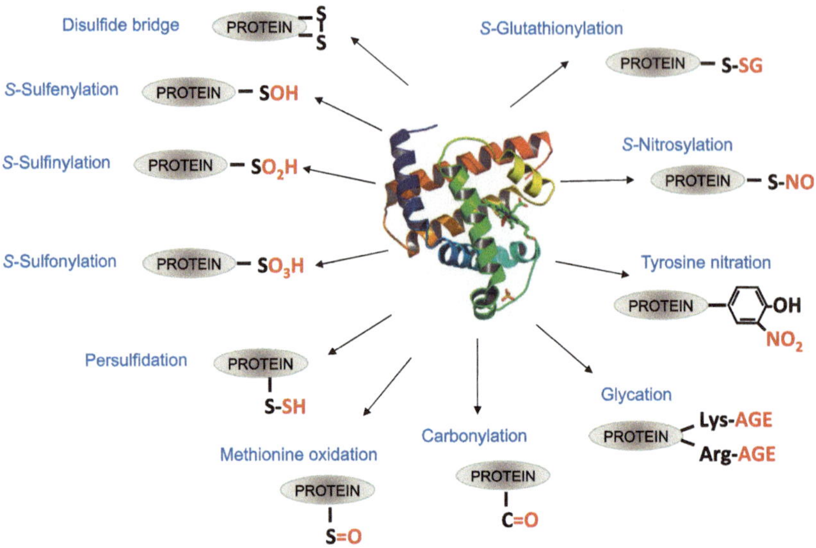

Fig. 3 Redox-dependent PTMs of proteins from legume nodules. Oxidation of thiols of cysteine residues (—SH/—S$^-$) leads to the formation of sulfenic acid (—SOH), which may react with another —SH to form disulfides (—S—S—). This modification can be reverted by Trxs and Grxs. The —SOH group can be further oxidized to sulfinic acid (—SO_2H) and sulfonic acid (—SO_3H). The —SO_2H group may be reduced back to —SH by sulfiredoxins, whereas —SO_3H formation is irreversible. The —SH group can be also persulfidated to —SSH. *S*-nitrosylation (—SNO) may be mediated by NO, NO^+, nitrogen oxides and *trans*-nitrosylating agents such as *S*-nitrosoglutathione (GSNO). Glutathionylation occurs by three main mechanisms: reaction of the target protein with GSNO, —SH/—SS— exchange with glutathione disulfide and reaction of GSH with —SOH. $ONOO^-$-derived radicals modify tyrosine residues. Tyrosyl radicals generated by hydroxyl (•OH) and carbonate (CO_3^-•) radicals react with NO_2 yielding NO_2—Tyr. Methionine is typically oxidized to a sulfoxide. The *S* and *R* stereoisomers are specifically reduced back to methionine by methionine sulfoxide reductases A and B, respectively. Direct oxidation of lysine, arginine, proline and threonine by •OH incorporates the carbonyl moiety into proteins. Alternatively, oxidation of polyunsaturated fatty acid produces unstable lipid hydroperoxides that decompose to secondary products. These react with amino acid side chains and generate carbonyl derivatives. Moreover, arginine and lysine residues may react with reducing sugars or α-dicarbonyls such as glyoxal and methylglyoxal, generating glycation products that are readily oxidized yielding relatively stable advanced glycation end products (AGEs).

4.2 Sulfenylation

Along with methionine, the chemical characteristics of the sulfur atom make cysteine residues major targets of oxidation in proteins. However, not all the cysteines show the same reactivity, which depends on the protein microenvironment and the residue pK_a value. Usually, only thiols with low pK_a play key roles in catalysis and serve as important sites for PTM (Go, Chandler, & Jones, 2015). Deprotonated thiolates are prone to oxidation to form disulfide bridges (S—S) and sulfenic (R—SOH), sulfinic (R—SO_2H) and sulfonic (R—SO_3H) acids. Sulfonic acid formation seems irreversible but the other modifications can be reversed by Trxs, Grxs and sulfiredoxins (Meyer et al., 2009; Sevilla et al., 2015).

Thiol redox modifications may alter the structure, localization and activity of enzymes and transcription factors (Waszczak et al., 2015). In legumes, the use of chemical and genetic probes combined with mass spectrometry analyses allowed the identification of sulfenylated proteins in inoculated roots and mature nodules of *M. truncatula* (Oger, Marino, Guigonis, Pauly, & Puppo, 2012). Most proteins (~80%) were of plant origin. Interestingly, the pattern of sulfenylation differed at different stages of the symbiosis. Proteins involved in redox signaling constituted the largest group during the establishment of the symbiosis, whereas in mature nodules most proteins were related to amino acid and carbohydrate metabolism, protein synthesis, folding, modification and degradation. This is consistent with the well-recognized importance of redox regulation for nodule development (Frendo, Matamoros, Alloing, & Becana, 2013; Ramu, Peng, & Cook, 2002; Rubio et al., 2004). The sulfenylated proteins identified in the bacteroids include the iron protein (NifH) and the iron-molybdenum protein (NifK) of nitrogenase.

4.3 *S*-nitrosylation

Thiol-containing proteins can also undergo *S*-nitrosylation. This is an important NO-dependent PTM in which NO is reversibly incorporated to a reactive cysteine residue (Astier et al., 2012). In Arabidopsis, a study aimed at identifying endogenously *S*-nitrosylated proteins showed that this PTM is a regulatory mechanism in photosynthesis, carbohydrate metabolism and stress responses (Hu et al., 2015). In mature nodules of *M. truncatula*, 80 proteins of bacterial or plant origin were identified as *S*-nitrosylated and 27 proteins were also sulfenylated (Puppo, Pauly, Boscari, Mandon, & Brouquisse, 2013). Most proteins susceptible to both PTMs participate in carbon and nitrogen metabolism and energy production, which underlines the importance of redox regulation in nodule metabolism.

Although the list of *S*-nitrosylated proteins is long, the effect of this PTM on protein function is known only in a few cases. For example, ascorbate peroxidase activity is up-regulated by *S*-nitrosylation of a cysteine residue located in the ascorbate binding site (Begara-Morales et al., 2014). In *L.*

japonicus, LjGpx1 and LjGpx3 are highly expressed in nodules and are regulated by *S*-nitrosylation of the peroxidatic cysteine *in vitro* and *in vivo* (Matamoros et al., 2015). In contrast to ascorbate peroxidase, the modification of Gpxs inhibits their enzymatic activities. In *M. truncatula, in vitro* studies showed that GS isoenzymes can be differently regulated by NO. Plastid-located MtGS2a activity is inhibited by *S*-nitrosylation. MtGS1a is not affected by this PTM but is inactivated by tyrosine nitration (see below; Melo, Silva, Ribeiro, Seabra, & Carvalho, 2011), thus showing the versatility of NO-mediated regulation of protein function.

4.4 *S*-glutathionylation

S-glutathionylation is the reversible addition of glutathione to a reactive cysteine residue of a protein. This PTM usually occurs in response to increases in ROS and NO levels and protects cysteine residues against further deleterious oxidation. As for other PTMs, *S*-glutathionylation may induce functional changes in the target protein and regulate signal transduction and metabolic pathways (Zaffagnini, Bedhomme, Lemaire, & Trost, 2012). Whereas this modification may occur via nonenzymatic mechanisms, deglutathionylation is usually carried out by Grxs (Meyer et al., 2009). *S*-glutathionylation has been widely studied in animal systems (Zhang, Ye, Singh, Townsend, & Tew, 2018) and is also emerging as an important mechanism of redox regulation in plants. In Arabidopsis, a number of *S*-glutathionylated proteins were identified in response to oxidative stress (Dixon, Skipsey, Grundy, & Edwards, 2005). This PTM regulates the activity of Trx*f* in chloroplasts and glycine decarboxylase in mitochondria and is therefore involved in the regulation of carbon fixation and photorespiration (Zaffagnini et al., 2012). *S*-glutathionylation is also expected to be a redox signaling mechanism in the onset and functioning of the rhizobium-legume symbiosis. This is suggested by the observations that SmGrx1 displays deglutathionylation activity and that the corresponding bacterial mutant strain has an impaired symbiotic phenotype (Benyamina et al., 2013). To our knowledge, there is no information about the occurrence of this PTM in nodules.

4.5 Persulfidation

Besides sulfenylation, nitrosylation and glutathionylation, the thiol group of cysteine residues may be modified in yet other ways. Hydrogen sulfide (H_2S) has emerged as a novel signaling molecule playing an important role in many physiological and pathological processes in plants and animals (Aroca, Gotor, & Romero, 2018). There are three main routes by which H_2S exerts its biological effects: metal center interactions, ROS and RNS scavenging, and persulfidation. This PTM is accepted as the main mechanism by which H_2S transmits its signaling capacity. In this process, a thiol (R—SH) is converted into a perthiol (R—SSH, also called a persulfide), and this can alter protein structure and function because of the decrease in the pK_a and the increase in nucleophilicity of the persulfide group (Ono et al., 2014). In Arabidopsis, >2000 proteins involved in key processes of plant biology have been recently identified as targets of persufidation (Aroca et al., 2018). Of these, ~25% have been found to be also *S*-nitrosylated and ~3% are *S*-glutathionylated, showing that many proteins may be regulated by these redox-dependent mechanisms. To gain insight into the possible roles of protein persufidation in nodules, a proteomic study is underway with bean nodules at different stages of development (Matamoros et al., unpublished data). Preliminary data show that >1000 proteins of nodule host cells and bacteroids are persulfidated *in vivo* under physiological conditions.

4.6 Nitration

This PTM consists in the covalent addition of a nitro group (—NO_2) to one of the two equivalent *ortho* carbons in the aromatic ring of tyrosine residues to form 3-nitrotyrosine (NO_2—Tyr) (Kolbert, Feigl, Bordé, Molnár, & Erdei, 2017). Tyrosine nitration requires the presence of $ONOO^-$ or NO_2 because NO itself is not reactive enough to nitrate tyrosine residues. $ONOO^-$ is formed by the reaction of NO and superoxide radicals. Radicals derived from $ONOO^-$ breakdown oxidize tyrosine residues to tyrosyl radicals, which react with NO_2, produced also from $ONOO^-$ decomposition, to yield NO_2—Tyr. Alternatively, NO_2 may be generated by the oxidation of NO to NO_2 in the presence of H_2O_2 and peroxidases (Radi, 2018). In nodules another mechanism has been described: H_2O_2 oxidizes Lb to ferryl Lb and this, in turn, oxidizes NO_2^- to NO_2 and tyrosine to tyrosyl radicals; then both radicals react to form NO_2—Tyr (Sainz et al., 2015).

Tyrosine nitration causes a decrease of the residue pK_a, enhances its hydrophobicity and provokes steric restrictions because NO_2—Tyr is larger than tyrosine. In plant cells tyrosine nitration generally leads to protein loss-of-function, although there are some exceptions. In nodules tyrosine nitration may have an important regulatory role because two key proteins for nodule functioning, GS and Lb, are targets of this PTM. MtGS1a is inactivated by tyrosine nitration and the level of nitrated protein increases under conditions in which SNF is impaired, such as in ineffective nodules or in nodules of plants fed with NO_3^- (Melo et al., 2011). Lb is susceptible to nitration in both the heme and globin. Navascués et al. (2012) identified green Lb derivatives in senescing soybean nodules. These modified Lbs have identical globins to the parent red Lbs but their hemes are nitrated in a vinyl group. In a follow-up study, it was found that one tyrosine residue located in the distal heme pocket was the major target of nitration of the globin moiety (Sainz et al., 2015). However, the amount of nitrated globin decreased during senescence, suggesting that heme and globin nitration occurs through different mechanisms and/or that globin nitration, but not heme nitration, makes the protein prone to degradation by nodule proteases as observed for other plant proteins (Castillo et al., 2015).

The significance of protein nitration in redox signaling is still poorly defined. This stable PTM was categorized as irreversible, but recent research in animals has identified denitrase mechanisms that could be also operative in plants (Kolbert et al., 2017).

4.7 Carbonylation and glycation

In cells, metal-catalyzed oxidation occurs when free Fe^{2+} or Cu^{+} reacts with H_2O_2 and generates hydroxyl radicals through the Fenton reaction (Halliwell, 2006). These radicals can irreversibly oxidize amino acid side chains and introduce the carbonyl moiety in proteins (Møller, Rogowska-Wrzesinska, & Rao, 2011). Carbonyl groups may also be generated indirectly by Michael addition of lipid peroxidation decomposition products to arginine, cysteine, histidine and lysine residues (Matamoros et al., 2018; Møller et al., 2011). Protein carbonylation contributes to cellular damage caused by stress conditions and age-associated diseases in animals (Höhn, König, & Grune, 2013) and plants (Matamoros et al., 2013; Sun et al., 2014). It is also likely, however, that irreversible protein carbonylation has a role in the regulation of protein function, thus contributing to redox signaling (Oracz et al., 2007; Winger, Taylor, Heazlewood, Day, & Millar, 2007).

Protein glycation occurs when arginine and lysine residues react with reducing sugars, generating Amadori and Heyns compounds (Heyns & Noack, 1962). These glycation products are readily oxidized, yielding relatively stable advanced glycation end products. Alternatively, these glycation products can be formed by the reaction of arginine and lysine residues with α-dicarbonyls (eg. glyoxal and methylglyoxal) generated by monosaccharide auto-oxidation under oxidative conditions. In humans, the formation of glycation end products accompanies atherosclerosis and diabetes (Höhn et al., 2013). However, very little is known about protein glycation in plants.

The Arabidopsis proteome modified by advanced glycation products, as well as the age-dependent increase of glycation at specific sites, have been recently reported (Bilova et al., 2017).

An extensive study on protein oxidation in nodules allowed the identification of 238 and 131 plant and bacterial carbonylated proteins, respectively, and the carbonylation sites were determined. The study revealed that carbonylation occurs under normal growth conditions and that lipid peroxidation-derived products are the major contributors to protein carbonylation in nodule cells (Matamoros et al., 2018). The same study also uncovered major effects of carbonylation on two key nodule proteins, malate dehydrogenase and Lb. Malate dehydrogenase is essential for SNF because malate is the primary source of carbon transported to the bacteroids, and its activity is negatively correlated to the carbonylation level. Carbonylation also induced Lb aggregation, probably rendering the protein inactive and more susceptible to degradation by cell proteases. Other numerous glycated proteins were identified *in vivo*, including three nodule proteins that are central to carbon and nitrogen metabolism: sucrose synthase, GS and glutamate synthase. Label-free quantification identified 10 plant proteins and 18 bacterial proteins as age-specifically glycated, although the functional implications of these modifications are unknown.

5. Conclusions

Extensive transcriptomic work has been carried out to elucidate the mechanisms of redox control in legume nodules. This has not been in pace, however, with parallel studies at the proteomic and metabolomic levels. An increasing number of nodule proteins bearing redox-dependent PTMs have been identified, but in most cases the precise effects of such modifications on protein function remain unknown. Because the same PTM may cause distinct effects on various proteins, an individual study of each protein is necessary. The results available so far suggest that the selective redox-dependent modification of enzymes, transcriptional regulators and components of signaling pathways may constitute a major control mechanism of the infection process and of nodule metabolism, development and senescence. In the next years, considerable effort will be required to identify ROS and RNS sensors and to establish the protein interactions that convey specific signals for the functioning of symbiosis and its response to abiotic and biotic stress. Understanding redox signaling in plant and nodule cells will constitute a solid base to manipulate nodule activity and improve adaptation of legumes to a changing environment.

Acknowledgments

We are very grateful to Carmen Pérez-Rontomé for drawing the figures. Work from our laboratory has been funded by grant AGL2017-85775-R from the Ministry of Economy and Competiveness—European Regional Development Fund and grant A09_17R from the Government of Aragón (Spain).

References

Alesandrini, F., Mathis, R., Van de Sype, G., Hérouart, D., & Puppo, A. (2003). Possible roles of a cysteine protease and hydrogen peroxide in soybean nodule development and senescence. *New Phytologist, 158*, 131—138.

Alkhalfioui, F., Renard, M., Frendo, P., Keichinger, C., Meyer, Y., Gelhaye, E., et al. (2008). A novel type of thioredoxin dedicated to symbiosis in legumes. *Plant Physiology, 48*, 424—435.

Appleby, C. A. (1984). Leghemoglobin and *rhizobium* respiration. *Annual Review of Plant Physiology, 35*, 443—478.

Aroca, A., Gotor, C., & Romero, L. C. (2018). Hydrogen sulfide signaling in plants: Emerging roles of protein persulfidation. *Frontiers in Plant Science, 9*, 1369.

Arthikala, M. K., Montiel, J., Sánchez-López, R., Nava, N., Cárdenas, L., & Quinto, C. (2017). Respiratory burst oxidase homolog gene A is crucial for *Rhizobium* infection and nodule maturation and function in common bean. *Frontiers in Plant Science, 8*, 2003.

Astier, J., Kulik, A., Koen, E., Besson-Bard, A., Bourque, S., Jeandroz, S., et al. (2012). Protein *S*-nitrosylation: what's going on in plants? *Free Radical Biology & Medicine, 53*, 1101—1110.

Baudouin, E., Pieuchot, L., Engler, G., Pauly, N., & Puppo, A. (2006). Nitric oxide is formed in *Medicago truncatula—Sinorhizobium meliloti* functional nodules. *Molecular Plant-Microbe Interactions, 19*, 970—975.

Becana, M., Dalton, D. A., Moran, J. F., Iturbe-Ormaetxe, I., Matamoros, M. A., & Rubio, M. C. (2000). Reactive oxygen species and antioxidants in legume nodules. *Physiologia Plantarum, 109*, 372—381.

Becana, M., & Klucas, R. V. (1992). Oxidation and reduction of leghemoglobin in root nodules of leguminous plants. *Plant Physiology, 98*, 1217—1221.

Becana, M., Matamoros, M. A., Ramos, J., Rubio, M. C., & Sainz, M. (2014). Reactive oxygen/nitrogen species and antioxidant defenses in *Lotus japonicus*. In S. Tabata, & J. Stougaard (Eds.), *The Lotus japonicus genome, compendium of plant genomes* (pp. 137—147). Berlin, Heidelberg: Springer-Verlag.

Becana, M., Matamoros, M. A., Udvardi, M., & Dalton, D. A. (2010). Recent insights into antioxidant defenses of legume root nodules. *New Phytologist, 188*, 960—976.

Begara-Morales, J. C., Chaki, M., Valderrama, R., Sánchez-Calvo, B., Mata-Pérez, C., Padilla, M. N., et al. (2018). Nitric oxide buffering and conditional nitric oxide release in stress response. *Journal of Experimental Botany, 69*, 3425—3438.

Begara-Morales, J. C., Sánchez-Calvo, B., Chaki, M., Valderrama, R., Mata-Pérez, C., Lopez-Jaramillo, J., et al. (2014). Dual regulation of cytosolic ascorbate peroxidase (APX) by tyrosine nitration and *S*-nitrosylation. *Journal of Experimental Botany, 65*, 527—538.

Benyamina, S. M., Baldacci-Cresp, F., Couturier, J., Chibani, K., Hopkins, J., Bekki, A., et al. (2013). Two *Sinorhizobium meliloti* glutaredoxins regulate iron metabolism and symbiotic bacteroid differentiation. *Environmental Microbiology, 15*, 795–810.

Bilova, T., Paudel, G., Shilyaev, N., Schmidt, R., Brauch, D., Tarakhovskaya, E., et al. (2017). Global proteomic analysis of advanced glycation end products in the *Arabidopsis* proteome provides evidence for age-related glycation hot spots. *Journal of Biological Chemistry, 292*, 15758–15776.

Boscari, A., Meilhoc, E., Castella, C., Bruand, C., Puppo, A., & Brouquisse, R. (2013). Which role for nitric oxide in symbiotic N_2-fixing nodules: Toxic by-product or useful signaling/metabolic intermediate? *Frontiers in Plant Science, 4*, 384.

Bruand, C., & Meilhoc, E. (2019). Nitric oxide in plants: Pro- or anti-senescence. *Journal of Experimental Botany, 70*, 4419–4427.

Bueno, P., Soto, M. J., Rodríguez-Rosales, M. P., Sanjuan, J., Olivares, J., & Donaire, J. P. (2001). Time-course of lipoxygenase, antioxidant enzyme activities and H_2O_2 accumulation during early stages of *Rhizobium*-legume symbiosis. *New Phytologist, 152*, 91–96.

Bustos-Sanmamed, P., Tovar-Méndez, A., Crespi, M., Sato, S., Tabata, S., & Becana, M. (2011). Regulation of nonsymbiotic and truncated hemoglobin genes of *Lotus japonicus* in plant organs and in response to nitric oxide and hormones. *New Phytologist, 189*, 765–776.

Calvo-Begueria, L., Rubio, M. C., Martínez, J. I., Pérez-Rontomé, C., Delgado, M. J., Bedmar, E. J., et al. (2018). Redefining nitric oxide production in legume nodules through complementary insights from electron paramagnetic resonance spectroscopy and specific fluorescent probes. *Journal of Experimental Botany, 69*, 3703–3714.

Cam, Y., Pierre, O., Boncompagni, E., Hérouart, D., Meilhoc, E., & Bruand, C. (2012). Nitric oxide (NO): A key player in the senescence of *Medicago truncatula* root nodules. *New Phytologist, 196*, 548–560.

Cárdenas, L., Martínez, A., Sánchez, F., & Quinto, C. (2008). Fast, transient and specific intracellular ROS changes in living root hair cells responding to Nod factors (NFs). *The Plant Journal, 56*, 802–813.

Castillo, M.-C., Lozano-Juste, J., González-Guzmán, M., Rodriguez, L., Rodriguez, P. D., & León, J. (2015). Inactivation of PYR/PYL/RCAR ABA receptors by tyrosine nitration may enable rapid inhibition of ABA signaling by nitric oxide in plants. *Science Signaling, 8*, ra89.

Cheng, G., Karunakaran, R., East, A. K., Munoz-Azcarate, O., & Poole, P. S. (2017). Glutathione affects the transport activity of *Rhizobium leguminosarum* 3841 and is essential for efficient nodulation. *FEMS Microbiology Letters, 364*.

Clemente, M. R., Bustos-Sanmamed, P., Loscos, J., James, E. K., Pérez-Rontomé, C., Navascués, J., et al. (2012). Thiol synthetases of legumes: Immunogold localization and differential gene regulation by phytohormones. *Journal of Experimental Botany, 63*, 3923–3934.

Dalton, D. A., Langeberg, L., & Treneman, N. (1993). Correlations between the ascorbate-glutathione pathway and effectiveness in legume root nodules. *Physiologia Plantarum, 87*, 365–370.

Dalton, D. A., Russell, S. A., Hanus, F. J., Pascoe, G. A., & Evans, H. J. (1986). Enzymatic reactions of ascorbate and glutathione that prevent peroxide damage in soybean root nodules. *Proceedings of the National Academy of Sciences of the United States of America, 83*, 3811–3815.

Dam, S., Dyrlund, T. F., Ussatjuk, A., Jochimsen, B., Nielsen, K., Goffard, N., et al. (2014). Proteome reference maps of the *Lotus japonicus* nodule and root. *Proteomics, 14*, 230–240.

del Giudice, J., Cam, Y., Damiani, I., Fung-Chat, F., Meilhoc, E., Bruand, C., et al. (2011). Nitric oxide is required for an optimal establishment of the *Medicago truncatula-Sinorhizobium meliloti* symbiosis. *New Phytologist, 191*, 405–417.

Dietz, K. J. (2011). Peroxiredoxins in plants and cyanobacteria. *Antioxidants & Redox Signaling, 15*, 1129–1159.
Dixon, D. P., Skipsey, M., Grundy, N. M., & Edwards, R. (2005). Stress-induced protein S-glutathionylation in *Arabidopsis*. *Plant Physiology, 138*, 2233–2244.
Dupont, L., Alloing, G., Pierre, O., El Msehli, S., Hopkins, J., Hérouart, D., et al. (2012). The legume root nodule: From symbiotic nitrogen fixation to senescence. In T. Nagata (Ed.), *Senescence* (pp. 137–168). Rijeka: InTech Europe.
El Msehli, S., Lambert, A., Baldacci-Cresp, F., Hopkins, J., Boncompagni, E., Smiti, S. A., et al. (2011). Crucial role of (homo)glutathione in nitrogen fixation in *Medicago truncatula* nodules. *New Phytologist, 192*, 496–506.
Erickson, J. R., Joiner, M. L., Guan, X., Kutschke, W., Yang, J., Oddis, C. V., et al. (2008). A dynamic pathway for calcium-independent activation of CaMKII by methionine oxidation. *Cell, 133*, 462–474.
Evans, P. J., Gallesi, D., Mathieu, C., Hernández, M. J., de Felipe, M. R., Halliwell, B., et al. (1999). Oxidative stress occurs during soybean nodule senescence. *Planta, 208*, 73–79.
Foyer, C. H., & Noctor, G. (2011). Ascorbate and glutathione: The heart of the redox hub. *Plant Physiology, 155*, 2–18.
Frendo, P., Gallesi, D., Turnbull, R., Van de Sype, G., Hérouart, D., & Puppo, A. (1999). Localisation of glutathione and homoglutathione in *Medicago truncatula* is correlated to a differential expression of genes involved in their synthesis. *The Plant Journal, 17*, 215–219.
Frendo, P., Harrison, J., Norman, C., Hernández, M. J., Van de Sype, G., Gilabert, A., et al. (2005). Glutathione and homoglutathione play a critical role in the nodulation process of *Medicago truncatula*. *Molecular Plant-Microbe Interactions, 18*, 254–259.
Frendo, P., Hernández, M. J., Mathieu, C., Duret, L., Gallesi, D., Can de Sype, G., et al. (2001). A *Medicago truncatula* homoglutathione synthetase is derived from glutathione synthetase by gene duplication. *Plant Physiology, 126*, 1706–1715.
Frendo, P., Matamoros, M. A., Alloing, G., & Becana, M. (2013). Thiol-based redox signaling in the nitrogen-fixing symbiosis. *Frontiers in Plant Science, 4*, 376.
Fukudome, M., Calvo-Begueria, L., Kado, T., Osuki, K. I., Rubio, M. C., Murakami, E., et al. (2016). Hemoglobin LjGlb1-1 is involved in nodulation and regulates the level of nitric oxide in the *Lotus japonicus–Mesorhizobium loti* symbiosis. *Journal of Experimental Botany, 67*, 5275–5283.
Fukudome, M., Watanabe, E., Osuki, K., Imaizumi, R., Aoki, T., Becana, M., et al. (2019). Stably transformed *Lotus japonicus* plants overexpressing phytoglobin LjGlb1-1 show decreased nitric oxide levels in roots and nodules as well as delayed nodule senescence. *Plant and Cell Physiology, 60*, 816–825.
Gardner, P. R. (2012). Hemoglobin: A nitric-oxide dioxygenase. *Scientifica, 2012*, 1–34.
Go, Y. M., Chandler, J. D., & Jones, D. P. (2015). The cysteine proteome. *Free Radical Biology & Medicine, 84*, 227–245.
Groten, K., Vanacker, H., Dutilleul, C., Bastian, F., Bernard, S., Carzaniga, R., et al. (2005). The roles of redox processes in pea nodule development and senescence. *Plant, Cell & Environment, 28*, 1293–1304.
Halliwell, B. (2006). Reactive species and antioxidants: Redox biology is a fundamental theme of aerobic life. *Plant Physiology, 141*, 312–322.
Hardin, S. C., Larue, C. T., Oh, M. H., Jain, V., & Huber, S. C. (2009). Coupling oxidative signals to protein phosphorylation via methionine oxidation in *Arabidopsis*. *Biochemical Journal, 422*, 305–312.
Harrison, J., Jamet, A., Muglia, C. I., Van de Sype, G., Aguilar, O. M., Puppo, A., et al. (2005). Glutathione plays a fundamental role in growth and symbiotic capacity of *Sinorhizobium meliloti*. *Journal of Bacteriology, 187*, 168–174.

Hebelstrup, K. H., Shah, J. K., & Igamberdiev, A. U. (2013). The role of nitric oxide and hemoglobin in plant development and morphogenesis. *Physiologia Plantarum, 148*, 457–469.

Heyns, K., & Noack, H. (1962). Die umsetzung von D-fructose mit L-lysine und L-arginin und deren beziehung zu nichtenzymatischen bräunungsreaktionen. *Chemische Berichte, 95*, 720–727.

Hill, R. D. (2012). Non-symbiotic haemoglobins – What's happening beyond nitric oxide scavenging? *AoB Plants, 2012*, pls004.

Hill, R. D., Hargrove, M., & Arredondo-Peter, R. (2016). Phytoglobin: A novel nomenclature for plant globins accepted by the globin community at the 2014 XVIII conference on oxygen-binding and sensing proteins. *F1000Research, 5*, 212.

Höhn, A., König, J., & Grune, T. (2013). Protein oxidation in aging and the removal of oxidized proteins. *Journal of Proteomics, 92*, 132–159.

Horchani, F., Prévot, M., Boscari, A., Evangelisti, E., Meilhoc, E., Bruand, C., et al. (2011). Both plant and bacterial nitrate reductases contribute to nitric oxide production in *Medicago trucatula* nitrogen-fixing nodules. *Plant Physiology, 155*, 1023–1036.

Hu, J., Huang, X., Chen, L., Sun, X., Lu, C., Zhang, L., et al. (2015). Site-specific nitrosoproteomic identification of endogenously *S*-nitrosylated proteins in *Arabidopsis*. *Plant Physiology, 167*, 1731–1746.

Iñigo, S., Durand, A. N., Ritter, A., Le Gall, S., Termathe, M., Klassen, R., et al. (2016). Glutaredoxin GRXS17 associates with the cytosolic iron–sulfur cluster assembly pathway. *Plant Physiology, 172*, 858–873.

Innocenti, G., Pucciariello, C., Le Gleuher, M., Hopkins, J., de Stefano, M., Delledonne, M., et al. (2007). Glutathione synthesis is regulated by nitric oxide in *Medicago truncatula* roots. *Planta, 225*, 1597–1602.

Iturbe-Ormaetxe, I., Heras, B., Matamoros, M. A., Ramos, J., Moran, J. F., & Becana, M. (2002). Cloning and functional characterization of a homoglutathione synthetase from pea nodules. *Physiologia Plantarum, 115*, 69–73.

Iturbe-Ormaetxe, I., Matamoros, M. A., Rubio, M. C., Dalton, D. A., & Becana, M. (2001). The antioxidants of legume nodule mitochondria. *Molecular Plant-Microbe Interactions, 14*, 1189–1196.

Jamet, A., Mandon, K., Puppo, A., & Hérouart, D. (2007). H_2O_2 is required for optimal establishment of the *Medicago sativa/Sinorhizobium meliloti* symbiosis. *Journal of Bacteriology, 189*, 8741–8745.

Jamet, A., Sigaud, S., Van de Sype, G., Puppo, A., & Hérouart, D. (2003). Expression of the bacterial catalase genes during *Sinorhizobium meliloti-Medicago sativa* symbiosis and their crucial role during the infection process. *Molecular Plant-Microbe Interactions, 16*, 217–225.

Kiirika, L. M., Bergmann, H. F., Schikowsky, C., Wimmer, D., Korte, J., Schmitz, U., et al. (2012). Silencing of the Rac1 GTPase *MtROP9* in *Medicago truncatula* stimulates early mycorrhizal and oomycete root colonizations but negatively affects rhizobial infection. *Plant Physiology, 159*, 501–516.

Klapheck, S. (1988). Homoglutathione: Isolation, quantification, and occurrence in legumes. *Physiologia Plantarum, 74*, 727–732.

Knuesting, J., Riondet, C., Maria, C., Kruse, I., Bécuwe, N., König, N., et al. (2015). *Arabidopsis* glutaredoxin S17 and its partner NF-YC11/NC2α contribute to maintenance of the shoot apical meristem under long-day photoperiod. *Plant Physiology, 167*, 1643–1658.

Kolbert, Z., Feigl, G., Bordé, A., Molnár, A., & Erdei, L. (2017). Protein tyrosine nitration in plants: Present knowledge, computational prediction and future perspectives. *Plant Physiology and Biochemistry, 113*, 56–63.

Lindermayr, C. (2017). Crosstalk between reactive oxygen species and nitric oxide in plants: Key role of *S*-nitrosoglutathione reductase. *Free Radical Biology & Medicine, 122*, 110–115.
Lorenzo, C., Lucas, M. M., Vivo, A., & de Felipe, M. R. (1990). Effect of nitrate on peroxisome ultrastructure and catalase activity in nodules of *Lupinus albus* L. cv. Multolupa. *Journal of Experimental Botany, 41*, 1573–1578.
Loscos, J., Matamoros, M. A., & Becana, M. (2008). Ascorbate and homoglutathione metabolism in common bean nodules under stress conditions and during natural senescence. *Plant Physiology, 146*, 1282–1292.
Marino, D., Dunand, C., Puppo, A., & Pauly, N. (2012). A burst of plant NADPH oxidases. *Trends in Plant Science, 17*, 9–15.
Martí, M. C., Olmos, E., Calvete, J. J., Díaz, I., Barranco-Medina, S., Whelan, J., et al. (2009). Mitochondrial and nuclear localization of a novel pea thioredoxin: Identification of its mitochondria target proteins. *Plant Physiology, 150*, 646–657.
Matamoros, M. A., Clemente, M. R., Sato, S., Asamizu, E., Tabata, S., Ramos, J., et al. (2003). Molecular analysis of the pathway for the synthesis of thiol tripeptides in the model legume *Lotus japonicus*. *Molecular Plant-Microbe Interactions, 16*, 1039–1046.
Matamoros, M. A., Dalton, D. A., & Becana, M. (2017). Ascorbate metabolism and nitrogen fixation in legumes. In M. Hossain, S. Munné-Bosch, D. Burritt, P. Diaz-Vivancos, M. Fujita, & A. Lorence (Eds.), *Ascorbic acid in plant growth, development and stress tolerance* (pp. 471–490). Springer.
Matamoros, M. A., Fernández-García, N., Wienkoop, S., Loscos, J., Saiz, A., & Becana, M. (2013). Mitochondria are an early target of oxidative modifications in senescing legume nodules. *New Phytologist, 197*, 873–885.
Matamoros, M. A., Kim, A., Peñuelas, M., Ihling, C., Griesser, E., Hoffmann, R., et al. (2018). Protein carbonylation and glycation in legume nodules. *Plant Physiology, 177*, 1510–1528.
Matamoros, M. A., Loscos, J., Coronado, M. J., Ramos, J., Sato, S., Testillano, P. S., et al. (2006). Biosynthesis of ascorbic acid in legume root nodules. *Plant Physiology, 141*, 1068–1077.
Matamoros, M. A., Moran, J. F., Iturbe-Ormaetxe, I., Rubio, M. C., & Becana, M. (1999). Glutathione and homoglutathione synthesis in legume root nodules. *Plant Physiology, 121*, 879–888.
Matamoros, M. A., Saiz, A., Peñuelas, M., Bustos-Sanmamed, P., Mulet, J. M., Barja, M. V., et al. (2015). Function of glutathione peroxidases in legume root nodules. *Journal of Experimental Botany, 66*, 2979–2990.
Matamoros, M. A., Cutrona, M. C., Wienkoop, S., Begara-Morales, J. C., Sandal, N., Orera, I., et al. (2019). Altered plant and nodule development and protein S-nitrosylation in *Lotus japonicus* mutants deficient in *S*-nitrosoglutathione reductases. *Plant and Cell Physiology*. https://doi.org/10.1093/pcp/pcz182.
Mathieu, C., Moreau, S., Frendo, P., Puppo, A., & Davies, M. J. (1998). Direct detection of radicals in intact soybean nodules: Presence of nitric oxide-leghemoglobin complexes. *Free Radical Biology & Medicine, 24*, 1242–1249.
Meilhoc, E., Blanquet, P., Cam, Y., & Bruand, C. (2013). Control of NO level in rhizobium-legume root nodules: Not only a plant globin story. *Plant Signaling & Behavior, 8*, e25923.
Melo, P. M., Silva, L. S., Ribeiro, I., Seabra, A. R., & Carvalho, H. G. (2011). Glutamine synthetase is a molecular target of nitric oxide in root nodules of *Medicago truncatula* and is regulated by tyrosine nitration. *Plant Physiology, 157*, 1505–1517.
Meyer, Y., Buchanan, B. B., Vignols, F., & Reichheld, J. P. (2009). Thioredoxins and glutaredoxins: Unifying elements in redox biology. *Annual Review of Genetics, 43*, 335–367.

Mhamdi, A., Queval, G., Chaouch, S., Vanderauwera, S., Van Breusegem, F., & Noctor, G. (2010). Catalase function in plants: A focus on *Arabidopsis* mutants as stress-mimic models. *Journal of Experimental Botany, 61*, 4197–4220.

Minchin, F. R., James, E. K., & Becana, M. (2008). Oxygen diffusion, production of reactive oxygen and nitrogen species, and antioxidants in legume nodules. In M. J. Dilworth, E. K. James, J. I. Sprent, & W. E. Newton (Eds.), *Nitrogen-fixing leguminous symbioses* (pp. 321–362). Dordrecht: Springer.

Mishina, T. E., Lamb, C., & Zeier, J. (2007). Expression of a nitric oxide degrading enzyme induces a senescence programme in *Arabidopsis*. *Plant, Cell & Environment, 30*, 39–52.

Møller, I. M., Rogowska-Wrzesinska, A., & Rao, R. S. P. (2011). Protein carbonylation and metal-catalyzed protein oxidation in a cellular perspective. *Journal of Proteomics, 74*, 2228–2242.

Montiel, J., Arthikala, M., Cárdenas, L., & Quinto, C. (2016). Legume NADPH oxidases have crucial roles at different stages of nodulation. *International Journal of Molecular Sciences, 17*, 680.

Moseler, A., Aller, I., Wagner, S., Nietzel, T., Przybyla-Toscano, J., Mühlenhoff, U., et al. (2015). The mitochondrial monothiol glutaredoxin S15 is essential for iron-sulfur protein maturation in *Arabidopsis thaliana*. *Proceedings of the National Academy of Sciences of the United States of America, 44*, 13735–13740.

Muglia, C., Comai, G., Spegazzini, E., Riccillo, P. M., & Aguilar, O. M. (2008). Glutathione produced by *Rhizobium tropici* is important to prevent early senescence in common bean nodules. *FEMS Microbiology Letters, 286*, 191–198.

Murakami, E., Nagata, M., Shimoda, Y., Kucho, K., Higashi, S., Abe, M., et al. (2011). Nitric oxide production induced in roots of *Lotus japonicus* by lipopolysaccharide from *Mesorhizobium loti*. *Plant and Cell Physiology, 52*, 610–617.

Nagata, M., Murakami, E., Shimoda, Y., Shimoda-Sasakura, F., Kucho, K., Suzuki, A., et al. (2008). Expression of a class 1 hemoglobin gene and production of nitric oxide in response to symbiotic and pathogenic bacteria in *Lotus japonicus*. *Molecular Plant-Microbe Interactions, 21*, 1175–1183.

Navascués, J., Pérez-Rontomé, C., Gay, M., Marcos, M., Yang, F., Walker, F. A., et al. (2012). Leghemoglobin green derivatives with nitrated hemes evidence production of highly reactive nitrogen species during aging of legume nodules. *Proceedings of the National Academy of Sciences of the United States of America, 109*, 2660–2665.

Noctor, G., Mhamdi, A., Chaouch, S., Han, Y., Neukermans, J., Marquez-Garcia, B., et al. (2012). Glutathione in plants: An integrated overview. *Plant, Cell & Environment, 35*, 454–484.

Oger, E., Marino, D., Guigonis, J. M., Pauly, N., & Puppo, A. (2012). Sulfenylated proteins in the *Medicago truncatula-Sinorhizobium meliloti* symbiosis. *Journal of Proteomics, 75*, 4102–4113.

Oldroyd, G. E. D. (2013). Speak, friend, and enter: Signaling systems that promote beneficial symbiotic associations in plants. *Nature Reviews Microbiology, 11*, 252–263.

Ono, K., Akaike, T., Sawa, T., Kumagai, Y., Wink, D. A., Tantillo, D. J., et al. (2014). Redox chemistry and chemical biology of H_2S, hydropersulfides, and derived species: Implications of their possible biological activity and utility. *Free Radical Biology & Medicine, 77*, 82–94.

Oracz, K., El-Maarouf Bouteau, H., Farrant, J. M., Cooper, K., Belghazi, M., Job, C., et al. (2007). ROS production and protein oxidation as a novel mechanism for seed dormancy alleviation. *The Plant Journal, 50*, 452–465.

Ott, T., van Dongen, J. T., Günther, C., Krusell, L., Desbrosses, G., Vigeolas, H., et al. (2005). Symbiotic leghemoglobins are crucial for nitrogen fixation in legume root nodules but not for general plant growth and development. *Current Biology, 15*, 531–535.

Passaia, G., & Margis-Pinheiro, M. (2015). Glutathione peroxidases as redox sensor proteins in plant cells. *Plant Science, 234*, 22–26.

Pérez Guerra, J. C., Coussens, G., De Keyser, A., De Rycke, R., De Bodt, S., Van de Velde, W., et al. (2006). Comparison of developmental and stress-induced nodule senescence in *Medicago truncatula*. *Plant Physiology, 152*, 1574–1584.

Puppo, A., Groten, K., Bastian, F., Carzaniga, R., Soussi, M., Lucas, M. M., et al. (2005). Legume nodule senescence: Roles for redox and hormone signalling in the orchestration of the natural aging process. *New Phytologist, 165*, 683–701.

Puppo, A., Pauly, N., Boscari, A., Mandon, K., & Brouquisse, R. (2013). Hydrogen peroxide and nitric oxide: Key regulators of the legume-*Rhizobium* and mycorrhizal symbioses. *Antioxidants & Redox Signaling, 18*, 2202–2219.

Puppo, A., Rigaud, J., & Job, D. (1981). Role of superoxide anion in leghemoglobin autoxidation. *Plant Science Letters, 22*, 353–360.

Radi, R. (2018). Oxygen radicals, nitric oxide, and peroxynitrite: Redox pathways in molecular medicine. *Proceedings of the National Academy of Sciences of the United States of America, 115*, 5839–5848.

Ramu, S. K., Peng, H. M., & Cook, D. R. (2002). Nod factor induction of reactive oxygen species production is correlated with expression of the early nodulin gene *rip1* in *Medicago truncatula*. *Molecular Plant-Microbe Interactions, 15*, 522–528.

Rey, P., Becuwe, N., Tourrette, S., & Rouhier, N. (2017). Involvement of *Arabidopsis* glutaredoxin S14 in the maintenance of chlorophyll content. *Plant, Cell & Environment, 40*, 2319–2332.

Ribeiro, C. W., Baldacci-Cresp, F., Pierre, O., Larousse, M., Benyamina, S., Lambert, A., et al. (2017). Regulation of differentiation of nitrogen-fixing bacteria by microsymbiont targeting of plant thioredoxin s1. *Current Biology, 27*, 250–256.

Robson, R. L., & Postgate, J. R. (1980). Oxygen and hydrogen in biological nitrogen fixation. *Annual Review of Microbiology, 34*, 183–207.

Rouhier, N., & Jacquot, J. P. (2005). The plant multigenic family of thiol peroxidases. *Free Radical Biology & Medicine, 38*, 1413–1421.

Roux, B., Rodde, N., Jardinaud, M. F., Timmers, T., Sauviac, L., Cottret, L., et al. (2014). An integrated analysis of plant and bacterial gene expression in symbiotic root nodules using laser-capture microdissection coupled to RNA sequencing. *The Plant Journal, 77*, 817–837.

Rubio, M. C., Becana, M., Sato, S., James, E. K., Tabata, S., & Spaink, H. P. (2007). Characterization of genomic clones and expression analysis of the three types of superoxide dismutases during nodule development in *Lotus japonicus*. *Molecular Plant-Microbe Interactions, 20*, 262–275.

Rubio, M. C., James, E. K., Clemente, M. R., Bucciarelli, B., Fedorova, M., Vance, C. P., et al. (2004). Localization of superoxide dismutases and hydrogen peroxide in legume root nodules. *Molecular Plant-Microbe Interactions, 17*, 1294–1305.

Sainz, M., Calvo-Begueria, L., Pérez-Rontomé, C., Wienkoop, S., Abián, J., Staudinger, C., et al. (2015). Leghemoglobin is nitrated in functional legume nodules in a tyrosine residue within the heme cavity by a nitrite/peroxide-dependent mechanism. *The Plant Journal, 81*, 723–735.

Sánchez, C., Cabrera, J. J., Gates, A. J., Bedmar, E. J., Richardson, D. J., & Delgado, M. J. (2011). Nitric oxide detoxification in the rhizobia-legume symbiosis. *Biochemical Society Transactions, 39*, 184–188.

Sánchez, C., Gates, A. J., Meakin, G. E., Uchiumi, T., Girard, L., Richardson, D. J., et al. (2010). Production of nitric oxide and nitrosylleghemoglobin complexes in soybean nodules in response to flooding. *Molecular Plant-Microbe Interactions, 23*, 702–711.

Santos, R., Hérouart, D., Sigaud, S., Touati, D., & Puppo, A. (2001). Oxidative burst in alfalfa—*Sinorhizobium meliloti* symbiotic interaction. *Molecular Plant-Microbe Interactions, 14*, 86—89.

Sanz-Luque, E., Ocaña-Calahorro, F., de Montaigu, A., Chamizo Ampudia, A., Llamas, A., Galván, A., et al. (2015). THB1, a truncated hemoglobin, modulates nitric oxide levels and nitrate reductase activity. *The Plant Journal, 81*, 467—479.

Sevilla, F., Camejo, D., Ortiz-Espín, A., Calderón, A., Lázaro, J. J., & Jiménez, A. (2015). The thioredoxin/peroxiredoxin/sulfiredoxin system: Current overview on its redox function in plants and regulation by reactive oxygen and nitrogen species. *Journal of Experimental Botany, 66*, 2945—2955.

Shimoda, Y., Shimoda-Sasakura, F., Kucho, K., Kanamori, N., Nagata, M., Suzuki, A., et al. (2009). Overexpression of class 1 plant hemoglobin genes enhances symbiotic nitrogen fixation activity between *Mesorhizobium loti* and *Lotus japonicus*. *The Plant Journal, 57*, 254—263.

Smagghe, B. J., Blervacq, A. S., Blassiau, C., Decottignies, J. P., Jacquot, J. P., Hargrove, M. S., et al. (2007). Immunolocalization of non-symbiotic hemoglobins during somatic embryogenesis in chicory. *Plant Signaling and Behavior, 2*, 43—49.

Smagghe, B. J., Hoy, J. A., Percifield, R., Kundu, S., Hargrove, M. S., Sarath, G., et al. (2009). Correlations between oxygen affinity and sequence classifications of plant hemoglobins. *Biopolymers, 91*, 1083—1096.

Smirnoff, N. (2018). Ascorbic acid metabolism and functions: A comparison of plants and mammals. *Free Radical Biology & Medicine, 122*, 116—129.

Stadtman, E. R., Moskovitz, J., & Levine, R. L. (2003). Oxidation of methionine residues of proteins: Biological consequences. *Antioxidants & Redox Signaling, 5*, 577—582.

Sun, C., Lu, L., Liu, L., Liu, W., Yu, Y., Liu, X., et al. (2014). Nitrate reductase-mediated early nitric oxide burst alleviates oxidative damage induced by aluminum through enhancement of antioxidant defenses in roots of wheat (*Triticum aestivum*). *New Phytologist, 201*, 1240—1250.

Tang, G., Li, N., Liu, Y., Yu, L., Yan, J., & Luo, L. (2018). *Sinorhizobium meliloti* glutathione reductase is required for both redox homeostasis and symbiosis. *Applied and Environmental Microbiology, 84*. e01937-17.

Tarrago, L., Laugier, E., & Rey, P. (2009). Protein-repairing methionine sulfoxide reductases in photosynthetic organisms: Gene organization, reduction mechanisms, and physiological roles. *Molecular Plant, 2*, 202—217.

Torres-Jerez, I., Huertas-Ruz, R., Lara-Dampier, V., Dalton, D., & Udvardi, M. (2017). Enhancing SNF and stress tolerance via over-expression of ascorbate biosynthesis genes. In *20th International nitrogen fixation conference, Granada, Spain*.

Tovar-Méndez, A., Matamoros, M. A., Bustos-Sanmamed, P., Dietz, K. J., Cejudo, F. J., Rouhier, N., et al. (2011). Peroxiredoxins and NADPH-dependent thioredoxin systems in the model legume *Lotus japonicus*. *Plant Physiology, 156*, 1535—1547.

Udvardi, M., & Poole, P. S. (2013). Transport and metabolism in legume-rhizobia symbioses. *Annual Review of Plant Biology, 64*, 781—805.

Umbreen, S., Lubega, J., Cui, B., Pan, Q., Jiang, J., & Loake, G. J. (2018). Specificity in nitric oxide signaling. *Journal of Experimental Botany, 69*, 3439—3448.

Van de Velde, W., Pérez Guerra, J. C., De Keyser, A., De Rycke, R., Rombauts, S., Maunoury, N., et al. (2010). Aging in legume symbiosis. A molecular view on nodule senescence in *Medicago truncatula*. *Plant Physiology, 141*, 711—720.

Wang, L., Rubio, M. C., Xin, X., Zhang, B., Fan, Q., Wang, Q., et al. (2019). CRISPR/Cas9 knockout of leghemoglobin genes in *Lotus japonicus* uncovers their synergistic roles in symbiotic nitrogen fixation. *New Phytologist, 224*, 818—832.

Wang, J., Sun, P., Li, Y., Liu, Y., Yu, J., Ma, X., et al. (2017). Hierarchically aligning 10 legume genomes establishes a family-level genomics platform. *Plant Physiology, 174*, 284–300.

Waszczak, C., Akter, S., Jacques, S., Huang, J., Messens, J., & Van Breusegem, F. (2015). Oxidative post-translational modifications of cysteine residues in plant signal transduction. *Journal of Experimental Botany, 66*, 2923–2934.

Waszczak, C., Carmody, M., & Kangasjärvi, J. (2018). Reactive oxygen species in plant signaling. *Annual Review of Plant Biology, 69*, 209–236.

Winger, A. M., Taylor, N. L., Heazlewood, J. L., Day, D. A., & Millar, A. H. (2007). The cytotoxic lipid peroxidation product 4-hydroxy-2-nonenal covalently modifies a selective range of proteins linked to respiratory function in plant mitochondria. *Journal of Biological Chemistry, 282*, 37436–37447.

Wittenberg, J. B., Bolognesi, M., Wittenberg, B. A., & Guertin, M. (2002). Truncated hemoglobins: A new family of hemoglobins widely distributed in bacteria, unicellular eukaryotes, and plants. *Journal of Biological Chemistry, 277*, 871–874.

Yuan, S. L., Li, R., Chen, H. F., Zhang, C. J., Chen, L. M., et al. (2017). RNA-Seq analysis of nodule development at five different developmental stages of soybean (*Glycine max*) inoculated with *Bradyrhizobium japonicum* strain 113-2. *Scientific Reports, 7*, 42248.

Zaffagnini, M., Bedhomme, M., Lemaire, S. D., & Trost, P. (2012). The emerging roles of protein glutathionylation in chloroplasts. *Plant Science, 185–186*, 86–96.

Zhang, J., Ye, Z. W., Singh, S., Townsend, D. M., & Tew, K. D. (2018). An evolving understanding of the *S*-glutathionylation cycle in pathways of redox regulation. *Free Radical Biology & Medicine, 120*, 204–216.

CHAPTER FOUR

Rhizobia infection, a journey to the inside of plant cells

Rafael E. Venado, Juan Liang and Macarena Marín*
Genetics, Faculty of Biology, Ludwig Maximilians University Munich, Germany
*Corresponding author: E-mail: m.marin@biologie.uni-muenchen.de

Contents

Abstract

Diazotrophic rhizobia infect legume plants in a highly regulated manner. This infection process can be divided in three steps in which bacteria cross the root epidermis, spread through the cortex and are released inside the cells of a growing nodule primordium. The way bacteria transverse the different layers is intimately controlled by the host. However, different hosts have developed different strategies (root hair, crack-entry,

Advances in Botanical Research, Volume 94
ISSN 0065-2296
https://doi.org/10.1016/bs.abr.2019.09.007

intercellular) to guide the infection of bacteria. In this chapter, we review the molecular mechanisms that govern the infection of root nodules leading to the intracellular accommodation of bacteria and discuss the commonalities and differences between the different strategies in terms of distribution, architecture, and regulation.

1. Introduction

The nitrogen fixing root nodule symbiosis is a mutualistic association between plants of the Fabales, Fagales, Rosales, and Curcubitales orders and diazotrophic bacteria (Soltis et al., 1995). Among these, the Fabales are of great economical relevance, as they include crops such as soybean, peanut, and pea. Such legume crops are considered pivotal to reach sustainability in agriculture, because of their high nutrient content, and their low dependency on synthetic nitrogen fertilisation due to the symbiosis that they establish with nitrogen-fixing rhizobia (Considine, Siddique, & Foyer, 2017). These convert atmospheric nitrogen into ammonia in a reaction catalyzed by the nitrogenase enzyme. This reaction takes place within cells of specialised plant organs called nodules (Oldroyd, Murray, Poole, & Downie, 2011). The accommodation of bacteria inside of plant cells is key to the success of the symbiosis and one of the most unique features of this mutualistic association. How rhizobia reach the inside of these specialised cells (symbiotic infection) is the topic of this chapter.

Compatible symbiotic partners recognise each other through a signal crosstalk that activates two distinct, but interconnected programs. Cell divisions are initiated in the distal plant root cortex leading to nodule organogenesis (Soyano & Hayashi, 2014), while, rhizobia move from the root surface into the interior of developing nodule cells in a process known as infection. Symbiotic infection follows at least three physical steps: (1) crossing of the epidermis, (2) cortical spreading, and (3) intracellular uptake of rhizobia (Fig. 1). This is attained differently depending on the host legume. For example, Papilionoids plants are often colonised through root hair infection in which tubular structures known as infection threads form in growing root hairs, while Genistoids and Dalbergioids are infected via cracks on the epidermis (Sprent, 2007). Nevertheless, in all cases a tightly regulated genetic program governs the infection process and ensures the controlled colonisation of plant cells (Madsen et al., 2010).

Decades of work have identified components of the signaling cascade that initiates infection and the transcriptional network responsible of the

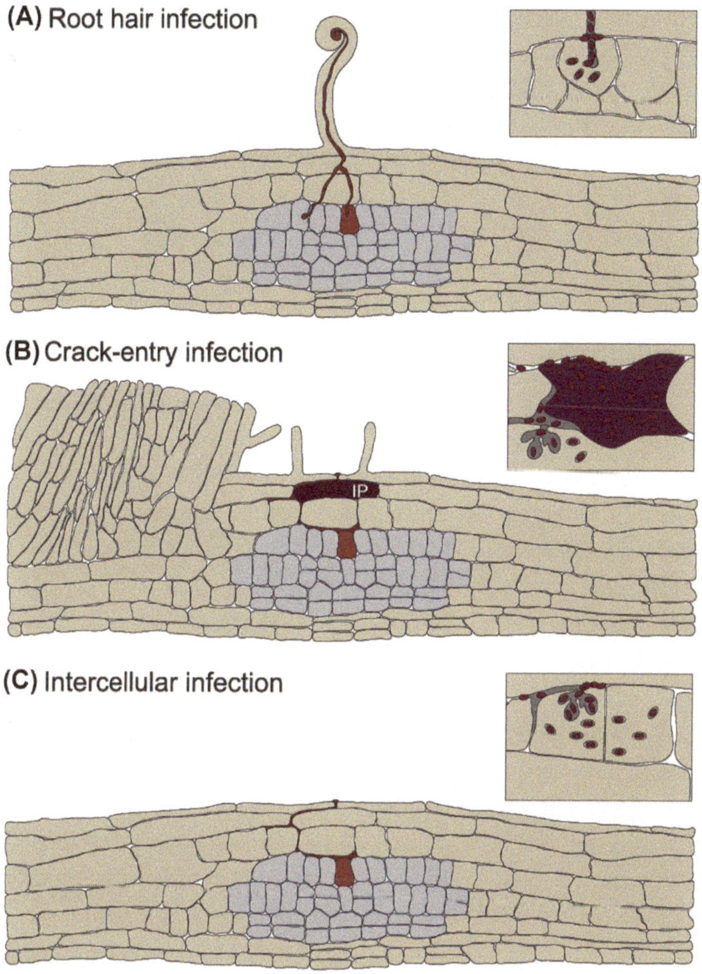

Fig. 1 Schematic representation of infection modes. (A) Root hair infection as in *Lotus*. Rhizobia are entrapped in a curled root hair and progress via an infection thread. Upon arrival to the dividing primordium (in blue), rhizobia are internalised (in red). (B) "Crack-entry" infection as in *Arachis*. Rhizobia enter through cracks at the lateral root base and accumulate in an infection pocket (IP) in the sub-epidermis. From there rhizobia spread and infect other cells. (C) Intercellular infection as in *Lupinus*. Rhizobia penetrate epidermis between cells and are directly internalised via "infection pegs". Infected cells divide and disperse the rhizobia.

reprograming of host cells. However, still little is known about how the infection progresses through the cortex, and how rhizobia ultimately enter plant cells. This is partially, because the epidermal and cortical programs are genetically intertwined (Madsen et al., 2010). This chapter aims to present a general overview about the different steps of the infection with a special emphasis on how rhizobia enter plant cells and the similarities and differences between the infection mechanisms in different legumes. We argue that exploring the wide diversity of symbiotic infection phenotypes observed in different legumes will help not only in identifying the fundamental requirements for infection, but also expand our understanding of how this process is regulated.

2. Signaling

The scope of this section is to summarise the signaling framework that controls infection: from perception to transcriptional reprogramming. We will cover exclusively the so-called common symbiotic signaling pathway downstream Nodulation (Nod) factor perception, and some of the transcriptional responses. We will not review the role of other factors, such as hormones (Liu, Zhang, Yang, Yu, & Wang, 2018), reactive oxygen species (Damiani, Pauly, Puppo, Brouquisse, & Boscari, 2016), and extracellular polysaccharides (Ghosh & Maiti, 2016).

2.1 Early perception and signal transduction

The crosstalk between symbionts begins in the rhizosphere. Legumes secrete flavonoids that act as chemo-attractants and induce in rhizobia the biosynthesis and secretion of Nod factors (Cooper, 2004). Each rhizobia strain produces a specific cocktail of these lipochitooligosaccharides with different modifications, which partially determine the recognition specificity (Fliegmann & Bono, 2015). Nod factors are perceived by LysM receptor-like kinases located at the plasma membrane of root epidermal cells. These receptors are called Nod Factor Receptor (NFR)1 and NFR5 in *Lotus* (Madsen et al., 2003; Radutoiu et al., 2003), and LysM Receptor Kinase 3 (LYK3) and Nod Factor Perception (NFP) in *Medicago* (Arrighi et al., 2006; Limpens et al., 2003). A third member of the perception complex is the leucine-rich repeat SYMbiosis Receptor-like Kinase (SYMRK) (Stracke et al., 2002) and Does not Make Infection 2 (DMI2) in *Lotus* and *Medicago*, respectively (Endre et al., 2002). Although SYMRK interacts with NFR1 and NFR5 (Ried, Antolin-Llovera, & Parniske, 2014), it is not directly involved in the binding of Nod factors and its molecular function is yet to be fully elucidated. Several proteins, such as the SYMbiotic REMorin (SYMREM1) (Lefebvre et al., 2010; Toth et al., 2012), and the E3 ubiquitin ligase Seven IN Absentia (SINA) (Den Herder, Yoshida, Antolín-Llovera, Ried, & Parniske, 2012), interact with members of the perception complex. These proteins most likely form a protein hub at the intracellular leaflet of the plasma membrane involved in receptor regulation and/or downstream signal transduction (Ott, 2017).

How the signal generated at the plasma membrane is transduced downstream into nuclear calcium oscillations and later into transcriptional reprogramming is not fully understood (Fig. 2). One hypothesis is that mevalonate or derivatives act as secondary messengers, as exogenous application induces calcium spiking (Venkateshwaran et al., 2015). This isoprenoid is produced by a 3-hydroxy-3-methyglutaryl-CoA reductase (HMGR) that interacts with DMI2 (Kevei et al., 2007; Venkateshwaran et al., 2015). However, if mevalonate, its derivatives, or a different secondary messenger can directly activate nuclear cation channels remains to be elucidated.

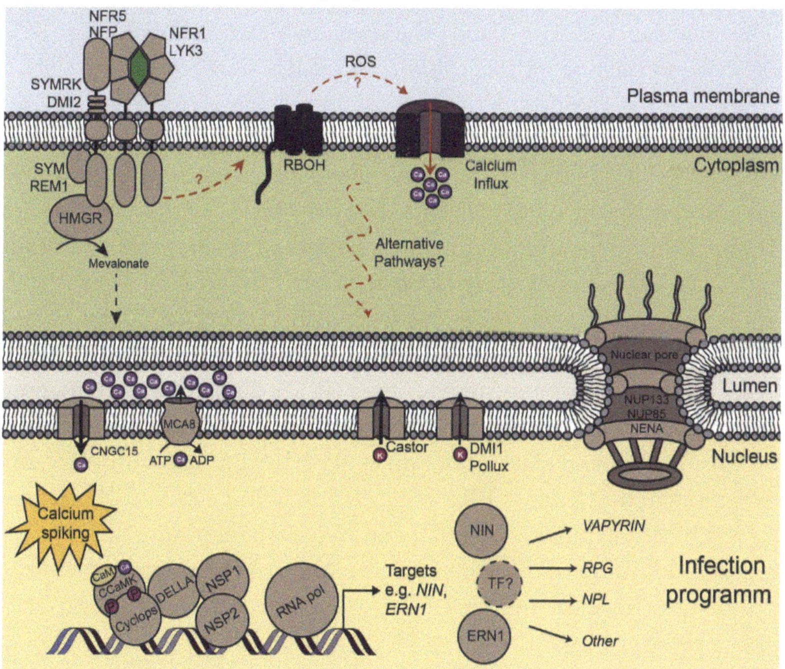

Fig. 2 Root nodule symbiotic signaling. Nod factors are perceived by surface receptors, which triggers a downstream signaling cascade that is transduced into calcium spiking in the nucleus. Decoding of these calcium oscillations leads to transcriptional activation of transcription factors and other symbiotic genes. Evidence suggests that alternative pathways exist, but these (in red) are not discussed in this review.

2.2 Calcium spiking and transcriptional reprogramming

One of the most defining features of the symbiotic signaling pathway is the oscillation in the calcium concentration between the nuclear and perinuclear spaces, the so-called calcium spiking (Oldroyd, 2013). Mathematical models that describe these oscillations have been proposed (Capoen et al., 2011; Charpentier, Vaz Martins, Granqvist, Oldroyd, & Morris, 2013). These include at least three components: a cation channel (LjCASTOR/LjPOLLUX (Charpentier et al., 2008), MtDMI1 (Ané et al., 2004)), a calcium pump (MCA8 (Capoen et al., 2011)), and a calcium channel (the cyclic nucleotide-gated channels CNGC15a, b, and c (Charpentier et al., 2016)). Genetic evidence has established that the nuclear pore complex proteins NUP133, NUP85, and NENA are required for the generation of calcium spiking. Based on their subcellular localisation and the function of orthologs, it is believed that this complex regulates the movement of components required for the spiking (Groth et al., 2010; Kanamori et al., 2006; Saito et al., 2007).

The nuclear Calcium- and CalModulin-dependent protein Kinase (LjCCaMK (Tirichine et al., 2006), MtDMI3 (Lévy et al., 2004)) has been proposed as the decoder of the calcium spiking. An active CCaMK form phosphorylates a coiled-coil domain containing transcription factor called CYCLOPS (Yano et al., 2008) (MtIPD3 (Messinese et al., 2007)). The phosphorylation of CYCLOPS in two specific serine residues leads to its functional activation. This form activates the promoter of Nodule inception (NIN) (Singh, Katzer, Lambert, Cerri, & Parniske, 2014), the first identified symbiotic transcription factor (Schauser, Roussis, Stiller, & Stougaard, 1999). This promoter is regulated at different levels, and evidence suggests that other proteins, such as, the GRAS-domain-type transcription factors Nodulation Signaling Pathway (NSP)1, NSP2, and DELLA, also play a role (Heckmann et al., 2006; Jin et al., 2016). Along with NIN, other transcription factors, such as ERF Required for Nodulation (ERN)1 (Cerri et al., 2012; Middleton et al., 2007), and NF-YA1 (Laporte et al., 2013) are required for symbiotic infection. Therefore there is an emerging picture of a complex transcriptional network controlling infection (Fig. 2) (Soyano & Hayashi, 2014).

The root hair transcriptome of *nin* and *ern1* mutants revealed that NIN has a large impact on the symbiotic transcriptional network, as it controls the expression of hundreds of genes. In contrast, ERN1 regulates a limited but distinct set of targets (Liu et al., 2019). Genetic studies have revealed a long list of genes downstream of this transcriptional network, which encode proteins that execute cell remodeling. Mutants in these genes show impaired infection phenotypes. Examples include genes encoding E3 ubiquitin ligases, such as CERBERUS (Yano et al., 2009), and PUB1 (Mbengue et al., 2010), a coiled-coil protein of unknown function called RPG (Arrighi et al., 2008), and a major sperm protein involved in membrane trafficking VAPYRIN (Murray et al., 2011). Around 10% of the NIN targets are related to cell wall remodeling (Liu et al., 2019). One example is the Nodule Pectate Lyase (NPL), which is involved in the formation of infection threads (Liu et al., 2019; Xie et al., 2012). In the following sections, we will elaborate on the phenotypes exhibited by some of these mutants and the molecular functions of the key players.

3. Infection stages: from the epidermis to the inside of plant cells

Legumes allow controlled colonisation by symbiotic rhizobia through diverse infection modes. These can be dissected depending on how rhizobia cross the epidermis and spread in the cortex, and if they induce the formation of infection threads or not. Despite this large diversity, most studies have been conducted in hosts of the Papilionoideae subfamily, which contains model organisms such as *Medicago*, *Lotus*, and *Pisum* (Sprent, Ardley, & James, 2017). These plants develop infection threads in both the epidermis and the cortex. In this section we will illustrate fundamental principles of symbiotic infection reviewing how infection threads facilitate the colonisation of plant cells.

3.1 Epidermal program: Infection initiation

3.1.1 Genetic control of epidermal infection

The transcriptional reprogramming downstream of the common symbiotic signaling pathway changes the fate of the root hair cells that interact with the rhizobia. Some of these cells undergo major cell biological changes that enable the formation of infection threads. One of the first responses to Nod factors, as early as 3 to 5 min post perception, is the *NFR1* and *NFR5*-dependent deformation of root hairs, e.g. swelling, branching, and curling (Madsen et al., 2003; Miwa, Sun, Oldroyd, & Downie, 2006; Radutoiu et al., 2003). Likewise, *nfp* mutants in the *Medicago* ortholog show a similar phenotype, while mutants in *lyk3* retain root hair deformations

(Arrighi et al., 2006; Smit et al., 2007). Furthermore, *symrk* and *dmi2* mutants maintain deformations such as swelling and branching, but completely block the entrapment of rhizobia into curled root hairs and the formation of infection foci (Endre et al., 2002; Miwa et al., 2006; Stracke et al., 2002). A similar phenotype, is observed in mutants of other genes required for the calcium spiking, such as the ion channels *castor, pollux/dmi1,* the nucleoporins *nup85,* and *nup133*, and the kinase *ccamk* (Ané et al., 2004; Charpentier et al., 2008; Kanamori et al., 2006; Miwa et al., 2006; Saito et al., 2007; Tirichine et al., 2006). Altogether these results indicate that the perception of the Nod factor and the calcium spiking are essential for the formation of epidermal infection threads.

Interruption of genes immediately downstream of the calcium spiking leads to infection arrest at the curling or micro-colony stage. For instance, *nin* mutants usually display an excessive curling phenotype although they occasionally form infection foci (Marsh et al., 2007; Schauser et al., 1999). The arrest at this stage coincides with the initiation of the infection chamber remodeling. Fournier et al. proposed a two-step model for the development of infection threads. In this model, rhizobia multiplication after root hair curling leads to the enlargement of the infection chamber, followed by elongation of the infection thread. Cell wall and exocytosis-associated markers, accumulate around rhizobia foci. These are likely to facilitate the cell wall plasticity and the delivery of materials required for remodeling (Fournier et al., 2015).

A group of mutants is impaired in the radial and polar growth of the chamber and as a consequence these develop truncated infection threads. Interestingly, thread development in these mutants is arrested at different points. Mutants in *cerberus/lin* (Kuppusamy et al., 2004; Yano et al., 2009), *cyclops* (Yano et al., 2008), and *vapyrin* (Murray et al., 2011) entrap the rhizobia at the root curl with no further thread progression, whereas in *crinckle* (Tansengco, Hayashi, Kawaguchi, Imaizumi-Anraku, & Murooka, 2003), *nap1*, and *pir1* (Yokota et al., 2009) mutants, rhizobia progress until the base of the root hair, but not to the sub-epidermal cells. Other mutants such as *rpg* (Arrighi et al., 2008), *pub1* (Mbengue et al., 2010), and *nf-ya1* (Laporte et al., 2013) have thick infection threads. Several of these genes participate in cell biological processes, such as actin and cell wall remodeling, membrane trafficking, and polar growth. However, how these processes are coordinated in a spatio-temporal manner remains to be elucidated.

3.1.2 Nanodomains of infection

Membrane nanodomains are organisational units for proteins at the plasma membrane. They have been postulated to act as platforms that facilitate the spatio-temporal organisation of a symbiotic signaling hub (Ott, 2017). Two protein families, flotillins and remorins, have been described as nanodomain markers in plants (Jarsch et al., 2014). The flotillin FLOT4 is required for infection thread-elongation and localises at membrane nanodomains at the tip of elongating root hairs and on infection threads (Haney & Long, 2010; Liang et al., 2018). The remorin SYMREM1 also localises to the infection thread membrane and is required for correct thread elongation (Lefebvre et al., 2010; Liang et al., 2018). A model has been proposed in which FLOT4 and SYMREM1 form a symbiosis specific nanodomain (Liang et al., 2018). In this model, FLOT4 together with actin forms the nanodomain core. Upon perception of rhizobia, SYMREM1 is expressed and localises to the nanodomain. This remorin acts then as a secondary scaffold, which recruits the ligand-activated receptor LYK3 and stabilises it within the nanodomain. This stabilisation prevents receptor endocytosis and is indispensable for optimal infection (Liang et al., 2018). Guidance of the infection thread toward the base of the root hair requires localised delivery of cell wall and membrane materials. Thus it is tempting to speculate that this symbiotic nanodomain is involved in this process.

3.2 Cortical program: a network of trans-cellular infection threads

The progression of the infection from the epidermis to the cortex is a major milestone in the infection process. Depending on the species, a variable number of infection threads abort in the epidermis (Nutman, 1962) suggesting regulation at this stage. Before the arrival of the growing infection thread to the base of the root hair cell, local weakening of the cell wall occurs and the sub-epidermal cell layer forms a pre-infection thread (van Brussel et al., 1992). This process continues throughout multiple cell layers as a tree-like network that allows the progression of rhizobia into cells of the developing nodule. Excellent reviews describe the growth and topology of this network (Gage, 2004; Monahan-Giovanelli, Pinedo, & Gage, 2006).

3.2.1 The epidermal and cortical programs are linked, but can be genetically uncoupled

A number of mutants in which epidermal infection threads develop fully, but rhizobia do not progress to the sub-epidermal cell layer, evidence that the transition between the epidermis and the cortex is genetically controlled. For instance, the *Pisum* mutant *brz* that has a defect in hormonal control (Guinel & Larue, 1992) and the *Lotus vag1-1* that has impaired endoreduplication and cortical cell proliferation (Suzaki et al., 2014) form normal epidermal infection threads, but these do not progress into the cortex. Contrastingly, in mutants, such as *nena-1* (Groth et al., 2010) and *symrk-14* (Kosuta et al., 2011) nodules contain trans-cellular threads in the absence of epidermal threads. In both mutants calcium spiking is impaired. Elegant complementation studies using epidermal specific promoters have demonstrated that expression in the epidermis of genes involved in the generation of the calcium spiking, such as *CASTOR*, *POLLUX*, *NUP85*, and *NUP133* is sufficient to complement the formation of fully infected nodules (Hayashi et al., 2014). In contrast, epidermal expression of *CCaMK*, *CYCLOPS*, *NSP1*, and *NSP2* only rescues the epidermal, but not the cortical infection. Notably, components upstream of the spiking are also required for the cortical program, as epidermal complementation of *nfr5-2*, but not *nfr1-4* only partially restores nodule colonisation. Perhaps most interestingly, epidermal complementation is not sufficient to fully restore epidermal infection in both mutants (Hayashi et al., 2014). This suggests that the expression of both receptors in both regions is relevant for infection, especially of NFR1. These observations open interesting questions, such as what is the role of the Nod factor receptors in the cortex?

3.3 Internalisation program

The last stage of the infection is arguably its most unique step. Cells of the developing nodule actively internalise rhizobia in a process that resembles cell exocytosis (Ivanov et al., 2012). In the interface of the cell entry, so called infection droplets are formed and rhizobia are released in a localised manner forming symbiosomes surrounded by a host-derived membrane (Coba de la Pena, Fedorova, Pueyo, & Lucas, 2017). The infected cell undergoes profound changes that include cytoskeleton rearrangements and vacuole degradation.

3.3.1 Requirements for intracellular accommodation of rhizobia

The symbiosis between legumes and rhizobia is one of the few associations between plants and bacteria that occur within living plant cells. The ability to accommodate bacteria intracellularly is hypothesised to be a feature of the common ancestor of the nodulating plants (Parniske, 2018). But what are the characteristics shared by nodulating plants that enable intracellular accommodation? Geurts et al. postulated that induction of plant cell division is one of these pre-requisites, as the wall of dividing cells is less rigid and this probably increases their susceptibility to infection (Geurts, Xiao, & Reinhold-Hurek, 2016). However, two plant cell characteristics still limit the entry of bacteria: 1) the remaining plant cell wall, and 2) the turgor pressure. Lysis of the cell wall would not only release damage associated molecular patterns, but would also disrupt the integrity of the protoplast. It has been proposed that the turgor pressure of neighboring cells and the localised secretion of an electron-dense matrix might act as compensatory measures to maintain cell integrity (Parniske, 2018).

3.3.2 Perception at the internalisation interface

The localised release of rhizobia and the confined secretion of this matrix suggest a perception mechanism controlling cell entry. Consistently, Nod synthesis genes are expressed in the pre-fixation zone (Roux et al., 2014) and Nod factors accumulate strongly in infection threads, but drop drastically after cell entry (Marie, Plaskitt, & Downie, 1994; Timmers, Auriac, de Billy, & Truchet, 1998). In addition, the NFP and LYK3 receptors accumulate in the periphery of cells located in a narrow space at the border between the meristematic and the infection zones. Both receptors, especially NFP, gather on the membrane of infection threads growing in this area. Specific down-regulation of NFP in nodules impairs the release of bacteria in cells derived from the meristem (Moling et al., 2014). This together with the abovementioned requirement of NFR1 in the cortical program of *Lotus* suggests that the perception of Nod factors plays an important role in the internalisation of rhizobia.

3.3.3 Role of the common symbiotic signaling pathway in the internalisation of rhizobia

Moling et al. postulated that components of the common symbiotic signaling pathway are not only required for epidermal infection, but also play a role in the internalisation of rhizobia (Moling et al., 2014). In *Medicago*, some components such as *DMI1*, *DMI2*, *DMI3*, and *IDP3* are expressed in the nodule apex (Limpens et al., 2005; Roux et al., 2014). DMI2 locates to the membrane surrounding the infection threads and its down-regulation causes the formation of enlarged infection threads and impaired rhizobia release (Limpens et al., 2005). Similarly, down-regulation of the orthologue *SrSYMRK* in *Sesbania rostrata* impairs rhizobial release (Capoen, Goormachtig, De Rycke, Schroeyers, & Holsters, 2005).

Arguably, RNAi mediated down-regulation of transcripts can be unspecific and thus result in silencing of off-targets. Nevertheless, these results at minimum suggest a role for this gene or of a close orthologue in internalisation. The *Mtsym1-1* mutant allele of *IDP3* shows enlarged infection threads and no rhizobia release, which correlates with specific down-regulation of IDP3 target genes that localise to the infection thread and symbiosome membranes (Ovchinnikova et al., 2011). Altogether these results evidence that the common symbiotic signaling pathway also plays a role during cell entry.

3.3.4 Remodeling of the symbiotic interface

The delivery of new membrane material is essential for the local increase in membrane area at the infection site (Grefen, Honsbein, & Blatt, 2011), as the plasma membrane is an inelastic surface (Apodaca, 2002). It has been hypothesised that the membrane tension caused by the growing rhizobia "creates a spatial vector for membrane fusion to the interface membrane ensuring the enlargement of the symbiotic interface" (Gavrin, Kulikova, Bisseling, & Fedorova, 2017). Although the internalisation of single rhizobia into symbiosomes resembles an endocytosis-like process, all cell biological evidence suggests that this process is exocytotic. Exocytosis markers, such as components of the Soluble NSF Attachment Protein Receptor (SNARE) complex, locate to the internalisation interface (Ivanov et al., 2012). Examples of these markers are members of the vesicle associated membrane protein (VAMP) family and of the Syntaxin family (Han, Pluhackova, & Bockmann, 2017). For instance, *Medicago* VAMP72d and VAMP72e localise at infection sites in dot-like structures and are required for the formation of the symbiotic interface. Silencing of their expression causes the formation of abnormal infection droplets that retain a thin layer of cell wall. The presence of this layer likely obstructs the bacteria release and the formation of symbiosomes (Ivanov et al., 2012). Another example is the symbiosis specific SYntaxin of Plants SYP132 that is alternatively spliced. Both splice variants localise at the infection thread membrane, infection droplets, and symbiosomes, but only specific down-regulation of the alpha form impairs the differentiation of released rhizobia (Huisman et al., 2016). Interestingly, several of these factors have been shown to have a shared function in the arbuscular mycorrhiza symbiosis (Huisman et al., 2016; Ivanov et al., 2012). Altogether, these examples emphasize the role of membrane remodeling in shaping the symbiotic interface.

The release of rhizobia into plant cells requires the formation of unwalled infection droplets. However, how the cell wall is locally modified is not fully understood. Immunocytological studies have revealed that infection droplets are devoid of cellulose, xyloglucan, methyl-esterified pectin, and non-esterified pectin, which are components of the infection thread cell wall (Rae, Bonfante-Fasolo, & Brewin, 1992). This specific depletion could be explained by blocking of the delivery of cell wall materials to the infection droplets or the localised targeting of cell wall degrading enzymes to these sites. The silencing of *Glycine* GmVAMP721a and d blocks the release of rhizobia, which remain in clusters surrounded by a matrix of methyl-esterified and de-methyl-esterified pectin. GmVAMP721d partly co-localises with GmNPL1 in vesicles that do not transport methyl-esterified pectin. This suggests that these vesicles might deliver pectin-modifying enzymes to infection droplets (Gavrin et al., 2016). Future studies, should address the cell wall composition of the internalisation interface.

3.3.5 Cell infection and cellular rearrangement

The intracellular accommodation of rhizobia not only causes the redirection of vesicular traffic and the remodeling of the membrane interface, but also the reorganisation of the cytoskeleton and the endomembranes of the host cell. Cytoskeletal disorganisation has been observed upon infection of cells in e.g. *Medicago sativa* (Timmers et al., 1998) and *Pisum sativum* (Davidson & Newcomb, 2001). During symbiosome maturation actin microfilaments undergo drastic changes, as they form a network surrounding mature symbiosomes (Gavrin, Jansen, Ivanov, Bisseling, & Fedorova, 2015). In *Medicago*, silencing of actin related protein (ARP) 3 ARP3, which regulates actin branching (Pollard, 2007), hinders the maturation and life span of symbiosomes (Gavrin et al., 2015). It has been postulated that an abnormal actin organisation reduces vesicle trafficking toward symbiosomes and that this network could help blocking symbiosome fusion with lytic vacuoles (Gavrin et al., 2015).

During symbiosome maturation, the peri-bacteroid membrane composition changes from having a plasma membrane identity to gaining late endosomal or tonoplast markers (Limpens et al., 2009). In parallel, vacuoles contract, their pH increases, and tonoplast proteins, such as the aquaporin TIPg are retargeted to the peri-bacteroid membrane. These changes have been postulated to permit the expansion and maturation of symbiosomes, as interruption of this process arrests their development (Gavrin et al., 2014) and is likely to cause the arrest of cell divisions after internalisation, as observed in *Pisum*, *Medicago*, and *Trifolium* (Brewin, 1991). Remarkably, division of infected cells has been reported in other species, such as *Lupinus* (Gonzalez-Sama, Lucas, de Felipe, & Pueyo, 2004), *Arachis* (Chandler, 1978), and *Stylosanthes* (Chandler, Date, & Roughley, 1982). These observations illustrate the diversity of strategies in the legume-rhizobia symbiosis, highlighted by in a process as essential as the intracellular accommodation of rhizobia.

4. Unifying principles of alternative mechanisms of infection

4.1 Crossing of the root epidermis

The root epidermis acts as the first physical barrier to the colonisation of nodules. How rhizobia cross this barrier mainly depends on the plant species (Fig. 1). For instance, several plants from the Mimosoide, Cassiinaee, Caesalpinieae clades mostly develop infection threads in epidermal root hairs. Also Papilionoid plants from the Milletioids, Robinioids, and IRLC clades are infected via root hairs. In contrast, in Genistoids and Dalbergioids legumes, rhizobia often invade the roots in zones close to natural cracks or wounds ("crack-entry") (Sprent et al., 2017).

Plants infected via "crack-entry" develop nodules almost exclusively at the base of their lateral roots (Ibanez, Wall, & Fabra, 2017). Lateral root emergence sites are characterised by a disrupted epidermis and bacterial accumulation. The exact entry point is often located between axillary root hairs. For example, rhizobia infect *Arachis* between an enlarged root hair cell and the neighboring cell (Chandler, 1978). Similarly, *Aeschynomene indica* is infected between enlarged axillary root hairs (Bonaldi et al., 2011). In contrast, root hairs play no significant role in some aquatic legumes, such as *Neptunia natans* (James, Sprent, Stuherland, Mcinroy, & Minchin, 1992; Subba-Rao et al., 1995) or in *S. rostrata* grown under flooded conditions (Ndoye, de Billy, Vasse, Dreyfus, & Truchet, 1994).

In a third mode of infection, rhizobia penetrate the epidermis through the middle lamella between intact epidermis cells (Fig. 1). Although widespread among Actinorhizal plants of the order Rosales (Pawlowski & Demchenko, 2012), this form of infection is limited in legumes. *Mimosa scabrella* (de Faria, Hay, & Sprent, 1988) and *Lupinus albus* (Gonzalez-Sama et al., 2004) are among the few legumes in which this intercellular infection mode has been observed. In *M. scabrella* however rhizobia seem to also invade the primary cell wall (de Faria et al., 1988).

4.2 Cortical spreading and internalisation

The spreading and the internalisation of rhizobia are distinct from the epidermal program. For instance, under flooding conditions *S. rostrata* crosses the epidermis via "crack-entry", but in the nodule it forms trans-cellular threads. This resembles the phenotypes of some of the *Lotus* mutants described in section three. The diversity of thread-independent phenotypes is striking. After crossing the epidermis, rhizobia can either directly enter a sub-epidermal cell or remain in the intercellular space. Often the infected cell dies and an infection pocket is formed from which rhizobia can spread further. For example, in *Stylosanthes*, early-infected cells die and collapse, as neighboring cells squeeze them into a narrow pocket (Chandler et al., 1982). A similar phenotype is observed in *Aeschynomene indica* in which wall-thickening reminiscent of plant defense surrounds the infected cell (Bonaldi et al., 2011). An exception to this is *L. albus* in which rhizobia do not kill the primary infected cell, but they proliferate inside and are distributed through the division of host cells (Gonzalez-Sama et al., 2004). In *Arachis* and *Aeschynomene*, secondary infected cells also divide and thereby spread the intracellular rhizobia (Bonaldi et al., 2011; Chandler, 1978). As a consequence in these plants almost every cell in the inner tissue of the nodule is infected. In contrast, nodules infected via trans-cellular infection threads display normally both uninfected and infected cells (Sprent, 2007).

As discussed in section three, one of the prerequisites for cell infection is the localised production of an electron-dense matrix at entry sites. This is most clearly observed in infection thread independent infection modes. Host cells internalise rhizobia either directly from the apoplast or from dead cell pockets. Here rhizobia are embedded in the matrix from which they are later released into the host cell. These irregularly shaped structures are often called "infection pegs", however across the literature there is no consensus in the nomenclature. Such structures have been described for instance in *M. scabrella* (de Faria et al., 1988), *N. natans* (Subba-Rao et al., 1995), *Lotus burttii* (Liang et al., 2019), and *L. albus* (Gonzalez-Sama et al., 2004).

5. Conclusions

The natural diversity of symbiotic infection mechanisms can help us identify the minimal set of requirements. Based on our current understanding these can be summarised as follows: (1) activation of cell divisions in the host, (2) induction of a weak spot on the epidermis, (3) creation of local accumulations of rhizobia below the epidermis, (4) formation of an electron-dense matrix to facilitate cell entry, and (5) rearrangement of endomembrane trafficking and cytoskeleton organisation. However, natural diversity is also a so-far untapped source that can be used to study the regulation of the symbiotic infection, as there is large variation, in for example the number and shape of infection threads, or the cell biological rearrangements. The progress in sequencing and genome editing technologies will facilitate in the near future the study of the genetic basis of alternative symbiotic infection strategies in non-model organisms.

References

Ané, J.-M., Kiss, G. B., Riely, B. K., Penmetsa, R. V., Oldroyd, G. E. D., Ayax, C., ... Kalo, P. (2004). *Medicago truncatula* DMI1 required for bacterial and fungal symbioses in legumes. *Science, 303*(5662), 1364–1367.

Apodaca, G. (2002). Modulation of membrane traffic by mechanical stimuli. *American Journal of Physiology - Renal Physiology, 282*(2), F179–F190.

Arrighi, J.-F., Barre, A., Amor, B. B., Bersoult, A., Soriano, L. C., Mirabella, R., ... Huguet, T. (2006). The *Medicago truncatula* lysine motif-receptor-like kinase gene family includes NFP and new nodule-expressed genes. *Plant Physiology, 142*(1), 265–279.

Arrighi, J.-F., Godfroy, O., de Billy, F., Saurat, O., Jauneau, A., & Gough, C. (2008). The RPG gene of *Medicago truncatula* controls *Rhizobium*-directed polar growth during infection. *Proceedings of the National Academy of Sciences of the U S A, 105*(28), 9817–9822.

Bonaldi, K., Gargani, D., Prin, Y., Fardoux, J., Gully, D., Nouwen, N., ... Giraud, E. (2011). Nodulation of *Aeschynomene afraspera* and *A. indica* by photosynthetic *Bradyrhizobium* sp. strain ORS285: The Nod-dependent versus the Nod-independent symbiotic interaction. *Molecular Plant-Microbe Interactions, 14*(11), 1359–1371.

Brewin, N. J. (1991). Development of the legume root nodule. *Annual Review of Cell Biology, 7*(1), 191–226.

van Brussel, A. A. N., Bakhuizen, R., van Spronsen, P. C., Spaink, H. P., Tak, T., Lugtenberg, B. J. J., et al. (1992). Induction of pre-infection thread structures in the leguminous host plant by mitogenic lipo-oligosaccharides of *Rhizobium*. *Science, 257*(5006), 70–72.

Capoen, W., Goormachtig, S., De Rycke, R., Schroeyers, K., & Holsters, M. (2005). SrSymRK, a plant receptor essential for symbiosome formation. *Proceedings of the National Academy of Sciences of the U S A, 102*(29), 10369–10374.

Capoen, W., Sun, J., Wysham, D., Otegui, M. S., Venkateshwaran, M., Hirsch, S., ... Ané, J.-M. (2011). Nuclear membranes control symbiotic calcium signaling of legumes. *Proceedings of the National Academy of Sciences of the U S A, 108*(34), 14348–14353.

Cerri, M. R., Frances, L., Laloum, T., Auriac, M. C., Niebel, A., Oldroyd, G. E., ... de Carvalho-Niebel, F. (2012). *Medicago truncatula* ERN transcription factors: Regulatory interplay with NSP1/NSP2 GRAS factors and expression dynamics throughout rhizobial infection. *Plant Physiology, 160*(4), 2155–2172.

Chandler, M. (1978). Some observations on infection of *Arachis hypogaea* L. by *Rhizobium*. *Journal of Experimental Botany, 29*(110), 749–755.

Chandler, M., Date, R. A., & Roughley, R. J. (1982). Infection and root nodule development in *Stylosanthes* species by *Rhizobium*. *Journal of Experimental Botany, 33*(132), 47–57.

Charpentier, M., Bredemeier, R., Wanner, G., Takeda, N., Schleiff, E., & Parniske, M. (2008). *Lotus japonicus* CASTOR and POLLUX are ion channels essential for perinuclear calcium spiking in legume root endosymbiosis. *The Plant Cell, 20*(12), 3467–3479.

Charpentier, M., Sun, J., Martins, T. V., Radhakrishnan, G. V., Findlay, K., Soumpourou, E., … Morris, R. J. (2016). Nuclear-localized cyclic nucleotide–gated channels mediate symbiotic calcium oscillations. *Science, 352*(6289), 1102–1105.

Charpentier, M., Vaz Martins, T., Granqvist, E., Oldroyd, G. E. D., & Morris, R. J. (2013). The role of DMI1 in establishing Ca^{2+} oscillations in legume symbioses. *Plant Signaling & Behavior, 8*(2). e22894-e22894.

Coba de la Pena, T., Fedorova, E., Pueyo, J. J., & Lucas, M. M. (2017). The symbiosome: Legume and rhizobia co-evolution toward a nitrogen-fixing organelle? *Frontiers of Plant Science, 8*, 2229.

Considine, M. J., Siddique, K. H. M., & Foyer, C. H. (2017). Nature's pulse power: Legumes, food security and climate change. *Journal of Experimental Botany, 68*(8), 1815–1818.

Cooper, J. E. (2004). *Multiple responses of rhizobia to flavonoids during legume root infection* (Vol. 41, pp. 1–62). Elsevier.

Damiani, I., Pauly, N., Puppo, A., Brouquisse, R., & Boscari, A (2016). Reactive oxygen species and nitric oxide control early steps of the legume - rhizobium symbiotic interaction. *Frontiers of Plant Science, 7*, 454.

Davidson, A. L., & Newcomb, W. (2001). Organization of microtubules in developing pea root nodule cells. *Canadian Journal of Botany, 79*, 767–776.

Den Herder, G., Yoshida, S., Antolín-Llovera, M., Ried, M. K., & Parniske, M. (2012). *Lotus japonicus* E3 ligase SEVEN IN ABSENTIA4 destabilizes the symbiosis receptor-like kinase SYMRK and negatively regulates rhizobial infection. *The Plant Cell, 24*(4), 1691–1707.

Endre, G., Kereszt, A., Kevei, Z., Mihacea, S., Kaló, P., & Kiss, G. B. (2002). A receptor kinase gene regulating symbiotic nodule development. *Nature, 417*(6892), 962-962.

de Faria, S. M., Hay, G. T., & Sprent, J. I. (1988). Entry of rhizobia into roots of *Mimosa scabrella* Bentham occurs between epidermal cells. *Journal of General Microbiology, 143*, 2291–2296.

Fliegmann, J., & Bono, J. J. (2015). Lipo-chitooligosaccharidic nodulation factors and their perception by plant receptors. *Glycoconjugate Journal, 32*(7), 455–464.

Fournier, J., Teillet, A., Chabaud, M., Ivanov, S., Genre, A., Limpens, E., … Barker, D. G. (2015). Remodeling of the infection chamber before infection thread formation reveals a two-step mechanism for rhizobial entry into the host legume root hair. *Plant Physiology, 167*(4), 1233–1242.

Gage, D. J. (2004). Infection and invasion of roots by symbiotic, nitrogen-fixing rhizobia during nodulation of temperate legumes. *Microbiology and Molecular Biology Reviews, 68*(2), 280–300.

Gavrin, A., Chiasson, D., Ovchinnikova, E., Kaiser, B. N., Bisseling, T., & Fedorova, E. E. (2016). VAMP721a and VAMP721d are important for pectin dynamics and release of bacteria in soybean nodules. *New Phytologist, 210*(3), 1011–1021.

Gavrin, A., Jansen, V., Ivanov, S., Bisseling, T., & Fedorova, E. (2015). ARP2/3-mediated Actin nucleation associated with symbiosome membrane is essential for the development of symbiosomes in infected cells of *Medicago truncatula* root nodules. *Molecular Plant-Microbe Interactions, 28*(5), 605–614.

Gavrin, A., Kaiser, B. N., Geiger, D., Tyerman, S. D., Wen, Z., Bisseling, T., et al. (2014). Adjustment of host cells for accommodation of symbiotic bacteria: Vacuole defunctionalization, HOPS suppression, and TIP1g retargeting in *Medicago*. *The Plant Cell, 26*(9), 3809–3822.

Gavrin, A., Kulikova, O., Bisseling, T., & Fedorova, E. E. (2017). Interface symbiotic membrane formation in root nodules of *Medicago truncatula*: The role of synaptotagmins MtSyt1, MtSyt2 and MtSyt3. *Frontiers of Plant Science, 8*, 201.
Geurts, R., Xiao, T. T., & Reinhold-Hurek, B. (2016). What does it take to evolve a nitrogen-fixing endosymbiosis? *Trends in Plant Science, 21*(3), 199–208.
Ghosh, P. K., & Maiti, T. K. (2016). Structure of extracellular polysaccharides (EPS) produced by rhizobia and their functions in legume–bacteria symbiosis:—a Review. *Achievements in the Life Sciences, 10*(2), 136–143.
Gonzalez-Sama, A., Lucas, M. M., de Felipe, M. R., & Pueyo, J. J. (2004). An unusual infection mechanism and nodule morphogenesis in white lupin (*Lupinus albus*). *New Phytologist, 163*(2), 371–380.
Grefen, C., Honsbein, A., & Blatt, M. R. (2011). Ion transport, membrane traffic and cellular volume control. *Current Opinion in Plant Biology, 14*(3), 332–339.
Groth, M., Takeda, N., Perry, J., Uchida, H., Draxl, S., Brachmann, A., ... Parniske, M. (2010). NENA, a *Lotus japonicus* homolog of Sec13, is required for rhizodermal infection by arbuscular mycorrhiza fungi and rhizobia but dispensable for cortical endosymbiotic development. *The Plant Cell, 22*(7), 2509–2526.
Guinel, F. C., & Larue, T. A. (1992). Ethylene inhibitors partly restore nodulation to pea mutant E 107 (brz). *Plant Physiology, 99*(2), 515–518.
Haney, C. H., & Long, S. R. (2010). Plant flotillins are required for infection by nitrogen-fixing bacteria. *Proceedings of the National Academy of Sciences of the U S A, 107*(1), 478–483.
Han, J., Pluhackova, K., & Bockmann, R. A. (2017). The multifaceted role of SNARE proteins in membrane fusion. *Frontiers in Physiology, 8*, 5.
Hayashi, T., Shimoda, Y., Sato, S., Tabata, S., Imaizumi-Anraku, H., & Hayashi, M. (2014). Rhizobial infection does not require cortical expression of upstream common symbiosis genes responsible for the induction of Ca($^{2+}$) spiking. *The Plant Journal,* 77(1), 146–159.
Heckmann, A. B., Lombardo, F., Miwa, H., Perry, J. A., Bunnewell, S., Parniske, M., ... Downie, J. A. (2006). *Lotus japonicus* nodulation requires two GRAS domain regulators, one of which is functionally conserved in a non-legume. *Plant Physiology, 142*(4), 1739–1750.
Huisman, R., Hontelez, J., Mysore, K. S., Wen, J., Bisseling, T., & Limpens, E. (2016). A symbiosis-dedicated SYNTAXIN OF PLANTS 13II isoform controls the formation of a stable host-microbe interface in symbiosis. *New Phytologist, 211*(4), 1338–1351.
Ibanez, F., Wall, L., & Fabra, A. (2017). Starting points in plant-bacteria nitrogen-fixing symbioses: Intercellular invasion of the roots. *Journal of Experimental Botany, 68*(8), 1905–1918.
Ivanov, S., Fedorova, E. E., Limpens, E., De Mita, S., Genre, A., Bonfante, P., et al. (2012). *Rhizobium*-legume symbiosis shares an exocytotic pathway required for arbuscule formation. *Proceedings of the National Academy of Sciences of the U S A, 109*(21), 8316–8321.
James, E. K., Sprent, J. I., Stuherland, J. M., Mcinroy, S. G., & Minchin, F. R. (1992). The structure of nitrogen fixing root nodule on the aquatic Mimosoid legume *Neptunia plena*. *Annals of Botany, 69*, 173–180.
Jarsch, I. K., Konrad, S. S., Stratil, T. F., Urbanus, S. L., Szymanski, W., Braun, P., ... Ott, T. (2014). Plasma membranes are subcompartmentalized into a plethora of coexisting and diverse microdomains in *Arabidopsis* and *Nicotiana benthamiana*. *The Plant Cell, 26*(4), 1698–1711.
Jin, Y., Liu, H., Luo, D., Yu, N., Dong, W., Wang, C., & Wang, E. (2016). DELLA proteins are common components of symbiotic rhizobial and mycorrhizal signalling pathways. *Nature Communications,* 7, 12433-12433.

Kanamori, N., Madsen, L. H., Radutoiu, S., Frantescu, M., Quistgaard, E. M. H., Miwa, H., ... Haaning, L. L. (2006). A nucleoporin is required for induction of Ca^{2+} spiking in legume nodule development and essential for rhizobial and fungal symbiosis. *Proceedings of the National Academy of Sciences of the U S A, 103*(2), 359–364.

Kevei, Z., Lougnon, G., Mergaert, P., Horváth, G. V., Kereszt, A., Jayaraman, D., ... Kiss, G. B. (2007). 3-Hydroxy-3-methylglutaryl coenzyme A reductase1 interacts with NORK and is crucial for nodulation in *Medicago truncatula. Plant Cell, 19*(12), 3974–3989.

Kosuta, S., Held, M., Hossain, M. S., Morieri, G., Macgillivary, A., Johansen, C., ... Szczyglowski, K. (2011). *Lotus japonicus symRK-14* uncouples the cortical and epidermal symbiotic program. *The Plant Journal, 67*(5), 929–940.

Kuppusamy, K. T., Endre, G., Prabhu, R., Penmetsa, R. V., Veereshlingam, H., Cook, D. R., ... Vandenbosch, K. A. (2004). LIN, a *Medicago truncatula* gene required for nodule differentiation and persistence of rhizobial infections. *Plant Physiology, 136*(3), 3682–3691.

Laporte, P., Lepage, A., Fournier, J., Catrice, O., Moreau, S., Jardinaud, M.-F., ... Gamas, P. (2013). The CCAAT box-binding transcription factor NF-YA1 controls rhizobial infection. *Journal of Experimental Botany, 65*(2), 481–494.

Lefebvre, B., Timmers, T., Mbengue, M., Moreau, S., Hervé, C., Tóth, K., ... Godiard, L. (2010). A remorin protein interacts with symbiotic receptors and regulates bacterial infection. *Proceedings of the National Academy of Sciences of the U S A, 107*(5), 2343–2348.

Lévy, J., Bres, C., Geurts, R., Chalhoub, B., Kulikova, O., Duc, G., ... Bisseling, T. (2004). A putative Ca^{2+} and calmodulin-dependent protein kinase required for bacterial and fungal symbioses. *Science, 303*(5662), 1361–1364.

Liang, J., Klingl, A., Lin, Y. Y., Boul, E., Thomas-Oates, J., & Marín, M. (2019). A subcompatible rhizobium strain reveals infection duality in *Lotus. Journal of Experimental Botany, 70*(6), 1903–1913.

Liang, P., Stratil, T. F., Popp, C., Marín, M., Folgmann, J., Mysore, K. S., ... Ott, T. (2018). Symbiotic root infections in *Medicago truncatula* require remorin-mediated receptor stabilization in membrane nanodomains. *Proceedings of the National Academy of Sciences of the U S A, 115*(20), 5289–5294.

Limpens, E., Franken, C., Smit, P., Willemse, J., Bisseling, T., & Geurts, R. (2003). LysM domain receptor kinases regulating rhizobial Nod factor-induced infection. *Science, 302*(5645), 630–633.

Limpens, E., Ivanov, S., van Esse, W., Voets, G., Fedorova, E., & Bisseling, T. (2009). *Medicago* N_2-fixing symbiosomes acquire the endocytic identity marker Rab7 but delay the acquisition of vacuolar identity. *Plant Cell, 21*(9), 2811–2828.

Limpens, E., Mirabella, R., Fedorova, E., Franken, C., Franssen, H., Bisseling, T., et al. (2005). Formation of organelle-like N_2-fixing symbiosomes in legume root nodules is controlled by DMI2. *Proceedings of the National Academy of Sciences of the U S A, 102*(29), 10375–10380.

Liu, C.-W., Breakspear, A., Guan, D., Cerri, M. R., Abbs, K., Jiang, S., ... Bone, C. (2019). NIN acts as a network hub controlling a growth module required for rhizobial Infection. *Plant Physiology*, 01572.

Liu, H., Zhang, C., Yang, J., Yu, N., & Wang, E. (2018). Hormone modulation of legume-rhizobial symbiosis. *Journal of Integrative Plant Biology, 60*(8), 632–648.

Madsen, E. B., Madsen, L. H., Radutoiu, S., Olbryt, M., Rakwalska, M., Szczyglowski, K., ... Sandal, N. (2003). A receptor kinase gene of the LysM type is involved in legume-perception of rhizobial signals. *Nature, 425*(6958), 637-637.

Madsen, L. H., Tirichine, L., Jurkiewicz, A., Sullivan, J. T., Heckmann, A. B., Bek, A. S., ... Stougaard, J. (2010). The molecular network governing nodule organogenesis and infection in the model legume *Lotus japonicus. Nature Communications, 1*, 10.

Marie, C., Plaskitt, K. A., & Downie, J. A. (1994). Abnormal bacteroid development in nodules induced by a glucosamine synthase mutant of *Rhizobium ieguminosarum*. *Molecular Plant-Microbe Interactions,* 7(4), 482–487.

Marsh, J. F., Rakocevic, A., Mitra, R. M., Brocard, L., Sun, J., Eschstruth, A., … Oldroyd, G. E. (2007). *Medicago truncatula* NIN is essential for rhizobial-independent nodule organogenesis induced by autoactive calcium/calmodulin-dependent protein kinase. *Plant Physiology, 144*(1), 324–335.

Mbengue, M., Camut, S., de Carvalho-Niebel, F., Deslandes, L., Froidure, S., Klaus-Heisen, D., … Hervé, C. (2010). The *Medicago truncatula* E3 ubiquitin ligase PUB1 interacts with the LYK3 symbiotic receptor and negatively regulates infection and nodulation. *The Plant Cell, 22,* 3474–3488. tpc-110.

Messinese, E., Mun, J.-H., Yeun, L. H., Jayaraman, D., Rougé, P., Barre, A., & Cook, D. R. (2007). A novel nuclear protein interacts with the symbiotic DMI3 calcium-and calmodulin-dependent protein kinase of *Medicago truncatula. Molecular Plant-Microbe Interactions, 20*(8), 912–921.

Middleton, P. H., Jakab, J., Penmetsa, R. V., Starker, C. G., Doll, J., Kaló, P., … Kereszt, A. (2007). An ERF transcription factor in *Medicago truncatula* that is essential for Nod factor signal transduction. *The Plant Cell, 19*(4), 1221–1234.

Miwa, H., Sun, J., Oldroyd, G. E., & Downie, J. A. (2006). Analysis of Nod-factor-induced calcium signaling in root hairs of symbiotically defective mutants of *Lotus japonicus. Molecular Plant-Microbe Interactions, 19*(8), 914–923.

Moling, S., Pietraszewska-Bogiel, A., Postma, M., Fedorova, E., Hink, M. A., Limpens, E., … Bisseling, T. (2014). Nod factor receptors form heteromeric complexes and are essential for intracellular infection in *Medicago* nodules. *The Plant Cell, 26*(10), 4188–4199.

Monahan-Giovanelli, H., Pinedo, C. A., & Gage, D. J. (2006). Architecture of infection thread networks in developing root nodules induced by the symbiotic bacterium *Sinorhizobium meliloti* on *Medicago truncatula. Plant Physiology, 140*(2), 661–670.

Murray, J. D., Muni, R. R. D., Torres-Jerez, I., Tang, Y., Allen, S., Andriankaja, M., … Wen, J. (2011). Vapyrin, a gene essential for intracellular progression of arbuscular mycorrhizal symbiosis, is also essential for infection by rhizobia in the nodule symbiosis of *Medicago truncatula. The Plant Journal, 65*(2), 244–252.

Ndoye, I., de Billy, F., Vasse, J., Dreyfus, B., & Truchet, G. (1994). Root nodulation of *Sesbania rostrata. Journal of Bacteriology, 176*(4), 1060–1068.

Nutman, P. S. (1962). The relation between root hair infection by rhizobium and nodulation in *Trifolium* and *Vicia. Proceedings of the Royal Society of London B Biological Sciences, 156*(962), 122–132.

Oldroyd, G. E. (2013). Speak, friend, and enter: Signalling systems that promote beneficial symbiotic associations in plants. *Nature Reviews Microbiology, 11*(4), 252–263.

Oldroyd, G. E., Murray, J. D., Poole, P. S., & Downie, J. A. (2011). The rules of engagement in the legume-rhizobial symbiosis. *Annual Review of Genetics, 45,* 119–144.

Ott, T. (2017). Membrane nanodomains and microdomains in plant–microbe interactions. *Current Opinion in Plant Biology, 40,* 82–88.

Ovchinnikova, E., Journet, E. P., Chabaud, M., Cosson, V., Ratet, P., Duc, G., … Limpens, E. (2011). IPD3 controls the formation of nitrogen-fixing symbiosomes in pea and *Medicago* spp. *Molecular Plant-Microbe Interactions, 24*(11), 1333–1344.

Parniske, M. (2018). Uptake of bacteria into living plant cells, the unifying and distinct feature of the nitrogen-fixing root nodule symbiosis. *Current Opinion in Plant Biology, 44,* 164–174.

Pawlowski, K., & Demchenko, K. N. (2012). The diversity of actinorhizal symbiosis. *Protoplasma, 249*(4), 967–979.

Pollard, T. D. (2007). Regulation of actin filament assembly by Arp2/3 complex and formins. *Annual Review of Biophysics and Biomolecular Structure, 36,* 451–477.

Radutoiu, S., Madsen, L. H., Madsen, E. B., Felle, H. H., Umehara, Y., Grønlund, M., … Sandal, N. (2003). Plant recognition of symbiotic bacteria requires two LysM receptor-like kinases. *Nature, 425*(6958), 585-585.

Rae, A. L., Bonfante-Fasolo, P., & Brewin, N. J. (1992). Structure and growth of infection threads in the legume symbiosis with *Rhizobium leguminosarum*. *The Plant Journal, 2*(3), 385−395.

Ried, M. K., Antolin-Llovera, M., & Parniske, M. (2014). Spontaneous symbiotic reprogramming of plant roots triggered by receptor-like kinases. *Elife, 3*.

Roux, B., Rodde, N., Jardinaud, M. F., Timmers, T., Sauviac, L., Cottret, L., … Gamas, P. (2014). An integrated analysis of plant and bacterial gene expression in symbiotic root nodules using laser-capture microdissection coupled to RNA sequencing. *The Plant Journal, 77*(6), 817−837.

Saito, K., Yoshikawa, M., Yano, K., Miwa, H., Uchida, H., Asamizu, E., … Umehara, Y. (2007). NUCLEOPORIN85 is required for calcium spiking, fungal and bacterial symbioses, and seed production in *Lotus japonicus*. *The Plant Cell, 19*(2), 610−624.

Schauser, L., Roussis, A., Stiller, J., & Stougaard, J. (1999). A plant regulator controlling development of symbiotic root nodules. *Nature, 402*(6758), 191-191.

Singh, S., Katzer, K., Lambert, J., Cerri, M., & Parniske, M. (2014). CYCLOPS, a DNA-binding transcriptional activator, orchestrates symbiotic root nodule development. *Cell Host & Microbe, 15*(2), 139−152.

Smit, P., Limpens, E., Geurts, R., Fedorova, E., Dolgikh, E., Gough, C., et al. (2007). Medicago LYK3, an entry receptor in rhizobial nodulation factor signaling. *Plant Physiology, 145*(1), 183−191.

Soltis, D. E., Soltis, P. S., Morgan, D. R., Swensen, S. M., Mullin, B. C., Dowd, J. M., et al. (1995). Chloroplast gene sequence data suggest a single origin of the predisposition for symbiotic nitrogen fixation in angiosperms. *Proceedings of the National Academy of Sciences of the U S A, 92*(7), 2647−2651.

Soyano, T., & Hayashi, M. (2014). Transcriptional networks leading to symbiotic nodule organogenesis. *Current Opinion in Plant Biology, 20*, 146−154.

Sprent, J. I. (2007). Evolving ideas of legume evolution and diversity: A taxonomic perspective on the occurrence of nodulation. *New Phytologist, 174*(1), 11−25.

Sprent, J. I., Ardley, J., & James, E. K. (2017). Biogeography of nodulated legumes and their nitrogen-fixing symbionts. *New Phytologist, 215*(1), 40−56.

Stracke, S., Kistner, C., Yoshida, S., Mulder, L., Sato, S., Kaneko, T., … Szczyglowski, K. (2002). A plant receptor-like kinase required for both bacterial and fungal symbiosis. *Nature, 417*(6892), 959.

Subba-Rao, N. S., Mateos, P. F., Baker, D., Pankratz, H. S., Palma, J., Dazzo, F. B., et al. (1995). The unique root-nodule symbiosis between *Rhizobium* and the aquatic legume, *Neptunia natans* (L. f.) Druce. *Planta, 196*, 311−320.

Suzaki, T., Ito, M., Yoro, E., Sato, S., Hirakawa, H., Takeda, N., et al. (2014). Endoreduplication-mediated initiation of symbiotic organ development in *Lotus japonicus*. *Development, 141*(12), 2441−2445.

Tansengco, M. L., Hayashi, M., Kawaguchi, M., Imaizumi-Anraku, H., & Murooka, Y. (2003). crinkle, a novel symbiotic mutant that affects the infection thread growth and alters the root hair, trichome, and seed development in *Lotus japonicus*. *Plant Physiology, 131*(3), 1054−1063.

Timmers, A. C., Auriac, M. C., de Billy, F., & Truchet, G. (1998). Nod factor internalization and microtubular cytoskeleton changes occur concomitantly during nodule differentiation in alfalfa. *Development, 125*(3), 339−349.

Tirichine, L., Imaizumi-Anraku, H., Yoshida, S., Murakami, Y., Madsen, L. H., Miwa, H., … Kawaguchi, M. (2006). Deregulation of a Ca^{2+}/calmodulin-dependent kinase leads to spontaneous nodule development. *Nature, 441*(7097), 1153-1153.

Toth, K., Stratil, T. F., Madsen, E. B., Ye, J., Popp, C., Antolin-Llovera, M., & Ott, T. (2012). Functional domain analysis of the remorin protein LjSYMREM1 in *Lotus japonicus*. *PLoS One,* 7(1), e30817.

Venkateshwaran, M., Jayaraman, D., Chabaud, M., Genre, A., Balloon, A. J., Maeda, J., … Ane, J. M. (2015). A role for the mevalonate pathway in early plant symbiotic signaling. *Proceedings of the National Academy of Sciences of the U S A, 112*(31), 9781–9786.

Xie, F., Murray, J. D., Kim, J., Heckmann, A. B., Edwards, A., Oldroyd, G. E. D., et al. (2012). Legume pectate lyase required for root infection by rhizobia. *Proceedings of the National Academy of Sciences of the U S A, 109*(2), 633–638.

Yano, K., Shibata, S., Chen, W. L., Sato, S., Kaneko, T., Jurkiewicz, A., … Kojima, T. (2009). CERBERUS, a novel U-box protein containing WD-40 repeats, is required for formation of the infection thread and nodule development in the legume–*Rhizobium* symbiosis. *The Plant Journal, 60*(1), 168–180.

Yano, K., Yoshida, S., Muller, J., Singh, S., Banba, M., Vickers, K., … Parniske, M. (2008). CYCLOPS, a mediator of symbiotic intracellular accommodation. *Proceedings of the National Academy of Sciences of the U S A, 105*(51), 20540–20545.

Yokota, K., Fukai, E., Madsen, L. H., Jurkiewicz, A., Rueda, P., Radutoiu, S., … Morieri, G. (2009). Rearrangement of actin cytoskeleton mediates invasion of *Lotus japonicus* roots by *Mesorhizobium loti*. *The Plant Cell, 21*(1), 267–284.

CHAPTER FIVE

Nodule diversity, evolution, organogenesis and identity

Shengbin Liu, Pascal Ratet* and Kévin Magne
Institute of Plant Sciences Paris Saclay (IPS2), CNRS, INRA, Univ Paris Diderot, Univ Paris-Sud, Univ d'Evry, Université Paris-Saclay, Gif-sur-Yvette, France
*Corresponding author: E-mail: pascal.ratet@ips2.universite-paris-saclay.fr

Contents

Abstract

Some plants can engage symbiosis with nitrogen fixing soil bacteria, allowing them to grow when soils are poor in nitrogen. This symbiotic association results in the formation of a unique plant root organ called the nodule. In these nodules, bacteria can convert atmospheric dinitrogen into ammonia that can be used by the plant. Establishment of a successful symbiosis requires the two symbiotic partners to be compatible with each other throughout the process of symbiotic development. Nodulating plants include actinorhizal plants, *Parasponia* and the economically important legume family. In this chapter, we will describe the diversity of nodule shapes, the molecular mechanisms underlying the inception of nodule primordia, as well as the development and the maintenance of the symbiotic organ. We will also describe the molecular actors involved in these processes with an emphasis on the *NOOT-BOP-COCH-LIKE (NBCL)* genes controlling the identity of the symbiotic organ.

Advances in Botanical Research, Volume 94
ISSN 0065-2296
https://doi.org/10.1016/bs.abr.2019.09.009

1. Symbiotic association engaging plants and nitrogen-fixing bacteria

The term symbiosis (*Symbiotismus*) was introduced to define all cases in which two different species live on or in another, irrespective of the role of individuals. In the case of the plant-bacteria nitrogen fixing symbiosis, the atmospheric dinitrogen (N_2) is reduced to ammonia (NH_3) by soil nitrogen-fixing *rhizobia* or *frankia* (*actinobacteria*) entering symbiosis with plants. Following a molecular dialogue between the two symbiotic partners, the plant-bacteria association results in the formation of specific organs called nodules on the roots or, in few cases, on the stems of the host plant (Eaglesham & Szalay, 1983; Fernandez-Lopez, Goormachtig, Gao, D'Haeze, van Montagu, Van, & Holsters, 1998; Ndoye, de Billy, Vasse, Dreyfus, & Truchet, 1994). The nodulation ability is confined to the Rosid I clade including actinorhizal plants, *Parasponia* and legumes (Soltis et al., 1995). Currently, nitrogen-fixing root nodule symbiosis is believed to have evolved only once, around 100 million years ago (MYA), using components of the signaling pathway from the ancestral and widespread arbuscular mycorrhiza symbiosis which itself appeared 400 MYA (Doyle, 2011; van Velzen, Doyle, & Geurts, 2019; Werner, Cornwell, Sprent, Kattge, & Kiers, 2014). Nodule organs are specialized for N_2 reduction into NH_3, which benefits to the host plant and is assimilated into organic compounds such as amino acids and nucleotides (Frendo, Matamoros, Alloing, & Becana, 2013). In return, symbiotic bacteria receive photosynthetic carbohydrates and benefit of the nodule shelter (Gresshoff, Mens, Ferguson, Li, & Haaima, 2018; Sprent, Ardley, & James, 2013).

Rhizobia are unicellular Gram-negative bacteria belonging to the widespread *Proteobacteria* division, whereas *frankia* strains are filamentous, branching and Gram-positive bacteria. The alpha-proteobacterial genera (*Agrobacterium, Allorhizobium, Azorhizobium, Bradyrhizobium, Mesorhizobium, Rhizobium, Sinorhizobium, Devosia, Methylobacterium, Ochrobactrum* and *Phyllobacterium*) and the beta-proteobacterial genera (*Burkholderia* and *Cupriavidus*) contain nodule-forming bacteria (Andrews & Andrews, 2017;Lindstrom & Martinez-Romero, 2007, Table 1). However, the majority of soil *rhizobia* strains are not able to form nodules on plants, often because they lack Nod factor (NF) producing genes needed to induce nodulation (Sachs, Ehinger, & Simms, 2010). Most *actinobacteria* cannot be cultivated, making them less characterized than *rhizobia* (Pawlowski & Demchenko, 2012).

Table 1 General nodule features in nodulating-plant species among *Fagales, Cucurbitales, Rosales* and *Fabales* orders confined in the Rosid I clade.

Order	Family	Sub-family	Sub-clade	Example of nodulating tribes - genera	Symbionts	Symbiont compartimentation	Infection mechanism	Infected tissues characteristics	Origin of first cell divisions	Vasculature position	Nodule determinacy
Fagales	*Casuarinaceae*			*Casuarina, Allocasuarina, Ceuthostoma, Gymnostoma*	mainly *Frenkia* clade I few *Frankia* clade III	vesicles	root hair	infected + uninfected cells	cortex, pericycle	central	indeterminate
	Betulaceae			*Alnus*							
	Myricaceae			*Comptonia, Myrica*							
Cucurbitales	*Datiscaceae*			*Datisca*	*Frankia* clade II	vesicles	intercellular	entirely filled of infected cells	pericycle	central	indeterminate
	Coriariaceae			*Coriaria*							
Rosales	*Rosaceae*			*Dryas, Rubus, Purshia, Cowania, Cercocarpus,*	Frankia clade III or *Franka* clade II	vesicles	intercellular	variable depending plant species	pericycle	central	indeterminate
	Elaeagnaceae			*Elaeagnus, Hippophae, Shepherdia*							
	Rhamnaceae			*Adolphia, Colletia, Discaria, Kentrothamnus, Retanilla,*							
	Cannabaceae			*Parasponia*	*Bradyrhizobia, Rhizobia*	fixation threads	crack entry + IT	infected + uninfected cells	cortex, pericycle		indeterminate

(Continued)

Table 1 General nodule features in nodulating-plant species among *Fagales, Cucurbitales, Rosales* and *Fabales* orders confined in the Rosid I clade.

Order	Family	Sub-family	Sub-clade	Example of nodulating tribes - genera	Symbionts	Symbiont compartimentation	Infection mechanism	Infected tissues characteristics	Origin of first cell divisions	Vasculature position	Nodule determinacy
Fabales	*Leguminosae*	*Duparquetioideae*		Absent							
		Cercidoideae									
		Detarioideae		Uncertain							
		Dialioideae									
		Caesalpinioideae including mimosoid clade or *Mimosoideae - Caesalpinieae - Cassieae* (MCC) clade		***Caesalpinioideae*** : *Dimorphandra, Campsiandra, Sclerolobium, Tachigali, Vouacapoua, Chamaecrista, Peltophorum, Moldenhawera, Cassia, Erythrophleum, Melanoxylon*	***Caesalpinioideae:*** *Bradyrhizobia*	***Caesalpinioideae:*** fixation threads with exceptions	***Caesalpinioideae:*** root hair + IT	infected + uninfected cells	***Caesalpinioideae:*** outer cortex	peripheral	indeterminate
				Mimosoideae : *Parkia, Pentaclethra, Acacia, Anadenanthera, Entada, Mimosa, Stryphnodendron, Abarema, Albizia, Calliandra, Cedrelinga, Enterolobium, Inga, Pithecellobium, Zygia, Chidlowia, Desmanthus*	***Mimosoideae:*** *Bradyrhizobia , Burkholderia, Rhizobia, Ensifer, Cupriavidus, Mesorhizobium, Ochrobactrum*	***Mimosoideae:*** symbiosomes	***Mimosoideae:*** mainly root hair + IT or epidermal		***Mimosoideae:*** Inner cortex with exceptions		
		Papilionoideae	outside the 50-kb inversion clade	**Swartzieae** *(Swartzia, Ateleia, Bobgunnia, Cyathostegia)*	*Bradyrhizobia*	NA	NA	NA	NA	peripheral	indeterminate
			50-kb inversion clade	**Andiraeae** *(Andira)*	*Bradyrhizobia*	fixation threads	crack entry	infected + uninfected cells	NA		indeterminate
			50-kb inversion clade	**Brongniartieae** *(Poecilanthe, Cyclolobium, Hovea, Templetonia)*	*Bradyrhizobia*	fixation threads	NA	infected + uninfected cells	NA		indeterminate
			50-kb inversion clade	**Leptolobieae** *(Diplotropis)*	NA	NA	NA	NA	NA		indeterminate
			50-kb inversion clade	**Sophoreae** *(Sophora, Ctathrotropis, Ormosia)*	*Ensifer, Mesorhizobium, Phyllobacterium, Rhizobium, Bradyrhizobium, Burkholderia*	symbiosomes with expeptions	root hair + IT	infected + uninfected cells	cortex		indeterminate
			50-kb inversion clade	**Podalyrieae** *(Cyclopia, Podalyria, Virgilia)*	*Burkholderia, Rhizobium*	symbiosomes	root hair + IT	infected + uninfected cells	NA		indeterminate
			50-kb inversion clade	**Genistoid** or **Lupinoid** *(Genista, Lupinus, Cytisus, Lotononis)*	*Bradyrhizobia, Phyllobacteria, Ensifer, Mesorhizobium, Ochrobactrum, Microvirga*	symbiosomes	crack entry / no IT	entirely filled of infected cells	outer cortex	peripheral	indeterminate (many lateral meristems)
			50-kb inversion clade	**Crotalarieae** *(Crotalaria, Listia, Aspalathus, Lebeckia)*	*Methylobacteria, Mesorhizobia Rhizobium, Microvirga sometimes Burkholderia*	symbiosomes	epidermal / no IT	entirely filled of infected cells	NA		indeterminate
			50-kb inversion clade	**Dalbergioid** *(Arachis, Andira, Dalbergia, Etaballia, Machaerium, Platymiscium, Stylosanthes, Centrolobium, Pterocarpus, Zornia)* **Aeschynomenoid** *(Aschynomene)* **Adesmieae** *(adesmia)*	*Bradyrhizobia, Rhizobium*	symbiosomes	crack entry / no IT	entirely filled of infected cells	outer or inner cortex depending on plant species		determinate (developing at lateral root axil)
			NPAAA clade	**Mirbelioid** *(Aotus, Gompholobium)*	*wide range of α - and β-rhizobia*	symbiosomes	root hair * + IT	infected + uninfected cells	NA		indeterminate
			NPAAA clade	**Indigoferoid (Indigofera)**	*Burkholderia, Bradyrhizobium, Ensifer, Microvirga*	symbiosomes	root hair + IT	infected + uninfected cells	outer cortex		indeterminate desmodioid
			NPAAA clade	**Millettioid** *(Millettia, Derris, Tephrosia, Dahlstedtia, Lonchocarpus)*	Bradyrhizobium, Rhizobium, Ensifer	symbiosomes with exceptions	epidermal + IT	infected + uninfected cells	outer cortex *		indeterminate and determinate
			NPAAA clade	**Phaseoleae** *(Glycine, Phaseolus, Vigna, Cajanus, Clitoria, Dioclea, Erythrina, Amphicarpaea, Bolusafra, Canavalia, Centrosema, Dipogon, Lablab, Neonotonia, Pachyrhizus, Pachyrhizus, Pueraria, Rhynchosia)*	*Mesorhizobia, Rhizobia, Bradyrhizobia, Burkholderia, Ensifer*	symbiosomes	root hair + IT	infected + uninfected cells	outer cortex	peripheral	determinate with exception
			NPAAA clade	**Psoraleeae** *(Psoralea, Bituminaria, Otholobium)*	*Mesorhizobium, Burkholderia, Ensifer, Bradyrhizobium*	symbiosomes	root hair	infected + uninfected cells	outer cortex		determinate
			NPAAA clade	**Sesbanieae** *(Sesbania)*	*Azorhizobia, Ensifer, Rhizobium, Mesorhizobium, Bradyrhizobium*	symbiosomes	root hair + IT and crack entry	infected + uninfected cells	middle / outer cortex or inner / middle cortex		Both indeterminate / determinate
			NPAAA clade	**Loteae** *(Lotus, Tetragonolobus, Coronilla, Ornithopus)*	*Mesorhizobia, Rhizobia, Bradyrhizobia, Ensifer*	symbiosomes	root hair + IT and crack entry	infected + uninfected cells	outer cortex		determinate with exceptions
			NPAAA clade	**Robinieae** *(Gliricidia, Robinia)*	*Rhizobium, Ensifer, Mesorhizobium*	symbiosomes	root hair + IT	infected + uninfected cells	middle cortex		indeterminate
			IRLC	**Astragaleae / Hedysareae** *(Astragalus, Hedysarum)* **Cicereae** *(Cicer)* **Trifolieae** *(Trifolium, Medicago, Melilotus)* **Fabeae** *(Pisum, Vicia, Lathyrus, Lens)*	*Ensifer, Mesorhizobia* and *Rhizobia* somtimes *Bradyrhizobia, Neorhizobium, Phyllobacterium*	Symbiosomes (terminal differentiation)	root hair + IT	infected + uninfected cells	inner cortex	peripheral	indeterminate

NPAAA, Non-Protein Amino Acid Accumulating; *IRLC*, Inverted Repeat Lacking Clade; *IT*, Infection Threads; epidermal, bacterial entrance by epidermal cells dissolution; *NA*, information Not Available; *, probably. This table is based on the following publications (alphabetical order): Andrews and Andrews (2017), Ardley et al. (2013), Becking (1984), Behm et al. (2014), Berry and Sunell (1990), Boatwright, Wink and van Wyk (2011), Chen et al. (2013), Coba de la Peña, Fedorova, Pueyo, and Lucas (2018), Corby (1988), Dart (1977), Dawson (2008), de Faria, Sutherland and Sprent (1986), De souza Moreira, Da Silva and De Faria (1992), Dilworth, James, and Newton (2008), Doyle (2011), Fernandez-Lopez et al. (1998), Franche et al. (1998), Froussart et al. (2016), Gehlot et al. (2012), González-Sama, Lucas, De Felipe and Pueyo (2004), Guinel (2009), Hafeez et al. (1984a), (1984b), Hirsch (1992), Hocher et al. (2006), Ibáñez, Wall and Fabra (2017), Imanishi et al. (2014), Jiao et al. (2015), Kanu and Dakora (2017), Lalani Wijesundara, Van Holm, and Kulasooriya (2000), Laplaze et al. (2000), Lawrie (1983), Liu et al. (2014), LPWG (2013a), (2013b), LPWG (2017), Madsen et al. (2010), Ndoye et al. (1994), Newcomb and Pankhurst (1982), Op den Camp et al. (2012), Pawlowski and Bisseling (1996), Pawlowski and Demchenko (2012), Ren (2018), Renier et al. (2011), Soltis et al. (1995), Soltis et al. (2011), Sprent and James (2007), Sprent et al. (1987), Sprent et al. (2013), Sprent et al. (2017), Sprent (1988), Sprent (2000), Sprent (2007), Sprent (2008a), Sprent (2008b), Sprent (2009), Svistoonoff et al. (2014), Sy et al. (2001), Tajima, Abe, Lee, Morita and Lux (2008), Trinick (1973), Valverde and Wall (1999), van Velzen et al. (2019), Werner et al. (2014).

2. Nodule shapes and evolution in the Rosid I clade

Several types of nodule shapes have been reported (Corby, 1988; Doyle, 1998; Guinel, 2009; Sprent, 2008a, 2008b; Sprent, Ardley, & James, 2017; Sprent & James, 2007). In general, the morphology of nodules is based on the determinacy of their apical meristem (persistent/non-persistent), on the type of bacteroid compartmentation (vesicles/fixation-threads/symbiosomes), on the infection type (epidermal/crack entry/infection threads), on the presence or not of interstitial uninfected cells in infected tissues and on the position of the vascular tissues (peripheral/central). The shape of nodules appears to be mainly dependent of the host plant species, however genetic bases underlying the broad diversity of nodule morphologies are still poorly understood.

Actinorhizal plants represent approximately 260 species distributed among 24 genera in eight Angiosperm families of the *Cucurbitales*, *Fagales*, and *Rosales* orders. Actinorhizal plants are able to form nodules infected by filamentous Gram-positive *actinobacteria* commonly called *frankia* (Hocher et al., 2006; Soltis et al., 1995). *Actinobacteria* can infect their hosts by two ways, intracellularly, via root hair or intercellularly between epidermal cells, to finally form nitrogen-fixing vesicles (Berry & Sunell, 1990). Actinorhizal nodules formation tends to mainly originate from root pericycle cell divisions as for lateral root formation (Pawlowski & Bisseling, 1996). In addition, actinorhizal nodules present a single central vascular bundle that makes them resembling modified lateral roots (Franche, Laplaze, Duhoux, & Bogusz, 1998; Froussart, Bonneau, Franche, & Bogusz, 2016). Usually, mature actinorhizal nodules are indeterminate and branched, with each nodule lobe resembling a coralloid root (i.e. cycads/nostoc symbiotic modified lateral roots presenting a peripheral infected-root cortex, without a root cap; Costa & Lindblad, 2002). Indeed, mature actinorhizal nodules contain a single and central vascular bundle surrounded by infected-cortex cells, an apical meristem and a superficial periderm (Franche et al., 1998; Froussart et al., 2016; Laplaze et al., 2000; Pawlowski & Bisseling, 1996; Pawlowski & Demchenko, 2012, Table 1). In *Fagales* and *Rosales* nodules, infected cortical tissues present interstitial uninfected cells, while in *Cucurbitales* nodules, infected cortical tissues form an homogeneous and uninterrupted domain (Berg, Langenstein, & Silvester, 1999; Hafeez, Akkermans, & Chaudhary, 1984a, 1984b; Newcomb & Pankhurst, 1982, Table 1).

In the *Fabale* order, the *Leguminosae* family consists in ca. 751 genera and ca. 19,500 species. Legumes represent one of the most successful lineages in flowering plants (Legume Phylogeny Working Group, LPWG, 2013a). The LPWG recently proposed an updated legume classification based on both chloroplast and nuclear markers in which *Leguminosae* sub-families have been revised (LPWG, 2013b; 2017; Sprent et al., 2017). The *Leguminosae* family now consists in six sub-families: the *Duparquetioideae*, the *Detarioideae*, the *Cercidoideae*, the *Dialioideae*, the *Caesalpinioideae*, and the *Papilionoideae* (Table 1). Among these six *Leguminosae* sub-families, only the *Caesalpinioideae* (formerly called *Mimosoideae-Caesalpinieae-Cassieae* Clade, MCC) and the *Papilionoideae* contain legume species forming nodules. In the *Caesalpinioideae* and the *Papilionoideae* sub-families, nodules present central infected tissues and peripheral vascular tissues, surrounded by parenchyma, endodermis and cortex tissue layers (from the inside toward the outside of the nodule). Central infected tissues often contain a mixture of enlarged infected plant cells together with interstitial uninfected cells (Guinel, 2009; Sprent & James, 2007, Table 1).

Caesalpinioideae legumes occur primarily in tropical and sub-tropical regions and are mainly trees, some extremely large, as well as lianas and shrubs (LPWG, 2013a). Among these woody species, some *Caesalpinioid* genera are known to enter symbiosis with nitrogen-fixing bacteria and to form indeterminate branched nodules (Sprent et al., 2017). In *Caesalpinioid* species, the nitrogen-fixing bacteria are, in general, permanently retained within cell wall-bound structures called fixation threads (Table 1). These structures resemble infection threads (ITs) however, by contrast to ITs, bacteria hosted in fixation threads will never be released into symbiosomes, a plant membrane-derived organelle-like structure dedicated to host nitrogen-fixing bacteria into the plant cell.

Species belonging to the *Mimosoid* clade (formerly the *Mimosoideae* sub-family) are also generally woody. *Mimosoid* nodules infection mainly occurs via root hairs and involves ITs. *Mimosoid* nodules are all indeterminate, often branched, with bacteroids released into symbiosomes. While fixation threads are never observed, the central infected tissues display interstitial uninfected cells (Table 1).

In the *Papilionoideae* sub-family, a clade known as the Inverted Repeat Loss Clade (IRLC), was defined by Wojciechowski, Sanderson, Steele, and Liston (2000) on the basis of the loss of one copy of an inverted repeat in the plastid genome (Lavin, Doyle, & Palmer, 1990). The IRLC contains *Astragaleae/Hedysareae*, *Cicereae*, *Trifolieae*, *Fabeae* (formerly *Viceae*) tribes

(Table 1). These tribes are largely widespread from warm temperate to arctic climate regions and form indeterminate nodules with a persistent nodule meristem (NM) which is composed by an apical nodule central meristem (NCM) and multiple peripheral nodule vascular meristems (NVM, Couzigou et al., 2012; Franssen et al., 2015; Roux et al., 2014). In the IRLC, nodules are often branched, infected through root-hairs and develop ITs. Bacteroids are released into symbiosomes and undergo terminal differentiation, an irreversible plant-driven process consisting in bacterial cells enlargement, genome amplification and membrane permeabilization (Alunni & Gourion, 2016). The infected tissues of IRLC nodules contain a mixture of infected cells and interstitial uninfected cells (Sprent et al., 2017). Many economically important legumes such as lentils (*Lens culinaris*), clover (*Trifolium spp.*), vetches (*Vicia spp.*), garden pea (*Pisum sativum,* hereafter *P. sativum*), chickpea (*Cicer arietinum*) and alfalfa (*Medicago sativa,* hereafter *M. sativa*), as well as the model legume barrel medic (*Medicago truncatula,* hereafter *M. truncatula*), belong to the IRLC (Table 1).

Outside the IRLC but still in the monophyletic Non-Protein Amino Acid Accumulating (NPAAA) clade, characterized by the accumulation of a non-proteinogenic amino-acid called canavanine in seeds, species belonging to the *Phaseoloeae*, *Psoraleeae* and *Loteae* tribes form determinate nodules (also called desmodioid nodules) with determinate apical NCM and NVM meristems (Corby, 1988; Magne, George, Berbel Tornero, Broquet, Madueño, Andersen, et al. 2018, Table 1). In these tribes, nodule primordia tend to mostly originate from outer root cortical cell divisions (Hirsch, 1992). Due to the non-persistent NCM mitotic activity, these nodules have a determinate growth (Corby, 1988; Guinel, 2009; Sprent, 2008a, 2008b). Consequently, determinate nodules are usually spherical, in contrast to indeterminate nodules that have an elongated shape. Regarding infection features, infections occur mainly via root hairs, ITs are formed, infected tissues contain a mixture of infected and interstitial uninfected cells, and bacteroids are released within symbiosomes (Guinel, 2009; Sprent & James, 2007, Table 1). In addition, by contrast to indeterminate nodules, determinate nodules are unable to branch and often display lenticels, which are loosely packed clump of cells with large intercellular spaces that regulate the gas permeability of nodules. Lenticels usually draw a white star or stripes at the nodule surface and represent a typical feature of determinate nodules (Sprent, 2007). The NPAAA clade also integrates the *Mirbelioid, Indigoferoid, Millettioid, Sesbanieae* and *Robinieae* tribes. In these tribes, nodule characteristics are more variable. Nodules tend to be indeterminate even though many exceptions occur with species forming determinate nodules. Intermediate nodule structures, called indeterminate-desmodioid nodules, can also be found in the *Indigoferoid* tribe (Ren, 2018). Moreover, it also exists species able to form both indeterminate and determinate nodules, e.g. *Sesbania* (Fernandez-Lopez, Goormachtig, Gao, D'Haeze, van Montagu, Van, & Holsters, 1998; Ndoye et al., 1994). In *Mirbelioid, Indigoferoid, Millettioid, Sesbanieae* and *Robinieae* tribes, the infection mechanism involves root hair and the formation of ITs seems to be a dominant feature. The nodule vasculature is peripheral, central infected tissues consist in a mixture of infected and interstitial uninfected cells, and most species are able to release bacteroids into symbiosomes (Table 1).

In the *Papilionoideae* sub-family, outside the NPAAA clade but still in the 50-kb inversion clade (termed based on a 50-kb inversion in the chloroplast genome), the *Aeschynomenoid* (or *Dalbergioid*) nodule type is found in species belonging to the *Dalbergioid* tribe. In *Aeschynomenoid* nodules, the NCM is determinate, resulting in small and oblate nodules which develop at the upper root/lower stem transition, always in association with the axil of a lateral root. *Aeschynomenoid* nodules are often numerous and look-like determinate nodules, however they are generally smaller and without lenticels. *Aeschynomenoid* nodules infection occurs by crack-entry, does not present ITs and infected tissues do not present interstitial uninfected cells (Sprent & James, 2007, Table 1).

The *Genistoid* (or *Lupinoid*) nodule type is found in the genera *Genista, Lupinus, Cytisus* or *Lotononis* (Table 1). *Genistoid* nodules have lateral meristems, girdle the subtending root and display indeterminate growth. These nodules are roughly spherical, quite large and collar-shaped (Corby, 1988; Sprent, 2008a, 2008b). As for the *Aeschynomedoid* nodules, *Lupinoid* nodules are initiated by a crack-entry infection mechanism, ITs are absent and the central nodule tissue is entirely filled with infected cells (Sprent & James, 2007, Table 1).

The *Crotalarioid* tribe is phylogenetically close to the *Genistoid* tribe and contains both herbaceous and woody species. *Crotalarioid* nodules are indeterminate, laterally flattened, thin, and form fan-like structures (Corby, 1988). *Crotalarioid* nodules infection seems to occur without root hair infection nor ITs formation, and infected tissues do not contain interstitial uninfected cells, similarly to *Aeschynomenoid* and *Genistoid* nodules (Renier et al., 2011; Table 1).

In the 50-kb inversion clade, besides the *Aeschynomenoid, Genistoid* and *Crotalaroid* tribes, species from the *Andiraeae, Brongniartieae, Leptolobieae,*

Sophoreae and *Podalyrieae* families are also able to form indeterminate nodules with peripheral vasculatures. In these tribes, nodulating species are less studied but it seems that the less sophisticated fixation threads bacterial compartmentation is more frequent (Table 1).

Still in the *Papilionoideae* family, outside the 50-kb inversion clade, the basal *Swartzieae* group also contains few genera that have been described to form nodules, such as *Swartzia, Ateleia, Bobgunnia* or *Cyathostegia.* The nodulation traits in these groups are not well described but it seems that *Swartzieae* mostly associate with *bradyrhizobia* and form indeterminate nodules with peripheral vascular bundles (Table 1).

Outside the *Leguminosae* family, in the *Rosales, Parasponia sp.* belongs to the *Cannabaceae* family and represents the only known non-legume lineage having evolved a symbiosis with *rhizobia* (Behm, Geurts, & Kiers, 2014; Op den Camp et al., 2011; Op den Camp et al., 2012). The *Parasponia* genus is represented by five species and is phylogenetically close to the non-nodulating *Trema* genus (van Velzen et al., 2018). In *Parasponia*, nodules are indeterminate and root infection occurs via crack-entry, a less-sophisticated infection mode than root hair infection (Behm et al., 2014). In addition, *Parasponia* hosts its rhizobial partners in permanent fixation threads. Importantly, *Parasponia* nodules have a central vasculature surrounded by peripheral cortex infected-cells comparable to actinorhizal nodules. Infected tissues in *Parasponia* nodules consist in a mixture of infected and interstitial uninfected cells (Behm et al., 2014, Table 1).

3. Indeterminate *versus* determinate legume nodules

In legumes, nodules can harbor indeterminate or determinate meristems (Guinel, 2009; Hirsch, 1992). Indeterminate nodules are elongated due to the presence of a persistent meristem, and the indeterminate nodule primordium arises from root inner cortex cell divisions. By contrast, the transient activity of the determinate nodule meristem results in globular shaped-nodules. Determinate nodule primordia tend to mostly originate from root outer cortex cell divisions (Guinel, 2009; Hirsch, 1992, Table 1). *M. truncatula* and *P. sativum* nodules are among the best characterized indeterminate nodules, and typical determinate nodules are well described in *Lotus japonicus (L. japonicus), Glycine max (G. max)* and *Phaseolus vulgaris* (*P. vulgaris*). The development of these legume nodules is described below.

Xiao et al. (2014) established a cellular fate map for *M. truncatula* indeterminate nodules. This fate map precisely describes the root cellular origins of the different nodule tissues. In this pioneer work, authors highlighted that the NCM originates from the third root cortex cell layer and that once this NCM starts functioning and producing NCM derived-cells, it marks the transition between a nodule primordium toward a functional nodule. The continuously active NCM adds cells to the nodule all along its lifespan (Hadri, Spaink, Bisseling, & Brewin, 1998; Lotocka, Kopcińska, & Skalniak, 2012). There are mainly two types of NCM-derived cells: bacteria infected and uninfected cells. The central zone of the nodule presents a developmental gradient from the youngest cells adjacent to the NCM to the oldest cells near the root attachment point. In their study, Xiao et al. (2014) also showed that uninfected cell layers at the very base of nodules derived from root endodermis and pericycle cells divisions, and that the first infected cell layers at the base of the nodule derived from the inner fourth and fifth root cortical cell layers.

In indeterminate nodules, from the nodule apex to its base, six well defined histological zones can be observed: the zone I corresponds to the meristematic apical region producing cells of the central tissues of the nodule. The zone II corresponds to the ITs penetration zone. The interzone II—III corresponds to a region where the bacteria are released into plant cells and where they differentiate into bacteroids. The zone III corresponds to the nitrogen-fixing region. The zone IV corresponds to a region where the bacteroids start to senesce. Finally, the zone V corresponds to the oldest infected part of the nodule, where bacteria turn into a saprophytic behavior. The zone V then contains non-differentiated free-living *rhizobia* that live at the expense of the host plant (Lotocka et al., 2012). These central tissues are surrounded by multiple peripheral uninfected tissue layers, consisting from the inside toward the outside of the nodule in the nodule parenchyma, the nodule endodermis and the nodule cortex (Guinel, 2009; Lotocka et al., 2012; Vasse, de Billy, Camut, & Truchet, 1990). The nodule vascular bundles (NVB) are located between the nodule parenchyma and the nodule endodermis, and are ontologically related to root pericycle and endodermis cell layers (Couzigou et al., 2012; Magne, Couzigou, Schiessl, Liu, George, Zhukov, et al., 2018). NVBs develop through the activities of different apical NVMs (Franssen et al., 2015; Magne, Couzigou, Schiessl, Liu, George, Zhukov, et al., 2018; Roux et al., 2014).

Determinate nodules can be also broadly subdivided into central and peripheral tissues. The determinate nodule primordium originates from sub-epidermal cell divisions taking place in the outer root cortex. Later, cell divisions also occur in both the root pericycle and inner cortex, several cells layers away from the initial divisions (Gauthier-Coles, White, & Mathesius, 2019; Hirsch, 1992). Cells derived from the outer root cortex cell divisions will give rise to the *rhizobia*-infected central tissues. The determinate nodule expansion appears to be the consequence of bacteroids multiplication in infected cells (Guinel, 2009). Peripheral tissues of determinate nodules are similar to those of indeterminate nodules, even though the parenchyma, endodermis and cortex peripheral tissues completely surround the nodule central tissues in mature nodules. In addition, by contrast to the NVMs of indeterminate nodules, which are continuously active throughout the whole nodule lifespan, the NVMs of determinate nodules are also determinate (Corby, 1988; Magne, George, Berbel Tornero, Broquet, Madueño, Andersen, et al. 2018).

4. Nod factors signaling-dependent activation of nodule organogenesis

The current knowledge regarding the molecular mechanisms underlying legume-*rhizobia* symbiotic interaction, from partners recognition until mature nodule development has been principally established using two model legumes, *M. truncatula* and *L. japonicus* (Udvardi, Tabata, Parniske, & Stougaard, 2005; Young et al., 2011). Plant-bacteria early signaling and nodule organogenesis are less described for actinorhizal plants (Froussart et al., 2016; Pawlowski & Demchenko, 2012). Studies based on legumes have shown the importance of the bacterial NF perception for the symbiotic partners specific recognition and association (Dénarié, Debellé, & Promé, 1996; Dénarié Jean & Debellé Frédéric, 1996; Oldroyd, 2013; Oldroyd, Murray, Poole, & Downie, 2011; Wang, Liu, & Zhu, 2018). As the NF signaling cascade controls the symbionts recognition, the *rhizobial* infection, and the nodule organogenesis, it is difficult to precisely define the specific roles of NF signaling genes in organogenesis. However, the identification of spontaneous nodulation mutants, forming nodule-like structures in the absence of *rhizobia*, and symbiotic gene overexpression studies, clearly showed a role for several actors of the NF signaling pathway (see below) in the initiation of the nodule organogenesis (Murray et al., 2007; Suzaki & Kawaguchi, 2014; Tirichine, Imaizumi-Anraku et al., 2006; Tirichine, James, Sandal, & Stougaard, 2006; Tirichine et al., 2007; Truchet, Camut, de Billy, Odorico, and Vasse (1989).

NF are perceived by plant specific NF receptors/perceptors recognition complexes (e.g., the *Mt*LYSIN MOTIF RECEPTOR-LIKE KINASE3, *Mt*LYK3 and the *Mt*NOD FACTOR PERCEPTION, *Mt*NFP in *M. truncatula* or the *Lj*NOD FACTOR RECEPTOR 1 and 5 *Lj*NFR1/ *Lj*NFR5 in *L. japonicus*). These proteins are receptor kinases containing LysM-domain (Arrighi et al., 2006; Limpens, Franken, Smit, & Willemse, 2003; Radutoiu et al., 2003; Radutoiu et al., 2007; Smit et al., 2007). These LysM receptor kinases are known to associate with the LEUCINE RICH-REPEAT RECEPTOR-LIKE KINASE (LRR-RLK) *Mt*DOES NOT MAKE INFECTIONS 2 (*Mt*DMI2, *Lj*SYMRK in *L. japonicus*) and to induce calcium oscillations in the nucleus. These calcium oscillations are perceived by *Mt*DOES NOT MAKE INFECTIONS3 (*Mt*DMI3) a CALCIUM CALMODULIN-DEPENDENT PROTEIN KINASE (CCaMK; Lévy et al., 2004; Mitra et al., 2004; Tirichine, Imaizumi-Anraku et al., 2006). This CCaMK is able to phosphorylate *Mt*INTERACTING PROTEIN OF DMI3 (*Mt*IPD3, *Lj*CYCLOPS in *L. japonicus*) and to form a complex with this latter (Cerri et al., 2017; Horváth et al., 2011; Singh, Katzer, Lambert, Cerri, & Parniske, 2014; Yano et al., 2008). *Mt*IPD3/*Lj*CYCLOPS represent key symbiotic actors for nodule organogenesis and can directly bind the promoter of the RWP-RK-type TF *NODULE INCEPTION* (*NIN*) essential for nodule organogenesis (Marsh et al., 2007; Singh et al., 2014) as well as the promoter of the *APETALA2/ ETHYLENE RESPONSIVE FACTOR* (*AP2/ERF*) TF, *ERF REQUIRED FOR NODULATION 1,* to activate their expressions (*ERN1*; Andriankaja et al., 2007; Cerri et al., 2017; Middleton et al., 2007). *Mt*ERN1 and *Mt*ERN2 are essential for nodule organogenesis since primordia do not form in the *Mtern1ern2* double mutant (Cerri et al., 2016). In addition, GRAS-type TFs, NODULATION SIGNALING PATHWAY 1 and 2 (NSP1 and NSP2; Heckmann et al., 2006; Hirsch et al., 2009; Kaló et al., 2005; Murakami et al., 2006; Smit, Raedts, & Portyanko, 2005), *Ps*SCARECROW-LIKE13 INVOLVED IN NODULATION1 (*Ps*SIN1, Battaglia et al., 2014) and the gibberellin signaling DELLA proteins (Fonouni-Farde et al., 2016; 2017; Jin et al., 2016), also act downstream the calcium spiking signaling and might potentially associate with *Mt*IPD3/*Lj*CYCLOPS to form a heteromeric regulatory complex required for organogenesis. NIN activates the *MtNUCLEAR FACTOR-YA1* (*MtNFYA1*) subunit gene encoding a CCAAT box-binding TF that interacts with DELLA proteins (Fonouni-Farde et al., 2016). Both NIN and NF—Y play important roles in the reactivation of cortical cell divisions

(Combier et al., 2006; Laporte et al., 2014; Soyano, Kouchi, Hirota, & Hayashi, 2013). NIN also activates the cytokinin receptor *MtCYTOKININ RESPONSE 1* (*MtCRE1*) gene expression (*LjLHK1* in *L. japonicus*). The perception of cytokinins via *Mt*CRE1 activates cytokinin responses notably in the root cortex to promote nodule organogenesis (Tirichine et al., 2007; Vernié et al., 2015).

The *de novo* nodule organogenesis is promoted in the root cortex by the NF-induced cytokinin signaling pathway which involves the cytokinin receptor *Mt*CRE1/*Lj*LHK1 and regulates and/or requires the expression of the TFs NIN, NSP1, NSP2, DELLA, NF−Y subunits, SIN1, ERN1 and ERN2 (Andriankaja et al., 2007; Battaglia et al., 2014; Cerri et al., 2017; Diédhiou & Diouf, 2018; Heckmann et al., 2006; Kaló et al., 2005; Marsh et al., 2007; Smit et al., 2005; Soyano & Hayashi, 2014; Stougaard, Schauser, Roussis, & Stiller, 1999; Vernié et al., 2015). Some of these transcription factors are post-transcriptionally regulated by microRNAs, which therefore also participate in modulating nodule organogenesis (Hossain et al., 2019; Lelandais-Brière, Moreau, Hartmann, & Crespi, 2016). The reactivation of cell divisions following NF signaling pathway also involves the reactivation of the cell cycle machinery (Breakspear et al., 2014; Kuppusamy et al., 2009).

Plant hormones also play a major role in nodule organogenesis (see above for cytokinin perception and the role of the gibberellin signaling DELLA proteins) and their roles in nodule organogenesis were recently reviewed by Liu, Zhang, Yang, Yu, and Wang (2018) and in the dedicated chapter dealing on hormonal regulations of nodulation. Briefly, cytokinins play a pivotal and positive role in the initiation of legume nodule organogenesis (Gamas, Brault, Jardinaud, & Frugier, 2017; Plet, Wasson, Ariel, Le Signor, Baker, Mathesius et al., 2011; Tirichine et al., 2007). Strigolactones and local accumulation of auxins also participate in nodule organogenesis, the latter being regulated depending on the CRE1-dependent cytokinin signaling (Gauthier-Coles et al., 2019; Kohlen, Ng, Deinum, & Mathesius, 2018; Ng et al., 2015; Ng & Mathesius, 2018). Gibberellins through DELLA proteins may function as a hub to integrate different hormonal signaling pathways, especially cytokinins, to modulate nodule organogenesis (Fonouni-Farde et al., 2017; McAdam, Reid, & Foo, 2018). In contrast, auxins rather than cytokinins are necessary for the actinorhizal nodule organogenesis (Froussart et al., 2016; Gauthier-Coles et al., 2019) likely in relation to their root-like structure. Other hormones are likely involved in nodule organogenesis; even though specific knowledge to conclude specifically about this is mostly lacking.

5. Symbiotic organ identity regulation

Following nodule initiation stages, molecular mechanisms supporting nodule development and defining nodule identity remain poorly understood. The nodule identity directly influences nodule growth and shape, and might partly explain the diversity of nodule structures observed among nodulating species. Recent genetic studies using either bacterial or plant mutants pointed out the root origin of legume nodules. Assuming a common evolutionary origin for the nodulation capacity of the Rosid I plants, actinorhizal and *Parasponia* nodules, resembling more to modified lateral roots, might represent more ancients and less sophisticated nodule structures (Behm et al., 2014; Pawlowski & Bisseling, 1996; Santi, Bogusz, & Franche, 2013; Svistoonoff, Hocher, & Gherbi, 2014).

The limited information related to nodule identity regulation has been obtained using legume species belonging only to the NPAAA clade. The alteration of the nodule identity is often characterized by nodule-to-root conversions, or by the formation of calli structures. Such phenotypes have also been observed in *M. sativa* and *Trifolium sp.* indeterminate nodules by increasing the growing temperature conditions (Dart, 1977; Ferguson & Reid, 2005) or following hairy root transformation in *Arachis* (Sinharoy, Saha, Chaudhury, & DasGupta, 2009). The molecular bases of these organ identity changes are unknown. Nodule-to-root conversions were also observed in *G. max* determinate nodules inoculated with *Bradyrhizobium japonicum* mutant strains altered for PhyR and EcfG locus which encode response regulators involved in various stresses tolerance (*ΔphyR*, *ΔecfG*; Gourion et al., 2009). Similarly, in *P. vulgaris*, inoculation with *Rhizobium elti* mutant strains presenting an auxotrophy for lysin (*lysA*) also resulted in nodules forming ectopic roots (Ferraioli, Tatè, Rogato, Chiurazzi, & Patriarca, 2004). These bacterial mutants were thus classified as Root INDucer (RIND) bacterial mutants (Ferraioli et al., 2004). An altered infection process caused by these bacterial mutants may alters phytohormonal levels in nodules, probably the cytokinin/auxin balance, and results in an asynchronous development of NCM and NVM, altering the proliferation rate of the different nodule tissues (Ferraioli et al., 2004; Gourion et al., 2009). *M. truncatula* infected by the *Sinorhizobium meliloti exoY* mutant also develops root-like structures with central vasculature suggesting that a successful bacterial infection is required for a proper nodule development (Guan et al., 2013). Similar phenotypes were also observed using the *M.*

truncatula Mtlin-4 and *Mtvapyrin-2* infection deficient mutants (Guan et al., 2013).

Several genes participating in root development and identity are expressed during legume nodule development. This suggests that root-related genes have been recruited during evolution to regulate nodule organogenesis and identity. In *M. truncatula,* the *MtPLETHORA* genes (*MtPLT1-4*) are differentially expressed in both NCM and NVM, and probably function redundantly to coordinate the activities of the different nodule meristems and to allow nodule meristem maintenance (Franssen et al., 2015). Similarly, the root-related *WUSCHEL-RELATED HOMEOBOX 5* (*WOX5*) TF, probably together with the *CLAVATA1* receptor and the *CLAVATA3/EMBRYO-SURROUNDING REGION* (*CLE*) peptide encoding genes were recruited during nodule development (Osipova et al., 2012). In addition, Di Giacomo et al. (2016) and Azarakhsh et al. (2015) showed that in *M. truncatula,* three class II *KNOTTED1-LIKE HOMEOBOX* (*KNOX*) genes (*MtKNOX3, MtKNOX5* and *MtKNOX9*) belonging to the *MtKNAT3/4/5-LIKE* sub-clade were linked to nodule organogenesis. Class II KNOX likely participate to the definition of nodule boundaries, shape and size. These authors suggested that these class II KNOX control cytokinin-related regulatory modules required for nodule development.

Among the *NON-EXPRESSOR OF PATHOGENESIS-RELATED PROTEIN1-LIKE* (*NPR1-LIKE*) gene family, members of the *NOOT-BOP-COCH-LIKE* (*NBCL*) sub-clade are essential regulators of plant development and shape. This clade was called *NBCL* because it includes the *M. truncatula MtNODULE-ROOT* (*MtNOOT*), the *Arabidopsis thaliana AtBLADE-ON-PETIOLE1* and *AtBLADE-ON-PETIOLE2* (*AtBOP1/2*), and the *P. sativum PsCOCHLEATA* (*PsCOCH*) genes (Couzigou et al., 2012). Beside few exceptions, most legume species possess a second *NBCL* gene grouping in a legume-specific sub-clade called *NBCL2* (Magne, Couzigou, Schiessl, Liu, George, Zhukov, et al., 2018). Among legumes, *L. japonicus* represents one of theses exceptions because it possesses a single *NBCL* copy called *LjNBCL1* or *LjCOCH1* (Liu, Lei, Liu, Weng, Lei, Hu, et al., 2018; Magne, George, Berbel Tornero, Broquet, Madueño, Andersen, et al., 2018). Genes belonging to the *NBCL* clades encode transcriptional co-factors containing a Bric-a-Brac Tramtrack Broad complex/POx virus and Zinc finger (BTB/POZ) domain, as well as an ankyrin repeats domain. *NBCL* genes play pleiotropic roles in both dicots and grasses development because one of their major functions is to regulate the establishment of the meristem-to-organ boundaries (MTOB), as well as to promote both lateral organs differentiation and identity acquisition (Hepworth & Pautot, 2015; Khan, Xu, & Hepworth, 2014; Wang, Hasson, Rossmann, & Theres, 2016; Žádníková & Simon, 2014). In legume plants, *NBCL* functions have been recruited for nodule formation and appear particularly important for the establishment of a well-organized symbiotic organ as well as for the robust maintenance of the nodule identity (Couzigou et al., 2012; Ferguson & Reid, 2005).

In *M. truncatula* and *P. sativum* indeterminate nodules, the loss-of-function of *NBCL1* genes does not significantly impact the early nodule infection process, however, the development of the nodule is drastically altered (Couzigou et al., 2012; Ferguson & Reid, 2005; Magne, Couzigou, Schiessl, Liu, George, Zhukov, et al., 2018). *nbcl1* mutant plants are characterized by the outgrowth of ectopic roots arising from NVMs. This suggests that the control of the root identity at the NVM is impaired in the absence of *NBCL1* genes (Couzigou et al., 2012; Couzigou, Mondy, Sahl, Gourion, & Ratet, 2013; Ferguson & Reid, 2005). This phenotype is in agreement with the *MtNOOT1* spatial gene expression pattern which specifically associates to the developing part of the NVB and to the distal NVM regions. In *nbcl1* mutants, the nodule-to-root conversion is asynchronous and can occur randomly from the nodule primordium formation, to later developmental stages, and even until nodules are mature (Couzigou et al., 2012). In *M. truncatula*, the paralog of *MtNOOT1, MtNOOT2,* is mostly expressed in the NCM and surrounds the NVMs. The *Mtnoot2* mutant does not present any symbiotic phenotype, however, the loss-of-function of both *MtNOOT1* and *MtNOOT2* genes leads to the complete loss of the nodule identity. This complete loss of nodule identity is characterized by the presence of only non-functional nodules converted into roots which are unable to host *rhizobia*. This revealed that the success of *rhizobia* accomodation is fully dependent of the nodule identity and that both *MtNOOT1* and *MtNOOT2* genes participate to the nodule identity establishment and maintenance (Magne, Couzigou, Schiessl, Liu, George, Zhukov, et al., 2018). In addition, this complete loss of nodule identity is associated to drastic changes in gene expression. Indeed, symbiotic genes expression is drastically decreased while defense and root apical meristem (RAM) marker genes expression is increased (Magne, Couzigou, Schiessl, Liu, George, Zhukov, et al., 2018). As an example, the expression of the RAM markers *MtCRINKLY4, MtSHORT-ROOT* and *MtPLETHORA2* was significantly up-regulated in *Mtnoot1* and *Mtnoot1noot2* nodules in agreement with the nodule-to-root identity shift. In addition to these gene expression changes, the spatial expression pattern of the class II *MtKNOX9* identity gene was modified in *Mtnoot1* and *Mtnoot1noot2* converted nodules. Such a drastic spatial gene expression modification highlights a major role of NBCL in the definition of nodule domains as well as in nodule identity acquisition (Magne, Couzigou, Schiessl, Liu, George, Zhukov, et al., 2018).

More recently, using *L. japonicus* as a model to study the determinate nodule identity, it was shown that determinate nodule identity was also *NBCL*-dependent. Indeed, despite the absence of a persistent NCM in determinate nodule, the loss-of-function of the unique *LjNBCL1* gene triggers the outgrowth of nodule ectopic roots arising from the NVMs. The loss of nodule identity in *L. japonicus* nodules is associated with the reorganization of the NVB vessels. This phenomenon consists in a switch from the xylem/phloem collateral organization typical of NVB toward an alternate xylem/phloem organization reminiscent of a root vascular bundle organization. *NBCL1* functions are thus not only conserved between *M. truncatula* and *P. sativum* indeterminate nodules, but also in *L. japonicus* determinate nodules, for the maintenance of NVB and NVM identities as well as for nodule identity (Couzigou et al., 2012; Magne, George, Berbel Tornero, Broquet, Madueño, Andersen, et al., 2018).

Until now, from the plant side, these unique nodule-to-root homeotic events have only been reported for *nbcl1* mutants, making *NBCL1* genes key regulators of nodule development and identity.

6. Concluding remarks

During evolution, plants from the Rosid I clade acquired a unique ability to induce a *de novo* organogenesis from their root tissues resulting *in fine* in the formation of a new organ dedicated to host symbiotic bacteria.

To control nodule organogenesis, plants recruited pre-existing hormonal pathways together with regulatory elements involved in the development of already existing plant organs. Depending on nodule types (actinorhizal and *Parasponia vs.* legumes), root or shoot organogenesis programs tend to be used. In addition to the nodule initiation process, plants recruited other gene functions to control nodule architecture. The peripheral *versus* central vascularization depending on nodule types may, for example, allow a higher colonization of the nodule and favor exchanges between the symbiotic partners. Among the main actors known to participate in nodule vasculature organization, *VAPYRIN* and *LIN* gene functions seem to be essential for the central to peripheral switch of the NVBs position.

Among the most spectacular mutant nodule phenotypes, the loss of nodule identity is observed in the *nbcl1* mutant plants. Despite obvious differences existing among nodule types in term of meristematic determinacy, the role of *NBCL1* genes is conserved between indeterminate and determinate nodules for the control of nodule development and identity maintenance. This suggests that *NBCL1*-dependent nodule identity evolved earlier, in an ancestral nodule-forming species, from which *Papilionoideae* indeterminate and determinate nodule-forming species derived and evolved separately. The current trend of a single evolutionary origin of root nodule symbiosis suggests an original and common molecular pathway with many different additions (specificities around a single theme) resulting in different nodule shapes. Emphasizing on other nodule structures apart from the classic NPAAA indeterminate/determinate nodules might give crucial evolutionary information on the organogenesis programs allowing Rosid I plants to build different nodule shapes.

To date, the concept of *NBCL*-dependent nodule identity has only been tested and reported in legume species from the *Fabeae*, *Trifolieae* and *Loteae* tribes belonging to the NPAAA clade (Table 1). To better understand the role that *NBCL* genes played in the evolution of nodule shapes, it would be interesting to extend the concept of *NBCL*-dependent nodule identity to other *Papilionoideae* species outside the NPAAA clade. It would be particularly fruitful to focus on species belonging to the *Dalbergioid*, *Crotalarioid* or *Genistoid* tribes forming atypical nodule structures, as well as on the *Swartzieae* group which contains basal *Papilionoideae* species. Our understanding of the link existing between *NBCL* genes and the evolution of nodule shapes might also benefit from investigations focusing on the role of *NBCL* genes in species belonging to the *Caesalpinioideae* sub-family which represents the only other *Leguminosae* sub-family containing confirmed nodulating plants. Moreover, exploring the roles of *NBCL* genes in actinorhizal and *Parasponia* nodules would be of high interest because such species, belonging to the more distant *Fagales*, *Cucurbitales* and *Rosales* orders, form nodules with central vasculature. Actinorhizal and *Parasponia* nodules might represent more ancient, primitive and less-sophisticated nodule shapes. In such nodules, the NVB did not yet acquired the ability to dichotomize and to become peripheric. In this context, we can ask whether nodules with central or peripheral vasculature represent two choices in evolution or if the central vasculature predates the peripheral one. The peripheral vasculature option has been largely maintained among the *Fabales* order because it probably favored exchanges between the plant and the symbiotic bacteria. In addition, such NVB structural evolution might have allowed an increased volume capacity of the nitrogen-fixing zone, resulting in a more performant symbiotic association.

References

Alunni, B., & Gourion, B. (2016). Terminal bacteroid differentiation in the legume-rhizobium symbiosis: Nodule-specific cysteine-rich peptides and beyond. *New Phytologist, 211*(2), 411–417. https://doi.org/10.1111/nph.14025.

Andrews, M., & Andrews, M. E. (2017). Specificity in legume-rhizobia symbioses. *International Journal of Molecular Sciences, 18*(4). https://doi.org/10.3390/ijms18040705.

Andriankaja, A., Boisson-Dernier, A., Frances, L., Sauviac, L., Jauneau, A., Barker, D. G., et al. (2007). AP2-ERF transcription factors mediate Nod factor dependent Mt ENOD11 activation in root hairs via a novel cis-regulatory motif. *The Plant Cell, 19*(9), 2866–2885. https://doi.org/10.1105/tpc.107.052944.

Ardley, J. K., Reeve, W. G., O'Hara, G. W., Yates, R. J., Dilworth, M. J., & Howieson, J. G. (2013). Nodule morphology, symbiotic specificity and association with unusual rhizobia are distinguishing features of the genus Listia within the southern African crotalarioid clade Lotononis s.l. *Annals of Botany, 112*(1), 1–15. https://doi.org/10.1093/aob/mct095.

Arrighi, J.-F., Barre, A., Ben Amor, B., Bersoult, A., Soriano, L. C., Mirabella, R., et al. (2006). The Medicago truncatula lysin motif-receptor-like kinase gene family includes NFP and new nodule-expressed genes. *Plant Physiology, 142*(1), 265–279. https://doi.org/10.1104/pp.106.084657.

Azarakhsh, M., Kirienko, a. N., Zhukov, V.a., Lebedeva, M.a., Dolgikh, E.a., & Lutova, L.a. (2015). KNOTTED1-LIKE HOMEOBOX 3: A new regulator of symbiotic nodule development. *Journal of Experimental Botany, 66*(22), 7181–7195. https://doi.org/10.1093/jxb/erv414.

Battaglia, M., Ripodas, C., Clua, J., Baudin, M., Aguilar, O. M., Niebel, A., et al. (2014). A nuclear factor Y interacting protein of the GRAS family is required for nodule organogenesis, infection thread progression, and lateral root growth. *Plant Physiology, 164*(3), 1430–1442. https://doi.org/10.1104/pp.113.230896.

Becking, J. H. (1984). Identification of the endophypte of dryas and rubus (rosaceae). *78*, 105–128.

Behm, J. E., Geurts, R., & Kiers, E. T. (2014). Parasponia: A novel system for studying mutualism stability. *Trends in Plant Science, 19*(12), 757–763. https://doi.org/10.1016/J.TPLANTS.2014.08.007.

Berg, R. H., Langenstein, B., & Silvester, W. B. (1999). Development in the Datisca-Coriaria nodule type. *Canadian Journal of Botany, 77*(9), 1334–1350. https://doi.org/10.1139/b99-076.

Berry, A. M., & Sunell, L. A. (1990). The infection process and nodule development. *The Biology of Frankia and Actinorhizal Plants*, 61–81. https://doi.org/10.1016/B978-0-12-633210-0.50009-5.

Boatwright, J. S., Wink, M., & van Wyk, B.-E. (2011). The generic concept of Lotononis (crotalarieae, fabaceae): Reinstatement of the genera euchlora, leobordea and listia and the new genus ezoloba. *Taxon, 60*(1), 161–177.

Breakspear, A., Liu, C., Roy, S., Stacey, N., Rogers, C., Trick, M., et al. (2014). The root hair "infectome" of medicago truncatula uncovers changes in cell cycle genes and reveals a requirement for auxin signaling in rhizobial infection. *The Plant Cell Online, 26*(12), 4680–4701. https://doi.org/10.1105/tpc.114.133496.

Cerri, M. R., Frances, L., Kelner, A., Fournier, J., Middleton, P. H., Auriac, M.-C., et al. (2016). The symbiosis-related ERN transcription factors act in concert to coordinate rhizobial host root infection. *Plant Physiology, 171*(2), 1037–1054. https://doi.org/10.1104/pp.16.00230.

Cerri, M. R., Wang, Q., Stolz, P., Folgmann, J., Frances, L., Katzer, K., et al. (2017). The ERN1 transcription factor gene is a target of the CCaMK/CYCLOPS complex and controls rhizobial infection in Lotus japonicus. *New Phytologist, 215*(1), 323–337. https://doi.org/10.1111/nph.14547.

Chen, H., Chou, M., Wang, X., Liu, S., Zhang, F., & Wei, G. (2013). Profiling of differentially expressed genes in roots of Robinia pseudoacacia during nodule development using suppressive subtractive hybridization. *PLoS One, 8*(6), e63930. https://doi.org/10.1371/journal.pone.0063930.

Coba de la Peña, T., Fedorova, E., Pueyo, J. J., & Lucas, M. M. (2018). The symbiosome: Legume and rhizobia Co-evolution toward a nitrogen-fixing organelle? *Frontiers of Plant Science, 8*, 1–26. https://doi.org/10.3389/fpls.2017.02229.

Combier, J.-P., Frugier, F., de Billy, F., Boualem, A., El-Yahyaoui, F., Moreau, S., et al. (2006). MtHAP2-1 is a key transcriptional regulator of symbiotic nodule development regulated by microRNA169 in Medicago truncatula. *Genes & Development, 20*(22), 3084–3088. https://doi.org/10.1101/gad.402806.

Corby, H. D. L. (1988). Types of rhizobial nodules and their distribution among the leguminosae. *National Herbarium & Botanic, 13*(1), 53–123.

Costa, J.-L., & Lindblad, P. (2002). Cyanobacteria in symbiosis with cycads. In *Cyanobacteria in symbiosis* (pp. 195–205). https://doi.org/10.1007/0-306-48005-0_11.

Couzigou, J. M., Mondy, S., Sahl, L., Gourion, B., & Ratet, P. (2013). To be or noot to be: Evolutionary tinkering for symbiotic organ identity. *Plant Signaling & Behavior, 8*(8), 8–10. https://doi.org/10.4161/psb.24969.

Couzigou, J.-M., Zhukov, V., Mondy, S., Abu el Heba, G., Cosson, V., Ellis, T. H. N., et al. (2012). NODULE ROOT and COCHLEATA maintain nodule development and are legume orthologs of Arabidopsis BLADE-ON-PETIOLE genes. *The Plant Cell, 24*(11), 4498–4510. https://doi.org/10.1105/tpc.112.103747.

Dart, P. J. (1977). Infection and development of leguminous nodules. In R. W. F. Hardy, & W. S. Silver (Eds.), *A treatise on dinitrogen fixation. Section III, Biology* (pp. 367–472). John Wiley & Sons. A Treatise on Dinitrogen Fixation. Section III, Biology https://www.cabdirect.org/cabdirect/abstract/19771935874. Retrieved from.

Dawson, J. O. (2008). Ecology of actinorhizal plants. *Nitrogen-fixing Actinorhizal Symbioses*, 199–234. https://doi.org/10.1007/978-1-4020-3547-0_8.

De souza Moreira, F. M., Da Silva, M. F., & De Faria, S. M. (1992). Occurrence of nodulation in legume species in the Amazon region of Brazil. In *New phytol* (Vol. 121). Retrieved from https://nph.onlinelibrary.wiley.com/doi/pdf/10.1111/j.1469-8137.1992.tb01126.x.

Dénarié, J., Debellé, F., & Promé, J.-C. (1996). Rhizobium lipo-chitooligosaccharide nodulation factors: Signaling molecules mediating recognition and morphogenesis. *Annual Review of Biochemistry, 65*(1), 503–535. https://doi.org/10.1146/annurev.bi.65.070196.002443.

Di Giacomo, E., Laffont, C., Sciarra, F., Iannelli, M. A., Frugier, F., & Frugis, G. (2016). KNAT3/4/5-like class 2 KNOX transcription factors are involved in Medicago truncatula symbiotic nodule organ development. *New Phytologist, 213*(2), 822–837. https://doi.org/10.1111/nph.14146.

Diédhiou, I., & Diouf, D. (2018). Transcription factors network in root endosymbiosis establishment and development. *World Journal of Microbiology and Biotechnology, 34*(3), 37. https://doi.org/10.1007/s11274-018-2418-7.

Dilworth, M. J., James, E. K., & Newton, W. E. (2008). In M. J. Dilworth, E. K. James, & W. E. Newton (Eds.), *Nitrogen-fixing leguminous symbioses:Nitrogen fixation: Origins, applications, and research progress* (Vol. 7). https://doi.org/10.1007/978-1-4020-3548-7.

Doyle, J. J. (1998). Phylogenetic perspectives on nodulation: Evolving views of plants and symbiotic bacteria. *Trends in Plant Science, 3*(12), 473–478. https://doi.org/10.1016/S1360-1385(98)01340-5.

Doyle, J. J. (2011). Phylogenetic perspectives on the origins of nodulation. *Molecular Plant-Microbe Interactions, 24*(11), 1289–1295. https://doi.org/10.1094/mpmi-05-11-0114.

Eaglesham, A. R. J., & Szalay, A. A. (1983). Aerial stem nodules on Aeschynomene spp. *Plant Science Letters, 29*(2–3), 65–272. https://doi.org/10.1016/0304-4211(83)90151-7.

de Faria, S. M., Sutherland, J. M., & Sprent, J. I. (1986). A new type of infected cell in root nodules of Andira spp. (leguminosae). *Plant Science, 45*(2), 143–147. https://doi.org/10.1016/0168-9452(86)90050-6.

Ferguson, B. J., & Reid, J. B. (2005). Cochleata: Getting to the root of legume nodules. *Plant and Cell Physiology, 46*(9), 1583–1589. https://doi.org/10.1093/pcp/pci171.

Fernandez-Lopez, M., Goormachtig, S., Gao, M., D'Haeze, W., Van Montagu, M, M. Van, & Holsters, M. (1998). Ethylene-mediated phenotypic plasticity in root nodule development on Sesbania rostrata. *Proceedings of the National Academy of Sciences of the United States of America, 95*(21), 12724–12728. https://doi.org/10.1073/pnas.95.21.12724.

Ferraioli, S., Tatè, R., Rogato, A., Chiurazzi, M., & Patriarca, E. J. (2004). Development of ectopic roots from abortive nodule primordia. *Molecular Plant-Microbe Interactions, 17*(10), 1043–1050. https://doi.org/10.1094/mpmi.2004.17.10.1043.

Fonouni-Farde, C., Kisiala, A., Brault, M., Emery, R. J. N., Diet, A., & Frugier, F. (2017). DELLA1-Mediated gibberellin signaling regulates cytokinin-dependent symbiotic nodulation. *Plant Physiology, 175*(4), 1795–1806. https://doi.org/10.1104/pp.17.00919.

Fonouni-Farde, C., Tan, S., Baudin, M., Brault, M., Wen, J., Mysore, K. S., et al. (2016). DELLA-mediated gibberellin signalling regulates Nod factor signalling and rhizobial infection. *Nature Communications,* 7(1), 12636. https://doi.org/10.1038/ncomms12636.

Franche, C., Laplaze, L., Duhoux, E., & Bogusz, D. (1998). Actinorhizal symbioses: Recent advances in plant molecular and genetic transformation studies. *Critical Reviews in Plant Sciences, 17*(1), 1–28. https://doi.org/10.1016/S0735-2689(98)00356-6.

Franssen, H. J., Xiao, T. T., Kulikova, O., Wan, X., Bisseling, T., Scheres, B., et al. (2015). Root developmental programs shape the Medicago truncatula nodule meristem. *Development, 142*(17), 2941–2950. https://doi.org/10.1242/dev.120774.

Frendo, P., Matamoros, M.a., Alloing, G., & Becana, M. (2013). Thiol-based redox signaling in the nitrogen-fixing symbiosis. *Frontiers of Plant Science, 4,* 1–10. https://doi.org/10.3389/fpls.2013.00376.

Froussart, E., Bonneau, J., Franche, C., & Bogusz, D. (2016). Recent advances in actinorhizal symbiosis signaling. *Plant Molecular Biology, 90*(6), 613–622. https://doi.org/10.1007/s11103-016-0450-2.

Gamas, P., Brault, M., Jardinaud, M.-F., & Frugier, F. (2017). Cytokinins in symbiotic nodulation: When, where, what for? *Trends in Plant Science, 22*(9), 792–802. https://doi.org/10.1016/j.tplants.2017.06.012.

Gauthier-Coles, C., White, R. G., & Mathesius, U. (2019). Nodulating legumes are distinguished by a sensitivity to cytokinin in the root cortex leading to pseudonodule development. *Frontiers of Plant Science, 9,* 1901. https://doi.org/10.3389/fpls.2018.01901.

Gehlot, H. S., Panwar, D., Tak, N., Tak, A., Sankhla, I. S., Poonar, N., et al. (2012). Nodulation of legumes from the Thar desert of India and molecular characterization of their rhizobia. *Plant and Soil, 357*(1–2), 227–243. https://doi.org/10.1007/s11104-012-1143-5.
González-Sama, A., Lucas, M. M., De Felipe, M. R., & Pueyo, J. J. (2004). An unusual infection mechanism and nodule morphogenesis in white lupin (Lupinus albus). *New Phytologist, 163*(2), 371–380. https://doi.org/10.1111/j.1469-8137.2004.01121.x.
Gourion, B., Sulser, S., Frunzke, J., Francez-Charlot, A., Stiefel, P., Pessi, G., et al. (2009). The PhyR-σEcfG signalling cascade is involved in stress response and symbiotic efficiency in Bradyrhizobium japonicum. *Molecular Microbiology, 73*(2), 291–305. https://doi.org/10.1111/j.1365-2958.2009.06769.x.
Gresshoff, P. M., Mens, C., Ferguson, B. J., Li, D., & Haaima, L. E. (2018). Local and systemic effect of cytokinins on soybean nodulation and regulation of their isopentenyl transferase (IPT) biosynthesis genes following rhizobia inoculation. *Frontiers of Plant Science, 9*, 1–14. https://doi.org/10.3389/fpls.2018.01150.
Guan, D., Stacey, N., Liu, C., Wen, J., Mysore, K. S., Torres-Jerez, I., et al. (2013). Rhizobial infection is associated with the development of peripheral vasculature in nodules of medicago truncatula. *Plant Physiology, 162*(1), 107–115. https://doi.org/10.1104/pp.113.215111.
Guinel, F. C. (2009). Getting around the legume nodule: II. Molecular biology of its peripheral zone and approaches to study its vasculature. *Botany, 87*(12), 1139–1166. https://doi.org/10.1139/b09-075.
Hadri, A.-E., Spaink, H. P., Bisseling, T., & Brewin, N. J. (1998). Diversity of root nodulation and rhizobial infection processes. *The Rhizobiaceae*, 347–360. https://doi.org/10.1007/978-94-011-5060-6_18.
Hafeez, F., Akkermans, A. D. L., & Chaudhary, A. H. (1984a). Observations on the ultrastructure of *Frankia sp.* in root nodules of *Datisca cannabina L. Plant and Soil, 79*, 383–402. https://doi.org/10.2307/42934645.
Hafeez, F., Akkermans, A. D. L., & Chaudhary, A. H. (1984b). Observations on the ultrastructure of Frankia sp. in root nodules of Datisca cannabina L. *Plant and Soil, 79*(3), 383–402. https://doi.org/10.1007/BF02184330.
Heckmann, A. B., Lombardo, F., Miwa, H., Perry, J. A., Bunnewell, S., Parniske, M., et al. (2006). Lotus japonicus nodulation requires two GRAS domain regulators, one of which is functionally conserved in a non-legume. *Plant Physiology, 142*(4), 1739–1750. https://doi.org/10.1104/pp.106.089508.
Hepworth, S. R., & Pautot, V. A. (2015). Beyond the divide: Boundaries for patterning and stem cell regulation in plants. *Frontiers of Plant Science, 6*, 1–19. https://doi.org/10.3389/fpls.2015.01052.
Hirsch, A. M. (1992). Developmental biology of legume nodulation. *New Phytologist, 122*(2), 211–237. https://doi.org/10.1111/j.1469-8137.1992.tb04227.x.
Hirsch, S., Kim, J., Munoz, A., Heckmann, A. B., Downie, J. A., & Oldroyd, G. E. D. (2009). GRAS proteins form a DNA binding complex to induce gene expression during nodulation signaling in medicago truncatula. *The Plant Cell Online, 21*(2), 545–557. https://doi.org/10.1105/tpc.108.064501.
Hocher, V., Auguy, F., Argout, X., Laplaze, L., Franche, C., & Bogusz, D. (2006). Expressed sequence-tag analysis in *Casuarina glauca* actinorhizal nodule and root. *New Phytologist, 169*(4), 681–688. https://doi.org/10.1111/j.1469-8137.2006.01644.x.
Horváth, B., Yeun, L. H., Domonkos, Á., Halász, G., Gobbato, E., Ayaydin, F., et al. (2011). Medicago truncatula IPD3 is a member of the common symbiotic signaling pathway required for rhizobial and mycorrhizal symbioses. *Molecular Plant-Microbe Interactions, 24*(11), 1345–1358. https://doi.org/10.1094/MPMI-01-11-0015.

Hossain, M. S., Hoang, N. T., Yan, Z., Tóth, K., Meyers, B. C., & Stacey, G. (2019). Characterization of the spatial and temporal expression of two soybean miRNAs identifies SCL6 as a novel regulator of soybean nodulation. *Frontiers of Plant Science, 10,* 475. https://doi.org/10.3389/fpls.2019.00475.

Ibáñez, F., Wall, L., & Fabra, A. (2017). Starting points in plant-bacteria nitrogen-fixing symbioses: Intercellular invasion of the roots. *Journal of Experimental Botany, 68*(8). erw387 https://doi.org/10.1093/jxb/erw387.

Imanishi, L., Perrine-Walker, F. M., Ndour, A., VayssiÃ̃res, A., Conejero, G., Lucas, M., et al. (2014). Role of auxin during intercellular infection of Discaria trinervis by Frankia. *Frontiers of Plant Science, 5,* 1–9. https://doi.org/10.3389/fpls.2014.00399.

Jiao, Y. S., Liu, Y. H., Yan, H., Wang, E. T., Tian, C. F., Chen, W. X., et al. (2015). Rhizobial diversity and nodulation characteristics of the extremely promiscuous legume Sophora flavescens. *Molecular Plant-Microbe Interactions, 28*(12), 1338–1352. https://doi.org/10.1094/MPMI-06-15-0141-R.

Jin, Y., Liu, H., Luo, D., Yu, N., Dong, W., Wang, C., et al. (2016). DELLA proteins are common components of symbiotic rhizobial and mycorrhizal signalling pathways. *Nature Communications,* 7, 1–14. https://doi.org/10.1038/ncomms12433.

Kaló, P., Gleason, C., Edwards, A., Marsh, J., Mitra, R. M., Hirsch, S., et al. (2005). Nodulation signaling in legumes requires NSP2, a member of the GRAS family of transcriptional regulators. *Science, 308*(5729), 1786–1789. https://doi.org/10.1126/science.1110951.

Kanu, S. A., & Dakora, F. D. (2017). Symbiotic functioning, structural adaptation, and subcellular organization of root nodules from Psoralea pinnata (L.) plants grown naturally under wetland and upland conditions in the Cape Fynbos of South Africa. *Protoplasma, 254,* 137–145. https://doi.org/10.1007/s00709-015-0922-2.

Khan, M., Xu, H., & Hepworth, S. R. (2014). BLADE-ON-PETIOLE genes: Setting boundaries in development and defense. *Plant Science: An International Journal of Experimental Plant Biology, 215*–216, 157–171. https://doi.org/10.1016/j.plantsci.2013.10.019.

Kohlen, W., Ng, J. L. P., Deinum, E. E., & Mathesius, U. (2018). Auxin transport, metabolism, and signalling during nodule initiation: Indeterminate and determinate nodules. *Journal of Experimental Botany, 69*(2), 229–244. https://doi.org/10.1093/jxb/erx308.

Kuppusamy, K. T., Ivashuta, S., Bucciarelli, B., Vance, C. P., Gantt, J. S., & Vandenbosch, K. A. (2009). Knockdown of *CELL DIVISION CYCLE16* reveals an inverse relationship between lateral root and nodule numbers and a link to auxin in *Medicago truncatula*. *Plant Physiology, 151*(3), 1155–1166. https://doi.org/10.1104/pp.109.143024.

Lalani Wijesundara, T. I., Van Holm, L. H. J., & Kulasooriya, S. A. (2000). Rhizobiology and nitrogen fixation of some tree legumes native to Sri Lanka. *Biology and Fertility of Soils, 30*(5–6), 535–543. https://doi.org/10.1007/s003740050033.

Laplaze, L., Duhoux, E., Franche, C., Frutz, T., Svistoonoff, S., Bisseling, T., et al. (2000). *Casuarina glauca* prenodule cells display the same differentiation as the corresponding nodule cells. *Molecular Plant-Microbe Interactions, 13*(1), 107–112. https://doi.org/10.1094/mpmi.2000.13.1.107.

Laporte, P., Lepage, A., Fournier, J., Catrice, O., Moreau, S., Jardinaud, M.-F., et al. (2014). The CCAAT box-binding transcription factor NF-YA1 controls rhizobial infection. *Journal of Experimental Botany, 65*(2), 481–494. https://doi.org/10.1093/jxb/ert392.

Lavin, M., Doyle, J. J., & Palmer, J. D. (1990). Evolutionary significance of the loss of the chloroplast-DNA inverted repeat in the leguminosae subfamily Papilionoideae. *Evolution, 44*(2), 390. https://doi.org/10.2307/2409416.

Lawrie, A. C. (1983). Infection and nodule development in *Aotus ericoides* (vent.) G. Don, a woody native Australian legume. *Journal of Experimental Botany, 34*, 1168–1180. https://doi.org/10.2307/23690644.
Lelandais-Brière, C., Moreau, J., Hartmann, C., & Crespi, M. (2016). Noncoding RNAs, emerging regulators in root endosymbioses. *Molecular Plant-Microbe Interactions, 29*(3), 170–180. https://doi.org/10.1094/MPMI-10-15-0240-FI.
Lévy, J., Bres, C., Geurts, R., Chalhoub, B., Kulikova, O., Duc, G., et al. (2004). A putative Ca2+ and calmodulin-dependent protein kinase required for bacterial and fungal symbioses. *Science, 303*(5662), 1361–1364. https://doi.org/10.1126/science.1093038.
Limpens, E., Franken, C., Smit, P., & Willemse, J. (2003). LysM domain receptor kinases regulating rhizobial Nod factor. *Science, 302*, 630–633.
Lindstrom, K., & Martinez-Romero, E. (2007). International committee on systematics of prokaryotes; subcommittee on the taxonomy of agrobacterium and rhizobium: Minutes of the meeting, 23-24 july 2006, arhus, Denmark. *International Journal of Systematic and Evolutionary Microbiology, 57*(6), 1365–1366. https://doi.org/10.1099/ijs.0.65255-0.
Liu, Y. C., Lei, Y. W., Liu, W., Weng, L., Lei, M. J., Hu, X. H., et al. (2018). LjCOCH interplays with LjAPP1 to maintain the nodule development in Lotus japonicus. *Plant Growth Regulation, 85*(2), 267–279. https://doi.org/10.1007/s10725-018-0392-6.
Liu, W. Y. Y., Ridgway, H. J., James, T. K., James, E. K., Chen, W.-M., Sprent, J. I., et al. (2014). Burkholderia sp. induces functional nodules on the South African invasive legume dipogon lignosus (phaseoleae) in New Zealand soils. *Microbial Ecology, 68*(3), 542–555. https://doi.org/10.1007/s00248-014-0427-0.
Liu, H., Zhang, C., Yang, J., Yu, N., & Wang, E. (2018). Hormone modulation of legume-rhizobial symbiosis. *Journal of Integrative Plant Biology, 60*(8), 632–648. https://doi.org/10.1111/jipb.12653.
Lotocka, B., Kopcińska, J., & Skalniak, M. (2012). Review article: The meristem in indeterminate root nodules of Faboideae. *Symbiosis (Philadelphia, Pa, 58*(1–3), 63–72. https://doi.org/10.1007/s13199-013-0225-3.
LPWG. (2013a). Legume phylogeny and classification in the 21st century: Progress, prospects and lessons for other species-rich clades. *Taxon, 62*(2), 217–248. https://doi.org/10.5167/uzh-78167.
LPWG. (2013b). Towards a new classification system for legumes: Progress report from the 6th International Legume Conference. *South African Journal of Botany, 89*, 3–9. https://doi.org/10.1016/j.sajb.2013.07.022.
LPWG. (2017). A new subfamily classification of the leguminosae based on a taxonomically comprehensive phylogeny. *Taxon, 66*(1), 44–77. https://doi.org/10.12705/661.3.
Madsen, L. H., Tirichine, L., Jurkiewicz, A., Sullivan, J. T., Heckmann, A. B., Bek, A. S., et al. (2010). The molecular network governing nodule organogenesis and infection in the model legume Lotus japonicus. *Nature Communications, 1*(1). https://doi.org/10.1038/ncomms1009.
Magne, K., Couzigou, J.-M., Schiessl, K., Liu, S., George, J., Zhukov, V., et al. (2018). MtNODULE ROOT1 and MtNODULE ROOT2 are essential for indeterminate nodule identity. *Plant Physiology, 178*(September), 00610.2018 https://doi.org/10.1104/pp.18.00610.
Magne, K., George, J., Berbel Tornero, A., Broquet, B., Madueño, F., Andersen, S. U., et al. (2018). Lotus japonicus NOOT-BOP-COCH-LIKE1 is essential for nodule, nectary, leaf and flower development. *The Plant Journal, 94*(5), 880–894. https://doi.org/10.1111/tpj.13905.
Marsh, J. F., Rakocevic, A., Mitra, R. M., Brocard, L., Sun, J., Eschstruth, A., et al. (2007). Medicago truncatula NIN is essential for rhizobial-independent nodule organogenesis induced by autoactive calcium/calmodulin-dependent protein kinase. *Plant Physiology, 144*(1), 324–335. https://doi.org/10.1104/pp.106.093021.

McAdam, E. L., Reid, J. B., & Foo, E. (2018). Gibberellins promote nodule organogenesis but inhibit the infection stages of nodulation. *Journal of Experimental Botany, 69*(8), 2117–2130. https://doi.org/10.1093/jxb/ery046.

Middleton, P. H., Jakab, J., Penmetsa, R. V., Starker, C. G., Doll, J., Kalo, P., et al. (2007). An ERF transcription factor in medicago truncatula that is essential for Nod factor signal transduction. *The Plant Cell Online, 19*(4), 1221–1234. https://doi.org/10.1105/tpc.106.048264.

Mitra, R. M., Gleason, C. A., Edwards, A., Hadfield, J., Downie, J. A., Oldroyd, G. E. D., et al. (2004). A Ca2+/calmodulin-dependent protein kinase required for symbiotic nodule development: Gene identification by transcript-based cloning. *Proceedings of the National Academy of Sciences, 101*(13), 4701–4705. https://doi.org/10.1073/pnas.0400595101.

Murakami, Y., Miwa, H., Imaizumi-Anraku, H., Kouchi, H., Downie, J. A., Kawaguchi, M., et al. (2006). Positional cloning identifies Lotus japonicus NSP2, a putative transcription factor of the GRAS family, required for NIN and ENOD40 gene expression in nodule initiation. *DNA Research, 13*(6), 255–265. https://doi.org/10.1093/dnares/dsl017.

Murray, J. D., Karas, B. J., Sato, S., Tabata, S., Amyot, L., & Szczyglowski, K. (2007). A cytokinin perception mutant colonized by rhizobium in the absence of nodule organogenesis. *Science, 315*, 101–104.

Ndoye, I., de Billy, F., Vasse, J., Dreyfus, B., & Truchet, G. (1994). Root nodulation of Sesbania rostrata. *Journal of Bacteriology, 176*(4), 1060–1068. https://doi.org/10.1128/jb.176.4.1060-1068.1994.

Newcomb, W., & Pankhurst, C. E. (1982). Fine structure of actinorhizal nodules of Coriaria arborea (Coriariaceae). *New Zealand Journal of Botany, 20*(2), 93–103. https://doi.org/10.1080/0028825X.1982.10426409.

Ng, J. L. P., Hassan, S., Truong, T. T., Hocart, C. H., Laffont, C., Frugier, F., et al. (2015). Flavonoids and auxin transport inhibitors rescue symbiotic nodulation in the medicago truncatula cytokinin perception mutant cre1. *The Plant Cell, 27*(8), 2210–2226. https://doi.org/10.1105/TPC.15.00231.

Ng, J. L. P., & Mathesius, U. (2018). Acropetal auxin transport inhibition is involved in indeterminate but not determinate nodule formation. *Frontiers of Plant Science, 9*, 169. https://doi.org/10.3389/fpls.2018.00169.

Oldroyd, G. E. D. (2013). Speak, friend, and enter: Signalling systems that promote beneficial symbiotic associations in plants. *Nature Reviews Microbiology, 11*(4), 252–263. https://doi.org/10.1038/nrmicro2990.

Oldroyd, G. E. D., Murray, J. D., Poole, P. S., & Downie, J. A. (2011). The rules of engagement in the legume-rhizobial symbiosis. *Annual Review of Genetics, 45*(1), 119–144. https://doi.org/10.1146/annurev-genet-110410-132549.

Op den Camp, R. H. M., Polone, E., Fedorova, E., Roelofsen, W., Squartini, A., Op den Camp, H. J. M., et al. (2012). Non-legume Parasponia andersonii deploys a broad rhizobium host range strategy resulting in largely variable symbiotic effectiveness. *Molecular Plant-Microbe Interactions, 25*(7), 954–963. https://doi.org/10.1094/mpmi-11-11-0304.

Op den Camp, R., Streng, A., De Mita, S., Cao, Q., Polone, E., Liu, W., et al. (2011). LysM-type mycorrhizal receptor recruited for rhizobium symbiosis in non-legume Parasponia. *Science, 331*(6019), 909–912. https://doi.org/10.1126/science.1198181.

Osipova, M. A., Mortier, V., Demchenko, K. N., Tsyganov, V. E., Tikhonovich, I. A., Lutova, L. A., et al. (2012). WUSCHEL-RELATED HOMEOBOX5 gene expression and interaction of CLE peptides with components of the systemic control add two pieces to the puzzle of autoregulation of nodulation. *Plant Physiology, 158*(3), 1329–1341. https://doi.org/10.1104/pp.111.188078.

Pawlowski, K., & Bisseling, T. (1996). Rhizobial and actinorhizal symbioses: What are the shared features? *The Plant Cell, 8*(10), 1899–1913. https://doi.org/10.2307/3870238.
Pawlowski, K., & Demchenko, K. N. (2012). The diversity of actinorhizal symbiosis. *Protoplasma, 249*(4), 967–979. https://doi.org/10.1007/s00709-012-0388-4.
Plet, J., Wasson, A., Ariel, F., Le Signor, C., Baker, D., Mathesius, U., et al. (2011). MtCRE1-dependent cytokinin signaling integrates bacterial and plant cues to coordinate symbiotic nodule organogenesis in *Medicago truncatula. The Plant Journal, 65*(4), 622–633. https://doi.org/10.1111/j.1365-313X.2010.04447.x.
Radutoiu, S., Madsen, L. H., Madsen, E. B., Felle, H. H., Umehara, Y., Grønlund, M., et al. (2003). *Plant recognition of symbiotic bacteria* (pp. 585–592).
Radutoiu, S., Madsen, L. H., Madsen, E. B., Jurkiewicz, A., Fukai, E., Quistgaard, E. M. H., et al. (2007). LysM domains mediate lipochitin-oligosaccharide recognition and Nfr genes extend the symbiotic host range. *The EMBO Journal, 26*(17), 3923–3935. https://doi.org/10.1038/sj.emboj.7601826.
Ren, G. (2018). *The evolution of determinate and indeterminate nodules within the Papilionoideae subfamily*. Retrieved from https://www.semanticscholar.org/paper/The-evolution-of-determinate-and-indeterminate-the-Ren/b81480b240d67d7636ed7f2224551ddd306ea79e.
Renier, A., De Faria, S. M., Jourand, P., Giraud, E., Dreyfus, B., Rapior, S., et al. (2011). Nodulation of Crotalaria podocarpa DC. by Methylobacterium nodulans displays very unusual features. *Journal of Experimental Botany, 62*(10), 3693–3697. https://doi.org/10.1093/jxb/err083.
Roux, B., Rodde, N., Jardinaud, M. F., Timmers, T., Sauviac, L., Cottret, L., et al. (2014). An integrated analysis of plant and bacterial gene expression in symbiotic root nodules using laser-capture microdissection coupled to RNA sequencing. *The Plant Journal, 77*(6), 817–837. https://doi.org/10.1111/tpj.12442.
Sachs, J. L., Ehinger, M. O., & Simms, E. L. (2010). Origins of cheating and loss of symbiosis in wild Bradyrhizobium. *Journal of Evolutionary Biology, 23*(5), 1075–1089. https://doi.org/10.1111/j.1420-9101.2010.01980.x.
Santi, C., Bogusz, D., & Franche, C. (2013). Biological nitrogen fixation in non-legume plants. *Annals of Botany, 111*, 743–767. https://doi.org/10.1093/aob/mct048.
Singh, S., Katzer, K., Lambert, J., Cerri, M., & Parniske, M. (2014). CYCLOPS, A DNA-binding transcriptional activator, orchestrates symbiotic root nodule development. *Cell Host & Microbe, 15*(2), 139–152. https://doi.org/10.1016/j.chom.2014.01.011.
Sinharoy, S., Saha, S., Chaudhury, S. R., & DasGupta, M. (2009). Transformed hairy roots of Arachis hypogea: A tool for studying root nodule symbiosis in a non–infection thread legume of the aeschynomeneae tribe. *Molecular Plant-Microbe Interactions, 22*(2), 132–142. https://doi.org/10.1094/MPMI-22-2-0132.
Smit, P., Limpens, E., Geurts, R., Fedorova, E., Dolgikh, E., Gough, C., et al. (2007). Medicago LYK3, an entry receptor in rhizobial nodulation factor signaling. *Plant Physiology, 145*(1), 183–191. https://doi.org/10.1104/pp.107.100495.
Smit, P., Raedts, J., & Portyanko, V. (2005). NSP1 of the GRAS protein family is essential for rhizobial Nod factor. *Science, 08250*, 1789–1792.
Soltis, D. E., Smith, S. A., Cellinese, N., Wurdack, K. J., Tank, D. C., Brockington, S. F., et al. (2011). Angiosperm phylogeny: 17 genes, 640 taxa. *American Journal of Botany, 98*(4), 704–730. https://doi.org/10.3732/ajb.1000404.
Soltis, D. E., Soltis, P. S., Morgan, D. R., Swensen, S. M., Mullin, B. C., Dowd, J. M., et al. (1995). Chloroplast gene sequence data suggest a single origin of the predisposition for symbiotic nitrogen fixation in angiosperms. *Proceedings of the National Academy of Sciences of the United States of America, 92*(7), 2647–2651. https://doi.org/10.1073/pnas.92.7.2647.

Soyano, T., & Hayashi, M. (2014). Transcriptional networks leading to symbiotic nodule organogenesis. *Current Opinion in Plant Biology, 20*, 146–154. https://doi.org/10.1016/j.pbi.2014.07.010.

Soyano, T., Kouchi, H., Hirota, A., & Hayashi, M. (2013). NODULE INCEPTION directly targets NF-Y subunit genes to regulate essential processes of root nodule development in Lotus japonicus. *PLoS Genetics, 9*(3), e1003352. https://doi.org/10.1371/journal.pgen.1003352.

Sprent, J. I., Sutherland, J. M., de Faria, S. M., Dilworth, M. J., Corby, H. D. L., et al. (1987). A century of nitrogen fixation research: Present status and future prospects. *Philosophical Transactions of the Royal Society Society of London. Series B: Biological Science, 317*(1184), 67–67 https://doi.org/10.1098/rstb.1987.0048.

Sprent, J. I. (1988). Which steps are essential for the formation of functional legume nodules? *New Phytologist, 111*(2), 129–153. https://doi.org/10.1111/j.1469-8137.1989.tb00675.x.

Sprent, J. (2000). Nodulation as a taxonomic tool. In P. S. Herendeen, & A. Bruneau (Eds.), *Advances in legume systematics*. Retrieved from https://www.press.uchicago.edu/ucp/books/book/distributed/A/bo9854731.html.

Sprent, J. I. (2007). Evolving ideas of legume evolution and diversity: A taxonomic perspective on the occurrence of nodulation: Tansley review. *New Phytologist, 174*(1), 11–25. https://doi.org/10.1111/j.1469-8137.2007.02015.x.

Sprent, J. I. (2008a). 60Ma of legume nodulation. What's new? What's changing? *Journal of Experimental Botany, 59*(5), 1081–1084. https://doi.org/10.1093/jxb/erm286.

Sprent, J. I. (2008b). Evolution and diversity of legume symbiosis. In *Nitrogen-fixing leguminous symbioses* (pp. 1–21). https://doi.org/10.1007/978-1-4020-3548-7_1.

Sprent, J. I. (2009). *Legume nodulation : A global perspective*. Retrieved from https://books.google.fr/books?id=c-DeQ_wQr3MC&pg=PA174&lpg=PA174&dq=nodulation+as+a+taxonomic+tool&source=bl&ots=n_Lfxi4XpO&sig=ACfU3U2f41GjwFvyZvoy5OHP5Bk41NA4IA&hl=fr&sa=X&ved=2ahUKEwjmjof81oLhAhX_gM4BHYgnCUsQ6AEwA3oECAUQAQ#v=onepage&q=nodulation as a taxonomic tool&f=false.

Sprent, J. I., Ardley, J. K., & James, E. K. (2013). From north to south: A latitudinal look at legume nodulation processes. *South African Journal of Botany, 89*, 31–41. https://doi.org/10.1016/j.sajb.2013.06.011.

Sprent, J. I., Ardley, J., & James, E. K. (2017). Biogeography of nodulated legumes and their nitrogen-fixing symbionts. *New Phytologist, 215*(1), 40–56. https://doi.org/10.1111/nph.14474.

Sprent, J. I., & James, E. K. (2007). Legume evolution: Where do nodules and mycorrhizas fit in? *Plant Physiology, 144*(2), 575–581. https://doi.org/10.1104/pp.107.096156.

Stougaard, J., Schauser, L., Roussis, A., & Stiller, J. (1999). A plant regulator controlling development of symbiotic root nodules. *Nature, 402*(6758), 191–195. https://doi.org/10.1038/46058.

Suzaki, T., & Kawaguchi, M. (2014). Root nodulation: A developmental program involving cell fate conversion triggered by symbiotic bacterial infection. *Current Opinion in Plant Biology, 21*, 16–22. https://doi.org/10.1016/j.pbi.2014.06.002.

Svistoonoff, S., Hocher, V., & Gherbi, H. (2014). Actinorhizal root nodule symbiosis: What is signalling telling on the origins of nodulation? *Current Opinion in Plant Biology, 20*, 11–18. https://doi.org/10.1016/j.pbi.2014.03.001.

Sy, A., Giraud, E., Jourand, P., Garcia, N., Willems, A., Lajudie, P. de, et al. (2001). Methylotrophic Methylobacterium bacteria nodulate and fix nitrogen in symbiosis with legumes. *Journal of Bacteriology, 183*(1), 214–220. https://doi.org/10.1128/JB.183.1.214-220.2001.

Tajima, R., Abe, J., Lee, O. N., Morita, S., & Lux, A. (2008). Developmental changes in peanut root structure during root growth and root-structure modification by nodulation. *Annals of Botany, 101*(4), 491–499. https://doi.org/10.1093/aob/mcm322.

Tirichine, L., Imaizumi-Anraku, H., Yoshida, S., Murakami, Y., Madsen, L. H., Miwa, H., et al. (2006). Deregulation of a Ca 2+/calmodulin-dependent kinase leads to spontaneous nodule development. *Nature, 441*(7097), 1153–1156. https://doi.org/10.1038/nature04862.

Tirichine, L., James, E. K., Sandal, N., & Stougaard, J. (2006). Spontaneous root-nodule formation in the model legume Lotus japonicus: A novel class of mutants nodulates in the absence of rhizobia. *Molecular Plant Microbe Interaction, 19*, 373–382. https://doi.org/10.1094/MPMI-19-0373.

Tirichine, L., Sandal, N., Madsen, L. H., Radutoiu, S., Albrektsen, A. S., Sato, S., et al. (2007). A gain-of-function mutation in a cytokinin receptor triggers spontaneous root nodule organogenesis. *Science, 2680*, 104–107. https://doi.org/10.1126/science.1132397.

Trinick, M. J. (1973). Symbiosis between Rhizobium and the non-legume, Trema aspera. *Nature, 244*, 459–460.

Truchet, G., Camut, S., de Billy, F., Odorico, R., & Vasse, J. (1989). The Rhizobium-legume symbiosis. Two methods to discriminate between nodules and other root-derived structures. *Protoplasma, 149*, 82–88. https://doi.org/10.1007/BF01322980.

Udvardi, M. K., Tabata, S., Parniske, M., & Stougaard, J. (2005). Lotus japonicus: Legume research in the fast lane. *Trends in Plant Science, 10*(5), 222–228. https://doi.org/10.1016/j.tplants.2005.03.008.

Valverde, C., & Wall, L. G. (1999). Regulation of nodulation in Discaria trinervis (Rhamnaceae)-Frankia symbiosis. *Canadian Journal of Botany, 77*(9), 1302–1310. https://doi.org/10.1139/b99-072.

Vasse, J., de Billy, F., Camut, S., & Truchet, G. (1990). Correlation between ultrastructural differentiation of bacteroids and nitrogen fixation in alfalfa nodules. *Journal of Bacteriology, 172*(8), 4295–4306. https://doi.org/10.1128/jb.172.8.4295-4306.1990.

van Velzen, R., Doyle, J. J., & Geurts, R. (2019). A resurrected scenario: Single gain and massive loss of nitrogen-fixing nodulation. *Trends in Plant Science, 24*(1), 49–57. https://doi.org/10.1016/j.tplants.2018.10.005.

van Velzen, R., Holmer, R., Bu, F., Rutten, L., van Zeijl, A., Liu, W., et al. (2018). Comparative genomics of the nonlegume Parasponia reveals insights into evolution of nitrogen-fixing rhizobium symbioses. *Proceedings of the National Academy of Sciences, 115*(20), E4700–E4709. https://doi.org/10.1073/pnas.1721395115.

Vernié, T., Kim, J., Frances, L., Ding, Y., Sun, J., Guan, D., et al. (2015). The NIN transcription factor coordinates diverse nodulation programs in different tissues of the *Medicago truncatula* root. *The Plant Cell, 27*(12), 3410–3424. https://doi.org/10.1105/tpc.15.00461.

Wang, Q., Hasson, A., Rossmann, S., & Theres, K. (2016). Divide et impera: Boundaries shape the plant body and initiate new meristems. *New Phytologist, 209*(2), 485–498. https://doi.org/10.1111/nph.13641.

Wang, Q., Liu, J., & Zhu, H. (2018). Genetic and molecular mechanisms underlying symbiotic specificity in legume-rhizobium interactions. *Frontiers of Plant Science, 9*, 1–8. https://doi.org/10.3389/fpls.2018.00313.

Werner, G. D.a., Cornwell, W. K., Sprent, J. I., Kattge, J., & Kiers, E. T. (2014). A single evolutionary innovation drives the deep evolution of symbiotic N2-fixation in angiosperms. *Nature Communications, 5*, 1–9. https://doi.org/10.1038/ncomms5087.

Wojciechowski, M. F., Sanderson, M. J., Steele, K. P., & Liston, A. (2000). In P. S.H., & A. Bruneau (Eds.), *Molecular phylogeny of the "temperate herbaceous tribes" of papilionoid legumes: A supertree approach* (p. 398). Retrieved from http://citeseerx.ist.psu.edu/viewdoc/download?doi=10.1.1.
Xiao, T. T., Schilderink, S., Moling, S., Deinum, E. E., Kondorosi, E., Franssen, H., et al. (2014). Fate map of Medicago truncatula root nodules. *Development, 141*(18), 3517–3528. https://doi.org/10.1242/dev.110775.
Yano, K., Yoshida, S., Muller, J., Singh, S., Banba, M., Vickers, K., et al. (2008). CYCLOPS, a mediator of symbiotic intracellular accommodation. *Proceedings of the National Academy of Sciences, 105*(51), 20540–20545. https://doi.org/10.1073/pnas.0806858105.
Young, N. D., Debellé, F., Oldroyd, G. E. D., Geurts, R., Cannon, S. B., Udvardi, M. K., et al. (2011). The Medicago genome provides insight into the evolution of rhizobial symbioses. *Nature, 480*(7378), 520–524. https://doi.org/10.1038/nature10625.
Žádníková, P., & Simon, R. (2014). How boundaries control plant development. *Current Opinion in Plant Biology, 17*, 116–125. https://doi.org/10.1016/j.pbi.2013.11.013.

CHAPTER SIX

Differentiation of symbiotic nodule cells and their rhizobium endosymbionts

Peter Mergaert*
Institute for Integrative Biology of the Cell, UMR9198, CEA, CNRS, Université Paris-Saclay, Gif-sur-Yvette, France
*Corresponding author: E-mail: peter.mergaert@i2bc.paris-saclay.fr

Contents

Abstract

The symbiotic nodule cells are at the heart of the rhizobium-legume symbiosis. They harbor thousands of nitrogen-fixing rhizobia, called bacteroids, within symbiosome vesicles. The symbiotic cells differentiate from meristem cells and this process is driven by endoreduplication and a transcriptome switch, establishing an optimal environment for the maintenance and the metabolic integration of the nitrogen-fixing bacteroids. The differentiating symbiotic cells uptake rhizobia by an endocytosis-like process and the resulting symbiosomes are matured by a specific exocytotic pathway. Finally, the rhizobia within the symbiosomes differentiate into bacteroids involving a metabolic transition controlled by the low oxygen conditions in the symbiotic cells. Depending on the host plant, the bacteroid differentiation involves in addition an irreversible morphological and cytological transition, which is driven by host-produced peptides, called Nodule-specific Cysteine-Rich peptides or NCRs. The parallel differentiation of the host cells and their bacterial endosymbionts results in a unique cellular configuration that is extremely efficient for symbiotic nitrogen fixation.

Advances in Botanical Research, Volume 94
ISSN 0065-2296
https://doi.org/10.1016/bs.abr.2019.09.008

1. Introduction

In response to nitrogen starvation and the presence in the nearby soil environment of specific compatible nitrogen-fixing rhizobium bacteria, legumes will engage into a symbiosis with these bacteria. The plants will form symbiotic organs on their roots, called nodules which are infected by the microsymbionts (Fig. 1). In the nodules, the rhizobia form in large quantities ammonia from nitrogen gas and transfer this reduced form of nitrogen to the plant that incorporates it in its metabolism and uses it for growth. Nodulated

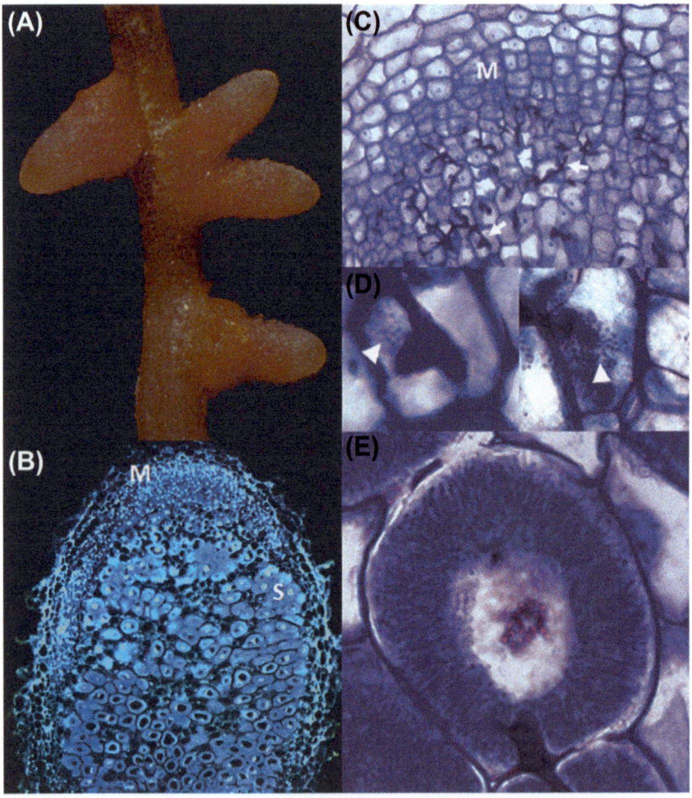

Fig. 1 *Medicago truncatula* root nodules. (A) Nodules on a root of a *M. truncatula* plant. (B) A nodule section stained with DAPI showing small dividing cells with small nuclei in the meristem (M), and symbiotic cells (S) with large nuclei in the central zone of the nodule. (C) Close-up of a nodule meristem and the underlying differentiating symbiotic cells that are crossed by infection threads (arrows). (D) Differentiating symbiotic cells with freshly released rhizobia (arrowheads). (E) A symbiotic nodule cell filled with radially aligned bacteroids.

legumes obtain most, if not all, of their nitrogen needs from their nitrogen-fixing symbionts. Nitrogen is one of the major elemental constituents of plant biomass, forth after carbon, hydrogen and oxygen (about 1%–2% of the dry biomass is nitrogen, compared to 20%–40% of carbon). Thus to sustain plant growth, the symbiosis must be capable of providing massive amounts of reduced nitrogen. This is achieved by producing large numbers of nodules in which very high amounts of bacteria are maintained (Fig. 1). At the heart of the nodules are the symbiotic nodule cells. These cells carry intracellularly the nitrogen-fixing rhizobia, called bacteroids (Fig. 1). A typical nodule will contain several thousands of symbiotic cells (Fig. 1B) and each symbiotic cell is infected with several thousand bacteroids (Fig. 1E). Thus, each nodule contains a few tens of millions of bacteria, the equivalent of a large plate-grown colony.

The heavily infected symbiotic nodule cells are among the most distinguishing features of nodules. Their formation, and also the overall nodule organogenesis, is based on similar processes in different legume lineages but it also diverges in quite some details. Here, I focus on the formation of symbiotic nodule cells in *Medicago truncatula* and closely related legumes of the Inverted Repeat-Lacking Clade (IRLC) but comparison with other legumes is made when relevant. In *M. truncatula*, the symbiotic nodule cells are formed from dividing nodule primordium cells or nodule meristem cells. Primordium or meristem cells that engage into the symbiotic cell formation pathway stop dividing and undergo endoreduplication that drives the differentiation process. Specific structures called infection threads conduct the rhizobia, starting from root hairs in the root epidermis, inside the tissues of the growing nodule (Fig. 1C) and release rhizobia inside the differenti ating symbiotic cells through an endocytotic process (Fig. 1D). The endocytotic uptake does not release rhizobia freely in the cytosol of growing symbiotic cells but inside organelle-like vesicles, called symbiosomes. Within the symbiosomes, the rhizobia grow and differentiate into bacteroids, the nitrogen-fixing form of the rhizobia. This latter process is enabled by the exocytosis pathway, which traffics specific proteins to the symbiosomes. The maintenance of thousands of bacterial endosymbiont-containing symbiosomes within individual plant cells necessitates the acquisition of a specific vesicular identity of the symbiosomes and avoidance of a vacuolar identity that would lead to the elimination of the symbiont. In addition, the symbiotic cells have a specific machinery that suppresses innate immunity responses against the large number of endosymbionts within them. The host cell-endosymbiont homeostasis also involves the production of symbiosis-specific peptides called Nodule-specific Cysteine-Rich peptides (NCRs), which target the bacteroids and keep them under check by forcing them into a terminally differentiated state. Finally, the functioning of the symbiotic cells requires the integration of the bacterial metabolism with the metabolism of the host cells. Major features of the metabolism of the symbiotic cells is the supply of carbon derived from shoot photosynthesis to the energy-devouring bacteroids, the maintenance of a low oxygen concentration to avoid the irreversible oxidation of the nitrogenase enzyme and the assimilation of ammonium provided by the nitrogen-fixing bacteroids. The plant forces the bacteroids to uncouple nitrogen-fixation from the bacterial nitrogen stress metabolism so that the rhizobia generate ammonia in large excess to their own needs, contrary to what they would do as autonomous bacteria. They release this ammonia to the benefit of the plant. This specific metabolism requires the transcriptional activation of a suite of enzymes and transporters and the massive production of leghemoglobin to bind oxygen. This general outline of the symbiotic nodule cell formation will be described in detail in the following paragraphs in which I will try to provide an integrated view of the differentiation of the symbiotic nodule cells, the symbiosomes and the bacteroids they contain.

2. Endoreduplication and infection driven differentiation of the symbiotic cells

Nodules are made *de novo* from fully differentiated root cells that, following the recognition of the rhizobial Nod factor signal, dedifferentiate and engage into the new developmental program of nodule organogenesis. In *M. truncatula*, the radial organization of roots consists, from the outside to the inside, in an epidermal cell layer, five cortical cell layers named C1 to C5, the endodermal cell layer, the pericycle cell layer and the vascular tissue. Based on sequential longitudinal sections of roots and nodule primordia, on the use of marker genes that allow the distinction of cells originating from different root tissues and on the use of nodulation mutants, a nodule cell fate map was constructed that precisely describes the origin of the different nodule tissues (Fig. 2A) (Xiao et al., 2014). This analysis showed that upon perception of the rhizobia and the initiation of infection in the epidermis, the first cell divisions occur in the pericycle and then spread into the cortex. Divisions are first anticlinal (perpendicular to the root axis adding cells within the same cell layer) and then, periclinal divisions happen that add new cell layers provoking the outgrowth of the incipient organ or

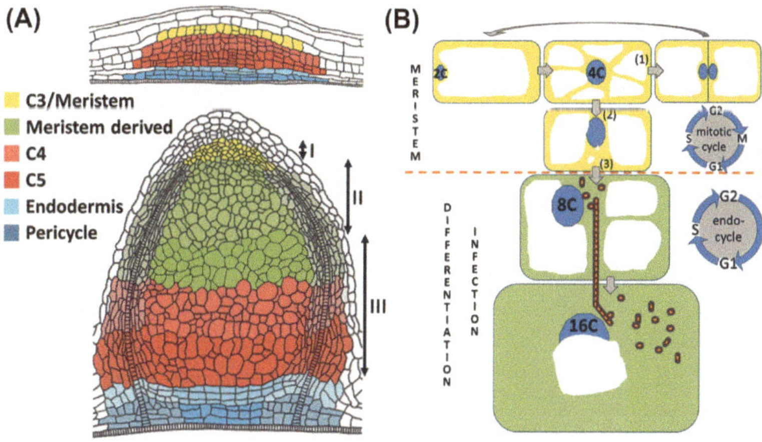

Fig. 2 Nodule organogenesis and differentiation of symbiotic cells. (A) Top: Primordium stage in which anticlinal and periclinal divisions occurred in the C3, C4, C5 cell layers and in the endodermis and pericycle, resulting in the first new cell layers and a visible outgrowth of the incipient organ. Bottom: Mature nodule with the six basal uninfected cell layers derived from the pericycle and endodermis, the eight most basal infected cell layers derived from the C4 and C5 root cortical layers and the remaining infected cells derived from the meristem. The color key shows the cell fate map of the primordium and nodule tissues. (B) Differentiation of symbiotic cells at the margin of the nodule meristem. Meristem cells that have replicated their genome (passed through the S phase of the cell cycle) have two possible fates: either they complete the cell cycle, go through mitosis (M phase) and divide to generate two new meristem cells (1) or they prepare for infection and differentiation into a symbiotic nodule cell (2). Entry into the differentiation pathway is triggered by the penetration of an infection thread and the endocytotic release of rhizobia, a process that is regulated by Nod factor production by the rhizobia in the infection threads and by the plant common symbiosis signaling pathway (3). The differentiation requires a cell cycle switch driven by the Ccs52A protein (4). The cell cycle switch produces polyploid and large symbiotic cells. Panel A is republished with permission of "The Company of Biologists Ltd", from Xiao et al., 2014.

primordium. While these divisions are ongoing, the infection threads cross the epidermis and the outer cortex and ramify as the branches of a tree. The C4 and C5 have formed about eight cell layers when the infection thread network reaches them and releases rhizobia in them. At this point, these cells stop dividing, switch to an endoreduplication cycle, and start the symbiotic cell differentiation process. The pericycle and endodermis have formed six to eight cell layers but these cell layers do not become infected. C3-derived cells that have not been infected form a multi-layered meristem. Next, the meristem starts to add cells to the nodule tissues, which on their turn become infected by infection threads and released rhizobia, upon which they switch to the endoreduplication-driven differentiation process. Several vascular bundles are established at the periphery of the incipient nodule, starting from the pericycle and endodermis cell layers. From this stage, the primordium has become a nodule. Thus in a mature nodule, the six to eight uninfected basal cell layers are derived from the pericycle and endodermis. The about eight most basal infected cell layers are derived from the C4 and C5 root cortical cell layers. The remaining infected cell layers are derived from the indeterminate nodule meristem and continuously increase in number with the nodule age due to the persistent meristematic activity (Fig. 2A).

Three-dimensional reconstructions of infection threads inside *M. truncatula* showed that the transition from a primordium stage to a mature nodule is also reflected in the architecture of the infection thread network (Monahan-Giovanelli, Pinedo, & Gage, 2006). The infection threads form tree-like networks in both primordia and mature nodules. These infection thread networks have relative constant characteristics in terms of branch lengths and branch density in nodules of different developmental stages but the overall direction of growth of the networks changes as nodules develop. In primordia, the network grows from the epidermis toward the inner C4- and C5-derived cell layers. However, sub-regions of this network are repolarizing their direction of development in response to the establishment of the nodule meristem. In mature nodules, the network is entirely reoriented and its branches grow now toward the meristem and away from the nodule base.

The central tissues of a mature *M. truncatula* nodule is usually described as partitioned in distinct zones: zone I is the meristem; zone II, derived from the meristem, is the symbiotic cell differentiation and infection zone where the infection thread network grows; and zone III is the fixation zone where the symbiotic cells are mature. The zone III is thus in fact composed of cells with two distinct origins, the most proximal cells are derived from the C4 and C5 cell layers of the primordium and the more distal cells are derived from the zone II (Fig. 2A) (Xiao et al., 2014).

The nodule meristem is organized in at least two subdomain types, a central meristem and several peripheral vascular meristems in which different sets of marker genes are expressed and in which the levels of the auxin and cytokinin phytohormones are distinct (Franssen et al., 2015; Magne et al., 2018). These subdomains give rise to the different tissues and cell types that compose the nodule. Cells located at the edge of the nodule meristem engender the peripheral tissues of nodules, the parenchyma, endodermis and cortex, and the vascular meristems generate vascular bundles that are located in the nodule parenchyma and that surround the central nodule tissue like the fingers of a hand holding a ball. The central meristem is the source of the symbiotic cells that constitute together with specialized uninfected cells the central zone of the nodules. The differentiation into symbiotic cells requires that the cells at the edge of the central meristem stop dividing and switch the mitotic cell cycle into an endoreduplication cycle, a modified cell cycle in which no mitosis and cytokinesis is taking place but repeated genome duplications are happening, forming polyploid cells (Fig. 2B) (Vinardell et al., 2003). Endoreduplication is triggered very early on, as soon as bacteria are released in the cells (Vinardell et al. 2003; Maunoury et al., 2010). The signal that triggers the cell cycle switch is unknown but the analysis of mutants in both *M. truncatula* and *Sinorhizobium meliloti*, its rhizobium symbiont, suggest that the trigger is related to infection. Indeed, the *vapyrin* and *lin M. truncatula* and *exoY S. meliloti* mutants form nodule-like structures without endoreduplication and in which infection threads are arrested in the epidermal cell layer (Maunoury et al., 2010; Guan et al. 2013; Xiao et al. 2014). Remarkably, the nodule-like structures formed by these mutants have a central vasculature instead of the normal peripheral bundles. Thus, it is as if the differentiation of the symbiotic cells inhibits the initiation of a vascular bundle in the pericyle and endodermis cells located immediately beneath them and shifts the vascular bundles to the periphery. At a later stage, in mature nodules, the NOOT1 and NOOT2 transcriptional regulators control the interaction between the central nodule meristem and the vascular meristems. The central meristem vanishes in the absence of these regulators and ectopic roots arise from the nodule vascular meristems (Magne et al., 2018). Another *M. truncatula* mutant, affected in the *IPD3* gene, connects the cell cycle switch of differentiation with the infection process. This mutant can form nodules with a meristem, peripheral vascular bundles and an extensive network of infection threads in its central zone without however releasing bacteria from them in the nodule cells. The cells of the central zone of *ipd3* nodules do not display endoreduplication either and are mostly in the 4C stage (four genome complements). Thus, in these nodules the central zone cells are arrested in the cell cycle but do not progress into an endoreduplication cycle. These cells do also not activate the characteristic transcriptome of the symbiotic cells (Maunoury et al., 2010). It seems thus from this, that penetration of an infection thread arrests division of meristem cells and that the endocytotic release of rhizobia is the trigger for endoreduplication, differentiation and the characteristic gene expression of symbiotic nodule cells. Some cells originating from the meristem in wild type nodules are not penetrated by the infection threads and thus never become infected by the rhizobia. These cells do not go into the endoreduplication cycles and can be seen as small cells in-between the large infected cells in the zone III (Fig. 1B). These uninfected cells have a specialized function in the metabolite transport to and from the infected cells (Limpens et al., 2013).

At the molecular level, several actors in the process of endoreduplication and endocytotic uptake have been identified. One of them is the cell cycle switch protein CCS52A. Master regulators of the eukaryotic cell cycle are cyclins that are active in different phases of the cell cycle and promote the transitions from one phase to the next. CCS52A is an activator of the anaphase promoting complex, an E3 ubiquitin ligase that targets cyclins for degradation by the 26S proteasome. CCS52A is expressed in the post-meristematic cells and promotes the degradation of mitosis-stimulating cyclins, thereby forcing the ongoing cell cycle in these cells into the endocycle (Fig. 2B) (Vinardell et al., 2003). When the *ccs52A* gene is knocked-down in nodules, the symbiotic cell formation is drastically affected, with lower ploidy levels, reduced cell size, inefficient invasion, and no maturation of bacteroids. Thus, endoreduplication is an integral part of normal symbiotic cell formation.

The so-called common symbiosis signaling pathway (CSSP) is a module of regulators that control the initial stages of nodulation (Geurts, Xiao, & Reinhold-Hurek, 2016). This module makes the plant response to the bacterial Nod factor signals possible in the epidermal root cells and is composed of a receptor complex and a signal transduction cascade that activate a downstream transcription factor (TF) network. Interestingly, using RNAi knock-down strategies that leaves the functioning of the *SymRK* and *NFP* components of the Nod factor receptor complex of the pathway unaffected in the root epidermis, allowing nodules to form, but that reduces strongly the activity of the genes at later stages, shows that the CSSP is required for the endocytotic uptake of rhizobia in the symbiotic cells (Limpens et al., 2005; Moling et al., 2014). Moreover, as mentioned above, mutants in the *IPD3* gene, encoding a TF of the pathway, can also not release rhizobia in the post-meristematic cells. The *ipd3* mutation does not block the epidermal function of the CSSP because of the presence of a second *IPD3* gene in *M. truncatula* that is sufficient for epidermal signaling but not for rhizobium uptake (Jin et al., 2018). From these observations follows that Nod factors produced by rhizobia in the infection threads activate the

CSSP to trigger rhizobium uptake (Moling et al., 2014). Thus, the CSSP seems to be a multipurpose regulatory module that is used to control nodulation in different stages of nodulation, including the first responses to Nod factors in the root epidermis and the endocytosis of rhizobia in the infection zone of the nodule adjacent to the meristem.

As discussed in detail in the next paragraph, the endocytotic release of rhizobia further requires exocytotic vesicle trafficking to the site of infection as well as a specific architecture of the actin cytoskeleton for the guidance of the vesicles. Exocytosis and the actin cytoskeleton are subsequently also primordial in the maturation of the symbiosomes. The exocytosis process and the actin cytoskeleton could constitute a link between infection and the cell cycle switch. The cell division, and in particular the cell plate assembly, as well as the symbiosome formation and maturation depend on the exocytosis pathway (Kanazawa & Ueda, 2017) and at least in part on the same protein players, rendering these processes possibly incompatible with each other. In addition, the actin cytoskeleton architecture required for rhizobium uptake and subsequent symbiosome maturation on the one hand and for cell division on the other hand might also be incompatible with each other. Thus, possibly cells cannot execute simultaneously division and symbiosome formation. However, in certain legumes as for example in *Aeschynomene* spp., nodule founder cells in the root cortex are infected by direct uptake of intercellular rhizobia, without the involvement of infection threads. These infected cells subsequently divide repeatedly to form the symbiotic nodule cells of the central zone of the nodules (Bonaldi et al., 2011). This infection process is thus fundamentally different from the infection thread-dependent process. It is further interesting to note that these symbiotic nodule cells in *Aeschynomene* spp. are not endoreduplicated.

3. Exocytosis pathway in symbiotic nodule cells

Infection threads are transcellular structures, delineated by a plasma membrane, which is aligned in its interior by a cell wall and a matrix imbedding the rhizobia. To release bacteria inside a symbiotic nodule cell, unwalled infection droplets containing rhizobia are budding off next to the tip of an infection thread (Fig. 3). These infection droplets release rhizobia enclosed in the symbiosome vesicles that have a membrane derived from the plasma membrane. Therefore, the interior of infection threads and symbiosomes are equivalent to the apoplastic compartment despite the

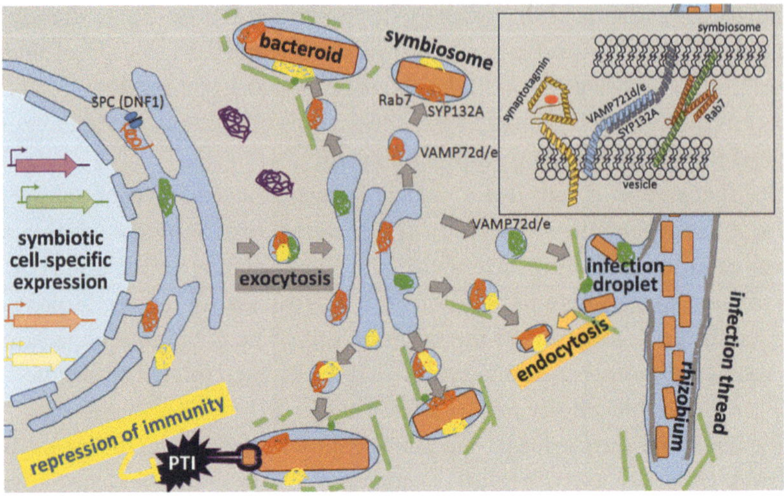

Fig. 3 Differentiation of bacteroids in symbiotic nodule cells. An infection thread releases rhizobia from unwalled infection droplets by endocytosis. The trafficking of secretory pathway vesicles to infection droplets and symbiosomes delivers nodule-specific cargo proteins (green, orange, and yellow globules). For example, cell wall degrading enzymes (green globules) are targeted to the droplets. The NCR bacteroid differentiation factors (orange globules) are targeted to the symbiosomes. Immune suppressor proteins (yellow) are targeted to the ER or the symbiosomes containing differentiating bacteroids. The actin cytoskeleton (green filaments) guide the secretory vesicles to the target organelles. The symbiotic cells produce also nodule specific soluble proteins (purple globules), for example leghemoglobin and enzymes for ammonia assimilation or carbon delivery to the symbionts. The inset shows a detailed view of the identified SNARE and Rab proteins involved in vesicle targeting to symbiosomes. The synaptotagmins bind calcium (red) to control the membrane fusion mediated by the SNARE proteins. The exocyst tethering complex is depicted in green. *Adapted with permission of American Society of Plant Biologists, from Mergaert, Kereszt, & Kondorosi, 2019; permission conveyed through Copyright Clearance Center, Inc.*

intracellular localization of these structures. Bacteria are repeatedly released in each symbiotic nodule cell, the symbiosomes initially also divide and in addition, they strongly enlarge before becoming mature (see below). The outcome is a symbiotic nodule cell that is completely filled with radially aligned symbiosomes (Fig. 1E). Thus, the proliferation of the symbiosomes in a symbiotic nodule cell is so extensive that it requires considerable cellular investment in trafficking proteins and lipids to these organelles. Circumstantial evidence strongly suggested that the secretory pathway orchestrates this massive colonization of the symbiotic nodule cells by the microsymbionts. Electron microscopy imaging for instance shows the abundant presence of endoplasmic reticulum (ER), Golgi bodies as well as small vesicles associated and fusing with symbiosomes (Hirsch, Bang, & Ausubel, 1983; Robertson & Lyttleton, 1982). Fluorescence immunolocalization of the ER shows a very strong signal in the developing and young symbiotic cells (Maunoury et al., 2010). Transcriptome analysis shows furthermore that symbiotic cells express a very high number of secretory proteins (Maunoury et al., 2010). Together, these observations show thus that the symbiotic nodule cells have a very strong secretory activity targeting the symbiosomes. Genetic and reverse genetic experiments combined with cell biology analyses reinforce this assumption and highlight the essential functions in symbiosome formation and maturation for multiple proteins of the secretory pathway (Fig. 3).

One of the first steps in handling secretory proteins is the cleavage of their signal peptide sequence in the ER by the signal peptidase complex (SPC) (Fig. 3). A forward genetic screen identified the mutant *dnf1* (*defective in nitrogen fixation 1*) of *M. truncatula* forming non-functional nodules (Wang et al., 2010). *DNF1* encodes SPC22, one of the subunits of the SPC. The *dnf1* mutation only affected nodule formation and no other phenotypes in the mutant plants were reported, despite the fact that the SPC is essential for viability in eukaryotes. The reason is that a second SPC22-encoding gene is present in the genome. Two additional components, SPC12 and SPC25, are also encoded by duplicated genes. Strikingly, *DNF1* and one copy of the SPC12-and SPC25-encoding genes are nearly exclusively expressed in nodules and are specifically active in the infection zone were symbiosomes are formed and grow. The other gene copies on the other hand have a ubiquitous expression explaining the absence of additional phenotypes in *dnf1*. The nodules of the *dnf1* mutant initially develop normally and release bacteria in symbiotic nodule cells, which become nearly completely filled with symbiosmes. However, these symbiosomes and the bacteria they contain remain unmodified and do not differentiate into their enlarged nitrogen-fixing form. At the molecular level, secretory proteins that are targeted to the symbiosomes in the wild type nodules, retain their signal peptide and are trapped in the ER in the *dnf1* nodules (Van de Velde et al., 2010). The *dnf1* mutant provides thus a genetic proof that the secretory pathway is crucial for the formation of functional symbiotic nodule cells in *M. truncatula*.

Additional genes encoding ER-localized components of the secretory pathway are upregulated in nodules and are co-expressed with the SPC genes in the in the infection zone of nodules (Maunoury et al., 2010). Among them are a gene encoding a signal peptide peptidase (SPP), an enzyme that degrades signal peptides after they have been cleaved off by the SPC; a protein disulfide isomerase (PDI) and ER oxidoreductin 1 (Ero1), both involved in the introduction of disulfide bridges in secretory proteins; a small family of nodule-specific secretory prolyl-4-hydroxylases possibly involved in modifications of prolyl residues of secretory proteins; and an ER lumen protein retaining receptor for sorting and retaining ER luminal proteins.

Exocytotic trafficking of proteins through the ER might have multiple final destinations, including the plasma membrane and the extracellular space, vacuoles, and in the symbiotic cells, thus also the symbiosomes. How the endomembrane trafficking process of a cell in general distinguish between possible destinations is determined by membrane identity markers that mediate specific vesicle delivery to corresponding target membranes. These proteins include the exocyst tethering factors and Rab (Ras-associated binding) GTPase proteins that promote the complex formation between vesicles and target membranes, and v-SNARE (Soluble N-ethylmaleimide-sensitive-factor Attachment protein REceptor) and t-SNARE complexes that mediate the fusion itself of the vesicle membrane with the target membrane, resulting also in the delivery of the vesicle cargo into the target compartment. Symbiosomes must thus have specific factors that contribute to the exact delivery of secretory proteins.

The *M. truncatula Syntaxin 132* (*SYP132*) gene encodes two isoforms of the t-SNARE protein SYP132, resulting from alternative splicing (Pan et al., 2016; Huisman et al., 2016). One of these transcripts, *SYP132C*, has a ubiquitous expression while *SYP132A* is most strongly expressed in symbiotic nodule cells. Both protein isoforms are nearly identical and differ only in their last 60 amino acids. The SYP132C isoform is mostly found on the cell periphery, in agreement with the known plasma membrane localization of the SYP132 t-SNARE in plants. In contrast, SYP132A is localized to the symbiosome membrane (Fig. 3) (Pan et al., 2016; but see also Huisman et al., 2016). RNAi experiments specifically targeting one or the other alternative transcript revealed that *SYP132C* is essential for plant growth while the knock-down of the *SYP132A* transcript resulted in the formation of symbiotic cells containing undifferentiated bacteroids, a phenotype very similar to the *dnf1* mutant. Presumably, the lack of SYP132A on the symbiosome membrane precluded vesicle fusion and thus provoked a blockage in the transport of the nodule-specific secretory proteins to the symbiosomes needed for symbiosome and bacteroid maturation.

Also on the side of the vesicles that target the symbiotic compartment, key v-SNARE proteins have been identified. In plants, v-SNAREs belong to the VAMP72 family. By genome mining seven VAMP72 proteins were identified in *M. truncatula* and phylogenetic analysis comparing the family in *M. truncatula* with the corresponding family in the non-symbiotic Arabidopsis plant combined with transcriptional analysis, restricted the candidate VAMP72s to the closely related VAMP721d and VAMP721e (Ivanov et al., 2012). Contrary to the *dnf1* mutant and *SYP132A* knock-down, knock-down of the *VAMP721d* and *VAMP721e* genes blocked nearly completely symbiosome formation. Ultrastructural analysis showed that this was because infection droplets in these RNAi lines contained a cell wall, hampering the release of the bacteria from infection droplets (Ivanov et al., 2012). Probably this is the consequence of the inability to deliver VAMP721d/e-vesicles containing cell wall degrading enzymes to the droplets (Gavrin et al., 2016). Immunolocalization and confocal or transmission electron microscopy further showed that the VAMP721d/e proteins are located on vesicles that are in the vicinity of infection droplets as well as symbiosomes, consistent with a targeted VAMP721d/e-mediated delivery of exocytotic vesicles to droplets but also to symbiosomes (Fig. 3). Interestingly, MtVAMP721d/e accumulated at the site of cell plate formation in meristem cells, in agreement with the earlier proposed common machinery for cytokinesis in dividing cells and for infection in symbiotic nodule cells (Kanazawa & Ueda, 2017).

Strikingly, VAMP72a, closely related to VAMP72d/e, is repressed in symbiotic nodule cells thanks to the symbiotic cell-specific transcriptional repressor RSD (Sinharoy et al., 2013). This repression is essential for the symbiosome maintenance because in an *rsd* mutant, which has elevated *VAMP72a* expression in the symbiotic cells, bacteroids are not fully developed and strong defense reactions are induced against them. Possibly, VAMP721a-bearing vesicles contain cargo different from that of VAMP721d/e-vesicles, and their fusion with symbiosomes in the *rsd* mutant delivers compounds that mount an immune response against the bacteria. In line with this hypothesis, Arabidopsis VAMP721/722, the closest homologs VAMP721a, are part of the secretory apparatus that targets vesicles containing defense compounds to the plasma membrane at sites of fungal pathogen infection (Kwon et al., 2008).

Other characterized components involved in vesicle fusion on the symbiotic interface are the calcium-sensing synaptotagmins that regulate the calcium-dependent vesicle fusions mediated by v-SNAREs and t-SNAREs

(Fig. 3) (Gavrin, Kulikova, Bisseling, & Fedorova, 2017). The *M. truncatula* synaptotagmins were localized nearby the tips of infection threads, unwalled droplets and symbiosomes and their knock-down by RNAi retarded rhizobium release and blocked symbiosome maturation. The Rab GTPase Rab7 is also located on symbiosomes and could thus be involved in the tethering of secretory vesicles to the symbiosomes. Accordingly, knock-down of *Rab7* gene expression hampers symbiosome and bacteroid differentiation (Limpens et al., 2009).

In summary (Fig. 3), in the infection process of the symbiotic cells, the v-SNAREs VAMP72d/e are required before SYP132A and Rab7, for the formation of unwalled droplets but subsequently all three proteins cooperate in the focal vesicle delivery to symbiosomes. Thus, the t-SNARE and Rab proteins on the infection droplet remain to be described and are likely to be different from SYP132A and Rab7. Synaptotagmins are likely involved in regulating the fusion of vesicles on the unwalled droplets and the symbiosomes. So far, no symbiosome-specific exocyst tethering complex has been identified. The *M. truncatula* genome contains multiple genes (42), encoding the Exo70 subunit that is anchored on the target membrane and the other subunits of the octameric exocyst complex are duplicated, complicating the identification of the symbiosome-specific subunits even if some of them have a nodule-enhanced expression.

The secretory pathway uses filamentous actin as a scaffold for vesicle transport. Thus, the actin cytoskeleton in the symbiotic cells must be of primordial importance to guide vesicles of the exocytosis pathway toward the ensemble of symbiosis organelles in the cell. In agreement, the actin cytoskeleton architecture reorganizes during the differentiation of the symbiotic nodule cells (Gavrin, Jansen, Ivanov, Bisseling, & Fedorova, 2015; Zhang et al., 2019). Dense filamentous actin clusters around growing infection threads and infection droplets; then the actin filaments reorganizes around immature symbiosomes, forming a network around them, until when short actin filaments and actin dots form around the mature and radially aligned symbiosomes. Strikingly, the microtubular cytoskeleton closely parallels the changes in the actin cytoskeleton during infection thread growth, droplet formation, bacterial release, and symbiosome positioning (Kitaeva, Demchenko, Tikhonovich, Timmers, & Tsyganov, 2016; Timmers, Auriac, de Billy, & Truchet, 1998). A key factor in the actin dynamics during the formation of the symbiotic cells is the ARP2/3 actin nucleation complex, which associates with maturing symbiosomes and might anchor the actin filaments to them. Silencing of the *ARP3* gene abolishes the differentiation of the symbiosomes (Gavrin et al., 2015). Another intriguing factor that might affect the actin cytoskeleton in the symbiotic nodule cells is queuosine, a modified nucleoside, produced by the bacteroids (Marchetti et al., 2013). Indeed, in a search for a suitable cellular experimental model system for bacterial uptake, it was found that *S. meliloti* can induce a strong actin cytoskeleton reorganization in human HeLa cells, and this response depends on bacterial queuosine production since *S. meliloti* mutants in the biosynthesis of this compound fail to induce the actin rearrangement. Interestingly, these mutants in queuosine synthesis form nitrogen-fixing nodules in which however the radial organization of the symbiosomes is lost, presumably because of an effect on the actin cytoskeleton of the host cell.

How the symbiosome organelle relates to lytic vacuoles is also partially understood thanks to the characterization of specific vacuolar marker proteins. Concomitantly with their gradual filling with symbiosomes, the volume of the vacuoles of the symbiotic nodule cells contracts temporarily and vacuoles are very small in young mature symbiotic cells but they become then larger again in older symbiotic nodule cells (Fig. 1) (Gavrin et al., 2014). This collapse of the vacuole permits the expansion of the symbiosomes and is most likely due to the suppression of the expression of the *VPS11* and *VPS39* genes encoding HOPS vacuolar tethering complex proteins. It could be of interest to check if this downregulation is mediated by the RSD repressor that downregulates VAMP72a. Moreover, the vacuoles of the infected nodule cells lose their functionality and are not acidic contrary to the vacuoles of uninfected cells. This defunctionalization of the vacuoles is further accompanied with retargeting of the aquaporin TIP1g from the tonoplast membrane to the symbiosome membrane. The relocalization of TIP1g appears to be essential for the maturation of symbiosomes because they did not differentiate in RNAi knock-down lines of *TIP1g*. Thus taken together, the symbiotic nodule cells adjust the vacuole features as a mechanism that facilitates symbiosome formation. Later on in the existence of nodules, during senescence, lytic compartments appear in the symbiotic nodule cells. Accordingly, *VPS11* and *VPS39* are expressed again in these senescing symbiotic cells. Concomitantly, vacuolar t-SNAREs, the syntaxins SYP22 and VTI11, start to accumulate on the symbiosome membrane, indicating that the identity of the symbiosome compartment is changing toward a lytic compartment heralding the degeneration of the bacteroids they contain (Limpens et al., 2009).

The ensemble of these studies starts to draw a picture of the mechanisms that allow the symbiotic cells to sort out the symbiosomes within the endomembrane system of the cell and deliver them the resources they need for their growth and later on for nitrogen fixation. But clearly the picture is still out of focus and many questions remain open. To name just one: we do not know how the cargo proteins for the symbiosomes are sorted from the cargo to be delivered to the extracellular space or other cellular compartments. The symbiosomes do not have a cell wall, contrary to the extracellular space and thus the cell wall synthesis machinery must be targeted specifically to the plasma membrane. Moreover, the plasma and symbiosome membranes have a different complement of small metabolite transporters and a large number of secretory proteins are transported specifically to the symbiosomes. As argued above, VAMP721d/e vesicles might contain a different cargo than the VAMP721a vesicles. We have at present no clue how these proteins are properly sorted and addressed to the right compartment. In this respect, it is intriguing that the *dnf1* mutation only affects symbiosome maturation but not the formation of the unwalled droplets or the cell wall. Thus, the DNF1 signal peptidase complex might be specifically involved in the processing of the secretory proteins that are destined to the symbiosomes, but how this can happen is a mystery. Maybe the DNF1 SPC complex is only localized in the ER complexes that are in vicinity of the symbiosomes.

Among the cargo proteins destined for the symbiosomes are a large number of proteins that are uniquely expressed in nodules, including thioredoxin s proteins, GRPs, NCRs, SNARPs, calmodulin-like proteins, Enod8, Enod16, Enod20, DNF2, SymCRK, nodulin 25, MtN22, MtN11, lipase/lipoxygenases (nodule-specific PLAT domain proteins or NPDs) and basic blue-like protein. The corresponding genes are mostly very highly expressed, suggesting that a major activity of the symbiotic cells is the protein trafficking to symbiosomes, in agreement with the highly developed ER in these cells. However, the function of most of these proteins is not or only poorly known. One of the exceptions is the NCR peptides that control the terminal differentiation of the bacteroids in the symbiosomes.

4. Terminal bacteroid differentiation

After their release from infection threads, the rhizobia within the symbiosomes become functional bacteroids implying an alteration in their physiology to adapt for nitrogen fixation. This physiological switch is triggered by the low oxygen levels prevailing in the symbiotic nodule cells and is made possible by a complete rewiring of the bacterial transcription affecting a major fraction of the genes (Roux et al., 2014). Moreover, in *M. truncatula*, the physiological transformation of the bacteroids is also accompanied with the remarkable process of terminal differentiation. This bacterial differentiation process is irreversible or terminal, because the bacteroids are unable to resume growth when released from the nodules. The terminal bacteroids are characterized by a partially permeabilized cell membrane and by a very strong cell elongation leading to the formation of giant bacterial cells (Mergaert et al., 2006). Strikingly, these large bacteroids are polyploid, just like the symbiotic nodule cells in which they reside. Terminal differentiation of bacteroids is typical for particular legume clades, *e.g.* the IRLC to which *M. truncatula* belongs, or the Dalbergioid clade with *Aeschynomene* spp. However, it is not common to all legumes. In these other legumes, bacteroids are in a reversible state, not displaying any of the distinctive features of terminally differentiated bacteroids. Their morphology, ploidy and membrane features are similar to the free-living rhizobia. This implies that the process of terminal differentiation is not an absolute requirement for obtaining functional nitrogen-fixing bacteroids, but it was proposed to improve the efficiency of the symbiosis (Mergaert, 2018). Phylogenetic analyses and a range of experiments that compare symbiotic performances of terminally differentiated bacteroids with less or non-differentiated ones confirm this hypothesis (Kazmierczak et al., 2017; Lamouche et al., 2019a, 2019b; Oono & Denison, 2010; Oono, Schmitt, Sprent, & Denison, 2010; Sen & Weaver, 1984). However, it is unclear at present why these large, partially permeabilized, polyploid bacteroids would be more performant than the undifferentiated ones.

The terminal differentiation is triggered by the above-mentioned family of antimicrobial peptides (AMPs) produced by the symbiotic nodule cells, called the nodule-specific cysteine-rich (NCR) peptides (Van de Velde et al., 2010). Many lines of evidence support this. *NCR* genes are only found in the so-far analyzed legumes that have terminally differentiated bacteroids (Czernic et al., 2015; Mergaert et al., 2003). Within the IRLC clade, the extent of bacteroid differentiation varies depending on the plant and is correlated with the number of *NCR* genes in its genome, especially with the number of cationic ones (Montiel et al., 2017). In *M. truncatula* nodules, the *NCR* expression pattern overlaps with the differentiating and mature bacteroids (Guefrachi et al., 2014). Also in other IRLC legumes or in Dalbergioid legumes, tested *NCR* genes are only expressed in the bacteroid-containing nodule cells. Moreover, the NCR peptides accumulate in bacteroids. In agreement with this, the NCRs are indeed secretory peptides and their transport depends thus on the secretory pathway. In the *dnf1* mutant, the NCR peptides are retained in the ER and this correlates with the absence of bacteroid differentiation in this mutant as described above. In addition, mutants in some individual *NCR* genes are affected in bacteroid differentiation or in bacteroid persistence (see below). *In vitro* treatment of *S. meliloti* with synthetic NCRs resulted in bacteroid-like cells with permeabilized membranes, arrested division, cell elongation and DNA multiplication (Van de Velde et al., 2010). Most strikingly, the expression of *NCR* genes in *Lotus japonicus* plants, a legume that does not possess NCR genes and normally has reversible bacteroids, leads to bacteroids in the transgenic plants with features of terminal differentiation. And finally, rhizobium mutants that are affected in their response to NCR peptides are inept to differentiate into bacteroids (see below).

The *NCR* gene family is notable in several ways. In *M. truncatula*, over 600 *NCR* genes were identified in the genome, and in several other species of the IRLC clade, many hundreds of them are present as well (Mergaert et al., 2003; Montiel et al., 2017). Also in *Aeschynomene* species, several tens of *NCR* genes are expressed in nodules (Czernic et al., 2015; Gully et al., 2018). A transcriptome analysis in *M. truncatula*, using data of a very large number of experiments, found that the *NCR* genes are only expressed in nodules where they are moreover among the most highly expressed genes (Guefrachi et al., 2014). In addition, all *NCR* genes are specifically expressed in the symbiotic cells (except for two). The orthologous *NCR* genes in different *M. truncatula* accessions showed a high degree of nucleotide variation with signatures of diversifying selection pressures (Nallu, Silverstein, Zhou, Young, & Vandenbosch, 2014), possibly reflecting an adaption to the diversity of possible rhizobium symbionts. This could be one of the underlying reasons for the surprising large size of the *NCR* gene family. But other factors might have contributed as well. One of them is that the *NCR* genes are not functionally equivalent and are expressed in different phases of the symbiotic cell and bacteroid differentiation processes. Moreover, many of them are only expressed when bacteroids are already mature and fix nitrogen, suggesting that these "late" genes are involved in the maintenance of the bacteroid state rather than in the formation of it. Thus, it is expected that NCR peptides provoke many physiological changes in the bacteroids. Finally, it is possible that deleterious side-effects on bacterial physiology of particular NCR peptides was compensated by additional peptides, also expanding the family (see below).

The NCR peptides have features shared with AMPs and at least some NCRs, and in particular the cationic ones, can *in vitro* kill or inhibit the growth of not only the rhizobium symbionts but also many other bacteria and even fungi (Farkas, Pap, Kondorosi, & Maróti, 2018). Their major antibacterial activity results from their capacity to disturb the integrity of the inner and outer membranes of bacteria, leading to a loss of the membrane potential (Van de Velde et al., 2010). In addition, NCR peptides have intracellular targets, binding to ribosomes and inhibiting translation (Farkas et al., 2014). However, eliminating the endosymbionts is evidently not what happens in the symbiotic nodule cells and the bacteroids remain metabolically active for very long despite the deployment of this impressive NCR arsenal. How the antimicrobial activity of the NCR peptides relates to their *in vivo* action and induces terminal bacteroid differentiation is still unclear. Possibly, many factors within the symbiotic nodule cells and symbiosomes contribute to modulate the antimicrobial activity of the peptides to a much more subtle but complex effect on the endosymbiont physiology.

Indeed, endosymbiont adaptations contribute to the bacteria's ability to endure exposure to the NCR peptides and to colonize stably the symbiotic cells. Among them is the BacA peptide transporter in *S. meliloti* (known as BclA in the bradyrhizobia infecting the NCR-producing *Aeschynomene*). BacA-deficient rhizobia are hypersensitive to the killing activity of NCR peptides. Such mutants induce nodules and infect symbiotic cells similarly as in the wild-type. However, in NCR-producing symbiotic cells, the bacteria die as soon as they are released in the symbiotic cells and this death can be avoided by blocking NCR transport to the infecting rhizobia in the *M. truncatula dnf1* mutant (Barrière et al., 2017; Guefrachi et al., 2015; Haag et al., 2011). On the other hand, the BacA or BclA proteins, which are widespread and found in all rhizobia and even in other non-symbiotic bacteria, have no function in symbiosis in NCR-lacking legumes. BacA and BclA proteins are ABC peptide transporters with a broad spectrum of peptide import activity and they can promote the uptake of NCR peptides, suggesting that the transport of NCR peptides through BacA or BclA provides resistance. Other bacterial envelope components, including outer membrane proteins, LPS, EPS, peptidoglycan and hopanoids, are additional potential factors that contribute to NCR resistance and terminal bacteroid differentiation (Mergaert, 2018). For example, rhizobial EPS interacts with NCR peptides and is proposed to form a shield that can reduce the harmful interaction of NCR peptides with the bacterial membrane (Arnold, Penterman, Shabab, Chen, & Walker, 2018; Montiel et al., 2017).

Moreover, transcriptome and Tn-seq (transposon sequencing) analyses of NCR-treated cells and NCR-protein interaction studies identified a whole suite of additional candidate NCR-responsive functions in *S. meliloti* (Arnold et al., 2017; Farkas et al., 2014; Penterman et al., 2014). Together, these bacterial functions may contribute to alleviating the stress on the bacteroids induced by the NCR peptides. This suggests that the NCR peptides can act as determinants of compatibility between the host plant and the endosymbiont and that the peptides can function in the elimination of unadapted rhizobium strains. That this is truly part of the role of NCR peptides is illustrated by two *NCR* genes, named *NFS1* and *NFS2* that both exist in two allelic forms in different *M. truncatula* accessions (Wang et al., 2017; Yang et al., 2017). In the presence of a particular allele of *NFS1* or *NFS2*, some *S. meliloti* strains are not able to differentiate into functional bacteroids while other strains are not sensitive to these *NCR* alleles and differentiate normally. On the other hand, the sensitive strains are fully competent when infecting *M. truncatula* lines expressing other alleles of the *NFS1* and *NFS2* genes. This suggests that the *NFS1* and *NFS2* alleles leading to incompatible interactions encode peptides that directly or indirectly kill unadapted rhizobia. The sensitive strains produce a slightly different EPS than the compatible ones and transferring EPS biosynthesis genes from a compatible strain to the incompatible one overcomes the incompatibility (Simsek, Ojanen-Reuhs, Stephens, & Reuhs, 2007). These *NCR* alleles thus highlight how the bacterial adaptation to the NCR-profile of a host plant determines the compatibility or incompatibility in a given symbiont-host genotype-genotype interaction. Some rhizobium species can even escape altogether the NCR-mediated reproductive inhibition and terminal bacteroid differentiation by producing the M16A family metallopeptidase HrrP capable of degrading NCR peptides (Price et al., 2015). HrrP expressing strains proliferate more abundantly within the nodules without fixing nitrogen for the host and have thus shifted from a cooperative to a parasitic behavior.

Well-adapted rhizobia resist the NCR challenge and establish chronically within the symbiotic nodule cells. They nevertheless do not emerge unaffected and rather undergo an irreversible physiological shift. The NCR peptides provoke a cell cycle transition, resulting from an inhibition of the bacterial cytokinesis, while at the same time DNA replication is maintained leading to polyploidy (Mergaert et al., 2006). The central cell cycle regulator CtrA has been proposed as a key target for the NCR peptides that promote this cell cycle switch (Penterman et al., 2014; Pini et al.,

2015). Besides the cell cycle, many other cellular components were found to interact with NCR peptides, including the chaperone GroEL, which is also required for bacteroid differentiation, the metabolic enzyme pyruvate dehydrogenase, the F-ATP synthase and the nitrogenase enzyme complex (Farkas et al., 2014). These observations are compatible with the view that the NCR peptides interfere with a broad panel of bacterial functions. It is possible that the differentiation of the bacteroids by cell enlargement and polyploidization is a secondary effect induced by the AMP stress, while the primary effect is a manipulation of the metabolism of the endosymbiont in order to mold the bacterial metabolism for optimal nitrogen fixation. Two *M. truncatula* mutants, *dnf4* and *dnf7* (*defective in nitrogen fixation 4* and *7*), disrupted in the NCR211 and NCR169 encoding genes, respectively, form non-functional nodules in which bacteroids partially or even completely differentiate but they do not persist (Horváth et al., 2015; Kim et al., 2015). This was an unexpected finding, firstly because a high degree of redundancy was *a priori* expected in the very large *NCR* gene family. And secondly, because in the NCR169 and NCR211 mutant plants bacteroids differentiate, or at least partially differentiate, one has to conclude that the function of these peptides is not related to the cell cycle switch. Because well-adapted rhizobium strains die in the NCR169 and NCR211-deficient plants, the function of these peptides cannot be the elimination of unadapted bacteria either, as the restrictive *NFS1/2* alleles are doing. Thus, their role should be related to the interference with another cellular process. How to explain mechanistically the phenotype of these two mutants is presently not possible, but one hypothesis is that the NCR169 and NCR211 peptides have essential antagonistic or compensating activities to other NCR peptides that are toxic when not counteracted. Identification of suppressor mutations in either *M. truncatula* or *S. meliloti* that restores the symbiotic activity of the *dnf4* and *dnf7* mutants could be helpful to settle this question.

This picture of NCR peptides interfering with a broad range of bacterial functions is based on the study of a few peptides only, suggesting that the full scale of interactions of the NCR armamentarium is enormous. Moreover, this complexity could even be enhanced by mechanisms that modify the NCR peptides or their activity. The conserved cysteine residues of the NCRs likely form disulfide bridges during their passage through the oxidizing environment of the ER. The earlier mentioned Ero1 and PDI-encoding genes that are upregulated in the symbiotic nodule cells could contribute to the oxidation of NCRs in the ER. On the other hand, the symbiotic cells produce also a set of particular thioredoxins, *i.e.* proteins that catalyze the reduction of disulfide bridges in proteins to the free cysteines. These nodule-specific thioredoxins are coexpressed with the NCR peptides and are secretory proteins targeted to the symbiosomes and localized in the bacterial cytosol. They reduce NCR peptides *in vitro* and knocking down their expression by RNAi reduces significantly the level of terminal bacteroid differentiation while expressing the thioredoxin gene in the *S. meliloti* symbiont complements the bacteroid defect in the RNAi lines (Ribeiro et al., 2017). This complementation supports the hypothesis that the thioredoxins reduce NCR peptides within the bacteroids and that the redox balance of NCR peptides is important. Accordingly, the redox state of the NCR peptides has *in vitro* a strong impact on the activity of the peptides. The antimicrobial activity, the activation or repression of gene expression, or the peptide stability toward the degradation by HrrP, were all affected by the NCR redox state (Haag et al., 2012; Shabab et al., 2016).

Many pathogenic bacteria inject "effector" proteins into host cells as part of their pathogenicity strategy. These effectors suppress host immune responses and subvert intracellular processes, thereby promoting bacterial colonization of the host. Pathogens usually inject a large diversity of effectors, up to over a hundred different proteins. A picture that emerges from the above is that the NCR peptides are comparable to these effectors but acting in the opposite direction: they are produced by the host and modulate bacterial processes. The NCR effectors might be a strategy to limit the possibilities of the bacteroids to act against the interests of the host. On the other hand, the peptides should not overly act and kill the symbionts while the latter have possibilities for counteracting the effectors. Thus, a subtle balance between the manipulation of the bacterial physiology by the NCR effectors and the ability of the rhizobia to resist them is required for a functional symbiotic association. An imbalance, as seen in bacterial and plant mutants or natural variants, results in a breakdown of the symbiosis.

The large number of bacteroids in the symbiotic cells implies the massive presence of microbial molecules that can potentially act as pathogen-associated molecular patterns (PAMPs) and induce a strong PAMP Triggered Immunity (PTI). To prevent such a response, which would be destructive to the symbiosis, the symbiotic cells of *M. truncatula* nodules activate a specific machinery that enables the long-term maintenance of the rhizobia without mounting a PTI response (Fig. 3) (Berrabah et al., 2014; Bourcy et al., 2013; Domonkos et al., 2017; Wang et al., 2016). In the absence of the symbiotic cysteine-rich receptor-like kinase SymCRK or the phospholipase C-like

DNF2, both probably located in the symbiosomes, or the ER located small protein NAD1 (Nodules with Activated Defense 1), a strong immune response is induced, observable by the induction of marker genes of immunity, nodule necrosis and a complete halting of the symbiosis. The PTI response was reported to be dependent on ethylene signaling in the *symcrk* and *dnf2* (*defective in nitrogen fixation 2*) mutant nodules (Berrabah et al., 2018) and on a signaling cascade involving respiratory burst oxidase homologs (Rboh) and a calcium-dependent protein kinase (CDPK) in the *nad1* nodules (Yu et al., 2018). Both pathways are known regulators of PTI. At present, it is still unclear how the three nodule suppressors of immunity interfere with these pathways. However, it is unlikely that they act at the transcriptional level, given their localization in the ER or the symbiosomes. It is also unclear if other immunity pathways are suppressed and in how far the three suppressors act on overlapping or separate pathways. However, by combining the immunity suppressor mutants with the above-described *exoY*, *ipd3*, *lin*, *bacA* and *dnf1* plant and bacterial mutants affected in different stages of infection or bacteroid differentiation, it became clear that the suppressors, which are only expressed in the symbiotic nodule cells, only act after release of rhizobia from infection threads. Moreover, they act at different steps thereafter, at the bacterial uptake step for NAD1 and DNF2, and at the formation of mature bacteroids for SymCRK (Berrabah, Ratet, & Gourion, 2015; Domonkos et al., 2017). This suggests that depending on the stage of bacteroid differentiation, different PTI responses are activated and that there is a corresponding specialization of the three immunity repressors. Furthermore, the immune suppression by NAD1, DNF2 and SymCRK was proposed to be specific for the terminally differentiated bacteroids since equivalent mutants in *L. japonicus*, which has reversible bacteroids, have no phenotype whatsoever (Yu, Bao, Zhang, & Cao, 2019). Thus, the challenge with the antimicrobial NCR peptides possibly provokes the release of PAMPs, requiring the specific suppression of PTI.

5. Gene expression and regulation in the symbiotic cells

Genome-wide gene expression studies have been performed extensively in *M. truncatula*, revealing that nodule formation is accompanied by the differential expression of several thousand genes (reviewed in Mergaert, Kereszt, & Kondorosi, 2019). The analysis of the expression pattern of these nodule enhanced genes showed that a large proportion of them is extremely specific and only expressed in nodules (Guefrachi et al., 2014). The entire *NCR* gene family for instance is absolutely nodule specific as mentioned earlier. In addition, transcriptome studies providing spatial resolution at the cell or nodule-zone level, by using nodulation mutants or by combining laser-capture microdissection with RNA-seq, demonstrated that the large majority of the nodule-enhanced or nodule-specific genes is expressed in the differentiating or mature symbiotic cells (Maunoury et al., 2010; Roux et al., 2014).

Stimulation of the CSSP by Nod factors results in the activation of a well-characterized TF network that triggers the infection process and nodule organogenesis. However, how this TF network further connects to the downstream massive reprogramming in the symbiotic cells still has to be discovered. This regulation most likely involves a presently unknown downstream TF network. An important challenge for the future is to identify these TFs and their connection to the known upstream regulatory pathways.

Moreover, the strict expression of the nodule-specific genes suggests that regulation at the chromatin-level is important as well. Indeed, genes with high tissue- or condition-specific expression are often actively silenced by epigenetic mechanisms that are reversed when expression is required, giving TFs access to their target genes. Indications for such a key role of chromatin-based regulation in the massive gene activation during symbiotic cell differentiation were obtained by analyzing the chromatin features of the nodule-specific genes. Because of the coupling of differentiation to endoreduplication, symbiotic nodule cells multiply their genome from a 2C and 4C DNA content in meristem cells, over 8C and 16C in differentiating cells, till 32C and 64C in mature symbiotic cells (Vinardell et al., 2003). Using purified nodule nuclei of these various ploidy levels, a correlation was found between the ploidy level of the cells and the sequential activation of nodule specific genes, suggesting that the endoreduplication is implicated in the gene expression control (Nagymihály et al., 2017). Expression of the nodule-specific genes in the different ploidy-types further correlated with opening of the chromatin as well as with reduced levels of the chromatin mark H3K27me3 (Nagymihály et al., 2017). H3K27me3 (trimethylation of lysine 27 in histone H3) is a repressive chromatin mark, introduced by the Polycomb silencing complex. In another study, it was noticed that genes that are specifically expressed in the differentiating and mature symbiotic cells in *M. truncatula* often display local duplications and are clustered together with other types of nodule-specific genes (Pecrix et al., 2018). These genome clusters of nodule-specific genes were called symbiosis-related islands (SRIs). Almost 300 SRIs were identified and they contain on average more than seven nodule enhanced genes of more than five distinct gene families. The SRIs showed reduced H3K27me3 levels in nodules compared to roots. Thus, the activation of nodule-specific genes could be made possible by opening of the chromatin and release of H3K37me3-mediated repression during the differentiation of the symbiotic cells.

Promoter DNA methylation is another level of epigenetic gene silencing. It was found, by comparing the DNA methylation levels either in whole roots versus nodules or in nodule nuclei of different ploidy level, that many of the symbiotic nodule cell-specific genes, including genes encoding the NCRs, become demethylated in correlation with the activation of their expression (Nagymihály et al., 2017; Satgé et al., 2016). Remarkably, the SRIs displayed such a differential methylation pattern in roots versus nodules and the borders of the SRIs sharply delignated the limits of the differentially methylated regions while the SRI-flanking regions showed very low and non-differential methylation (Pecrix et al., 2018). This DNA demethylation of the nodule-specific genes is in large part mediated by the DNA demethylase DEMETER that is transcriptionally activated in the differentiating symbiotic cells (Satgé et al., 2016). Furthermore, the SRIs contain numerous expressed long non-coding RNAs, potentially also involved in the regulation of the SRI genes (Pecrix et al., 2018).

Together, these findings suggest thus that nodule-specific genes and the SRIs are silenced by epigenetic means, by DNA methylation and by histone modifications, and that endoreduplication in the differentiating symbiotic cells and the expression of long non-coding RNAs correlates with their epigenetic reprogramming in such a way that they become accessible and competent for transcription. However, the current data only establishes correlations between gene expression on the one hand and chromatin modifications or endoreduplication on the other hand. Clearly, a challenge for future studies will be to establish causal relations between these processes and to identify the underlying molecular mechanisms.

6. Conclusion: the integrated picture

The formation of the symbiotic nodule cells recruits existing and widely used plant processes. Indeed, meristems provide the cells for all plant tissues; endoreduplication is found in many plant cell types and depends on the same molecular mechanism in nodules and elsewhere; the tip growth of infection threads is shared with root hairs or pollen tubes; endocytotic and exocytotic vesicle trafficking and actin cytoskeleton rearrangements are common in all plant cells; AMPs are part of the innate immune system of plants; and the nitrogen and carbon metabolism of the symbiotic cells relies on enzymes participating in the central metabolism of plants. However, these common processes are integrated with completely new features not seen anywhere else in plant biology. The organization of the nodule meristem providing lateral vascular bundles and a central endoreduplicating tissue is unlike any other meristem. Infection threads are trans-cellular and subcellular structures that use cell cycle and polar growth-related processes but they are not autonomous cells like root hairs or pollen tubes. Endocytosis of whole bacteria is not known elsewhere in plants. The symbiosomes may resemble phagocytized pathogens in animals but in plants, such new organelle-like structures are unique. A heavy load of bacteria inside plant tissues as in the nodules would normally lead to an immune response or a collapse of the plant tissue and a unique deployment of immune effectors and immune suppression is therefore required. The metabolic integration of the bacterial endosymbionts needs the management of specific metabolic fluxes. Finally, the whole process involves the expression of a very large set of genes that are only expressed in the symbiotic cells. These genes arose from gene duplications combined with modifications leading to neofonctionalization (for example the oxygen transporting leghemoglobins, derived by replacement of a few amino acids from non-oxygen-transporting hemoglobins present in all plants; Sturms, Kakar, Trent, & Hargrove, 2010). Others arose from duplications whereby one paralog obtained a new, nodule-specific expression profile but without acquiring a new protein function (for example the secretory pathway genes or small metabolite transporters and metabolic enzyme genes). And still other symbiotic cell-specific genes were formed *de novo* and have no homologs (so-called orphan genes such as the *NCRs* as well as many others).

The net result of this tinkering with old and new parts is this particular cell type packed with nitrogen-fixing organelles that is very efficient for nitrogen and carbon exchange and that can produce sufficient reduced nitrogen to meet all the needs of the plant. All nitrogen-fixing nodule-forming plants, including legumes and actinorhizal plants, have some form of symbiotic nodule cells similar to the ones of *M. truncatula* described here, even if they might differ in many details (Pawlowski & Demchenko, 2012; Sprent, Ardley, & James, 2017). In the light of the ongoing discussion on the evolution of the nodule-forming symbiosis in legumes and actinorhizal plants (Parniske, 2018; van Velzen, Doyle, & Geurts, 2019), a fascinating question to think about is if the symbiotic nodule cell is ancestral in this symbiosis or if it appeared by convergent evolution in the different nodule-forming lineages. One can also imagine that, in case they are indeed ancestral, they have first existed in root cortical cells independently of organogenesis in early evolution of the symbiosis, in a similar form as the arbuscules in the mycorrhizal symbiosis (Lanfranco, Bonfante, & Genre, 2016). The nodule formation itself has then subsequently evolved by recruitment of cell divisions and the gradual appearance of additional cell types. Part of this evolutionary process could have taken place in the common ancestor of nodulating plants and part of it could have appeared in parallel in the different lineages.

Finally, it is interesting to note that cells with many features analogous to symbiotic nodule cells are found in insects that maintain nutritional endosymbionts in specific symbiotic cells called bacteriocytes (Baumann, 2005). These bacteria synthesize amino acids or vitamins for their host instead of the reduced nitrogen produced by the rhizobial symbionts of legumes. At least in the case of symbiotic nodule cells versus bacteriocytes, evolution was convergent. It has come up multiple times with a similar solution for the stable maintenance of a very large number of bacteria within a cell and for the establishment of an efficient metabolic interaction between eukaryotic cells and bacterial endosymbionts.

Acknowledgments

Research in my laboratory benefits from support from the Agence National de la Recherche (ANR), grants no. ANR-16-CE20-0013 and ANR-17-CE20-0011, and the Labex "Saclay Plant Sciences" (ANR-10-LABX-0040-SPS).

References

Arnold, M. F. F., Penterman, J., Shabab, M., Chen, E. J., & Walker, G. C. (2018). Important late-stage symbiotic role of the *Sinorhizobium meliloti* exopolysaccharide succinoglycan. *Journal of Bacteriology, 200*. e00665-17.

Arnold, M. F. F., Shabab, M., Penterman, J., Boehme, K. L., Griffitts, J. S., & Walker, G. C. (2017). Genome-wide sensitivity Analysis of the microsymbiont *Sinorhizobium meliloti* to symbiotically important, defensin-like host peptides. *mBio, 8*. e01060-17.

Barrière, Q., Guefrachi, I., Gully, D., Lamouche, F., Pierre, O., Fardoux, J., et al. (2017). Integrated roles of BclA and DD-carboxypeptidase 1 in *Bradyrhizobium* differentiation within NCR-producing and NCR-lacking root nodules. *Scientific Reports, 7*, 9063.

Baumann, P. (2005). Biology of bacteriocyte-associated endosymbionts of plant sap-sucking insects. *Annual Review of Microbiology, 59*, 155—189.

Berrabah, F., Balliau, T., Aït-Salem, E. H., George, J., Zivy, M., Ratet, P., et al. (2018). Control of the ethylene signaling pathway prevents plant defenses during intracellular accommodation of the rhizobia. *New Phytologist, 219*, 310–323.

Berrabah, F., Bourcy, M., Eschstruth, A., Cayrel, A., Guefrachi, I., Mergaert, P., et al. (2014). A nonRD receptor-like kinase prevents nodule early senescence and defense-like reactions during symbiosis. *New Phytologist, 203*, 1305–1314.

Berrabah, F., Ratet, P., & Gourion, B. (2015). Multiple steps control immunity during the intracellular accommodation of rhizobia. *Journal of Experimental Botany, 66*, 1977–1985.

Bonaldi, K., Gargani, D., Prin, Y., Fardoux, J., Gully, D., Nouwen, N., et al. (2011). Nodulation of *Aeschynomene afraspera* and *A. indica* by photosynthetic *bradyrhizobium* sp. strain ORS285: The nod-dependent versus the nod-independent symbiotic interaction. *Molecular Plant-Microbe Interactions, 24*, 1359–1371.

Bourcy, M., Brocard, L., Pislariu, C. I., Cosson, V., Mergaert, P., Tadege, M., et al. (2013). *Medicago truncatula* DNF2 is a PI-PLC-XD-containing protein required for bacteroid persistence and prevention of nodule early senescence and defense-like reactions. *New Phytologist, 197*, 1250–1261.

Czernic, P., Gully, D., Cartieaux, F., Moulin, L., Guefrachi, I., Patrel, D., et al. (2015). Convergent evolution of endosymbiont differentiation in Dalbergioid and Inverted Repeat-Lacking Clade legumes mediated by nodule-qpecific cysteine-rich peptides. *Plant Physiology, 169*, 1254–1265.

Domonkos, Á., Kovács, S., Gombár, A., Kiss, E., Horváth, B., Kováts, G. Z., et al. (2017). NAD1 controls defense-like responses in *Medicago truncatula* symbiotic nitrogen fixing nodules following rhizobial colonization in a BacA-independent manner. *Genes, 8*, E387.

Farkas, A., Maróti, G., Durgő, H., Györgypál, Z., Lima, R. M., & Medzihradszky, K. F. (2014). *Medicago truncatula* symbiotic peptide NCR247 contributes to bacteroid differentiation through multiple mechanisms. *Proceedings of the National Academy of Sciences of the United States of America, 111*, 5183–5188.

Farkas, A., Pap, B., Kondorosi, É., & Maróti, G. (2018). Antimicrobial activity of NCR plant peptides strongly depends on the test assays. *Frontiers in Microbiology, 9*, 2600.

Franssen, H. J., Xiao, T. T., Kulikova, O., Wan, X., Bisseling, T., Scheres, B., et al. (2015). Root developmental programs shape the *Medicago truncatula* nodule meristem. *Development, 142*, 2941–2950.

Gavrin, A., Chiasson, D., Ovchinnikova, E., Kaiser, B. N., Bisseling, T., & Fedorova, E. E. (2016). VAMP721a and VAMP721d are important for pectin dynamics and release of bacteria in soybean nodules. *New Phytologist, 210*, 1011–1021.

Gavrin, A., Jansen, V., Ivanov, S., Bisseling, T., & Fedorova, E. (2015). ARP2/3-mediated actin nucleation associated with symbiosome membrane is essential for the development of symbiosomes in infected cells of *Medicago truncatula* root nodules. *Molecular Plant-Microbe Interactions, 28*, 605–614.

Gavrin, A., Kaiser, B. N., Geiger, D., Tyerman, S. D., Wen, Z., Bisseling, T., et al. (2014). Adjustment of host cells for accommodation of symbiotic bacteria: Vacuole defunctionalization, HOPS suppression, and TIP1g retargeting in *Medicago*. *Plant Cell, 26*, 3809–3822.

Gavrin, A., Kulikova, O., Bisseling, T., & Fedorova, E. E. (2017). Interface symbiotic membrane formation in root nodules of *Medicago truncatula*: The role of synaptotagmins MtSyt1, MtSyt2 and MtSyt3. *Frontiers of Plant Science, 8*, 201.

Geurts, R., Xiao, T. T., & Reinhold-Hurek, B. (2016). What does it take to evolve a nitrogen-fixing endosymbiosis? *Trends in Plant Science, 21*, 199–208.

Guan, D., Stacey, N., Liu, C., Wen, J., Mysore, K. S., Torres-Jerez, I., et al. (2013). Rhizobial infection is associated with the development of peripheral vasculature in nodules of *Medicago truncatula*. *Plant Physiology, 162*, 107–115.

Guefrachi, I., Nagymihaly, M., Pislariu, C. I., Van de Velde, W., Ratet, P., Mars, M., et al. (2014). Extreme specificity of NCR gene expression in *Medicago truncatula*. *BMC Genomics, 15*, 712.

Guefrachi, I., Pierre, O., Timchenko, T., Alunni, B., Barrière, Q., Czernic, P., et al. (2015). *Bradyrhizobium* BclA is a peptide transporter required for bacterial differentiation in symbiosis with *Aeschynomene* legumes. *Molecular Plant-Microbe Interactions, 28*, 1155–1166.

Gully, D., Czernic, P., Cruveiller, S., Mahé, F., Longin, C., Vallenet, D., et al. (2018). Transcriptome profiles of Nod factor-independent symbiosis in the tropical legume *Aeschynomene evenia*. *Scientific Reports, 8*, 10934.

Haag, A. F., Baloban, M., Sani, M., Kerscher, B., Pierre, O., Farkas, A., et al. (2011). Protection of *Sinorhizobium* against host cysteine-rich antimicrobial peptides is critical for symbiosis. *PLoS Biology, 9*, e1001169.

Haag, A. F., Kerscher, B., Dall'Angelo, S., Sani, M., Longhi, R., Baloban, M., et al. (2012). Role of cysteine residues and disulfide bonds in the activity of a legume root nodule-specific, cysteine-rich peptide. *Journal of Biological Chemistry, 287*, 10791–10798.

Hirsch, A. M., Bang, M., & Ausubel, F. (1983). Ultrastructural analysis of ineffective alfalfa nodules formed by *nif*::Tn*5* mutants of *Rhizobium meliloti*. *Journal of Bacteriology, 155*, 367–380.

Horváth, B., Domonkos, Á., Kereszt, A., Szűcs, A., Ábrahám, E., Ayaydin, F., et al. (2015). Loss of the nodule-specific cysteine rich peptide, NCR169, abolishes symbiotic nitrogen fixation in the *Medicago truncatula dnf7* mutant. *Proceedings of the National Academy of Sciences of the United States of America, 112*, 15232–15237.

Huisman, R., Hontelez, J., Mysore, K. S., Wen, J., Bisseling, T., & Limpens, E. (2016). A symbiosis-dedicated SYNTAXIN OF PLANTS 13II isoform controls the formation of a stable host–microbe interface in symbiosis. *New Phytologist, 211*, 1338–1351.

Ivanov, S., Fedorova, E. E., Limpens, E., Mita, S. D., Genre, A., Bonfante, P., et al. (2012). *Rhizobium*–legume symbiosis shares an exocytotic pathway required for arbuscule formation. *Proceedings of the National Academy of Sciences of the United States of America, 109*, 8316–8321.

Jin, Y., Chen, Z., Yang, J., Mysore, K. S., Wen, J., Huang, J., et al. (2018). IPD3 and IPD3L function redundantly in rhizobial and mycorrhizal symbioses. *Frontiers of Plant Science, 9*, 267.

Kanazawa, T., & Ueda, T. (2017). Exocytic trafficking pathways in plants: Why and how they are redirected. *New Phytologist, 215*, 952–957.

Kazmierczak, T., Nagymihály, M., Lamouche, F., Barrière, Q., Guefrachi, I., Alunni, B., et al. (2017). Specific host-responsive associations between *Medicago truncatula* accessions and *Sinorhizobium* strains. *Molecular Plant-Microbe Interactions, 30*, 399–409.

Kim, M., Chen, Y., Xi, J., Waters, C., Chen, R., & Wang, D. (2015). An antimicrobial peptide essential for bacterial survival in the nitrogen-fixing symbiosis. *Proceedings of the National Academy of Sciences of the United States of America, 112*, 15238–15243.

Kitaeva, A. B., Demchenko, K. N., Tikhonovich, I. A., Timmers, A. C., & Tsyganov, V. E. (2016). Comparative analysis of the tubulin cytoskeleton organization in nodules of *Medicago truncatula* and *Pisum sativum*: Bacterial release and bacteroid positioning correlate with characteristic microtubule rearrangements. *New Phytologist, 210*, 168–183.

Kwon, C., Neu, C., Pajonk, S., Yun, H. S., Lipka, U., Humphry, M., et al. (2008). Co-option of a default secretory pathway for plant immune responses. *Nature, 451*, 835–840.

Lamouche, F., Bonadé-Bottino, N., Mergaert, P., & Alunni, B. (2019a). Symbiotic efficiency of spherical and elongated bacteroids in the *Aeschynomene-Bradyrhizobium* symbiosis. *Frontiers in Microbiology, 10*, 377.

Lamouche, F., Gully, D., Chaumeret, A., Nouwen, N., Verly, C., Pierre, O., et al. (2019b). Transcriptomic dissection of *Bradyrhizobium* sp. strain ORS285 in symbiosis with

Aeschynomene spp. inducing different bacteroid morphotypes with contrasted symbiotic efficiency. *Environmental Microbiology, 21*, 3244–3258.
Lanfranco, L., Bonfante, P., & Genre, A. (2016). The mutualistic interaction between plants and arbuscular mycorrhizal fungi. *Microbiology Spectrum, 4*, 4.
Limpens, E., Ivanov, S., van Esse, W., Voets, G., Fedorova, E., & Bisseling, T. (2009). *Medicago* N_2-fixing symbiosomes acquire the endocytic identity marker Rab7 but delay the acquisition of vacuolar identity. *Plant Cell, 21*, 2811–2828.
Limpens, E., Mirabella, R., Fedorova, E., Franken, C., Franssen, H., Bisseling, T., et al. (2005). Formation of organelle-like N_2-fixing symbiosomes in legume root nodules is controlled by DMI2. *Proceedings of the National Academy of Sciences of the United States of America, 102*, 10375–10380.
Limpens, E., Moling, S., Hooiveld, G., Pereira, P. A., Bisseling, T., Becker, J. D., et al. (2013). Cell- and tissue-specific transcriptome analyses of *Medicago truncatula* root nodules. *PLoS One, 8*, e64377.
Magne, K., Couzigou, J. M., Schiessl, K., Liu, S., George, J., Zhukov, V., et al. (2018). MtNODULE ROOT1 and MtNODULE ROOT2 are essential for indeterminate nodule identity. *Plant Physiology, 178*, 295–316.
Marchetti, M., Capela, D., Poincloux, R., Benmeradi, N., Auriac, M.-C., Le Ru, A., et al. (2013). Queuosine biosynthesis is required for *Sinorhizobium meliloti*-induced cytoskeletal modifications on HeLa cells and symbiosis with *Medicago truncatula. PLoS One, 8*, e56043.
Maunoury, N., Redondo-Nieto, M., Bourcy, M., Van de Velde, W., Alunni, B., Laporte, P., et al. (2010). Differentiation of symbiotic cells and endosymbionts are coupled to two transcriptome-switches in *Medicago truncatula* nodulation. *PLoS One, 5*, e9519.
Mergaert, P. (2018). Role of antimicrobial peptides in controlling symbiotic bacterial populations. *Natural Product Reports, 35*, 336–356.
Mergaert, P., Nikovics, K., Kelemen, Z., Maunoury, N., Vaubert, D., Kondorosi, A., et al. (2003). A novel family in *Medicago truncatula* consisting of more than 300 nodule-specific genes coding for small, secreted polypeptides with conserved cysteine motifs. *Plant Physiology, 132*, 161–173.
Mergaert, P., Uchiumi, T., Alunni, B., Evanno, G., Cheron, A., Catrice, O., et al. (2006). Eukaryotic control on bacterial cell cycle and differentiation in the *Rhizobium*-legume symbiosis. *Proceedings of the National Academy of Sciences of the United States of America, 103*, 5230–5235.
Mergaert, P., Kereszt, A., & Kondorosi, E. (2019). Gene expression in nitrogen-fixing symbiotic nodule cells in *Medicago turncatula* and other nodulating plants. *Plant Cell.* https://doi.org/10.1105/tpc.19.00494.
Moling, S., Pietraszewska-Bogiel, A., Postma, M., Fedorova, E., Hink, M. A., Limpens, E., et al. (2014). Nod factor receptors form heteromeric complexes and are essential for intracellular infection in *Medicago* nodules. *Plant Cell, 26*, 4188–4199.
Monahan-Giovanelli, H., Pinedo, C. A., & Gage, D. J. (2006). Architecture of infection thread networks in developing root nodules induced by the symbiotic bacterium *Sinorhizobium meliloti* on *Medicago truncatula. Plant Physiology, 140*, 661–670.
Montiel, J., Downie, J. A., Farkas, A., Bihari, P., Herczeg, R., Bálint, B., et al. (2017). Morphotype of bacteroids in different legumes correlates with the number and type of symbiotic NCR peptides. *Proceedings of the National Academy of Sciences of the United States of America, 114*, 5041–5046.
Nagymihály, M., Veluchamy, A., Györgypál, Z., Ariel, F., Jégu, T., Benhamed, M., et al. (2017). Ploidy-dependent changes in the epigenome of symbiotic cells correlate with specific patterns of gene expression. *Proceedings of the National Academy of Sciences of the United States of America, 114*, 4543–4548.

Nallu, S., Silverstein, K. A., Zhou, P., Young, N. D., & Vandenbosch, K. A. (2014). Patterns of divergence of a large family of nodule cysteine-rich peptides in accessions of *Medicago truncatula*. *The Plant Journal, 78*, 697–705.

Oono, R., & Denison, R. F. (2010). Comparing symbiotic efficiency between swollen versus nonswollen rhizobial bacteroids. *Plant Physiology, 154*, 1541–1548.

Oono, R., Schmitt, I., Sprent, J. I., & Denison, R. F. (2010). Multiple evolutionary origins of legume traits leading to extreme rhizobial differentiation. *New Phytologist, 187*, 508–520.

Pan, H., Oztas, O., Zhang, X., Wu, X., Stonoha, C., Wang, E., et al. (2016). A symbiotic SNARE protein generated by alternative termination of transcription. *Native Plants, 2*, 15197.

Parniske, M. (2018). Uptake of bacteria into living plant cells, the unifying and distinct feature of the nitrogen-fixing root nodule symbiosis. *Current Opinion in Plant Biology, 44*, 164–174.

Pawlowski, K., & Demchenko, K. N. (2012). The diversity of actinorhizal symbiosis. *Protoplasma, 249*, 967–979.

Pecrix, Y., Staton, S. E., Sallet, E., Lelandais-Brière, C., Moreau, S., Carrère, S., et al. (2018). Whole-genome landscape of *Medicago truncatula* symbiotic genes. *Native Plants, 4*, 1017–1025.

Penterman, J., Abo, R. P., De Nisco, N. J., Arnold, M. F., Longhi, R., Zanda, M., et al. (2014). Host plant peptides elicit a transcriptional response to control the *Sinorhizobium meliloti* cell cycle during symbiosis. *Proceedings of the National Academy of Sciences of the United States of America, 111*, 3561–3566.

Pini, F., De Nisco, N. J., Ferri, L., Penterman, J., Fioravanti, A., Brilli, M., et al. (2015). Cell cycle control by the master regulator CtrA in *Sinorhizobium meliloti*. *PLoS Genetics, 11*, e1005232.

Price, P. A., Tanner, H. R., Dillon, B. A., Shabab, M., Walker, G. C., & Griffitts, J. S. (2015). Rhizobial peptidase HrrP cleaves host-encoded signaling peptides and mediates symbiotic compatibility. *Proceedings of the National Academy of Sciences of the United States of America, 112*, 15244–15249.

Ribeiro, C. W., Baldacci-Cresp, F., Pierre, O., Larousse, M., Benyamina, S., Lambert, A., et al. (2017). Regulation of differentiation of nitrogen-fixing bacteria by microsymbiont targeting of plant thioredoxin s1. *Current Biology, 27*, 250–256.

Robertson, J. G., & Lyttleton, P. (1982). Coated and smooth vesicles in the biogenesis of cell walls, plasma membranes, infection threads and peribacteroid membranes in root hairs and nodules of white clover. *Journal of Cell Science, 58*, 63–78.

Roux, B., Rodde, N., Jardinaud, M. F., Timmers, T., Sauviac, L., Cottret, L., et al. (2014). An integrated analysis of plant and bacterial gene expression in symbiotic root nodules using laser-capture microdissection coupled to RNA sequencing. *The Plant Journal, 77*, 817–837.

Satgé, C., Moreau, S., Sallet, E., Lefort, G., Auriac, M.-C., Remblière, C., et al. (2016). Reprogramming of DNA methylation is critical for nodule development in *Medicago truncatula*. *Native Plants, 2*, 16166.

Sen, D., & Weaver, R. W. (1984). A basis for different rates of N_2-fixation by the same strains of *Rhizobium* in peanut and cowpea root nodules. *Plant Science Letters, 34*, 239–246.

Shabab, M., Arnold, M. F., Penterman, J., Wommack, A. J., Bocker, H. T., Price, P. A., et al. (2016). Disulfide cross-linking influences symbiotic activities of nodule peptide NCR247. *Proceedings of the National Academy of Sciences of the United States of America, 113*, 10157–10162.

Simsek, S., Ojanen-Reuhs, T., Stephens, S. B., & Reuhs, B. L. (2007). Strain-ecotype specificity in *Sinorhizobium meliloti-Medicago truncatula* symbiosis is correlated to succinoglycan oligosaccharide structure. *Journal of Bacteriology, 189*, 7733–7740.

Sinharoy, S., Torres-Jerez, I., Bandyopadhyay, K., Kereszt, A., Pislariu, C. I., Nakashima, J., et al. (2013). The C_2H_2 transcription factor REGULATOR OF SYMBIOSOME DIFFERENTIATION represses transcription of the secretory pathway gene *VAMP721a* and promotes symbiosome development in *Medicago truncatula*. *Plant Cell, 25*, 3584–3601.

Sprent, J. I., Ardley, J., & James, E. K. (2017). Biogeography of nodulated legumes and their nitrogen-fixing symbionts. *New Phytologist, 215*, 40–56.

Sturms, R., Kakar, S., Trent, J., III, & Hargrove, M. S. (2010). *Trema* and *Parasponia* hemoglobins reveal convergent evolution of oxygen transport in plants. *Biochemistry, 49*, 4085–4093.

Timmers, A. C., Auriac, M. C., de Billy, F., & Truchet, G. (1998). Nod factor internalization and microtubular cytoskeleton changes occur concomitantly during nodule differentiation in alfalfa. *Development, 125*, 339–349.

Van de Velde, W., Zehirov, G., Szatmari, A., Debreczeny, M., Ishihara, H., Kevei, Z., et al. (2010). Plant peptides govern terminal differentiation of bacteria in symbiosis. *Science, 327*, 1122–1126.

van Velzen, R., Doyle, J. J., & Geurts, R. (2019). A resurrected scenario: Single gain and massive loss of nitrogen-fixing nodulation. *Trends in Plant Science, 24*, 49–57.

Vinardell, J. M., Fedorova, E., Cebolla, A., Kevei, Z., Horvath, G., Kelemen, Z., et al. (2003). Endoreduplication mediated by the anaphase-promoting complex activator CCS52A is required for symbiotic cell differentiation in *Medicago truncatula* nodules. *Plant Cell, 15*, 2093–2105.

Wang, D., Griffitts, J., Starker, C., Fedorova, E., Limpens, E., Ivanov, S., et al. (2010). A nodule-specific protein secretory pathway required for nitrogen-fixing symbiosis. *Science, 327*, 1126–1129.

Wang, Q., Yang, S. M., Liu, J., Terecskei, K., Abraham, E., Gombar, A., et al. (2017). A host-secreted antimicrobial peptide enforces symbiotic selectivity in *Medicago truncatula*. *Proceedings of the National Academy of Sciences of the United States of America, 114*, 6854–6859.

Wang, C., Yu, H., Luo, L., Duan, L., Cai, L., He, X., et al. (2016). NODULES WITH ACTIVATED DEFENSE 1 is required for maintenance of rhizobial endosymbiosis in Medicago truncatula. *New Phytologist, 212*, 176–191.

Xiao, T. T., Schilderink, S., Moling, S., Deinum, E. E., Kondorosi, E., Franssen, H., et al. (2014). Fate map of *Medicago truncatula* root nodules. *Development, 141*, 3517–3528.

Yang, S., Wang, Q., Fedorova, E., Liu, J., Qin, Q., Zheng, Q., et al. (2017). Microsymbiont discrimination mediated by a host-secreted peptide in Medicago truncatula. *Proceedings of the National Academy of Sciences of the United States of America, 114*, 6848–6853.

Yu, H., Bao, H., Zhang, Z., & Cao, Y. (2019). Immune signaling pathway during terminal bacteroid differentiation in nodules. *Trends in Plant Science, 24*, 299–302.

Yu, H., Xiao, A., Dong, R., Fan, Y., Zhang, X., Liu, C., et al. (2018). Suppression of innate immunity mediated by the CDPK-Rboh complex is required for rhizobial colonization in Medicago truncatula nodules. *New Phytologist, 220*, 425–434.

Zhang, X., Han, L., Wang, Q., Zhang, C., Yu, Y., Tian, J., et al. (2019). The host actin cytoskeleton channels rhizobia release and facilitates symbiosome accommodation during nodulation in *Medicago truncatula*. *New Phytologist, 221*, 1049–1059.

CHAPTER SEVEN

Legume nodule senescence: a coordinated death mechanism between bacteria and plant cells

Théophile Kazmierczak[a,d], Li Yang[b], Eric Boncompagni[b], Eliane Meilhoc[c], Florian Frugier[a], Pierre Frendo[b], Claude Bruand[c], Véronique Gruber[a] and Renaud Brouquisse[b,*]

[a]Institute of Plant Sciences Paris Saclay (IPS2), CNRS, INRA, Univ Paris Diderot, Univ Paris-Sud, Univ d'Evry, Université Paris-Saclay, Gif-sur-Yvette, France
[b]INRA, CNRS, Université Côte d'Azur, ISA, France
[c]Laboratoire des Interactions Plantes-Microorganismes (LIPM), Université de Toulouse, INRA, CNRS, INSA, Castanet-Tolosan, France
[d]BIOtransfer, Montreuil, France
*Corresponding author: E-mail: renaud.brouquisse@inra.fr

Contents

Abstract

In the legume-rhizobia symbiosis, atmospheric dinitrogen (N_2) fixation by rhizobia takes place in a specific root organ, called the nodule. During biological N_2 fixation, the plant provides photosynthetic carbohydrates to bacteria and, in exchange, bacteria feed the plant with reduced nitrogen (N) under ammonia (NH_3) form. Like most other organs, the root nodule has a limited lifespan, and eventually enters a senescence process characterized by a decline of N_2 fixation and the coordinated death of both bacteria and plant cells. This senescence process characterized by a metabolic switch from a carbon sink to a nutrient source either results from nodule aging (developmental nodule senescence) or can be triggered prematurely by adverse environmental conditions. Accumulating evidence suggests that redox and hormone signaling contribute to

Advances in Botanical Research, Volume 94
ISSN 0065-2296
https://doi.org/10.1016/bs.abr.2019.09.013

the triggering of senescence. Indeed, nodule senescence is a developmentally programmed process in which antioxidants, reactive oxygen and N species (ROS/RNS), hormones and proteinases play key roles. This chapter reviews recent knowledge on nodule senescence and proposes models by which ROS, RNS, and likely hormones intervene in the orchestration of this process.

1. Introduction

Plant-microbe interactions represent a major aspect of the formation and of the development of ecosystems. Plants often establish symbiotic relationships with specific fungi and bacteria to acquire the nutrients they need to support growth and development (Smith & Smith, 2011; Van Der Heijden, De Bruin, Luckerhoff, Van Logtestijn, & Schlaeppi, 2016; Venkateshwaran, Volkening, Sussman, & Ané, 2013; Wang et al., 2017). Amongst the different nutrients, nitrogen (N) has a major importance as it represents a crucial constituent of proteins, nucleic acids and chlorophyll and is often in insufficient amounts in the soil under nitrate and/or ammonia forms. As a result, plants belonging to the legume family have developed a symbiotic association with soil bacteria collectively called rhizobia to increase their N nutrition via the biological fixation of atmospheric dinitrogen (N_2) (BNF). In BNF, the plant provides photosynthetic carbohydrates to the bacteria and, in exchange, bacteria feed the plant with reduced N under ammonia (NH_3) form. Legumes are a family of dicotyledonous plants of the *Fabales* order. It is the third most important family of flowering plants after the *Orchidaceae* and *Asteraceae* with more than 19,500 species (Christenhunsz & Byng, 2016). Economically, *Fabaceae* is the second largest family after *Poaceae* and is a valuable source of plant proteins for human food and animal feeding.

The success of the symbiotic interaction relies on the molecular dialogue between the plant and the bacteria (Oldroyd, 2013). The plant root exudates which contain flavonoids and isoflavonoids induce the production by rhizobia of lipochitooligosaccharides, called Nodulation Factors (NFs; Ferguson et al., 2010). NFs are usually essential for the bacterial infection process and the organogenesis of the root nodule, the organ which hosts the N_2-fixing bacteria. During the infection process, rhizobia attach to root hairs in the susceptible zone of the host roots. The recognition of the NFs by the plant induces the root hair curling and the formation of an infection pocket (Fig. 1). Then, bacteria enter into the root hairs through a

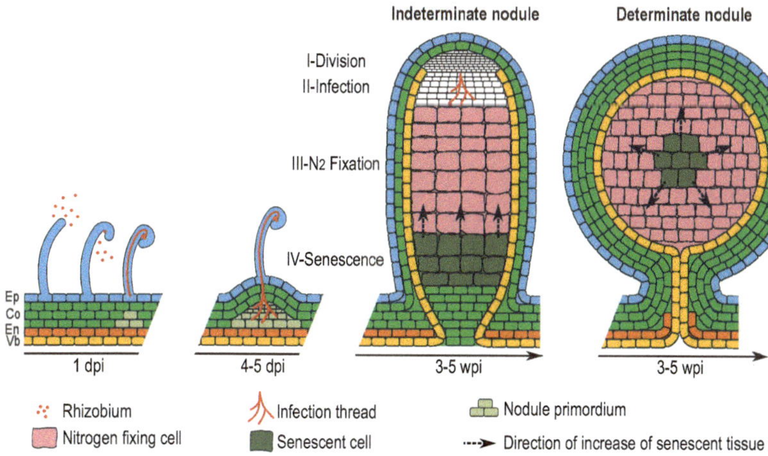

Fig. 1 Schematic representation of the establishment and of the development of the legume-rhizobium symbiosis, and of the structure of mature indeterminate and determinate nodules. *En*, endodermis; *Co*, cortex; *dpi*, day post-inoculation; *Ep*, epidermis; *Vb*, vascular bundle; *wpi*, week post-inoculation.

membrane invagination and the formation of tubular structures, the infection threads (Fig. 1), which allow the transit of the bacteria from the root surface to the cells of the neo-formed nodule (Timmers, Auriac, & Truchet, 1999). Later, the bacteria are endocytosed into the plant cell in a new organelle-like structure, called symbiosome. Bacteria inside the symbiosome, called bacteroids, become able to reduce N_2 to NH_3 thanks to the activity of a specific enzymatic complex, the nitrogenase (Peters, Fisher, & Dean, 1995).

Two predominant nodule types exist in legumes (Fig. 1). The nodule type depends on the host plant and is determined by the nature of the nodule meristem which could be either transient or persistent (Oldroyd, 2013). Determinate nodules, with a transient meristem, are spherical since cell divisions occur only during a short period of time. The final growth of the nodule is achieved by cell enlargement. In contrast, indeterminate nodules, with a persistent meristem, are cylindrical as they grow during the whole nodule life by increasing the number and the size of cells. Determinate nodules are generally formed in tropical legumes such as *Phaseolus*, *Vigna* and *Glycine*, whereas indeterminate nodules are found in temperate legumes such as *Medicago*, *Trifolium* or *Pisum*.

Beside the nodule types, the bacteroids possess also three different morphotypes named U, S and E (Kondorosi, Mergaert, & Kereszt, 2013;

Oono, Schmitt, Sprent, & Denison, 2010). The U-morphotype (for Undifferentiated) corresponds to bacteroids with minor morphological modifications compared to the free-living bacteria and the ability to reverse from the symbiotic lifestyle to the free-living state. The S− and E-morphotypes correspond to large Spherical and large Elongated bacteroids, respectively. These bacteroids cannot return to free-living life and are termed "terminally differentiated". This terminal differentiation is associated with the increase of permeability of the bacteroid membrane and of DNA content compared to free-living bacteria (Mergaert et al., 2006). The bacteroid differentiation is driven by the plant host via the production of a class of peptides called nodule-specific cysteine-rich peptides (NCRs; Van de Velde et al., 2010). To date, NCR peptides have been found to control the terminal differentiation of endosymbionts in inverted repeat-lacking clade (IRLC) legumes and in *Aeschynomene* spp. (Alunni et al., 2007; Czernic et al., 2015; Mergaert et al., 2003). At the functional level, NCRs induce membrane permeability (Lamouche, Bonadé-Bottino, Peter Mergaert, & Alunni, 2019) and have intracellular bacteroid targets such as the cell division protein FtsZ and the chaperone protein GroEL, suggesting that NCRs interact with the bacterial metabolism to induce the terminal differentiation (Farkas et al., 2014).

The cellular differentiation of both symbiotic partners is required to allow the symbiotic N_2 fixation. The production of the nitrogenase enzymatic complex in the bacteroid is a prerequisite for the reduction of N_2 to NH_3. This complex needs high amounts of energy, using sixteen ATP for each N_2 reduced to NH_3, and as a result demands a high respiration rate in the nodule. However, the nitrogenase activity is extremely sensitive to dioxygen (O_2) implying that free O_2 should be present at a very low level in bacteroids (Peters et al., 1995). Multiple strategies allow to deal with this paradox. To decrease the O_2 level in the infected cells, the nodule inner cortex cell layer acts as an oxygen barrier (Tjepkema & Yocum, 1974). In parallel, leghemoglobin (Lb), an efficient O_2-binding protein, present at millimolar concentration in the cytoplasm of infected cells (Appleby, 1992), delivers O_2 to the mitochondria and the bacteroids for respiration and decreases free O_2 to the nanomolar range in the N_2-fixing zone (Ott et al., 2005; Soupène, Foussard, Boistard, Truchet, & Batut, 1995). Finally, the bacteroid is able to function at a very low oxygen concentration thanks to a specific high affinity *cbb3* cytochrome oxidase (Preisig, Zufferey, Thony-Meyer, Appleby, & Hennecke, 1996). The reduction of N_2 to NH_3 is accompanied by the interuption of NH_3 assimilation into bacteroid amino acids. NH_3 is secreted from the bacteroid into the plant cytosol, for assimilation into the amides glutamine and asparagine, or into ureides (Day, Poole, Tyerman, & Rosendahl, 2001). In return, the plant provides carbon and energy sources to bacteroids in the form of dicarboxylic acids, particularly malate and succinate, which are produced from sucrose via sucrose synthase and glycolytic enzymes. Thus, the nodule presents multiple peculiarities, involving a plant-microbe crosstalk associated with a metabolism switch which needs a high energy level under microoxic conditions. These working conditions involve a specific cellular redox control in both symbiotic partners. The contribution of reactive oxygen and N species (ROS and RNS) and the importance of the antioxidant defense in the formation and the functioning of root nodules have been largely demonstrated (Alloing, Mandon, Boncompagni, Montrichard, & Frendo, 2018; Damiani, Pauly, Puppo, Brouquisse, & Boscari, 2016; Frendo, Matamoros, Alloing, & Becana, 2013; Puppo, Pauly, Boscari, Mandon, & Brouquisse, 2013).

Whereas formation and functioning of the root nodule has been extensively analyzed the rupture of the interaction leading to the nodule senescence has not been much characterized and only few reviews have been dedicated to nodule senescence (Bruand & Meilhoc, 2019; Dupont et al., 2012; Puppo et al., 2005; Serova & Tsyganov, 2014). Some investigators have used the initial decline in nitrogenase activity and BNF efficiency as an indication that nodule senescence has begun (Klucas, 1974). Others authors have defined nodule senescence as the period of rapid nodule decay which occurs at a plant physiological maturity stage, i.e. at leaf yellowing (Streeter, 1981), after flowering has finished (Serova & Tsyganov, 2014), or at pod filling (Bethlenfalvay & Phillips, 1977; Lawn & Brun, 1974). The nodule senescence represents a strong agronomical interest through the possible improvement of N_2 fixation efficiency under adverse environmental conditions to maintain or even increase the yield of legume crops. In this review, we summarize the different modifications of nodule functioning observed during the set up and the progress of nodule senescence, and we focus on the role of antioxidants, ROS/RNS, hormones and proteases in regulating this process.

2. The regulation of nodule senescence

2.1 Physiological and structural modifications

Although the lifespan of nodules of most legumes is 10–12 weeks, their ability to fix N starts to decline 3–5 weeks post-inoculation (Puppo et al., 2005). Nodule senescence is a highly organized and genetically controlled process that begins with metabolic changes. As nodules and leaves are tightly connected in a reciprocal sink-to-source relationship for carbon and N, any defect in photosynthetic activity will indeed impair the metabolic functions of the nodule associated with N assimilation and consequently provoke nodule senescence (Fig. 2). Developmental senescence (also known as natural senescence) occurs because of organ aging, but exposure to various environmental changes also induces nodule senescence (Fig. 2), e.g. when plants are submitted to drought (Aranjuelo et al., 2013; Gil-Quintana, Lyon, Staudinger, Wienkoop, & Gonzalez, 2015; Gogorcena et al., 1997), cadmium stress (Balestrasse, Gallego, & Tomaro, 2004; Redondo, de la Peña, Morcillo, Lucas, & Pueyo, 2009), photosynthesis inhibitors (Müller, Boller, & Wiemken, 2001), prolonged darkness (Hernández-Jiménez, Mercedes Lucas, & de Felipe, 2002; Matamoros et al., 1999; Perez-Guerra et al., 2010; Pierre et al., 2014), defoliation (Vance

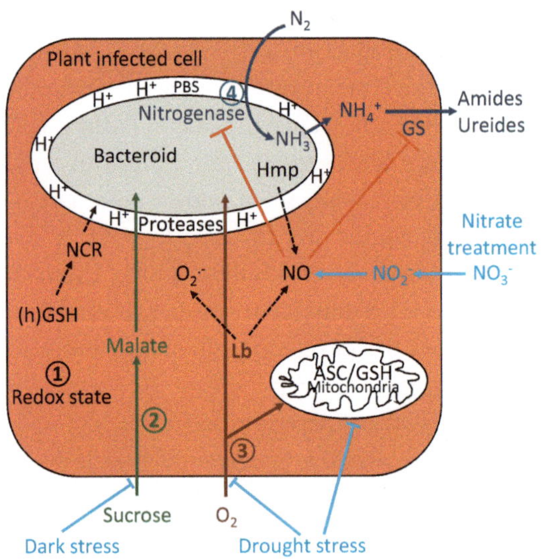

Fig. 2 Major processes triggering nodule senescence in response to environmental constraints. (1), Modification of the redox balance; (2), alteration of the bacteroid carbon nutrition; (3), alteration of O_2 homeostasis; (4), alteration of N_2 fixation (nitrogenase). Red lines indicate down-regulation. Black dotted lines indicate scavenging of NO or O_2 by Lb or Hmp proteins. The pink color of the plant cell cytoplasm illustrates the Lb content. *ASC*: Ascorbate; *Hmp*: Flavohemoglobin; *GS*: Glutamine; *(h)GSH*: (homo) Glutathione; synthetase; *Lb*: Leghemoglobin; *PBS*: Peribacteroid space; *Proteases*: Cysteine proteases; *NCR*: Nodule cysteine-rich peptides.

et al. al., 1979), or treatments with argon (Curioni, Hartwig, Nosberger, & Schuller, 1999), nitrate (Matamoros et al., 1999; Pierre et al., 2014) or abscisic acid (ABA; Gonzalez, Galvez, & Arrese-Igor, 2001). Alternately, senescence can result from a deficient recognition between the two partners (Banba, Siddique, Kouchi, Izui, & Hata, 2001). Depending on the type of triggering factor, the senescence process is more or less rapid and implement variable responses (Van de Velde et al., 2006), but a number of common typical features are found in most nodule senescence processes.

During developmental senescence, the first sign of nodule senescence is the drop in measurable N_2 fixation activity that occurs in parallel with a decrease in Lb content (Dalton, Russell, Hanus, Pascoe, & Evans, 1986; Lahiri, Chattopadhyay, Chatterjee, & Ghosh, 1993). However, during induced senescence, such as after dark treatment, drought stress, or ABA treatment, the rapid decline in N_2 fixation occurs well before a decrease in Lb transcripts or amounts is observed (Escuredo et al., 1996; Gogorcena et al., 1997; Gonzalez et al., 2001; Matamoros et al., 1999). The most visible sign of senescence is a change of the nodule color, from pink to green, caused by bilirubin and biliverdin accumulation after disruption of the Lb structure (Lehtovaara & Perttilä, 1978; Swaraj & Bishnoi, 1996). In senescent soybean nodules, the green color comes partially from Lb nitration, which highlights the critical role of ROS and RNS in the senescence process (Meakin et al., 2007; Navascués et al., 2012; Sainz et al., 2015).

In nodules of the determinate type, senescence develops radially (Fig. 1), starting from the center of the nodule and then progressively extending to the peripheral tissues in a few weeks (Puppo et al., 2005). In nodules of indeterminate type, the N_2 fixation zone is constantly renewed and the senescence zone (zone IV) grows as the nodule grows (Fig. 1); it increases from the root proximal zone toward the apical zone of the nodule (Montiel, Arthikala, Cardenas, & Quinto, 2016; Perez-Guerra et al., 2010; Puppo et al., 2005). Because this zonation is not strictly defined, a senescence inter-zone is sometimes described between zone III and zone IV. Histologically, early nodule senescent cells exhibit a less dense cytoplasm containing numerous vacuoles, altered peroxisomes, mitochondria, starch granules, and late senescent cells ultimately accumulate damaged "ghost" membranes and cell walls (Perez-Guerra et al., 2010; Puppo et al., 2005). In common bean nodule developmental senescence, mitochondria were found to be early targets of oxidative modifications and a potential source of ROS/RNS (Matamoros et al., 2013). A dynamic evolution of both plant and bacteroid cell structures is observed: bacteroids are first degraded whereas plant cells remain alive during the early senescence stage, allowing nutrient remobilization, even though at later stages plant cells are also degraded (Van de Velde et al., 2006; Vasse, de Billy, Camut, & Truchet, 1990). In senescent cells, the degradation of symbiosomes is correlated with the loss of the acidity of the peribacteroid space (Pierre et al., 2013), this acidification being an indicator of the bacteroid functioning in the nodule N fixing zone (Pierre et al., 2013).

The senescence process was analyzed at gene expression level in both leaves and nodules, particularly in *Medicago truncatula* (De Michele et al., 2009; Van de Velde et al., 2006). It appeared that nodule senescence has a significant degree of overlap with leaf senescence, arguing for the recruitment of similar pathways. A strong redox regulation, through antioxidants decrease and ROS/RNS accumulation, as well as hormonal regulations such as abscisic acid, ethylene or jasmonic acid, was evidenced to correlate with catabolic and remobilization processes during nodule senescence (Puppo et al., 2005; Van de Velde et al., 2006).

2.2 Redox state, ROS and RNS

Changes in the redox balance, which modify the antioxidant defense and the level of ROS and RNS, is directly associated with the transition between N_2 fixation and senescence in nodules (Fig. 2; Puppo et al., 2005).

Ascorbate (ASC) and glutathione (GSH) are the main reported soluble antioxidants within nodules and their regulation was supposed to control the organ lifespan (Becana, Matamoros, Udvardi, & Dalton, 2010; Loscos, Matamoros, & Becana, 2008). In soybean plants, drought-induced nodule senescence is characterized by a decrease in ASC reduction state and an increase in GSH content (Marquez-Garcia et al., 2015). In bean nodules, the progression of developmental senescence correlates with a decrease in both ASC and GSH concentrations and reduction states (Loscos et al., 2008). Moreover, the decrease in ASC and GSH concentrations in mitochondria correlates to an increase of oxidative modifications of lipids and proteins, supporting the hypothesis that the mitochondria is an early target of senescence in nodules (Matamoros et al., 2013). In contrast, an increased level of GSH in the N_2-fixing zone is correlated with a higher N_2 fixation efficiency (El Msehli et al., 2011). GSH content is notably involved in the control of glutathione *S*-transferases (GSTs) activities, which are enzymes involved in the cell protection from an oxidative damage, by catalyzing the conjugation of GSH with oxidative degradation products. The level of the GST9 protein, the most prevalent glutathione-S-transferase in soybean nodules, is accordingly increased during senescence (Dalton et al., 2009). A decrease in N_2 fixation associated with an increase of the oxidatively damaged protein content is observed in GST9-RNAi silenced soybean nodules, illustrating the role of GSTs in nodule senescence through the regulation of antioxidant defences (Dalton et al., 2009). In parallel, the importance of GSH during symbiosis of *Sinorhizobium meliloti* and *Rhizobium tropici* has been also demonstrated (Harrison et al., 2005; Muglia, Comai, Spegazzini, Riccillo, & Aguilar, 2008). The *S. meliloti* and *R. tropici* strains deficient in *gshB*, the gene encoding the enzyme catalyzing the second step of GSH synthesis, have a reduced N_2 fixation capacity associated with early nodule senescence. Similarly, mutation in the *S. meliloti* glutathione reductase gene, *gor*, impaired N_2 fixation efficiency (Tang et al., 2018).

During the legume-rhizobium symbiosis, both ROS and RNS, and especially nitric oxide (NO), appear to play crucial roles in the establishment and functioning of the interaction (Hichri et al., 2015; Puppo et al., 2013). ROS including superoxide radicals ($O_2^{\cdot-}$) and hydrogen peroxide (H_2O_2) are produced in high amounts in nodules due to elevated rates of respiration in bacteroids and plant mitochondria (Matamoros et al., 2013). Plant NADPH oxidases, known as Respiratory Burst Oxidase Homologues (RBOHs), are plasma membrane enzymes catalyzing the reduction of O_2 to generate $O_2^{\cdot-}$ using the reducing power of NADPH. Seven *Rboh* genes have been identified in legume genomes (Marino et al., 2011) and these oxidases are involved in legume-rhizobium symbioses (Arthikala et al., 2014; Marino et al., 2011; Montiel et al., 2016). Other possible sources for H_2O_2 are cell wall peroxidases, germin-like oxalate oxidases, and diamine oxidases (Smirnoff & Arnaud, 2019). On the other hand, NO is a reactive diffusible gas known to have important functions as inter- or intra-cellular signaling molecules in plant growth, development and defense against abiotic or biotic stresses (Brouquisse, 2019, and references therein). For a decade, NO has been also identified at all stages of legume-rhizobium symbiotic interactions (Berger, Boscari, Frendo, & Brouquisse, 2019; Hichri et al., 2016, 2015). The way NO is produced in legumes is not completely clear. The best characterized sources of NO are the nitrate reductase (NR) and the mitochondrial transfer chain (Horchani et al., 2011). Interestingly, it has been shown in soybean and *M. truncatula* nodules that the micro-symbiont accounts for 30% up to 90% of NO production thanks to a nitrite reductase (Nir) belonging to the denitrification pathway (Calvo-Begueria et al., 2018; Horchani et al., 2011; Sánchez et al., 2010). Even though ROS and NO are different types of molecules issuing from distinct synthesis pathways, some parallels can be established when looking at their occurrence during symbiosis. Strikingly, NO and ROS have comparable production patterns during symbiosis such as an early production burst in the root in response to rhizobia inoculation, later in the infection thread, then during nodule organogenesis and finally in mature nodules (Hichri et al., 2016; Montiel et al., 2016).

Interestingly, NO and ROS play a role in nodule senescence. At a late stage of the nodulation process (e.g. 6 weeks post-inoculation), H_2O_2 is detected in ultrathin sections of nodules (Puppo et al., 2013). At the ultrastructural level, the senescent nodules of *Glycine* max and *Pisum sativum* plants display alterations in the symbiosome structure accompanied by the presence of H_2O_2, supporting the idea that ROS have roles also during senescence (Montiel et al., 2016). In addition, it has been suggested that NO is present in higher amounts during the late symbiotic stages than at earlier stages (Bruand & Meilhoc, 2019). Indeed, NO detected by means of a bacterial biosensor is more abundant in the root-proximal part of *M. truncatula* nodules, where senescence is known to initiate (Cam et al., 2012). Similarly, NO detected by means of a fluorescent probe accumulates in old *Lotus japonicus* nodules and Lbs with nitrated heme groups are found in old soybean nodules, indicating the presence of NO in aging nodules (Chungopast, Duangkhet, Tajima, Ma, & Nomura, 2017; Navascués et al., 2012; Sainz et al., 2015). Remarkably, exogenous addition of a NO donor on mature nodules of *M. truncatula* lead to an accelerated senescence of the nodules (Cam et al., 2012). Also, several *S. meliloti* mutant strains, altered on *hmp*, *nnrS1*, *nnrS2* and *norB* genes (Table 1) and showing either increased or decreased NO degradation capacity compared to the wild-type, were used to modulate NO levels within nodules and, strikingly, strains deficient in NO degradation induced an early nodule senescence, whereas strains with higher degradation activity led to a delayed nodule senescence (Blanquet et al., 2015; Cam et al., 2012; Meilhoc, Blanquet, Cam, & Bruand, 2013). NO is also involved in stress-induced *M. truncatula* nodule senescence (Cam et al., 2012). Comparable results were recently obtained in determinate *L. japonicus* nodules (Fukudome et al., 2019). Indeed, overexpression of a phytoglobin gene *LjGlb1-1* (Table 1) decreased NO levels in nodules and delayed nodule senescence.

By analogy, Redondo et al. (2009, 2012) reported that overexpression in *S. meliloti* of a cyanobacterial flavodoxin involved in the response to oxidative stress in bacteria and cyanobacteria improves the oxidative balance, delays developmental nodule senescence in *Medicago sativa*, and enhances

Table 1 Genes involved in symbiotic nodule senescence.

Gene (mutant)	Species	Gene characteristics	Reference
LjIPT3	*Lotus japonicus*	Reduced nitrogen fixation observed in RNAi plants. Coding for an Isopenthyl-Transferase, a rate limiting step in cytokinin biosynthesis	Chen et al., 2014
MtCP6 and *MtVPE*	*Medicago truncatula*	RNAi plants are delayed in the apparition of senescence in nodule	Pierre et al., 2014
MtNAC969	*Medicago truncatula*	Accumulation of senescence marker transcripts in RNAi plants	de Zélicourt et al., 2012
Hmp (*hmp*$^{++}$)	*Sinorhizobium meliloti*	Hmp overexpression delays nodule senescence. Hmp encodes a flavohemoglobin involved in NO detoxication	Cam et al., 2012
nnrS1 *nnrS2*	*Sinorhizobium meliloti*	nnrS1/2 mutants induce early nodule senescence. nnrS1/2 encode for cooper-containing membrane proteins involved in NO detoxication	Blanquet et al., 2015
norB	*Sinorhizobium meliloti*	norB mutants induce early nodule senescence. norB encodes for a NO reductase in the denitrification pathway	Meilhoc et al., 2013
LjGlb1-1	*Lotus japonicus*	Glb1-1 overexpression delays nodule senescence. Glb1-1 encodes a Phytoglobin involved in NO detoxification	Fukudome et al., 2019
vapBC-5	*Sinorhizobium meliloti*	Delayed senescent phenotype. Coding an active bacteria toxin	Lipuma et al., 2014

IPT: IsoPentenyl Transferase, *CP*: Cystein Protease, *VPE*: Vacuolar Processing Enzyme, *NAC*:NAM ATAF CUC transcription factor, *Hmp*: flavohemoglobin, *nnrS*: nitrite and nitric oxide reductase S, *norB*: NO reductase B, *Glb1*: Class 1 Phytoglobulin, *vapBC*: toxin-antitoxin module.

tolerance to cadmium. Overexpression of this flavodoxin also permitted to avoid the decline in nitrogenase activity associated with a salinity stress whether the flavodoxin was expressed in *S. meliloti* or in *M. truncatula* (Coba de la Peña, Redondo, Manrique, Lucas, & Pueyo, 2010). In *S. meliloti*, various ROS scavenging enzymes (katA and katC catalases particularly) are expressed in the N_2-fixing zone (Roux et al., 2014). The *S. meliloti katA katC* double mutant displays a reduced N_2 fixation activity and bacteroids underwent faster senescence (Puppo et al., 2013). *Rhizobium etli* and *Bradyrhizobium japonicum* possess only one catalase gene and corresponding mutants were not affected in nodule functioning, but ROS scavenging enzymes other than catalases might reduce H_2O_2 (Puppo et al., 2013). This might also suggest a different role for ROS and antioxidant defense components in the senescence of determinate and indeterminate nodules.

A drastic reduction in N_2 fixation was observed in *Phaseolus vulgaris* nodules silenced in NADPH oxidase (*Rboh*) genes probably linked to an early senescence phenotype (Arthikala et al., 2014). By contrast, overexpression of *RbohB* delays senescence of the nodule. Hence, these elements suggest that ROS produced by RBOHs may have either a promoting, or an inhibiting role during nodule senescence in *P. vulgaris*. Two *Rboh* genes (*MtRbohA and MtRbohB)* were found specifically expressed in the N_2-fixing zone of *M. truncatula* nodules (Marino et al., 2011; Montiel et al., 2016; Roux et al., 2014). A decrease in *MtRbohA* expression via a RNA interference (RNAi) approach led to a reduction in the N_2 fixation capacity, showing again the link between MtRBOHA and nodule function (Marino et al., 2011). However, none of these genes displays any detectable promoter activity in the senescence zone of indeterminate nodules, suggesting that they do not directly act in nodule senescence. In addition, RBOHs might have a more prominent role in determinate nodules than in indeterminate nodules, even though the possibility that they participate in nodule senescence should not be eliminated before each gene has been studied in detail in different plants.

Hypotheses on mechanisms by which NO and ROS could trigger nodule senescence are multiple. These molecules can both act at the transcriptional and post-translational levels. Indeed, NO is known to activate the expression of bacterial as well as plant genes (Boscari et al., 2013a; Ferrarini et al., 2008; Meilhoc, Cam, Skapski, & Bruand, 2010). However, it is unknown yet which genes or transcription factors (TFs) could be involved at the onset of nodule senescence. One of the main downstream effects of NO is the post-translational modification of proteins such as

S-nitrosation of cysteine residues or nitration of tyrosine residues, which might cause conformational changes in the protein with concomitant changes in their activities, turnover or localization. About 80 proteins were found to be S-nitrosated in *M. truncatula* nodules; among them 20 and 8 plant and bacterial proteins, respectively, were also found to be sulfenylated by ROS (Puppo et al., 2013). These protein targets remain to be characterized to establish a potential link with the nodule senescence. However, the activity of many of them, involved either in the glycolysis, the tricarboxylic acid cycle, the energy regeneration or the N_2 assimilation, were found to be inhibited by NO, suggesting that an increased NO production could directly down-regulate N, carbon and energy metabolism in senescent nodules (Boscari et al., 2013b). Interestingly, the enzymatic activity of the plant glutamine synthetases (GS) involved in N assimilation decreases when GS is subjected to NO-mediated tyrosine nitration (Melo, Silva, Ribeiro, Seabra, & Carvalho, 2011). Nevertheless, methionine sulfoxidation of GS is also important for its regulation in response to ROS, showing again that both NO and ROS can be involved in the regulation of protein activity (Matamoros et al., 2013). It has also been shown that the nitrogenase activity is inhibited by NO (Trinchant & Rigaud, 1982).

NO and ROS are generated in cells at the same time and in response to the same cues, so that it is important to consider their possible interaction and associated consequences. One of the best known interaction between NO and ROS is the production of peroxynitrite ($ONOO^-$) from NO and superoxide anion ($O_2^{\cdot-}$). This has two potentially important outcomes: first, the reaction removes both $O_2^{\cdot-}$ and NO from the cell environment, thereby reducing their bioavailability; and second, there is a new compound produced ($ONOO^-$) which itself can act as a signaling/regulatory molecule, potentially giving different responses than that resulting from ROS or NO signaling alone. As an example, peroxynitrite is more potent than NO alone to trigger tyrosine nitration (Astier & Lindermayr, 2012).

NO/$ONOO^-$-dependent modifications may also alter signaling pathways mediated by other reactive signals. Indeed, Tyr nitration of superoxide dismutase can inhibit its activity and hence alter ROS signaling (Holzmeister et al., 2015). NO has also been proposed as a negative regulator of RBOH activity in nodulation (Damiani et al., 2016). Indeed, it has been reported that the AtRBOHD enzyme is S-nitrosated which thus reduces its ROS generating activity (Yun et al., 2011). Therefore, NO has a role in controlling ROS levels and hence their associated downstream signaling. In addition, the *B. japonicum* key regulator of the N_2-fixing enzymes $FixK_2$ is reversibly susceptible to oxidation, which might indirectly have an influence on the expression of bacterial genes involved in the NO metabolism regulated by $FixK_2$ (Mesa, Reutimann, Fischer, & Hennecke, 2009). Similarly, LsrB, a TF involved in the nodulation and redox regulation, is also regulated by its redox state and might be involved in the senescence process (Tang et al., 2017).

2.3 Hormones

Plant hormones are other obvious candidate regulatory molecules to control the transition between N_2 fixation and nodule senescence, some of these signals being previously associated with leaf senescence, such as cytokinins that delay leaf senescence (Zwack & Rashotte, 2013). In white lupin, expression of the cytokinin receptor *LaHK1*, closely related to *AHK4/CRE1* in *Arabidopsis thaliana*, is increased during nodule senescence (Coba de la Peña et al., 2008). In *L. japonicus*, expression of the *LjIPT3* (*IsoPentenyl-adenine Transferase 3,* Table 1) gene encoding a rate-limiting cytokinin biosynthetic enzyme, increases during nodule aging and is associated with vascular bundles of mature nodules. *LjIPT3* down-regulation by an RNAi approach triggers a decrease in N_2 fixation, consequently associated with premature nodule senescence (Chen et al., 2014). In addition, a maintained N_2 fixation associated with delayed nodule senescence was observed in drought-stressed nodules of alfalfa inoculated with an *S. meliloti* strain engineered to overexpress an *IPT* cytokinin biosynthesis gene and therefore producing increased levels of cytokinins (Xu, Li, & Luo, 2012). Cytokinin may therefore act as a positive regulator of N_2 fixation (Boivin et al., 2016; Reid, Heckmann, Novak, Kelly, & Stougaard, 2016), and possibly as a negative regulator of root nodule senescence as reported during leaf senescence (Zwack, Robinson, Risley, & Rashotte, 2013).

Besides cytokinins, several gene expression analyses revealed putative functions for other plant hormones (De Michele et al., 2009; Van de Velde et al., 2006). Among those, genes encoding enzymes of the ethylene biosynthesis pathway, such as an 1-aminocyclopropane-1-carboxylate (ACC) oxidase, are induced in senescent nodules (Van de Velde et al., 2006). Interestingly, the lack of an ACC deaminase activity in the *Sinorhizobium* strain sp. BL3 accelerates nodule senescence in mung bean, suggesting a positive role for ethylene in nodule senescence similarly to what is observed during leaf senescence (Tittabutr et al., 2015). Lipoxygenase genes associated with Jasmonic Acid (JA) biosynthesis and GA2-oxidase genes associated with Gibberellic Acid (GA) degradation are also up-regulated during nodule senescence, suggesting that JA may positively and GA negatively regulate nodule senescence (Benedito et al., 2008; Van de Velde et al., 2006). Abscisic acid (ABA) could be also involved in nodule senescence since some ABA responsive genes such as *ABR17* (*ABA Responsive 17*) are induced during nitrate-induced senescence (Liese, Schulze, & Cabeza, 2017). In addition, ABA, which is associated with various environmental stress responses, improves nodule N_2 fixation capacity in alfalfa salt-stressed nodules where antioxidant defenses mediated by the polyamine spermidine and spermine are induced (Palma, Lopez-Gomez, Tejera, & Lluch, 2014), and might therefore regulate senescence progression. Such putative negative effect of ABA on nodule senescence is however not observed in leaves where ABA is a positive regulator of senescence (Lee et al., 2011).

Considered together, if each of these studies suggests that hormones can intervene in the regulation of senescence, in most cases, functional analyses are lacking to clearly identify how these hormones regulate nodule senescence and more specifically how they may interact with ROS and RNS signaling pathways that are the most extensively linked to nodule senescence. These limited data may nevertheless suggest that common hormonal regulations between leaf and nodule senescence may exist, which could be surprising considering their differences in carbon versus N metabolic status.

2.4 Proteases associated with bacteroid and plant cell degradation

A hallmark of cell aging is the progressive decline in proteostasis and the accumulation of misfolded, oxidized and damaged proteins (López-Otín, Blasco, Partridge, Serrano, & Kroemer, 2013). Therefore, proteolytic activities are under strict regulation to ensure protease activation and degradation at appropriate time and space. In plants, proteases have been associated with global protein turnover, protein processing, degradation of regulatory proteins, cell growth and differentiation, stress response and developmental senescence, but to date functional evidence of their mode of action remains still limited (Buono, Hudecek, & Nowack, 2019). There are hundreds of protease-encoding genes; for example, more than 800 have been identified in Arabidopsis (Van der Hoorn, 2008). Such a large number of proteases can be grouped into families according to their catalytic residues, aspartic-, cysteine-, serine-, threonine-, glutamic-, asparagine-, or metalloproteases (Oda, 2012). A description of the protease classification can be retrieved in the MEROPS database (https://www.ebi.ac.uk/merops/; Rawlings, Barrett, & Bateman, 2010).

To date, many reports have elucidated the roles of cysteine proteases (CPs) in nodule senescence and in response to detrimental environments (Cilliers, van Wyk, van Heerden, Kunert, & Vorster, 2018; Perez-Guerra et al. 2010; Pierre et al. 2014; Pladys, Dimitrijevic, & Rigaud, 1991; Pladys & Vance, 1993; Van de Velde et al. 2006). The synthesis of CP precursors is orchestrated by a complex transcriptional regulation network. The post-translational activation of many CPs relies on either self-processing or on the interaction with other enzymes. Their activity is also modulated by inhibitors and co-factors. CPs encoding genes are among the most strongly and specifically induced nodule senescence genes, either developmentally or environmentally triggered (Cabeza et al., 2014; Chungopast et al., 2014; Seabra, Pereira, Becker, & Carvalho, 2012; Van de Velde et al., 2006). Their activity is directly correlated to the protein degradation occurring during senescence, and was documented in various legume plants including soybean (Alesandrini, Mathis, Van de Sype, Hérouart, & Puppo, 2003), *Astragalus sinicus* (Naito, Fujie, Usami, Murooka, & Yamada, 2000), pea (Kardailsky & Brewin, 1996), and *M. truncatula* (Pierre et al., 2014).

In soybean, 18 cysteine proteases were identified as actively transcribed in nodules during the developmental senescence (Van Wyk, Du Plessis, Cullis, Kunert, & Vorster, 2014). This expression was increased at specific time points of the senescence process with a majority of *CP* genes highly expressed in 14 week-old nodules. In *Trifolium repens*, the identified *CP* encoding genes were specifically expressed in the senescent zone of the nodule as revealed by a spatio-temporal analysis (Asp, Bowra, Borg, & Holm, 2004). In *A. sinicus* indeterminate nodules, the papain cysteine protease *AsNodF32* expressed in senescent tissues of the nodule was identified from a differential screening of a nodule cDNA library (Naito et al., 2000). An RNAi silencing of *AsNodF32* extended the period of the active N_2-fixation, delaying nodule senescence (Li et al., 2008). In *M. truncatula* nodules, the Cysteine Protease 6 (*MtCP6*), (MEROPS family C1A; https://www.ebi.ac.uk/merops/), and a Vacuolar Processing Enzyme (*MtVPE*; MEROPS family C13), belonging respectively to papain and legumain families, were strongly up-regulated in response to both developmental and environmentally-induced nodule senescence (Van de Velde et al., 2006; Pierre et al., 2014, Table 1). A detailed spatio-temporal analysis of the *MtCP6* and *MtVPE* expression patterns during nodule developmental senescence revealed an expression associated with the transition region between the N_2-fixing and senescence zones (Fig. 3A and B, Pierre et al., 2014). In nitrate-induced senescent nodules, *MtCP6* and *MtVPE* expression

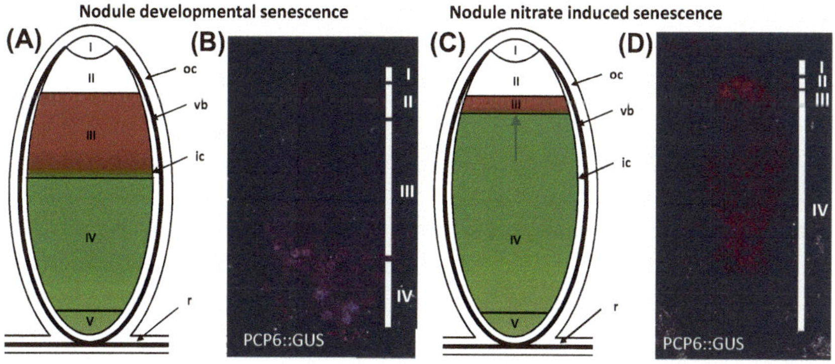

Fig. 3 Developmental and induced senescence in the *M. truncatula* nodule. (A) Schematic representation of developmental/aging senescence in indeterminate nodules. I, meristematic zone; II, rhizobial infection zone; III, nitrogen fixing zone (in red); IV, senescence zone (in green); V, saprophytic zone (in green). *oc*, outer cortex; *ic*, inner cortex; *vb*, vascular bundle; *r*, root. (B) Section of a nodule expressing the reporter gene *uidA* (GUS) under control of the *MtCP6* promoter 6 weeks post-inoculation. (C) Schematic representation of nitrate induced-senescence in indeterminate nodules. The arrow shows the extension of the senescent tissues revealed by the p*MtCP6*:GUS reporter. (D) Section of a nodule expressing the reporter gene *uidA* (GUS) under control of the *MtCP6* promoter 3 days after the nitrate treatment.

was detected in the whole nodule except in the basal senescence zone (Fig. 3C and D). Subcellular localization of both *Mt*CP6 and *Mt*VPE in senescent nodules was associated with vacuoles of cells non-infected by rhizobia and to symbiosomes of infected cells, suggesting a role of these CPs in symbiosomes and plant cell degradation (Pierre et al., 2014) Silencing of *MtCP6* or *MtVPE* genes delayed nodule senescence and increased N_2 fixation, while their ectopic expression in the N_2-fixing zone promoted nodule senescence and hindered N_2 fixation (Pierre et al., 2014). In pea, both *PsCyp1* and *PsCyp15a* (CP) genes transcript levels are increased during nodule aging. However, laser capture microdissection of pea nodule zones III and IV revealed a different pattern of expression: the *PsCyp1* gene is activated at an earlier stage of nodule senescence as transcripts are found exclusively in zone III, whereas the *PsCyp15a* gene is expressed in both zone III and senescent zone IV of pea nodule (Serova, Tikhonovich, & Tsyganov, 2017).

CPs were reported to interact with CP inhibitors called cystatins, which act as proteolysis regulators (Benchabane, Schluter, Vorster, Goulet, & Michaud, 2010; Kunert, van Wyk, Cullis, Vorster, & Foyer, 2015; Martínez, Cambra, Gonzalez-Melendi, Santamaria, & Diaz, 2012; Quain,

Makgopa, Cooper, Kunert, & Foyer, 2015; Van Wyk et al., 2014). In soybean developmental senescent nodules, three cystatin genes are co-expressed with CPs (Van Wyk et al., 2014). Remarkably, despite no direct evidence for a role of cystatins during nodule senescence was yet reported, an ectopic overexpression of a rice oryzacystatin in soybean leads to a strong decrease in cathepsin L-like CP activities that are linked to nodule senescence (Quain et al., 2015). The coordinated expression of transcripts encoding cystatins and CPs interacting partners suggests their involvement as a fine-tuning control of proteolysis events associated with nodule senescence. In *M. truncatula* nodules, serpins (serine protease inhibitors) have also been found to be activated in drought-induced senescence and might play a role as modulators of papain-like CPs (Dhanushkodi, Matthew, McManus, & Dijkwel, 2018).

Increasing data show that the symbiotic establishment relies on the suppression of plant immunity (Liang et al. 2013; Yu et al. 2018). Transcriptomic studies have revealed transient elicitation of legume plant defense related genes upon *rhizobium* inoculation (Libault et al. 2010; Lohar et al. 2006; Pucciariello et al., 2009), followed by their suppressions in the subsequent symbiotic stages (Aslam et al. 2008). Besides being known as an ending stage of organ development, nodule senescence is also proposed to involve a reversion of the plant innate immunity (Puppo et al. 2005). CPs have also been shown to be involved in salicylic acid induced plant defense to cope with microbial pathogens (Müeller, Ziemann, Treitschke, Aßmann, & Doehlemann, 2013; Shindo, Misas-Villamil, Hörger, Song, & van der Hoorn, 2012; Van der Linde et al., 2012). Thus, we could hypothesize that activation of CPs during nodule senescence might be linked to increasing plant defense responses linked to the declining nodule N_2 fixation capacity.

Genes encoding components of the SCF (skp1-cullin-F-Box protein) complex, also known as the E3 ubiquitin ligase complex, and genes encoding 26S proteasome AAA-ATPase subunits and the ubiquitin carrier protein 4 are up-regulated during *M. truncatula* developmental nodule senescence (De Michele et al., 2009; Perez-Guerra et al., 2010; Van de Velde et al., 2006). This suggests that the 26S proteasome pathway may be acting during nodule senescence.

3. Transcriptomic characterization of nodule senescence

To get insights into transcriptional regulations associated with symbiotic nodule senescence, several transcriptomic approaches were performed initially on *M. truncatula* nodules. First, a cDNA-AFLP (Amplified Fragment Length Polymorphism) technique was applied on developmental senescent nodules (Van de Velde et al., 2006). A second approach used microarrays on nodules treated for 8h with phosphinotricin, an inhibitor of GS activity and therefore affecting N_2 fixation (Seabra et al., 2012). More recently, an RNAseq was performed on nodules exposed to nitrate for 4 h or 28 h (Cabeza et al., 2014). Similar studies have been also performed in other legumes species than *M. truncatula*, such as in *L. japonicus* using microarrays and in soybean using RNAseq, at different time-points during nodule aging (Chungopast et al., 2014; Yuan et al., 2017).

Among the differentially expressed genes identified, the downregulation of specific N_2 fixation associated genes, such as Lb, late nodulins and NCR peptide encoding genes attests the inhibition of the N_2 fixation at the molecular level (Cabeza et al., 2014; Seabra et al., 2012; Van de Velde et al., 2006). Genes related to energy production, such as mitochondrial ATP synthase subunit encoding genes, or amino-acid biogenesis encoding genes, were accordingly repressed (Cabeza et al., 2014; Seabra et al., 2012). An upregulation of genes encoding proteins involved in macromolecule degradation, such as the CPs previously mentioned, and nutrient remobilization (phosphate, amino-acid, ion transporters and ABC [ATP-Binding Cassette] transporters) was observed, whereas ammonium, potassium and most of the sulfate transporters were down-regulated (Cabeza et al., 2014; Van de Velde et al., 2006). Genes associated with redox stress responses such as catalases, oxidoreductases, hydrolases, peroxidases, esterases, glutathione transferases were also upregulated, showing the relevance of redox regulations during nodule senescence (Seabra et al., 2012; Van de Velde et al., 2006), as previously described.

In order to distinguish at the molecular level environmentally induced-senescence from developmental senescence, selected genes up-regulated during a *M. truncatula* developmental nodule senescence were tested by qRT-PCR in nodules exposed to darkness for 3–5 days (Perez-Guerra et al., 2010; Van de Velde et al., 2006). Only 50% of the genes selected as up-regulated during developmental senescence were up-regulated during a dark-induced nodule senescence. Genes not regulated in the dark-induced nodule senescence were notably related to the 26S proteasome-mediated protein degradation (Perez-Guerra et al., 2010), suggesting either a less organized, or more probably an accelerated cellular degradation in dark-induced conditions. To refine the spatio-temporal expression profiles during nodule senescence, *M. truncatula* nodules were sectioned into apical versus basal zones and pea nodules were microdissected (Serova et al., 2017; Van de Velde et al., 2006). In this way, the *M. truncatula* or pea *ATB2* genes, encoding a bZIP (basic leucine ZIPper) transcription factor (TF), were identified as reaching a peak in fully senescent nodules (Van de Velde et al., 2006; D'haeseleer et al., 2010; Serova et al., 2017).

To evaluate if genes were specifically regulated during the senescence of nodule organs, a transcriptomic analysis was performed using a cDNA-AFLP approach in *M. truncatula* senescent leaves (De Michele et al., 2009). The overlap between nodule and leaf senescence associated genes was relatively low (23%). When excluding N-metabolism related genes, a higher degree of overlap was nevertheless observed in the early phase, including specific MAP kinases, ubiquitin cycle components, phosphatases, glutathione transferases, and transcription factors; as well as in the late phase, i.e. several CPs and components of the 26S proteasome (Van de Velde et al., 2006; De Michele et al., 2009).

Based on transcriptomic analyses previously described, transcription factors (TF) encoding genes differentially regulated during nodule senescence were identified, mainly from MYB (MYeloBlastosis), bHLH (basic Helix-Loop-Helix), and NAC (NAM ATAF CUC) families (Van de Velde et al., 2006; Seabra et al., 2012; Cabeza et al., 2014; Chungopast et al.; 2014; Yuan et al., 2017). Among those, only few TFs were selected for further functional analyses, with the aim of identifying regulatory pathways controlling the transition between N_2 fixation and nodule senescence, either during aging or in response to environmental stress conditions. The NAC TF family was previously largely associated with leaf senescence based on transcriptome and genetic studies performed in Arabidopsis (Breeze et al., 2011), and is also enriched in nodule senescence transcriptomes. Based on RNAseq or microarray experiments, approximately one-third of *NAC* genes expressed in nodules are induced by nitrate (16/59 in Cabeza et al. 2014; and 21/57 in Benedito et al., 2008) whereas over 30 out of 90 *NAC* genes are upregulated during *Arabidopsis* leaf senescence (Breeze et al., 2011). Expression of the *M. truncatula* TF *MtNAC969,* closely related to the well-known *Arabidopsis* leaf senescence associated gene *AtNAP* (*NAC-LIKE ACTIVATED BY AP3/PI*), is induced by nitrate and associated with the central region of symbiotic nodules including the N_2-fixation zone (de Zelicourt et al., 2012). In addition, both *AtNAP* and *MtNAC969* genes are induced during *Arabidopsis* and *M. truncatula* leaf senescence, respectively (De Michele et al., 2009; Balazadeh et al., 2010). An RNAi silencing of the *MtNAC969* TF promotes nodule senescence (Table 1), based on the increased expression of *CP* genes used as senescence markers, concomitantly to an accumulation of amyloplasts suggestive of an altered nodule metabolism (de Zelicourt et al., 2012). Expression of the *M. truncatula MtATB2* bZIP TF is increased during nodule aging senescence, as well as by sucrose but not by a nitrate treatment (Van de Velde et al., 2006; D'haeseleer et al., 2010). Despite this transcriptional regulation pattern, ectopic expression of *MtATB2* reduced the number of nodules formed but no senescence phenotype could be observed (D'haeseleer et al., 2010). This highlights that an increased expression in senescent nodules is not a sufficient criteria to identify genes functionally linked to nodule senescence. As *MtATB2* is expressed in the peripheral region of nodules containing the vascular bundles and not in the N_2 fixation or senescence central zones, and because of its regulation by sucrose and not by nitrate, this suggests that this TF might sense carbon sources availability and regulate remobilization processes linked to senescence (D'haeseleer et al., 2010). Finally, no TF was yet directly linked to the upregulation of genes encoding CPs or other cell degradation related enzymes, or to the downregulation of Lb or N-assimilation metabolic enzymes encoding genes, or to components of redox pathways. Similarly, a symbiotic bacterial contribution to the regulation of nodule senescence associated genes has not been explored even if some toxin/antitoxin bacterial systems, such as the vapBC-5 (Table 1), have been showed to alter nodule senescence progression (Oláhet al., 2001; Lipuma et al., 2014).

4. Concluding remarks - future prospects

Senescence affecting the legume-rhizobia symbiotic nodule organ during aging or in response to environmental conditions remains poorly documented. The best functionally characterized senescence mechanisms are related to the redox dependent inhibition of the N_2 fixation and assimilation. Physiological and molecular approaches show that nodule senescence is followed by a decrease in the level and/or reduction state of the main cellular antioxidants ASC and GSH, and by an increase in the production of ROS and RNS, such as H_2O_2 and NO. This results in an increase of oxidized/denatured proteins and subsequently in an inhibition of N, carbon and energy metabolisms. In addition from being harmful to cellular processes, both ROS and NO production are also important components of the signaling network used by legumes and their symbionts to communicate at the various stages of the symbiotic interaction. The evolution of efficient ROS/NO degrading systems has most likely enabled plant cells and bacteroids inside nodules to overcome ROS and NO toxicity. Diverse bacterial and plant systems contribute to the control of NO and ROS levels in nodules. Only when these systems are overcome by NO and ROS, these molecules can act as signals to trigger senescence (Puppo et al., 2013). Alternatively, senescence could be a consequence of NO-induced modifications of proteins essential for N_2 fixation. Further studies are needed to link the modifications of these proteins by NO or ROS to nodule senescence.

An increased proteolysis and the induction of plant vacuolar proteases such as CPs and VPEs during senescence are also well documented, and used as a hallmark to follow the senescence progression. Identifying how these nodule specific proteases are regulated and what are their targets remains a challenge to understand better nodule senescence. Both 26S and 20S proteasome proteolytic pathways are additionally induced during nodule senescence, and their contribution should be established. Finally, the nature of bacterial proteases as well as their contribution to the degradation of bacteroid proteins remains to be better documented.

Large-scale transcriptomic analyses from the last decade allowed identifying new candidate genes that may act in the senescence processes. These approaches were however focused on the plant transcriptome, and it would be interesting to analyze in parallel the transcriptional response of symbiotic bacteria to determine if they could contribute to the regulation of nodule senescence progression. In the model legume *M. truncatula*, partially overlapping plant regulatory pathways were identified between environmental, developmental and leaf senescence. This suggests that a core of common genes is conserved between the senescence of different organs, but also that a subset of genes may specifically regulate senescence in nodules, which may be notably related to the symbiotic and metabolic features of this organ hosting numerous N_2 fixing bacteria. Such putative plant-microbe gene network interactions remain however to be investigated. Only few regulatory genes directly regulating nodule senescence have been yet functionally characterized, such as TFs and hormone-related genes. A future prospect would be to integrate how the diversity of bacteria hosted inside the nodules, beyond rhizobia, could control senescence progression. Indeed, promising recent data (Rodrigues, Silveira, Bonifacio, & Figueiredo, 2013) reported that other soil bacteria than rhizobia might indeed modify the nodule lifespan, as cowpea nodules co-inoculated with *Bradyrhizobium* sp. and *Paenibacillus graminis* or *Paenibacillus durus* plant growth-promoting bacteria present an increased N availability and delayed nodule senescence. Finally, an applied perspective of studying nodule senescence would be, as achieved with "staygreen" leaves, to engineer "staypink" nodules that would have maintained, and even improved, capacities to fix N_2 under environmental stress conditions.

Acknowledgments

The three laboratories involved (IPS2, ISA and LIPM) are funded by a common "Agence Nationale de la Recherche" (ANR) project (STAYPINK ANR-15-CE20-0005). The IPS2 institute benefits from the support of the LabEx "Saclay Plant Sciences" (SPS) and of the Lidex "Plant Phenotyping Pipeline" (3P). T.K. was supported by Université Paris-Saclay and ANR fellowships. The ISA was supported by the Institut National de la Recherche Agronomique, the Center National de la Recherche Scientifique, the University of Nice–Sophia-Antipolis, and the French Government (National Research Agency, ANR) through the LABEX SIGNALIFE program (ANR-11-LABX-0028-01). L.Y. was supported by a doctoral fellowship from China Scholarship Council (CSC). The LIPM was supported by the French Laboratory of Excellence project TULIP (ANR-10-LABX-41). E.M. was supported by the National Institute for Applied Science (INSA-Toulouse).

References

Alesandrini, F., Mathis, R., Van de Sype, G., Hérouart, D., & Puppo, A. (2003). Possible roles for a cysteine protease and hydrogen peroxide in soybean nodule development and senescence. *New Phytologist, 158*, 131–138.

Alloing, G., Mandon, K., Boncompagni, E., Montrichard, F., & Frendo, P. (2018). Involvement of glutaredoxin and thioredoxin systems in the nitrogen-fixing symbiosis between legumes and rhizobia. *Antioxidants (Basel), 7*, 182.

Alunni, B., Kevei, Z., Redondo-Nieto, M., Kondorosi, A., Mergaert, P., & Kondorosi, E. (2007). Genomic organization and evolutionary insights on GRP and NCR genes, two large nodule-specific gene families in *Medicago truncatula. Molecular Plant-Microbe Interactions, 20*, 1138–1148.

Appleby, C. A. (1992). The origin and functions of haemoglobin in plants. *Science Progress, 76*, 365–398.

Aranjuelo, I., Tcherkez, G., Molero, G., Gilard, F., Avice, J. C., & Nogués, S. (2013). Concerted changes in N and C primary metabolism in alfalfa (*Medicago sativa*) under water restriction. *Journal of Experimental Botany, 64*, 885–897.

Arthikala, M. K., Sánchez-Lopez, R., Nava, N., Santana, O., Cardenas, L., & Quinto, C. (2014). RbohB, a Phaseolus vulgaris NADPH oxidase gene, enhances symbiosome number, bacteroid size, and nitrogen fixation in nodules and impairs mycorrhizal colonization. *New Phytologist, 202*, 886–900.

Aslam, S. N., Newman, M.-A., Erbs, G., Morrissey, K. L., Chinchilla, D., Boller, T., et al. (2008). Bacterial polysaccharides suppress induced innate immunity by calcium chelation. *Current Biology, 18*, 1078–1083.

Asp, T., Bowra, S., Borg, S., & Holm, P. B. (2004). Cloning and characterisation of three groups of cysteine protease genes expressed in the senescing zone of white clover (*Trifolium repens*) nodules. *Plant Science, 167*, 825–837.

Astier, J., & Lindermayr, C. (2012). Nitric oxide-dependent posttranslational modification in plants: An update. *International Journal of Molecular Sciences, 13*, 15193–15208.

Balazadeh, S., Siddiqui, H., Allu, A. D., Matallana-Ramirez, L. P., Caldana, C., Mehrnia, M., et al. (2010). A gene regulatory network controlled by the NAC transcription factor ANAC092/AtNAC2/ORE1 during salt-promoted senescence. *The Plant Journal, 62*, 250–264.

Balestrasse, K., Gallego, S., & Tomaro, M. (2004). Cadmium-induced senescence in nodules of soybean (*Glycine max* L.) plants. *Plant and Soil, 262*, 373–381.

Banba, M., Siddique, A. B., Kouchi, H., Izui, K., & Hata, S. (2001). *Lotus japonicus* forms early senescent root nodules with *Rhizobium etli*. *Molecular Plant-Microbe Interactions, 14*, 173–180.

Becana, M., Matamoros, M. A., Udvardi, M., & Dalton, D. A. (2010). Recent insights into antioxidant defences of legume root nodules. *New Phytologist, 188*, 960–976.

Benchabane, M., Schluter, U., Vorster, J., Goulet, M. C., & Michaud, D. (2010). Plant cystatins. *Biochimie, 92*, 1657–1666.

Benedito, V. A., Torres-Jerez, I., Murray, J. D., Andriankaja, A., Allen, S., Kakar, K., et al. (2008). A gene expression atlas of the model legume *Medicago truncatula*. *The Plant Journal, 55*, 504–513.

Berger, A., Boscari, A., Frendo, P., & Brouquisse, R. (2019). Nitric oxide signaling, metabolism and toxicity in nitrogen-fixing symbiosis. *Journal of Experimental Botany, 70*. https://doi.org/10.1093/jxb/erz159 (in press).

Bethlenfalvay, G. J., & Phillips, D. A. (1977). Ontogenetic interactions between photosynthesis and symbiotic nitrogen fixation in legumes. *Plant Physiology, 60*, 419–421.

Blanquet, P., Silva, L., Catrice, O., Bruand, C., Carvalho, H., & Meilhoc, E. (2015). *Sinorhizobium meliloti* controls nitric oxide-mediated post-translational modification of a *Medicago truncatula* nodule protein. *Molecular Plant-Microbe Interactions, 28*, 1353–1363.

Boivin, S., Kazmierczak, T., Brault, M., Wen, J., Gamas, P., et al. (2016). Different cytokinin histidine kinase receptors regulate nodule initiation as well as later nodule developmental stages in *Medicago truncatula*. *Plant, Cell and Environment, 39*, 2198–2209.

Boscari, A., Del Giudice, J., Ferrarini, A., Venturini, L., Zaffini, A.-L., Delledonne, M., et al. (2013a). Expression dynamics of the *Medicago truncatula* transcriptome during the symbiotic interaction with *Sinorhizobium meliloti*: Which role for nitric oxide? *Plant physiology, 161*, 425–439.

Boscari, A., Meilhoc, E., Castella, C., Bruand, C., Puppo, A., & Brouquisse, R. (2013b). Which role for nitric oxide in symbiotic N2-fixing nodules: Toxic by-product or useful signaling/metabolic intermediate? *Frontiers of Plant Science, 4*, 384.

Breeze, E., Harrison, E., McHattie, S., Hughes, L., Hickman, R., Hill, C., et al. (2011). High-resolution temporal profiling of transcripts during Arabidopsis leaf senescence reveals a distinct chronology of processes and regulation. *The Plant Cell, 23*, 873–894.

Brouquisse, R. (2019). Multifaceted roles of nitric oxide in plants. *Journal of Experimental Botany, 70*. https://doi.org/10.1093/jxb/erz352 (in press).

Bruand, C., & Meilhoc, E. (2019). Nitric oxide in plants: Pro- or anti-senescence. *Journal of Experimental Botany, 70*. https://doi.org/10.1093/jxb/erz117 (in press).

Buono, R. A., Hudecek, R., & Nowack, M. K. (2019). Plant proteases during developmental programmed cell death. *Journal of Experimental Botany, 70*, 2097–2112.

Cabeza, R., Koester, B., Liese, R., Lingner, A., Baumgarten, V., Dirks, J., et al. (2014). An RNA sequencing transcriptome analysis reveals novel insights into molecular aspects of the nitrate impact on the nodule activity of *Medicago truncatula*. *Plant Physiology, 164*, 400–411.

Calvo-Begueria, L., Rubio, M. C., Martinez, J. I., Perez-Rontome, C., Delgado, M. J., Bedmar, E. J., et al. (2018). Redefining nitric oxide production in legume nodules through complementary insights from electron paramagnetic resonance spectroscopy and specific fluorescent probes. *Journal of Experimental Botany, 69*, 3703–3714.

Cam, Y., Pierre, O., Boncompagni, E., Herouart, D., Meilhoc, E., & Bruand, C. (2012). Nitric oxide (NO): A key player in the senescence of *Medicago truncatula* root nodules. *New Phytologist, 196*, 548–560.

Chen, Y., Chen, W., Li, X., Jiang, H., Wu, P., Xia, K., et al. (2014). Knockdown of LjIPT3 influences nodule development in Lotus japonicus. *Plant and Cell Physiology, 55*, 183–193.

Christenhunsz, M. J., & Byng, J. W. (2016). The number of known plants species in the world and its annual increase. *Phytotaxa, 261*, 201–217.

Chungopast, S., Duangkhet, M., Tajima, S., Ma, J. F., & Nomura, M. (2017). Iron-induced nitric oxide leads to an increase in the expression of ferritin during the senescence of *Lotus japonicus* nodules. *J Plant Physiology, 208*, 40–46.

Chungopast, S., Hirakawa, H., Sato, S., Handa, Y., Saito, K., Kawaguchi, M., et al. (2014). Transcriptomic profiles of nodule senescence in *Lotus japonicus* and *Mesorhizobium loti* symbiosis. *Plant Biotechnology, 31*, 345–349.

Cilliers, M., van Wyk, S. G., van Heerden, P. D. R., Kunert, K. J., & Vorster, B. J. (2018). Identification and changes of the drought-induced cysteine protease transcriptome in soybean (*Glycine max*) root nodules. *Environmental and Experimental Botany, 148*, 59–69.

Coba de la Peña, T., Carcamo, C. B., Almonacid, L., Zaballos, A., Lucas, M. M., Balomenos, D., et al. (2008). A salt stress-responsive cytokinin receptor homologue isolated from *Medicago sativa* nodules. *Planta, 227*, 769–779.

Coba de la Peña, T., Redondo, F. J., Manrique, E., Lucas, M. M., & Pueyo, J. J. (2010). Nitrogen fixation persists under conditions of salt stress in transgenic *Medicago truncatula* plants expressing a cyanobacterial flavodoxin. *Plant Biotechnology Journal, 8*, 954–965.

Curioni, P. M., Hartwig, U. A., Nosberger, J., & Schuller, K. A. (1999). Glycolytic flux is adjusted to nitrogenase activity in nodules of detopped and argon-treated alfalfa plants. *Plant Physiology, 119*, 445–454.

Czernic, P., Gully, D., Cartieaux, F., Moulin, L., Guefrachi, I., Patrel, D., et al. (2015). Convergent evolution of endosymbiont differentiation in Dalbergioid and IRLC legumes mediated by nodule-specific cysteine-rich peptides. *Plant Physiology, 169*, 1254–1265.

D'haeseleer, K., De Keyser, A., Goormachtig, S., & Holsters, M. (2010). Transcription factor MtATB2: About nodulation, sucrose and senescence. *Plant and Cell Physiology, 51*, 1416–1424.

Dalton, D. A., Boniface, C., Turner, Z., Lindahl, A., Kim, H. J., Jelinek, L., et al. (2009). Physiological roles of glutathione s-transferases in soybean root nodules. *Plant Physiology, 150*, 521–530.

Dalton, D. A., Russell, S. A., Hanus, F. J., Pascoe, G. A., & Evans, H. J. (1986). Enzymatic reactions of ascorbate & glutathione that prevent peroxide damage in soybean root nodules. *Proceedings of the National Academy of Sciences, 83*, 3811–3815.

Damiani, I., Pauly, N., Puppo, A., Brouquisse, R., & Boscari, A. (2016). Reactive oxygen species and nitric oxide control early steps of the legume - rhizobium symbiotic interaction. *Frontiers of Plant Science, 7*, 454.

Day, D. A., Poole, P. S., Tyerman, S. D., & Rosendahl, L. (2001). Ammonia and amino acid transport across symbiotic membranes in nitrogen-fixing legume nodules. *Cellular and Molecular Life Sciences: CMLS, 58*, 61–71.

De Michele, R., Formentin, E., Todesco, M., Toppo, S., Carimi, F., Zottini, M., et al. (2009). Transcriptome analysis of *Medicago truncatula* leaf senescence: Similarities and differences in metabolic and transcriptional regulations as compared with Arabidopsis, nodule senescence and nitric oxide signalling. *New Phytologist, 181*, 563–575.

de Zelicourt, A., Diet, A., Marion, J., Laffont, C., Ariel, F., Moison, M., et al. (2012). Dual involvement of a *Medicago truncatula* NAC transcription factor in root abiotic stress response and symbiotic nodule senescence. *The Plant Journal, 70*, 220–230.

Dhanushkodi, R., Matthew, C., McManus, M. T., & Dijkwel, P. P. (2018). Drought-induced senescence of *Medicago truncatula* nodules involves serpin and ferritin to control proteolytic activity and iron levels. *New Phytologist, 220*, 196–208.

Dupont, L., Alloing, G., Pierre, O., El Msehli, S., Hopkins, J., Hérouart, D., et al. (2012). The legume root nodule: From symbiotic nitrogen fixation to senescence. In T. Nagata (Ed.), *Senescence* (Vol. 8, pp. 137–168). Shanghai, China: InTech.

El Msehli, S., Lambert, A., Baldacci-Cresp, F., Hopkins, J., Boncompagni, E., Smiti, S. A., et al. (2011). Crucial role of (homo)glutathione in nitrogen fixation in Medicago truncatula nodules. *New Phytologist, 192*, 496–506.

Escuredo, P. R., Minchin, F. R., Gogorcena, Y., Iturbe-Ormaetxe, I., Klucas, R. V., & Becana, M. (1996). Involvement of activated oxygen in nitrate-induced senescence of pea root nodules. *Plant Physiology, 110*, 1187–1195.

Farkas, A., Maroti, G., Durgo, H., Gyorgypal, Z., Lima, R. M., Medzihradszky, K. F., et al. (2014). *Medicago truncatula* symbiotic peptide NCR247 contributes to bacteroid differentiation through multiple mechanisms. *Proceedings of the National Academy of Sciences, 111*, 5183–5188.

Ferguson, B. J., Indrasumunar, A., Hayashi, S., Lin, M.-H., Lin, Y.-H., Reid, D. E., et al. (2010). Molecular analysis of legume nodule development and autoregulation. *Journal of Integrative Plant Biology, 52*, 61–76.

Ferrarini, A., de Stefano, M., Baudouin, E., Pucciariello, C., Polverari, A., Puppo, A., et al. (2008). Expression of *Medicago truncatula* genes responsive to nitric oxide in pathogenic and symbiotic conditions. *Molecular Plant-Microbe Interactions, 21*, 781–790.

Frendo, P., Matamoros, M. A., Alloing, G., & Becana, M. (2013). Thiol-based redox signaling in the nitrogen-fixing symbiosis. *Frontiers of Plant Science, 4*, 376.

Fukudome, M., Watanabe, E., Osuki, K. I., Imaizumi, R., Aoki, T., Becana, M., et al. (2019). Stably-transformed *Lotus japonicus* plants overexpressing phytoglobin LjGlb1-1 show decreased nitric oxide levels in roots and nodules as well as delayed nodule senescence. *Plant and Cell Physiology, 60*, 816–825.

Gil-Quintana, E., Lyon, D., Staudinger, C., Wienkoop, S., & Gonzalez, E. M. (2015). *Medicago truncatula* and *Glycine max*: Different drought tolerance and similar local response of the root nodule proteome. *Journal of Proteome Research, 14*, 5240–5251.

Gogorcena, Y., Gordon, A. J., Escuredo, P. R., Minchin, F. R., Witty, J. F., Moran, J. F., et al. (1997). N2 fixation, carbon metabolism, and oxidative damage in nodules of dark-stressed common bean plants. *Plant Physiology, 113*, 1193–1201.

Gonzalez, E. M., Galvez, L., & Arrese-Igor, C. (2001). Abscisic acid induces a decline in nitrogen fixation that involves leghaemoglobin, but is independent of sucrose synthase activity. *Journal of Experimental Botany, 52*, 285–293.

Harrison, J., Jamet, A., Muglia, C. I., Van de Sype, G., Aguilar, O. M., Puppo, A., et al. (2005). Glutathione plays a fundamental role in growth and symbiotic capacity of Sinorhizobium meliloti. *Journal of Bacteriology, 187*, 168–174.

Hernández-Jiménez, M. J., Mercedes Lucas, M., & de Felipe, M. R. (2002). Antioxidant defence and damage in senescing lupin nodules. *Plant Physiology and Biochemistry, 40*, 645–657.

Hichri, I., Boscari, A., Castella, C., Rovere, M., Puppo, A., & Brouquisse, R. (2015). Nitric oxide: A multifaceted regulator of the nitrogen-fixing symbiosis. *Journal of Experimental Botany, 66*, 2877–2887.

Hichri, I., Meilhoc, E., Boscari, A., Bruand, C., Frendo, P., & Brouquisse, R. (2016). Nitric oxide: Jack-of-All-Trades of the nitrogen-fixing symbiosis? *Advances in Botanical Research, 77*, 193–218.

Holzmeister, C., Gaupels, F., Geerlof, A., Sarioglu, H., Sattler, M., Durner, J., et al. (2015). Differential inhibition of Arabidopsis superoxide dismutases by peroxynitrite-mediated tyrosine nitration. *Journal of Experimental Botany, 66*, 989–999.

Horchani, F., Prevot, M., Boscari, A., Evangelisti, E., Meilhoc, E., Bruand, C., et al. (2011). Both plant and bacterial nitrate reductases contribute to nitric oxide production in Medicago truncatula nitrogen-fixing nodules. *Plant Physiology, 155*, 1023–1036.

Kardailsky, I. V., & Brewin, N. J. (1996). Expression of cysteine protease genes in pea nodule development and senescence. *Molecular Plant-Microbe Interactions, 9*, 689–695.

Klucas, R. V. (1974). Studies on soybean nodule senescence. *Plant Physiology, 54*, 612–616.

Kondorosi, E., Mergaert, P., & Kereszt, A. (2013). A paradigm for endosymbiotic life: Cell differentiation of rhizobium bacteria provoked by host plant factors. *Annual Review of Microbiology, 67*, 611–628.

Kunert, K. J., van Wyk, S. G., Cullis, C. A., Vorster, B. J., & Foyer, C. H. (2015). Potential use of phytocystatins in crop improvement, with a particular focus on legumes. *Journal of Experimental Botany, 66*, 3559–3570.

Lahiri, K., Chattopadhyay, S., Chatterjee, S., & Ghosh, B. (1993). Biochemical changes in nodules of *Vigna mungo* (L.) during vegetative and reproductive stages of plant growth in the field. *Annals of Botany, 71*, 485–488.

Lamouche, F., Bonadé-Bottino, N., Peter Mergaert, P., & Alunni, B. (2019). Symbiotic efficiency of spherical and elongated bacteroids in the aeschynomene-bradyrhizobium symbiosis. *Frontiers of Plant Science, 10*, 377.

Lawn, R. J., & Brun, W. A. (1974). Symbiotic nitrogen fixation in soybeans. I. Effect of photosynthetic source-sink manipulations1. *Crop Science, 14*, 11–16.

Lee, I. C., Hong, S. W., Whang, S. S., Lim, P. O., Nam, H. G., & Koo, J. C. (2011). Age-dependent action of an ABA-inducible receptor kinase, RPK1, as a positive regulator of senescence in Arabidopsis leaves. *Plant & Cell Physiology, 52*, 651–662.

Lehtovaara, P., & Perttilä, U. (1978). Bile-pigment formation from different leghaemoglobins. Methine-bridge specificity of coupled oxidation. *Biochemical Journal, 176*, 359–364.

Li, Y., Zhou, L., Li, Y., Chen, D., Tan, X., Lei, L., et al. (2008). A nodule-specific plant cysteine proteinase, AsNODF32, is involved in nodule senescence and nitrogen fixation activity of the green manure legume *Astragalus sinicus*. *New Phytologist, 180*, 185–192.

Liang, Y., Cao, Y., Tanaka, K., Thibivilliers, S., Wan, J., Choi, J., et al. (2013). Nonlegumes respond to rhizobial Nod factors by suppressing the innate immune response. *Science, 341*, 1384–1387.

Libault, M., Farmer, A., Brechenmacher, L., Drnevich, J., Langley, R. J., Bilgin, D. D., et al. (2010). Complete transcriptome of the soybean root hair cell, a single-cell model, and its alteration in response to Bradyrhizobium japonicum infection. *Plant Physiology, 152*, 541–552.

Liese, R., Schulze, J., & Cabeza, R. A. (2017). Nitrate application or P deficiency induce a decline in Medicago truncatula N2-fixation by similar changes in the nodule transcriptome. *Scientific Reports, 7*, 46264.

Lipuma, J., Cinege, G., Bodogai, M., Oláh, B., Kiers, A., Endre, G., et al. (2014). A vapBC-type toxin-antitoxin module of Sinorhizobium meliloti influences symbiotic efficiency and nodule senescence of Medicago sativa. *Environmental Microbiology, 16*, 3714–3729.

Lohar, D. P., Sharopova, N., Endre, G., Penuela, S., Samac, D., Town, C., et al. (2006). Transcript analysis of early nodulation events in *Medicago truncatula*. *Plant Physiology, 140*, 221–234.

López-Otín, C., Blasco, M. A., Partridge, L., Serrano, M., & Kroemer, G. (2013). The hallmarks of aging. *Cell, 153*, 1194–1217.

Loscos, J., Matamoros, M. A., & Becana, M. (2008). Ascorbate and homoglutathione metabolism in common bean nodules under stress conditions and during natural senescence. *Plant Physiology, 146*, 1282–1292.

Marino, D., Andrio, E., Danchin, E. G., Oger, E., Gucciardo, S., Lambert, A., et al. (2011). A Medicago truncatula NADPH oxidase is involved in symbiotic nodule functioning. *New Phytologist, 189*, 580–592.

Marquez-Garcia, B., Shaw, D., Cooper, J. W., Karpinska, B., Quain, M. D., Makgopa, E. M., et al. (2015). Redox markers for drought-induced nodule senescence, a process occurring after drought-induced senescence of the lowest leaves in soybean (*Glycine max*). *Annals of Botany, 116*, 497–510.

Martínez, M., Cambra, I., Gonzalez-Melendi, P., Santamaria, M. E., & Diaz, I. (2012). C1A cysteine-proteases and their inhibitors in plants. *Physiologia Plantarum, 145*, 85–94.

Matamoros, M. A., Baird, L. M., Escuredo, P. R., Dalton, D. A., Minchin, F. R., Iturbe-Ormaetxe, I., et al. (1999). Stress-induced legume root nodule senescence. Physiological, biochemical, and structural alterations. *Plant Physiology, 121*, 97–112.

Matamoros, M. A., Fernandez-Garcia, N., Wienkoop, S., Loscos, J., Saiz, A., & Becana, M. (2013). Mitochondria are an early target of oxidative modifications in senescing legume nodules. *New Phytologist, 197*, 873–885.

Meakin, G. E., Bueno, E., Jepson, B., Bedmar, E. J., Richardson, D. J., & Delgado, M. J. (2007). The contribution of bacteroidal nitrate and nitrite reduction to the formation of nitrosylleghaemoglobin complexes in soybean root nodules. *Microbiology, 153*, 411–419.

Meilhoc, E., Blanquet, P., Cam, Y., & Bruand, C. (2013). Control of NO level in rhizobium-legume root nodules: Not only a plant globin story. *Plant Signaling & Behavior, 8*. e25923.

Meilhoc, E., Cam, Y., Skapski, A., & Bruand, C. (2010). The response to nitric oxide of the nitrogen-fixing symbiont *Sinorhizobium meliloti*. *Molecular Plant-Microbe Interactions, 23*, 748–759.

Melo, P. M., Silva, L. S., Ribeiro, I., Seabra, A. R., & Carvalho, H. G. (2011). Glutamine synthetase is a molecular target of nitric oxide in root nodules of *Medicago truncatula* and is regulated by tyrosine nitration. *Plant Physiology, 157*, 1505–1517.

Mergaert, P., Nikovics, K., Kelemen, Z., Maunoury, N., Vaubert, D., Kondorosi, A., et al. (2003). A novel family in *Medicago truncatula* consisting of more than 300 nodule-specific genes coding for small, secreted polypeptides with conserved cysteine motifs. *Plant Physiology, 132*, 161–173.

Mergaert, P., Uchiumi, T., Alunni, B., Evanno, G., Cheron, A., Catrice, O., et al. (2006). Eukaryotic control on bacterial cell cycle and differentiation in the Rhizobium–legume symbiosis. *Proceedings of the National Academy of Sciences, 103*, 5230–5235.

Mesa, S., Reutimann, L., Fischer, H. M., & Hennecke, H. (2009). Posttranslational control of transcription factor FixK2, a key regulator for the *Bradyrhizobium japonicum*-soybean symbiosis. *Proceedings of the National Academy of Sciences, 106*, 21860–21865.

Montiel, J., Arthikala, M. K., Cardenas, L., & Quinto, C. (2016). Legume NADPH oxidases have crucial roles at different stages of nodulation. *International Journal of Molecular Sciences, 17*, 680.

Müeller, A. N., Ziemann, S., Treitschke, S., Aßmann, D., & Doehlemann, G. (2013). Compatibility in the Ustilago maydis–maize interaction requires inhibition of host cysteine proteases by the fungal effector Pit2. *PLoS Pathogens, 9*. e1003177.

Muglia, C., Comai, G., Spegazzini, E., Riccillo, P. M., & Aguilar, O. M. (2008). Glutathione produced by *Rhizobium tropici* is important to prevent early senescence in common bean nodules. *FEMS Microbiology Letters, 286*, 191–198.

Müller, J., Boller, T., & Wiemken, A. (2001). Trehalose becomes the most abundant non-structural carbohydrate during senescence of soybean nodules. *Journal of Experimental Botany, 52*, 943–947.

Naito, Y., Fujie, M., Usami, S., Murooka, Y., & Yamada, T. (2000). The involvement of a cysteine proteinase in the nodule development in Chinese milk vetch infected with *Mesorhizobium huakuii* subsp. rengei. *Plant Physiology, 124*, 1087–1096.

Navascués, J., Perez-Rontome, C., Gay, M., Marcos, M., Yang, F., Walker, F. A., et al. (2012). Leghemoglobin green derivatives with nitrated hemes evidence production of highly reactive nitrogen species during aging of legume nodules. *Proceedings of the National Academy of Sciences, 109*, 2660–2665.

Oda, K. (2012). New families of carboxyl peptidases: Serine-carboxyl peptidases and glutamic peptidases. *Journal of Biochemistry, 151*, 13–25.

Oláh, B., Kiss, E., Györgypál, Z., Borzi, J., Cinege, G., Csanádi, G., et al. (2001). Mutation in the ntrR gene, a member of the vap gene family, increases the symbiotic efficiency of *Sinorhizobium meliloti. Molecular Plant-Microbe Interactions, 14*, 887–894.

Oldroyd, G. E. (2013). Speak, friend, and enter: Signalling systems that promote beneficial symbiotic associations in plants. *Nature Reviews Microbiology, 11*, 252.

Oono, R., Schmitt, I., Sprent, J. I., & Denison, R. F. (2010). Multiple evolutionary origins of legume traits leading to extreme rhizobial differentiation. *New Phytologist, 187*, 508–520.

Ott, T., van Dongen, J. T., Gu, C., Krusell, L., Desbrosses, G., Vigeolas, H., et al. (2005). Symbiotic leghemoglobins are crucial for nitrogen fixation in legume root nodules but not for general plant growth and development. *Current Biology, 15*, 531–535.

Palma, F., Lopez-Gomez, M., Tejera, N. A., & Lluch, C. (2014). Involvement of abscisic acid in the response of *Medicago sativa* plants in symbiosis with *Sinorhizobium meliloti* to salinity. *Plant Science, 223*, 16–24.

Perez-Guerra, J. C., Coussens, G., De Keyser, A., De Rycke, R., De Bodt, S., Van De Velde, W., et al. (2010). Comparison of developmental and stress-induced nodule senescence in *Medicago truncatula. Plant Physiology, 152*, 1574–1584.

Peters, J. W., Fisher, K., & Dean, D. R. (1995). Nitrogenase structure and function: A biochemical-genetic perspective. *Annual Review of Microbiology, 49*, 335–366.

Pierre, O., Engler, G., Hopkins, J., Brau, F., Boncompagni, E., & Herouart, D. (2013). Peribacteroid space acidification: A marker of mature bacteroid functioning in *Medicago truncatula* nodules. *Plant, Cell and Environment, 36*, 2059–2070.

Pierre, O., Hopkins, J., Combier, M., Baldacci, F., Engler, G., Brouquisse, R., et al. (2014). Involvement of papain and legumain proteinase in the senescence process of *Medicago truncatula* nodules. *New Phytologist, 202*, 849–863.

Pladys, D., Dimitrijevic, L., & Rigaud, J. (1991). Localization of a protease in protoplast preparations of infected cells of French bean nodules. *Plant Physiology, 97*, 1174–1180.

Pladys, D., & Vance, C. P. (1993). Proteolysis during development and senescence of effective and plant gene-controlled ineffective alfalfa nodules. *Plant Physiology, 103*, 379–384.

Preisig, O., Zufferey, R., Thony-Meyer, L., Appleby, C. A., & Hennecke, H. (1996). A high-affinity cbb3-type cytochrome oxidase terminates the symbiosis-specific respiratory chain of *Bradyrhizobium japonicum. Journal of Bacteriology, 178*, 1532–1538.

Pucciariello, C., Innocenti, G., Van de Velde, W., Lambert, A., Hopkins, J., Clément, M., et al. (2009). (Homo)glutathione depletion modulates host gene expression during the symbiotic interaction between *Medicago truncatula* and *Sinorhizobium meliloti. Plant Physiology, 151*, 1186–1196.

Puppo, A., Groten, K., Bastian, F., Carzaniga, R., Soussi, M., Lucas, M. M., et al. (2005). Legume nodule senescence: Roles for redox and hormone signalling in the orchestration of the natural aging process. *New Phytologist, 165*, 683–701.

Puppo, A., Pauly, N., Boscari, A., Mandon, K., & Brouquisse, R. (2013). Hydrogen peroxide and nitric oxide: Key regulators of the legume-rhizobium and mycorrhizal symbioses. *Antioxidants and Redox Signaling, 18*, 2202–2219.

Quain, M. D., Makgopa, M. E., Cooper, J. W., Kunert, K. J., & Foyer, C. H. (2015). Ectopic phytocystatin expression increases nodule numbers and influences the responses of soybean (*Glycine max*) to nitrogen deficiency. *Phytochemistry, 112*, 179–187.

Rawlings, N. D., Barrett, A. J., & Bateman, A. (2010). MEROPS: The peptidase database. Nucleic Acids *Research, 38*, D227–D233.

Redondo, F. J., Coba de la Pena, T., Lucas, M. M., & Pueyo, J. J. (2012). Alfalfa nodules elicited by a flavodoxin-overexpressing *Ensifer meliloti* strain display nitrogen-fixing activity with enhanced tolerance to salinity stress. *Planta, 236*, 1687–1700.

Redondo, F. J., de la Peña, T. C., Morcillo, C. N., Lucas, M. M., & Pueyo, J. J. (2009). Overexpression of flavodoxin in bacteroids induces changes in antioxidant metabolism leading to delayed senescence and starch accumulation in alfalfa root nodules. *Plant Physiology, 149*, 1166–1178.

Reid, D. E., Heckmann, A. B., Novak, O., Kelly, S., & Stougaard, J. (2016). CYTOKININ OXIDASE/DEHYDROGENASE3 maintains cytokinin homeostasis during root and nodule development in *Lotus japonicus*. *Plant Physiology, 170*, 1060–1074.

Rodrigues, A. C., Silveira, J. A. G., Bonifacio, A., & Figueiredo, M. D. V. B. (2013). Metabolism of nitrogen and carbon: Optimization of biological nitrogen fixation and cowpea development. *Soil Biology and Biochemistry, 67*, 226–234.

Roux, B., Rodde, N., Jardinaud, M. F., Timmers, T., Sauviac, L., Cottret, L., et al. (2014). An integrated analysis of plant and bacterial gene expression in symbiotic root nodules using laser-capture microdissection coupled to RNA sequencing. *The Plant Journal, 77*, 817–837.

Sainz, M., Calvo-Begueria, L., Perez-Rontome, C., Wienkoop, S., Abian, J., Staudinger, C., et al. (2015). Leghemoglobin is nitrated in functional legume nodules in a tyrosine residue within the heme cavity by a nitrite/peroxide-dependent mechanism. *The Plant Journal, 81*, 723–735.

Sánchez, C., Gates, A. J., Meakin, G. E., Uchiumi, T., Girard, L., Richardson, D. J., et al. (2010). Production of nitric oxide and nitrosylleghemoglobin complexes in soybean nodules in response to flooding. *Molecular Plant-Microbe Interactions, 23*, 702–711.

Seabra, A. R., Pereira, P. A., Becker, J. D., & Carvalho, H. G. (2012). Inhibition of glutamine synthetase by phosphinothricin leads to transcriptome reprograming in root nodules of *Medicago truncatula*. *Molecular Plant-Microbe Interactions, 25*, 976–992.

Serova, T., & Tsyganov, V. (2014). Symbiotic nodule senescence in legumes: Molecular-genetic and cellular aspects (review). *Agricultural Biology, 5*, 3–15.

Serova, T. A., Tikhonovich, I. A., & Tsyganov, V. E. (2017). Analysis of nodule senescence in pea (*Pisum sativum* L.) using laser microdissection, real-time PCR, and ACC immunolocalization. *Journal of Plant Physiology, 212*, 29–44.

Shindo, T., Misas-Villamil, J. C., Hörger, A. C., Song, J., & van der Hoorn, R. A. (2012). A role in immunity for Arabidopsis cysteine protease RD21, the ortholog of the tomato immune protease C14. *PLoS One, 7*. e29317.

Smirnoff, N., & Arnaud, D. (2019). Hydrogen peroxide metabolism and functions in plants. *New Phytologist, 221*, 1197–1214.

Smith, S. E., & Smith, F. A. (2011). Roles of arbuscular mycorrhizas in plant nutrition and growth: New paradigms from cellular to ecosystem scales. *Annual Review of Plant Biology, 62*, 227–250.

Soupène, E., Foussard, M., Boistard, P., Truchet, G., & Batut, J. (1995). Oxygen as a key developmental regulator of Rhizobium meliloti N2-fixation gene expression within the alfalfa root nodule. *Proceedings of the National Academy of Sciences, 92*, 3759–3763.

Streeter, J. G. (1981). Seasonal distribution of carbohydrates in nodules and stem exudate from field-grown soya bean plants. *Annals of Botany, 48*, 441–450.

Swaraj, K., & Bishnoi, N. R. (1996). Physiological and biochemical basis of nodule senescence in legumes: A review. *Plant Physiology and Biochemistry, 23*, 105–116.

Tang, G., Li, N., Liu, Y., Yu, L., Yan, J., & Luo, L. (2018). *Sinorhizobium meliloti* glutathione reductase is required for both redox homeostasis and symbiosis. *Applied and Environmental Microbiology, 84*. e01937-17.

Tang, G., Xing, S., Wang, S., Yu, L., Li, X., Staehelin, C., et al. (2017). Regulation of cysteine residues in LsrB proteins from *Sinorhizobium meliloti* under free-living conditions and symbiotic oxidative stress. *Environmental Microbiology, 19*, 5130–5145.

Timmers, A. C. J., Auriac, M. C., & Truchet, G. (1999). Refined analysis of early symbiotic steps of the Rhizobium-Medicago interaction in relationship with microtubular cytoskeleton rearrangements. *Development, 126*, 3617–3628.

Tittabutr, P., Sripakdi, S., Boonkerd, N., Tanthanuch, W., Minamisawa, K., & Teaumroong, N. (2015). Possible role of 1-aminocyclopropane-1-carboxylate (ACC) deaminase activity of *Sinorhizobium* sp. BL3 on symbiosis with mung bean and determinate nodule senescence. *Microbes and Environments, 30*, 310–320.

Tjepkema, J. D., & Yocum, C. S. (1974). Measurement of oxygen partial pressure within soybean nodules by oxygen microelectrodes. *Planta, 119*, 351–360.

Trinchant, J. C., & Rigaud, J. (1982). Nitrite and nitric oxide as inhibitors of nitrogenase from soybean bacteroids. *Applied and Environmental Microbiology, 44*, 1385–1388.

Van Der Heijden, M. G., De Bruin, S., Luckerhoff, L., Van Logtestijn, R. S., & Schlaeppi, K. (2016). A widespread plant-fungal-bacterial symbiosis promotes plant biodiversity, plant nutrition and seedling recruitment. *The ISME Journal, 10*, 389.

Van Wyk, S. G., Du Plessis, M., Cullis, C. A., Kunert, K. J., & Vorster, B. J. (2014). Cysteine protease and cystatin expression and activity during soybean nodule development and senescence. *BMC Plant Biology, 14*, 294.

Vance, C. P., Heichel, G. H., Barnes, D. K., Bryan, J. W., & Johnson, L. E. (1979). Nitrogen fixation, nodule development, and vegetative regrowth of alfalfa (*Medicago sativa* L.) following harvest. *Plant Physiology, 64*, 1–8.

Van der Hoorn, R. A. (2008). Plant proteases: From phenotypes to molecular mechanisms. *Annual Review of Plant Biology, 59*, 191–223.

Van der Linde, K., Hemetsberger, C., Kastner, C., Kaschani, F., van der Hoorn, R. A., Kumlehn, J., et al. (2012). A maize cystatin suppresses host immunity by inhibiting apoplastic cysteine proteases. *The Plant Cell, 24*, 1285–1300.

Van de Velde, W., Guerra, J. C., De Keyser, A., De Rycke, R., Rombauts, S., Maunoury, N., et al. (2006). Aging in legume symbiosis. A molecular view on nodule senescence in *Medicago truncatula*. *Plant Physiology, 141*, 711–720.

Van de Velde, W., Zehirov, G., Szatmari, A., Debreczeny, M., Ishihara, H., Kevei, Z., et al. (2010). Plant peptides govern terminal differentiation of bacteria in symbiosis. *Science, 327*, 1122–1126.

Vasse, J., de Billy, F., Camut, S., & Truchet, G. (1990). Correlation between ultrastructural differentiation of bacteroids and nitrogen fixation in alfalfa nodules. *Journal of Bacteriology, 172*, 4295–4306.

Venkateshwaran, M., Volkening, J. D., Sussman, M. R., & Ané, J. M. (2013). Symbiosis and the social network of higher plants. *Current Opinion in Plant Biology, 16*, 118–127.

Wang, W., Shi, J., Xie, Q., Jiang, Y., Yu, N., & Wang, E. (2017). Nutrient exchange and regulation in arbuscular mycorrhizal symbiosis. *Molecular Plant, 10*, 1147–1158.

Xu, J., Li, S.-L., & Luo, L. (2012). Effects of engineered *Sinorhizobium meliloti* on cytokinin synthesis and tolerance of alfalfa to extreme drought stress. *Applied and Environmental Microbiology, 78*, 8056–8061.

Yu, H., Xiao, A., Dong, R., Fan, Y., Zhang, X., Liu, C., et al. (2018). Suppression of innate immunity mediated by the CDPK-Rboh complex is required for rhizobial colonization in *Medicago truncatula* nodules. *New Phytologist, 220*, 425–434.

Yuan, S. L., Li, R., Chen, H. F., Zhang, C. J., Chen, L. M., Hao, Q. N., et al. (2017). RNA-Seq analysis of nodule development at five different developmental stages of soybean (*Glycine max*) inoculated with *Bradyrhizobium japonicum* strain 113-2. *Scientific Reports, 7*, 42248.

Yun, B. W., Feechan, A., Yin, M., Saidi, N. B., Le Bihan, T., Yu, M., et al. (2011). S-nitrosation of NADPH oxidase regulates cell death in plant immunity. *Nature, 478*, 264–268.

Zwack, P. J., & Rashotte, A. M. (2013). Cytokinin inhibition of leaf senescence. *Plant Signaling & Behavior, 8*. e24737.

Zwack, P. J., Robinson, B. R., Risley, M. G., & Rashotte, A. M. (2013). Cytokinin response factor 6 negatively regulates leaf senescence and is induced in response to cytokinin and numerous abiotic stresses. *Plant and Cell Physiology, 54*, 971–981.

CHAPTER EIGHT

The genomes of rhizobia

Barney A. Geddes[a], Jason Kearsley[a], Richard Morton[a], George C. diCenzo[b] and Turlough M. Finan[a,]*
[a]Department of Biology, McMaster University, Hamilton, ON, Canada
[b]Department of Biology, Queen's University, Kingston, ON, Canada
*Correrspondiing author: E-mail: finan@mcmaster.ca

Contents

Abstract

The genomes of α- and β-rhizobia belonging to the α- and β-proteobacteria respectively contain genes for nitrogen fixation in symbiosis with host plants (encoded by *nif* and *fix* genes) and nodulation of their host plants, most often using Nod Factor (synthesized by *nod, noe, nol* gene products). In α-rhizobia symbiotic genes are often organized on plasmids of the *repABC* family, and symbiotic genes for the β-rhizobia are also plasmid borne. In some α-rhizobia, symbiotic genes are encoded on the chromosome within symbiotic islands that are integrative conjugal elements. In this chapter we give an overview of general features of rhizobial genomes including the symbiotic genes they contain and their organization. We review the current understanding of common features of rhizobial genomes and genomic backgrounds that support nodulation and nitrogen fixation and discuss genome deconstruction as a tool for uncovering new genes involved in symbiosis with legume plants.

Advances in Botanical Research, Volume 94
ISSN 0065-2296
https://doi.org/10.1016/bs.abr.2019.09.014

1. Introduction

Rhizobium bacteria form root nodules on leguminous plants and reduce atmospheric N_2 gas to ammonia. Many of these bacteria are members of the order Rhizobiales, one of nineteen orders of the alpha (α) class of proteobacteria, and are referred to as α-rhizobia. The α-rhizobia have a worldwide distribution. According to the NCBI taxonomy browser (https://www.ncbi.nlm.nih.gov/guide/taxonomy/), all of the α-rhizobia group within seven of the fourteen families in the order Rhizobiales. The rhizobia are found in but a few of the 84 genera within these seven families, including: *Azorhizobium* (family *Xanthobacteraceae*), *Bradyrhizobium* (*Bradyrhizobiaceae*), *Devosia* (*Hyphomicrobiaceae*), *Mesorhizobium* and *Phyllobacterium* (*Phyllobacteriaceae*), *Methylobacterium* (*Methylobacteriaceae*), *Ochrobactrum* (*Brucellaceae*), and *Rhizobium* and *Sinorhizobium* (syn. *Ensifer*) (*Rhizobiaceae*). These observations illustrate that while the α-rhizobia genera cluster within the Rhizobiales, many of the bacteria of this order do not form nodules. Indeed, bacteria in the Rhizobiales order are extremely diverse and possess a very wide range of metabolic capabilities such as photosynthesis (*Rhodobacter*), nodule formation (the rhizobia), and the use methanol or methane as a sole carbon source (*Methylobacterium)*. Rhizobiales also include pathogens of humans (*Brucella* and *Bartonella*) and plants (*Agrobacterium*), bacteria that divide by budding (*Caulobacter*), and bacteria that possess appendages called prosthecae (*Hyphomicrobium*). For the current status of the taxonomy of Rhizobia see the Capela and Tang chapter in this volume.

Another group of rhizobia are members of the beta (β) class of proteobacteria, and group in the family *Burkholderiaceae* within the order Burkholderiales. The β-rhizobia are found in three of the 21 genera of this family, including (*Para*)*Burkholderia* and *Cupriavidus* (ex *Ralstonia*). Some symbiotic species formerly classified as *Paraburkholderia* were recently reclassified to new genus, *Trinickia* (Estrada-de los Santos et al., 2018). The β-rhizobia form N_2-fixing root nodules on tropical and subtropical legumes and they are sometimes grouped according to whether they form nodules on *Papilionoideae* or *Mimosoideae* legume subfamilies (Chen, James, Prescott, Kierans, & Sprent, 2003; De Meyer et al., 2016; Gyaneshwar et al., 2011; Moulin, Munive, Dreyfus, & Boivin-Masson, 2001). As β-rhizobia have been recognized for only the past twenty years, it seems possible that the environments and hosts associated with the β-rhizobia will expand as more nodule isolates are investigated.

N_2-fixation has been studied for over 100 years and its importance in agriculture has long been recognized. However, concerns over availability of chemical fertilizers in the "oil crisis" of the early 1970s stimulated research interest and funding of N_2-fixation research and the ensuing 30 years resulted in the discovery of most of the major genes and pathways involved in free-living and symbiotic N_2-fixation (SNF). The introduction of DNA sequencing led to detailed molecular genetic analysis of gene structure and gene regulation in the 1980s—1990s and many of those studies focused on model representative organisms such as *Sinorhizobium* (*Ensifer*) *meliloti* 1021, *Bradyrhizobium japonicum* (*diazotrophicans*) USDA 110, *Rhizobium leguminosarium* RL3841, *Rhizobium tropici* CIAT 899 and *Mesorhizobium loti* (*japonicum*) MAFF 303099. The first bacterial genome sequences were reported in 1995 and now, some 25 years later, thousands of bacterial genomes are being sequenced each year thanks to advances in sequencing technology. Indeed, we have entered an era in which whole genomes rather than genes and operons are being deconstructed and synthetic bacterial genomes are being constructed (Gibson et al., 2010).

The bacterial genes involved in SNF processes are broadly characterized as nodulation (*nod, noe, nol*) and nitrogen fixation genes (*nif and fix*). The rhizobial *nod* genes are induced by flavones/flavonoids excreted by plant roots and the Nod proteins are responsible for the synthesis of lipochitooligosaccharides, called Nod factors (NFs), which induce legume root hair curling and nodule formation. NFs sensed by LysM-receptor-like kinases in the leguminous plant initiate a signal transduction cascade giving rise to the root nodule (Gage, 2004; Jones, Kobayashi, Davies, Taga, & Walker, 2007; Kelly, Radutoiu, & Stougaard, 2017). Part of the plant NF signaling pathway is also employed in the ancient interaction with Mycorrhizal fungi, which may have been a key event in the colonization of land by plants (Delaux et al., 2015). The *nif* genes encode proteins that are required for the synthesis, assembly, and control of the nitrogenase enzyme complex that reduces N_2 gas to ammonia. Most of these genes occur in both free-living and symbiotic diazotrophs. The *fix* genes are less defined; they are required for N_2-fixation in root nodules and are considered to be involved in the symbiotic process but they are not required for root nodulation. Some *fix* genes, particularly those involved in electron transfer to nitrogenase (*fixABCX*) and low oxygen affinity cytochrome oxidase (*fixNOQP*) are also found in aerobic N_2-fixing bacteria (Kopat et al., 2017; Ledbetter et al., 2017; Poudel et al., 2018).

In this chapter, we will attempt to review: general features of rhizobial genomes, rhizobia and secondary replicons; plasmids, pSym plasmids of α-rhizobia and β-rhizobia; symbiotic Islands – Integrative conjugative elements; Nod Factor Independent nodulation; symbiotic N_2-fixation (SNF) gene regions; and searches for common background genomic features required for SNF by rhizobia and rhizobia genome deconstruction.

2. General features of rhizobial genomes

Both α-rhizobia and β-rhizobia are aerobes and as free-living cells they grow well in both complex and defined media. Historically rhizobia were divided based on their growth rate into "fast-growing" and "slow-growing" classes (Stowers, 1985). Fast-growing rhizobia include the α-rhizobia, *Sinorhizobium* and *Rhizobium* strains that form colonies on complex media (TY or yeast mannitol agar) in 2–3 days; while slow-growing rhizobia included *Mesorhizobium* and *Bradyrhizobium* strains that generally take 4–7 days for visible colony formation. *Burkholderia phymatum* STM815^T forms colonies in 3–4 days at 28 °C (Bournaud et al., 2013; Moulin et al., 2014). General genomic features of representative bacteria from the α-rhizobia and β-rhizobia are shown in Table 1. We chose these particular strains as they are the most studied isolates of the various genera. In a previous analysis of 1708 bacterial genomes, the average and median genome sizes were found to be 3.65 Mb and 3.46 Mb respectively (diCenzo & Finan, 2017). As can be seen from Table 1, the α- and β-rhizobial genome sizes are often nearly twice or more times the average bacterial genome size. Presumably this reflects adaptation to complex environments and to the limited and diverse nutrients available to the rhizobia in the soil and rhizosphere environment (diCenzo, Checcucci et al., 2016). Many genomes of the α-rhizobia and all of the β-rhizobia are multipartite, i.e., genomes that are split between two or more large self replicating fragments (replicons). In all these cases, one of the replicons carries many (all) of the core essential genes and is equivalent to a typical chromosome of bacteria such as *Escherichia coli* or *Bacillus subtilis*. The additional replicons are referred to as second chromosomes, chromids, megaplasmids or plasmids. This multipartite genome structure is not unusual as it is found in approximately 10% of all bacteria and is not specific to rhizobia (diCenzo & Finan, 2017). The sizes of these replicons varies from ~100 kb to >2000 kb. Harrison and colleagues proposed the term chromid to describe a replicon that is a combination of chromosome and plasmid. Chromids carry some core

Table 1 Genome structure of selected rhizobia.

Species - strain	Chromosome (Mbp)		Mega-Plasmids (Mbp)		Plasmids (Mbp)							Reference	Bioproject Accession
Ensifer (*Sinorhizobium*) *meliloti* 1021	Chr (3.65)		pSymB (1.68)	pSymA (1.35)*#	-							Galibert et al., 2001	PRJNA19
Ensifer (*Sinorhizobium*) *medicae* WSM419	Chr (3.78)		pSMED01 (1.57)	pSMED02 (1.25)*#	pSMED03 (0.22)							Reeve, Chain, et al., 2010	PRJNA58549
Ensifer (*Sinorhizobium*) *fredii* NGR234	Chr (3.93)		pNGR234b (2.43)	pNGR234a (0.54)*#	-							Schmeisser et al., 2009	PRJNA59081
Rhizobium leguminosarum bv. *viciae* 3841	Chr (5.06)		-		pRL7 (0.15)	pRL8 (0.15)	pRL9 (0.35)	pRL10 (0.49)*#	pRL11 (0.68)	pRL12 (0.87)		Young et al., 2006	PRJNA344
Rhizobium leguminosarum bv. *trifolii* WSM1325	Chr (4.77)		-		pR132501 (0.83)*#	pR132502 (0.66)	pR132503 (0.52)	pR132504 (0.35)	pR132505 (0.29)			Reeve, O'Hara, et al., 2010	PRJNA20097
Rhizobium etli CFN 42	Chr (4.38)		-		p42a (0.19)	p42b (0.18)	p42c (0.25)	p42e (0.51)	p42f (0.64)	p42d (0.37)*#		González et al., 2006	PRJNA13932
Rhizobium tropici (*leucaenae*)CIAT 899	Chr (3.84)		-		pRtrCIAT899c (2.08)	pRtrCIAT899b (0.55)*#	pRtrCIAT899a (0.22)					Ormeño-Orrillo et al., 2012	PRJNA42391
Mesorhizobium japonicum (*loti*) MAFF 303099	Chr (7.04)*#		-		pMLa (0.35)	pMLb (0.21)						Kaneko et al., 2000	PRJNA18
Bradyrhizobium diazoeffi ciens(*japonicum*) USDA 110	Chr (9.11)*#		-		-							Kaneko et al., 2002	PRJNA282645
Bradyrhizobium sp. BTAi1	Chr (8.26)#		-		pBTAi1 (0.23)							Giraud et al., 2007	PRJNA16137
Bradyrhizobium sp. ORS278	Chr (7.46)#		-		-							Giraud et al., 2007	PRJNA19575
Azorhizobium caulinodans ORS 571	Chr (5.37)*#		-		-							Lee et al., 2008	PRJDA19267
Methylobacterium nodulans ORS 2060	Chr (7.77)*#		-		pMNOD01 (0.49)	pMNOD02 (0.46)	pMNOD03 (0.04)	pMNOD04 (0.04)	pMNOD05 (0.02)	pMNOD06 (0.01)	pMNOD07 (0.01)	Marx et al., 2012	PRJNA20477
Microvirga lupini Lut6	Chr (9.63)*#		-		None							Reeve et al., 2014	PRJNA66529 (scaffold)
Cupriavidus taiwanensis LMG 19424	Chr1 (3.42)	Chr2 (2.50)			pRALTA (0.56)*#							Amadou et al., 2008	PRJNA15733
Paraburkholderia (*Burkholderia*) *phymatum* STM815	Chr1 (3.48)	Chr2 (2.70)	pBPHY01 (1.90)		pBPHY02 (0.60)*#							Moulin et al., 2014	PRJNA17409

Representative strains were selected and are displayed in rows. Rows are colored according to family with the following scheme: Rhizobiaceae, blue; Phyllobacteriaceae, green; Bradyrhizobiaceae, purple; Xanthobacteraeae, orange; Methylobacteriaceae, brown; Burkholderiaceae, red. Replicons with size in megabases are indicated in columns. The replicon that contains *nod* (*) and *nif* (#) genes are marked with symbols and bolder coloring.

genes, and their nucleotide composition and codon usage are very similar to those of the chromosomes they are associated with (Harrison, Lower, Kim, & Young, 2010). They have plasmid replication and partitioning systems and the majority of their genes confer accessory functions. We have suggested that non-essential replicons that are <350 kb be referred to as a plasmid and that non-essential replicons >350 kb be referred to as a megaplasmid (diCenzo & Finan, 2017).

The term megaplasmid was first used in a 1981 publication (Rosenberg, Boistard, Dénarié, & Casse-Delbart, 1981) with reference to the location of symbiotic genes in *Sinorhizobium meliloti* 1021 – and as discussed below, many of the SNF genes (*nod*, *nif* and *fix*) are located on the 1.34 Mb pSymA megaplasmid in this strain. Similarly, in *Methylobacterium*, *Ochrobactrum*, *Rhizobium*, *Sinorhizobium*, and *Phyllobacterium* and for both of the β-rhizobia genera, *Burkholderia* and *Cupriavidus*, the SNF genes are located on plasmids or megaplasmids that are often referred to as the pSym plasmids or pSym megaplasmids.

Among the bacteria in Table 1, the *Bradyrhizobium*, *Mesorhizobium*, and *Azorhizobium* strains stand out because of their large single chromosomes (~5–9 Mb). For these strains the SNF regions are present on the chromosome in regions referred to as symbiotic islands that are integrative conjugative elements (ICEs) with structural similarities to pathogenecity islands in bacterial pathogens.

3. Rhizobia and secondary replicons: plasmids

Plasmids of the α-rhizobia have been the subject of several reviews concerning replication, conjugation, and other properties (Ding & Hynes, 2009; Pueppke, 1996). Following the recognition that some bacteria carry extra-chromosomal elements in the 1960s (episomes – plasmids), there were several reports linking nodule formation with the presence of plasmids. It was soon recognized that functions related to nodulation and nitrogen fixation were indeed located on plasmids in at least some rhizobia, and this insight was soon followed by reports in which nodule formation and nitrogen fixation was transferred between rhizobium strains and even into *Agrobacterium* (Beringer, 1974; Brewin, Beringer, & Johnston, 1980; Higashi, 1967; Hooykaas, Snijdewint, & Schilperoort, 1982; Truchet et al., 1984). Subsequent characterization of plasmids in the α-rhizobia revealed that invariably, these are low copy number plasmids that contain distinct

RepABC −cassette type replication systems. These loci include: *repC* that encodes the replication protein, *repA* and *repB* that encode ParAB-like proteins, an origin of replication (*oriV*), and a centromere-like element (*parS*) involved in plasmid partitioning (Cevallos, Cervantes-Rivera, & Gutiérrez-Ríos, 2008; Pinto, Pappas, & Winans, 2012). The α-rhizobia can carry as many as six of these RepABC plasmids (Ding & Hynes, 2009). To date, RepABC plasmids are found exclusively in α-proteobacteria and they reside in various other genera including the *Roseobacter* clade (order, *Rhodobacterales*) that are prominent marine bacterioplankton (Petersen, Brinkmann, & Pradella, 2009). Interestingly, analysis of phylogenetic trees for RepC with the adjacent RepA and RepB phylogenetic trees revealed a congruence in RepABC cassettes in the Rhodobacterales; however that is not the case in the Rhizobiales where it appears that frequent recombination events have occurred between *repC* and the partition *repAB* module (Cevallos et al., 2008; Petersen et al., 2009; Slater et al., 2009). It would seem that RepABC replication systems were present in α-proteobacteria prior to the evolution of SNF and that these were convenient systems to carry and transfer genes for SNF and other traits that allowed occupation on new environmental niches. Many of the genes carried on these plasmids appear to be associated with environmental niches involved in plant microbe interactions. These include the tumor inducing (Ti) and hair-root inducing (Ri) plasmids, other large plasmids (pAtC58), and the linear chromosome of *Agrobacterium* strains.

RepABC modules also contain conserved incompatibility (*inc*) regions: *incα,* and *incβ and/or incγ*; *incα* is highly conserved and is located between *repB* and *repC*; it encodes a small regulatory RNA and is transcribed in the opposite direction to *repABC* (Chai & Winans, 2005; MacLellan, Smallbone, Sibley, & Finan, 2005; Venkova-Canova, Soberón, Ramírez-Romero, & Cevallos, 2004). The *incγ* or *incβ* sequences consist of an ~16-bp palindromic sequences (*parS* site) that acts *in cis* as centromere-like sequences that are bound by RepB (Koper, Żebracki, Marczak, Skorupska, & Mazur, 2016; MacLellan, Zaheer, Sartor, MacLean, & Finan, 2006; Soberón, Venkova-Canova, Ramı;rez-Romero, Téllez-Sosa, & Cevallos, 2004). Unlike the highly conserved gene order of the *repABC* operon, the location of the *par*S is variable. Koper et al. have demonstrated specific RepB binding only to the parental *parS* elements and confirmed no cross-reactivity between individual RepB and *parS* from non-parental *repABC* cassettes (Koper et al., 2016). RepA shows a similar specificity for binding of the operator from its cognate *repABC* cassette (Żebracki, Koper,

Marczak, Skorupska, & Mazur, 2015). RepAB facilitate the faithful partitioning of these plasmids into daughter cells upon cell division (Bartosik, Szymanik, & Wysocka, 2001; Soberón, Venkova-Canova, Ramırez-Romero, Téllez-Sosa, & Cevallos, 2004). Indeed, RepABC-type plasmids are known to be quite stable and in the past, the curing (loss) of these plasmids was achieved by treating cells with agents such as ethidium bromide or high temperatures (Zurkowski, 1982). Hynes and colleagues successfully employed an approach where they marked plasmids with *sacB* and then directly identified those rare cells that lost the plasmid by selecting for colonies that grew in media containing high concentrations of sucrose (Baldani, Weaver, Hynes, & Eardly, 1992; Michael F Hynes, Quandt, O'Connell, & Pühler, 1989). More recently, direct procedures to select for the loss of specific RepABC plasmids are employing incompatibility with either whole *repABC* modules or *inc* regions *in trans* (e.g. MacLellan et al., 2005, Petersen et al., 2013). However, loss of these plasmids can result in cell death if; a) the plasmids carry toxin-antitoxin addiction modules such that daughter cells not receiving the replicon die due to the action of residual active toxin from the cytoplasm of mother cell (Milunovic, Morton, & Finan, 2014), or b) if the cured replicon carries genes that are required for cell viability or growth on the culture media (diCenzo, Milunovic, Cheng, & Finan, 2012; Landeta et al., 2011).

The location of SNF genes on plasmids allows their rapid transfer into other bacteria. This can occur as the plasmids may be self transmissible (have all of the genes required for conjugation) or the plasmid can be mobilized *in trans* by other conjugative systems in the donor cell; alternatively, transfer could occur via cointegration with another plasmid. Plasmid conjugal transfer systems are made of DNA transfer and replication (Dtr) and mating pair formation (Mpf) components. Plasmid transfer initiates at the *oriT* sequence and single stranded DNA transfers through the mating pore, a type IV secretion system, into the recipient cell. Among conjugative RepABC-type plasmids, at least four classes are recognized and the majority of plasmids are either Class 1 plasmids, where a bacterial cell density - quorum sensing (QS) regulates the conjugal transfer system, or Class 2 plasmids, where the conjugal transfer system is repressed by RctA/RctB (Ding & Hynes, 2009). For Class 2 plasmids, the nature of the signal responsible for regulating the RctA/RctB system is unknown. The *Agrobacterium tumefaciens* Ti plasmid is the best studied Class 1 plasmid and other members of this group include the symbiotic plasmids pRL1JI from *Rhizobium leguminosarum* and the pSym plasmid of *Sinorhizobium fredii* NGR234; Class 2

RctA regulated plasmids include the symbiotic plasmid pCFN42d from *Rhizobium etli* CFN42 and the pSymA megaplasmid from *S. meliloti* (Nogales, Blanca-Ordóñez, Olivares, & Sanjuán, 2013; Pérez-Mendoza et al., 2005). A recent analysis of eighteen Class 1 plasmids identified two groups based on the sites at which "cargo" DNA was located and the authors presented strong evidence that the quorum-sensing and transfer functions are coevolving within but not between these two groups (Wetzel, Olsen, Chakravartty, & Farrand, 2015). On the other hand, no coevolutionary relationship was seen between the quorum-sensing - transfer functions and the RepABC proteins although RepA proteins were found to coevolve with RepB proteins and as seen in other studies, there was little evidence for RepC coevolving with RepA or RepB (Wetzel et al., 2015). Overall this is consistent with the idea that RepAB modules combine with RepC to generate different RepABC replicons and that a unit containing the conjugation control and transfer functions are added to generate a conjugative plasmid.

4. pSym plasmids of α-rhizobia and β-rhizobia

The genes involved in nodulation and nitrogen fixation are carried on conjugal plasmids in the Rhizobiales family of α-rhizobia. Reports on the conjugation and symbiotic genes of several of these plasmids in the Rhizobiales family of α-rhizobia have been published and among these are pRL1JI (200 kb) from *R. leguminosarum* (Danino, Wilkinson, Edwards, & Downie, 2003), pSymA (1340 kb) from *S. meliloti* (Barnett et al., 2001), pSymNGR234 (536 kb) from *S. fredii* NGR234 (Schmeisser et al., 2009), and pRetCFN42d (371 kb) of *R. etli* CFN42 (González et al., 2006).

The *nod*, *nif* and *fix* genes are also located on plasmids in all of the β-rhizobia that have been examined to date. Detailed functional analysis of the replication and transmission of these plasmids have not been performed. *Burkholderia phymatum* STM 815^T was one of the first β-rhizobia to be isolated, and this forms nodules on species of the legume genus *Mimosa*, and a range of other legume genera. The STM 815^T genome is 8,676,562 bp and is composed of two chromosomes (3,479,187 and 2,697,374 bp), a megaplasmid (1,904,893 bp), and a plasmid (595,108 bp) that carries the symbiotic genes (Moulin et al., 2014). In addition to *nod*, *nif*, and *fix* genes, this Sym plasmid carries several genes coding for proteins indirectly linked to symbiotic interactions with plants such as indol acetic acid (*iaaHM*), ACC deaminase (*acdS*), and genes involved in the biosynthesis of rhizobitoxine (*rtxAC*-like) (Okazaki, Kaneko, Sato, & Saeki, 2013; Okazaki, Nukui, Sugawara, & Minamisawa, 2004). Interestingly, in a recent study of early symbiotic events, 23% (129 genes) of the genes on the STM 815^{T} pSym plasmid were found to be induced in response to root exudates of *Mimosa pudica*. These included *nod* genes and the above-mentioned *iaaHM*, *acdS*, and *rtxAC* genes, as well as a type IV secretion system possibly involved in transfer of this plasmid (Klonowska, Melkonian, Miché, Tisseyre, & Moulin, 2018).

The *Cupriavidus taiwanensis* LMG19424 genome consists of two circular chromosomes 3.42 Mb and 2.50 Mb size, and a symbiotic plasmid, pRalta of 560 kb, that includes one of the most compact SNF clusters (35-kb) known (Amadou et al., 2008). The pRalta plasmid carries a 57.9-kb region encompassing *pil*, *trb*, and *tra* loci related to pilus biogenesis and conjugative plasmid transfer; however, whether this is a self-transmissible plasmid remains unclear. Of 583 coding sequences present on the pRata plasmid, 207 are transposable element (TE) sequences. This extraordinary number is 20 times higher than the number of TE sequences present on either of the *C. taiwanensis* chromosomes. High numbers of insertion element sequences is a feature that is also seen for many of the pSym plasmids and ICEs of the α-rhizobia, which is consistent with the mobility of these regions and their inheritance via lateral transfer.

5. Symbiotic islands – integrative conjugative elements

The location of *nod*, *nif*, and *fix* genes on plasmids in many fast-growing rhizobia allows the horizontal transfer of these genes. It was curious to find that *nif* and *nod* genes in some rhizobia are located on the chromosome. In 1998, Ronson and colleagues reported that rhizobia can evolve by acquiring a 502-kb "symbiosis island" that integrates into a *phe-tRNA* gene on the chromosome of *M. loti* (Sullivan & Ronson, 1998). That report followed-up on earlier findings suggesting genetically diverse symbionts, isolated from root nodule of *Lotus corniculatus* plants at a remote field site, arose through horizontal transfer of a chromosomal symbiotic element from an inoculant strain of *M. loti* to non-symbiotic rhizobia in soil. In subsequent laboratory experiments, the symbiotic island was shown to be a member of a large group of mobile genetic elements, known as integrative and conjugative elements, or ICEs. These share features involved in integration, excision and conjugative transfer (type IV secretion system). ICEs integrate into the host chromosome (often at a tRNA) and replicate passively as part of the chromosome. Individual ICEs generally carry genes whose functions are related to a particular process such as antibiotic resistance (e.g. Tn916), pathogenesis (pathogenecity islands), degradative pathways, or symbiotic N_2-fixation (Johnson & Grossman, 2015). ICE - symbiotic islands have been found in *Mesorhizobium*, *Bradyrhizobium*, and *Azorhizobium* and vary in length from ~90 kb to 500 kb. As is the case for pSym plasmid transfer, the factors and environmental conditions that influence the excision, transfer and integration of these ICEs to other cells must be under considerable evolutionary selection and identifying these factors is of much interest and practical importance with respect to rhizobia and the application of commercial rhizobia inoculants.

In the most studied system, excision and transfer of the *M. loti* strain R7A symbiosis island, ICE*Ml*Sym^{R7A}, is under a sophisticated multi-level control. ICE*Ml*Sym^{R7A} excision is indirectly activated by a LuxRI-family autoinduction/quorum sensing (QS) system (TraRI) that responds to N-(3-oxo-hexanoyl)-L-homoserine lactone (3-oxo-C6-HSL) produced by TraI. Activated TraR stimulates transcription of the *msi172* and *msi171* genes and a +1 programmed ribosomal frameshift (PRF) fuses 4–13% of the *msi172* and *msi171* coding sequences to generate a protein designated as the frameshifted excision activator protein, FseA (Ramsay et al., 2013; Ramsay & Ronson, 2015). FseA in turn activates transcription of the recombination directionality factor *rdfS*, and RfdS stimulates excision of the ICE-*Ml*SymR7A island. However, in most cells this activation pathway is repressed by an anti-activator, QseM, that binds both TraR-3-oxo-C6-HSL and FseA to inhibit the transcriptional activation activity of both proteins. Thus in induced cells, ICE*Ml*Sym^{R7A} transfer is only observed at a frequency of $\sim 10^{-7}$. For a review of this work and its evolutionary implications see Ramsay and Ronson (2015). In addition to SNF *nod*, *nif*, and *fix* genes, and many genes yet to be characterized, ICE*Ml*Sym^{R7A} carries *dct* genes required for C_4-dicarboxylate transport, thiamine biosynthesis genes, and genes involved in biotin and nicotinate synthesis. The latter have been useful in detecting transfer of ICE*Ml*Sym^{R7A} transfer to non-symbiotic mesorhizobia that are Dct$^-$ and auxotrophic for thiamin and biotin (Sullivan, Brown, Yocum, & Ronson, 2001). In the case of *Azorhizobium caulinodans*, the ICE symbiotic island (87.6-kb) was found to excise, form a circular DNA element, and conjugatively transfer to a specific site of gly-tRNA gene of other rhizobial genera (Ling et al., 2016). Interestingly the transfer frequency increased in the rhizosphere in response to plant flavonoids that also induced the expression of nodulation genes (Ling et al., 2016). These conditions would seem to be ideally suited to the spread of the symbiosis island to non-symbiotic recipients such that they may then be able to nodulate host plants that are nearby.

A fascinating twist to the ICE symbiotic island story was uncovered with the discovery that some symbiosis ICEs exist as three separate chromosomal regions, as in *Mesorhizobium ciceri* WSM1271 (Haskett et al., 2017, 2016). As in the case of ICE*Ml*Sym^{R7A}, this discovery originated from field experiments – in this case in Western Australia in field trials of the pasture legume *Biserrula pelecinus* inoculated with *M. ciceri* WSM1271. In the field, the *M. ciceri* WSM1271 tripartite ICE (ICE^3) was found to transfer and integrate into other *Mesorhizobium* genomes and to regenerate the three separate chromosomal regions found in WSM1271 (Nandasena, O'Hara, Tiwari, Sezmiş, & Howieson, 2007). Each of the three ICE regions in the chromosome were bordered by two nonhomologous integrase attachment (*att*) sites, which together comprised three homologous pairs of *attL* and *attR* sites. Haskett, Ramsay and colleagues showed that prior to conjugative transfer, the three ICE^3 regions of *M. ciceri* WSM1271 recombine and excise to form a single circular element and that this assembly required three coordinated recombination events involving three site-specific recombinases (IntS, IntG, and IntM) and the sequential expression of three recombination directionality factors (excisionases) (Haskett et al., 2018). The expression of ICE^3 transfer and conjugation genes was induced by quorum sensing and the recombination directionality factors, such that the three excision events occur as stepwise sequence of recombination events that avoids chromosomal rearrangements. This unusual ICE^3 structure does not appear to be unique as nine additional tripartite ICEs were identified in diverse mesorhizobia and transfer was demonstrated for three of them (Haskett et al., 2017, 2016, 2018). Haskett et al., 2017 discuss factors that may influence the persistence of the ICE^3 structure such as enhanced host range, host stability and resistance to destabilization by tandem insertion of competing integrative elements.

6. Symbiotic N_2-fixation (SNF) gene regions

For most rhizobia, their ability to form N_2-fixing nodules is known to be defined by relatively conserved (but not identical) *nod-nif-fix* gene

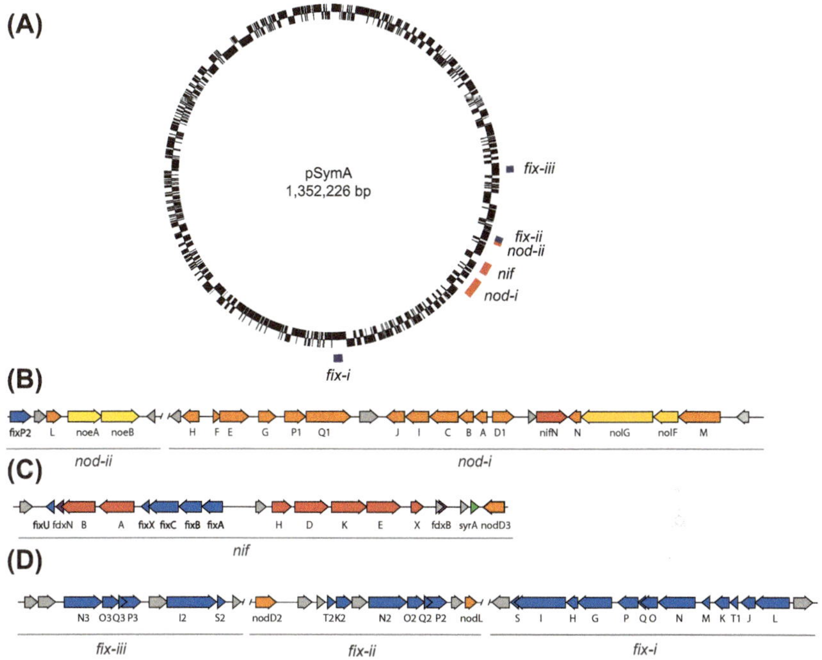

Fig. 1 ***The* nod, nif *and* fix *loci of* Sinorhizobium meliloti *Rm1021.*** (A) Map of the *S. meliloti* pSymA megaplasmid showing genes on the + and − strand created with CiVI (Overmars, van Hijum, Siezen, & Francke, 2015), and annotated with the location of *nod, nif* and *fix* loci. Genetic maps of the open reading frames in the *nod* (B), *nif* (C) and *fix* (D) loci are shown below. The *nif* genes are colored red, *fix* in blue, *nod* genes in orange, the *noe* genes in yellow, and the *nol* genes in brown. *Structure of genetic loci was obtained from IMG/M (Chen et al., 2019), and annotated with Adobe Illustrator.*

regions. These loci include genes responsible for N_2-fixation in nodules (*nif* and *fix*) and those involved in Nod factor (NF) synthesis (*nod, noe, nol*) (Fig. 1). We have summarized the occurrences of many of these genes among the rhizobia in Table 2 and present the organization of *nod* loci as an example of differences in gene arrangement in Fig. 2. We note that while many of the features and genes in this region are present in all rhizobia, as noted previously, there are considerable differences in the presence/absence of some of these genes among the rhizobia and how they are regulated and organized (Fischer, 1994; Masson-Boivin, Giraud, Perret, & Batut, 2009). Here we summarize features of the SNF genes with particular reference to the alfalfa symbiont, *S. meliloti* 1021. As illustrated in Fig. 1, in *S. meliloti* 1021, the *nod*, *nif*, and *fix* genes are dispersed in several clusters that encompass ~300 kb of the pSymA megaplasmid.

Table 2 Distribution of genes involved in nodulation and nitrogen fixation across select rhizobia.

	sme	smd	rhi	rtr	rle	ret	mlo	azc	mvl	mno	bja	bbt	bra	cti	bph
nodA				3											
nodB															
nodC															
nodD	3	3	2	5		3	2				2				2
nodE							2								
nodF							2								
nodG															
nodH															
nodI											2			2	2
nodJ											2			2	2
nodL					2		2								
nodM		2			2										
nodN	2	2			2		2				2			3	
nodO															
nodP	2	2													
nodQ	2	2		2										1	
nodS															
nodT					3	2						2		2	
nodU															
nodV															
nodW					2			2							
nodX															
nodY															
nodZ															
noeA															
noeB															
noeC							2								
noeD															
noeE											2				
noeH															
noeI															
noeJ															
noeK							2								
noeL						2	2								
noeO															
noeP															
nolA															
nolB/nopB															
nolE															
nolF															
nolG														2	

	sme	smd	rhi	rtr	rle	ret	mlo	azc	mvl	mno	bja	bbt	bra	cti	bph
nolK															
nolL															
nolM															
nolN															
nolO			2												
nolR	nf				2	2	2								
nolT															
nolU															
nolV															
nolW															
nolX/nopX															
nolY															
nolZ															
nopA															
nopC															
nopJ															
nopL															
nopM											2				
nopP											2				
nopT											2				
fixA															
fixB															
fixC															
fixF															
fixG					2	2	2		2		2	3	4		
fixH					2	2	2								
fixI	2	2	2		3		2								
fixJ															
fixK	3	3			2	2	2			2	5	5	2		
fixL															
fixM															
fixN	3	3	2		3	2	2			2					
fixO	3	3	2		3	2	2			2					
fixP	3	3	2		3	2	2			2					
fixQ	3	3			3	2	2			2					
fixR											2	2	2		
fixS	2	2	2		3	2	2								
fixT	3	2													
fixU/nifT															2
fixX															
fdxN															
nifA															
nifB															

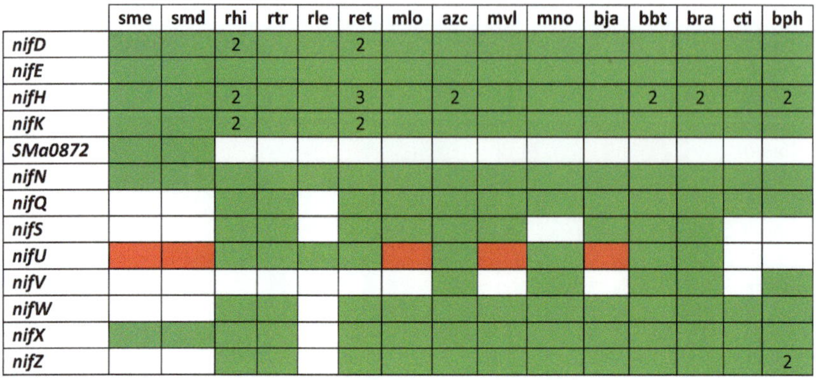

Colors indicate the presence or absence of genes based on the following scheme: **Green** — Gene is expected to be present, either due to annotation and/or a high degree of sequence similarity in conjunction with the genes' location relative to existing annotations. Each gene was analyzed separately, and homology was considered on a case-by-case basis using the Expect value. **Red** — Gene showed homology to annotated *nod*, *nif*, or *fix* genes present in another species, but its symbiotic relevance was inconclusive due to homology to other, non-symbiotic genes. **White** — The gene was concluded not to be present in this species due to a lack of annotations as well as sequence similarity. Numbers within the boxes represent multiple copies of the gene present within the species. Abbreviations of species in column titles are as follows: **sme** — *S. meliloti* 1021, **smd** — *Sinorhizobium medicae* WSM419, **rhi** — *S. fredii* NGR234, **rtr** — *Rhizobium tropici* CIAT 899, **rle** — *Rhizobium leguminosarum* bv. *viciae* 3841, **ret** — *Rhizobium etli* CFN 42, **mlo** — *M. loti* MAFF 303,099, **azc** — *Azorhizobium caulinodans* ORS 571, **mvl** — *Microvirga lupini* Lut6, **mno** — *Methylobacterium nodulans* ORS 2060, **bja** — *B. japonicum* USDA 110, **bbt** — *Bradyrhizobium* sp. BTAi1, **bra** — *Bradyrhizobium* sp. ORS278, **cti** — *C. taiwanensis* LMG 19,424, **bph** — *P. phymatum* STM815.
Data for expected gene presence in each species was obtained from a comparative analysis using BLAST-N to query sequenced genomes of select rhizobia in the Integrated Microbial Genomes & Microbiomes (IMG/M) system (I.-M. A. Chen et al., 2019).

***nif* genes** The *nif* genes are responsible for the synthesis and assembly of the nitrogenase enzyme which catalyzes the reduction of N_2 gas to ammonia. This reaction requires high levels of ATP and reductant to drive the reduction of N_2. Interestingly, while nitrogenase is inactivated by oxygen, N_2-fixation in rhizobia (bacteroids) within nodules occurs under microaerophilic conditions. In fact, leghaemoglobin facilitates the transfer of oxygen into the nodule to enable aerobic respiration in bacteroids while maintaining a low oxygen tension (Layzell et al., 1993). ATP is generated in bacteroids via oxidative phosphorylation employing an electron transport chain with a terminal cbb3 oxidase that has a high-affinity for O_2 (FixNOQP) (Preisig, Zufferey, Thöny-Meyer, Appleby, & Hennecke, 1996). The pathway(s) through which reductant is generated is still under investigation (Poudel et al., 2018). The *nif* genes are labeled according to the twenty homologs of *Klebsiella pneumonia (oxytoca)*. Of the eight *nif* gene in

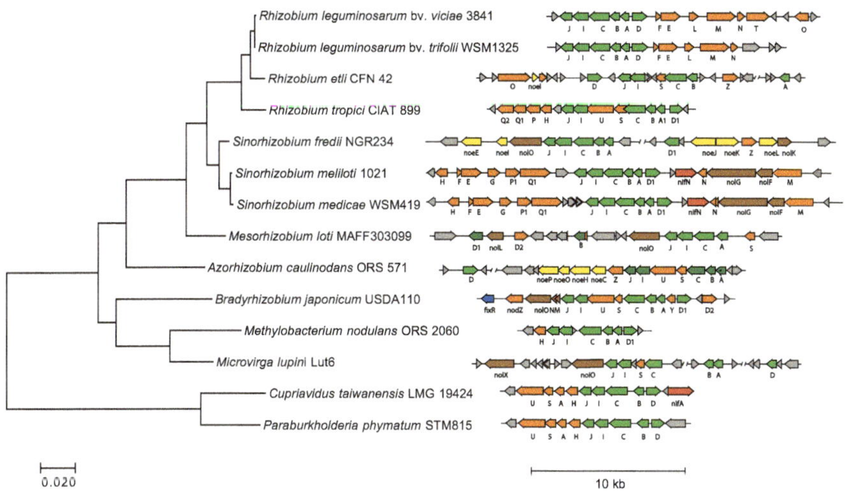

Fig. 2 ***Genetic organization* nod *loci in different rhizobia.*** Genetic maps of the *nod* loci surrounding the core nod genes *nodABCDIJ* are shown adjacent to the position of rhizobia in a phylogenetic tree constructed by maximum likelihood (Kumar, Stecher, & Tamura, 2016). Core *nod* genes are colored in green, accessary *nod* genes in orange, *noe* genes in yellow, and *nol* genes in brown. *Structure of genetic loci was obtained from IMG/M (Chen et al., 2019), and annotated with Adobe Illustrator.*

S. meliloti, *nifDK* encode the two subunits of the molybdenum nitrogenase protein and *nifH* the nitrogenase reductase. The *nifB*, *nifX*, *nifE*, and *nifN* genes encode proteins required for the biosynthesis of the iron molybdenum cofactor (FeMo-co) that is located at the active site of nitrogenase (Masson-Boivin et al., 2009; Rubio & Ludden, 2008). *S. meliloti* possesses another gene, *nifS*, on its chromosome that encodes a cysteine desulfurase possibly involved in FeMo-co synthesis (Capela et al., 2001). The *nifA* gene encodes a transcriptional regulator that is required for the *rpoN*-dependent (sigma-54) transcription of *nifB*, *nifN*, *nifHDKEX*, and *fixABCX* among others (Galardini, Pini, Bazzicalupo, Biondi, & Mengoni, 2013; Gong, He, Zhu, Yu, & Zou, 2006). Following a hiatus of several decades for research on rhizobial *nif* genes, there is now renewed interest in the functions of the different *nif* genes and this is driven in part by the prospect of directly engineering plants to fix N_2 through the direct introduction of *nif* genes (e.g. see Burén & Rubio, 2017; Wang et al.,. 2013), or transferring N_2-fixation to bacteria that effectively colonize cereal crops (Geddes et al., 2015; Mus et al., 2016). As noted previously, there are eight/nine *nif* genes present in *S. meliloti*, which is somewhat unusual among the rhizobia as many also have *nifQ*, *nifS*, *nifU*, *nifW*, and *nifZ*. These differences may reflect differences in the physiology of the organisms and or redundancy with other proteins yet to be uncovered (Masson-Boivin et al., 2009; Rubio & Ludden, 2008).

***fix* genes** The *fix* genes include those that are required to support N_2-fixation in nodules but are not required for nodule formation. Among the most important of these are the *fixA*, *fixB*, *fixC*, and *fixX* genes that form an operon in *S. meliloti*. These genes are present in many bacteria that fix nitrogen aerobically such as *Azotobacter vinelandii* and both the α- and β-rhizobia (Table 2) (Edgren & Nordlund, 2004; Ledbetter et al., 2017). Accumulating evidence suggests these proteins act via an electron-bifurcating electron transfer mechanism wherein two NADH are oxidized to yield a reduced ferredoxin and a reduced quinone, the former which serves as the electron donor for nitrogenase (Costas et al., 2017; Earl, Ronson, & Ausubel, 1987; Edgren & Nordlund, 2004; Ledbetter et al., 2017; Poudel et al., 2018). Mutagenesis has verified their importance for effective fixation of nitrogen in the nodule (Dusha, Kovalenko, Banfalvi, & Kondorosi, 1987; Ruvkun, Sundaresan, & Ausubel, 1982; Szeto, Zimmerman, Sundaresan, & Ausubel, 1984). FixA and FixB represent homologs of the *E. coli* electron-transfer flavoproteins (Etfs), while FixC and FixX are an oxidoreductase and a ferredoxin-like protein, respectively (Costas et al., 2017). Curiously, a *S. meliloti nif*-associated ferredoxin co-transcribed with *nifA* and *nifB* encoded by *fdxN* is essential for nitrogen fixation (Klipp, Reiländer, Schlüter, Krey, & Pühler, 1989) and FdxN was also proposed as an electron donor to nitrogenase in *Rhodospirillum rubrum* (Edgren & Nordlund, 2005).

In a symbiosis specific electron transport chain, a high-affinity *cbb*$_3$-type cytochrome *c* oxidase encoded by the *fixNOQP* genes (*ccoNOQP* orthologs) serves as the terminal oxidase functioning at the very low O_2 concentration (~10 nM) in legume root nodules (Preisig, Zufferey, Thöny-Meyer et al., 1996; Torres, Hidalgo-García, Bedmar, & Delgado, 2013). In *S. meliloti*, there are three copies of the *fixNOQP* genes on pSymA. The *fixNOQP1* and *fixNOQP2* copies share 95% homology between their sequences (Fig. 1, fix-I and fix-II) while third copy, *fixNOQP3*, only shares 61% homology with the other copies and its genes do not appear to be related to SNF (Torres, Hidalgo-García, Bedmar, & Delgado, 2013). A single functional copy of either *fixNOQP1* or *fixNOQP2* is necessary and sufficient for SNF in *S. meliloti* (Renalier et al., 1987) although a *fixN1* mutant shows a delay in wild type N_2-fixation (Torres et al., 2013). Downstream of *fixNOQP1* is another gene cluster, *fixGHI1S1* (*ccoGHIS*), whose products are required for the maturation of the FixNOQP cytochrome *cbb*$_3$-type oxidase complex (Kulajta, Thumfart, Haid, Daldal, & Koch, 2006; Preisig, Zufferey, & Hennecke, 1996). In other species such as *R. leguminosarum*, both copies of *fixNOQP* are required for optimal nitrogen fixation, hinting at the importance for the existence of multiple copies in *S. meliloti* (Schlüter et al., 1997). This was supported by another study that found knockout of either *fixN1* or *fixN2* in *S. meliloti* SM11 reduced SNF to approximately 50% of the wild type level after 39 days post inoculation (Reyes-González et al., 2016). The maintanence of multiple copies of *fixNOQP* the Rhizobiales is widespread (Kopat et al., 2017), but the mechanistic significance of the use of multiple copies for optimal SNF remains to be discovered.

The *fixL* and *fixJ* genes encode the two components of a histidine kinase - response regulator system that responds to low oxygen conditions and both are required for SNF in *S. meliloti* (Cosseau & Batut, 2004; Mesa et al., 2008; Reyes-González et al., 2016). Under microaerophilic conditions, FixJ activates transcription of the *nifA* regulator as well as another regulatory gene, *fixK1*, whose product, in turn, activates transcription of the *fixNOQP* and *fixGHIS* operons and other loci in a complex regulatory cascade (Bobik, Meilhoc, & Batut, 2006; Reyrat, David, Blonski, Boistard, & Batut, 1993). A second copy of *fixK2* is also present on pSymA which shares 97% amino acid identity with *fixK1* (Fig. 2 see fix-I and fix-II). Individual mutations do not appear to significantly alter expression of some of its targets, such as *fixN1* and *fixN2*, which hints at the possibility of redundancy in FixK activity (Reyes-González et al., 2016). Batut et al. (1989) demonstrated that at least one copy of *fixK*, *fixK1* in this case, is required for a Fix^+ phenotype to be observed with alfalfa (Batut et al., 1989). Indirect autoregulation of FixK occures via another protein, FixT. The role of FixT is to inhibit FixL autophosphorylation when its transcription is activated by FixK (Foussard et al., 1997; Garnerone, Cabanes, Foussard, Boistard, & Batut, 1999). Like *fixK*, two copies of *fixT* are present on pSymA – *fixT1* and *fixT2* – that bear 79% amino acid identity and occur within each of the reiterated *fix* regions (Garnerone et al., 1999; Renalier et al., 1987). The *fixT1* gene was determined to be necessary and sufficient for negative regulation of FixK1, but nodules induced by individual *fixT1* and *fixT2* mutants were Fix^+, as was a double mutant, indicating it is not essential for SNF (Foussard et al., 1997; Renalier et al., 1987).

Regulation of *fix* and *nif* genes in other α-rhizobia varies significantly. For example, while *R. leguminosarum* contains *fixK*, another similar gene, *fnrN* seems to be the main regulator of the *fixNOQP* operons (Schlüter et al., 1997). In *B. japonicum* a Redox responsive two-component system

(RegS/RegR) regulates *nifA* (Bauer, Kaspar, Fischer, & Hennecke, 1998), and in *M. loti*, a LacI type regulator controls *nifA* expression, possibly in response to an inositol-like molecule (Sullivan, Brown, & Ronson, 2013).

***nod* genes** The nodulation genes (*nod*, *nol*, *noe*) of *S. meliloti* are also located on the pSymA megaplasmid at two separate loci (Fig. 1 nod-i and nod-ii). Transcription of these genes is activated by a LysR-type transcriptional regulator, NodD, in response to flavonoids excreted by the plant host (Mulligan & Long, 1989; Spaink, 2000). The encoded proteins are responsible for the synthesis and secretion of lipochitooligosaccharidic Nod factors (NFs), with an acylated core backbone composed of 3–6 of N-acetylglucosamine residues synthesized by the NodABC proteins and the efficient secretion of NFs mediated by the NodIJ ABC-type transport system (Denarie, Debelle, & Prome, 1996). Representative arrangements of the loci surrounding core *nod* genes is presented in Fig. 2. All rhizobia that produce NFs contain the *nodABC*, *nodD*, and *nodIJ* genes; however, there is considerable variation in the presence or absence of other *nod*, *noe*, and *nol* genes between the various rhizobia (Table 2, Fig. 2). This reflects the fact that Nod factors from different rhizobia carry substituents and modifications that affect the plant host-specificity and the genes responsible for these decorations vary between the rhizobia (Denarie et al., 1996; Gage, 2004; Perret, Staehelin, & Broughton, 2000). For instance *nodL* and *nodH* are associated with acetylation of the nonreducing terminal residue and sulfation of the reducing terminal residue of *S. meliloti* NFs, respectively (Ardourel et al., 1994; Lerouge et al., 1990; Roche et al., 1991). Whereas *nodU* encodes an alternate N-terminal modification with a carbamoyl group in other rhizobia (D'Haeze, Van Montagu, Promé, & Holsters, 1999; Jabbouri et al., 1995), and a common alternative to sulfation of the nonreducing terminal residue is fucosylation encoded by *nodZ* (López-Lara et al., 1996; Mergaert et al., 1996) (Table 2, Fig. 2). As for some *nif* and *fix* genes, the number of copies of the nodulation genes also varies among rhizobial strains. For example, there are three *nodD* genes in *S. meliloti* (*nodD1*, *nodD2*, and *nodD3*; Fig. 1). Presumably this reflects gene rearrangements that result from the mobile nature of these gene regions and the high number of insertion elements in these regions. There may also be selection for such duplications as appears to be the case for *nodD* as of multiple *nodD* genes can extend the repertoire of flavone signal molecules that can be detected (Perret et al., 2000).

While relatively few functional studies have been performed on SNF gene functions in the β-rhizobia, the structure of Nod factors produced by *C. taiwanensis* have been determined and the *nod* genes associated with various substitutions were found on the genome; O-carbamoyl (NodU), N-methyl (NodS), and 6-O-sulfate (NodH) (Amadou et al., 2008). In a recent report, De Meyer and colleagues analyzed the symbiotic genes from eighteen β-rhizobia and concluded that most *nif/fix* genes were likely acquired from free-living diazotrophic β-proteobacteria with the exception of *nifA* that was likely aquired from the α-rhizobia (De Meyer et al., 2016). The phylogenetic profile for *nod* genes was more complex and inferences could not be made regarding their ancestry. While it is often assumed that *nod* genes originated in α-rhizobia, we note that on the bases of phylogenetic profiling, Aoki, Ito, and Iwasaki (2013) suggested *nodIJ* genes originated from gene duplication in a lineage of the β-proteobacteria (Aoki et al., 2013). Several reports have noted the absence of *fixNOQP* and *fixGHIS* genes in the β-rhizobial symbiotic gene clusters (Amadou et al., 2008; De Meyer et al., 2016; Moulin et al., 2014). However in *C. taiwanensis* a copy of the *ccoNOQP* operon encoding a *cbb3* type oxidase with a high affinity for oxygen is present on the chromosome (Amadou et al., 2008). No *cbb3* type oxidase was detected in *Paraburkholderia* rhizobial genomes but RNAseq analysis of *Phaseolus vulgaris* (common bean) nodules formed by *Paraburkholderia phymatum* STM817 revealed strong induction of genes encoding a potential cytochrome o ubiquinol oxidase (Lardi, Liu, Purtschert, Bolzan de Campos, & Pessi, 2017). Disruption of the genes encoding this system resulted in a reduction in SNF consistent with an important role of this oxidase in the nodule (Lardi et al., 2017).

7. Nod Factor independent nodulation

In 2007, Giraud et al., reported on a group of photosynthetic *Bradyrhizobium* strains (e.g., BTAi1 and ORS278) that lacked the *nodABC* genes yet form N_2-fixing root-nodules on some *Aeschynomene* species (Giraud et al., 2007). As the *nodABC* genes are required for Nod Factor (NF) production, this suggested that alternate symbiotic pathways could give rise to N_2-fixing root nodules on legumes and at least two such pathways are recognized. It was later shown that non-photosynthetic Bradyrhizobia can nodulate NF-independent *Aeschynomene* species using a pathway that employs effector proteins secreted by type three secretion systems (T3SS) to promote root nodule formation by directly activating the Nod factor signaling pathway of the plant (Miwa & Okazaki, 2017; Okazaki et al.,

2013, 2016). The NF independent pathway employed by photosynthetic *Bradyrhizobium* strains such as BTAi1 and ORS278 has yet to be determined; however, it appears to function independently of the T3SS-dependent pathway as a TTSS is absent from these strains (Okazaki et al., 2016). Mutants of *Lotus japonicus* that form nodules spontaneously have been used to uncouple nodule organogenesis from infection and to study the role of various host genes in these two processes (Madsen et al., 2010). Nod-factor perception was found to be required for epidermal root hair infection thread formation. These and other studies support the idea that rhizobial infection of the plant may have originated through intercellular invasion of the epidermis and subsequent evolution of crack entry and root hair invasions (Madsen et al., 2010). Okazaki et al. (2016) hypothesized that while the NF-independent symbiosis between bradyrhizobia and *Aeschynomene* involves the activation of a common symbiotic program in the plant, it also relies on a specific mechanism of microbe-associated molecular patterns (MAMPs) triggered immunity suppression that avoids the development of plant defense reactions during the interaction (Okazaki et al., 2016).

The use of type 3 secretion systems (T3SS) for host manipulation by some rhizobia is reminiscent of the use of these systems by many animal and plant pathogenic bacteria where T3SS are used to deliver effector proteins into host cells or the external environment in order to manipulate host cellular processes (Miwa & Okazaki, 2017). In rhizobia, T3SSs secreted proteins, called nodulation outer proteins (Nops) (Table 2), are involved in host-range determination and symbiotic efficiency (Deakin & Broughton, 2009). T3SS are broadly distributed among the α-rhizobia and as an example *S. fredii* HH103 secretes at least eight NOP proteins (Jiménez-Guerrero, Pérez-Montaño, Medina, Ollero, & López-Baena, 2017; Jiménez-Guerrero et al., 2017). In rhizobia and in bacterial pathogens, T3SS systems are often acquired by horizontal gene transfer via mobile genetic elements. It is therefore not surprising that T3SS often colocalize with other genes related to SNF. For example, the T3SS genes and effector proteins are located near *nod* and *nif* genes on the pSym plasmid of *S. fredii* NGR234 (Deakin & Broughton, 2009). In the photosynthetic *Bradyrhizobium* sp. ORS285 that nodulates *Aeschynomene*, a T3SS gene cluster and *nod* genes are present in a 262 kb symbiotic island that contains many transposable elements but not the *nif* genes which are located elsewhere on the chromosome (Gully et al., 2017). Consistent with their involvement in the interaction with the plant, transcription of several of the rhizobial T3SS gene clusters is activated by a transcription activator (*ttsI*) whose transcription is induced by plant secreted flavonoids via NodD and or other regulatory Nod proteins (Jiménez-Guerrero et al., 2017; Krause, Doerfel, & Göttfert, 2002; Wassem et al., 2008).

8. Searching for genes that are both common and specific to the α and β rhizobia

The availability of many complete genome sequences for nodulating α- and β-rhizobia, together with sequences for symbiotically effective *Bradyrhizobium* strains that lack canonical *nod* genes, allowed Amadou et al. (2008) to search for genes that were unique and shared by the α and β-rhizobia. As no such genes were identified, the authors inferred that there is not a single or unique biological strategy that is required for the symbiosis of rhizobia with legumes (Amadou et al., 2008). The authors followed this work with a review where they expand on the variety of mechanisms that rhizobia employ to develop a symbiosis with legumes (Masson-Boivin et al., 2009). In another study, Seshadri et al. (2015) searched 145 genomes belonging to seven genera of α—rhizobia, and 18 genomes from the two genera of β—rhizobia from diverse legume hosts from various biomes with the aim of identifying the genetic determinants of symbiotic association and plant growth promotion (Seshadri et al., 2015). Of 4896 Pfam protein families, 437 were over represented in the α— and β—rhizobia and of those, 180 corresponded to genes that were upregulated during SNF as deduced from at least one of four transcriptome studies (Seshadri et al., 2015). While several of these hits included known *nif* and *nod* genes, many other hits contained domains of unknown function and follow-up functional studies on the corresponding genes and on genes identified via similar approaches would be revealing (eg. Pini, Galardini, Bazzicalupo, & Mengoni, 2011).

8.1 Are there common background genomic features required for SNF by rhizobia?

The influence of the bacterial genetic background on symbiotic performance is unclear. However it would seem that several different background features affect the ability and efficiency (effectiveness) of SNF.

While the α-rhizobium isolates fall within the phylogenetically defined order Rhizobiales, these organisms show great phenotypic diversity including, for example, differences in growth rates, basic carbon utilization phenotypes (with substrates such as methanol, methane, organic acids, and glucose), and differences in lipopolysaccharide and exopolysaccharide composition and structure. The identification of β-rhizobia has further raised questions as to what genomic factors other than the SNF genes allow the formation of N_2-fixing nodules. On the other hand, despite the great phenotypic diversity of the α-rhizobia, there does appear to be clustering as rhizobia are only found in only a limited number of genera within the families of α-proteobacteria. In the future, as more nodule isolates are examined, it will be interesting to see if the number of genera increases dramatically. One possibility is that the rhizobia host background is of importance in avoiding the innate immune system of plants that recognize pathogen-associated molecular patterns (PAMPs) (Jones, Vance, & Dangl, 2016). Early in symbiosis a plant defense-like reaction can be observed that then subsides (Lohar et al., 2006), and much evidence indicates that plant immunity plays a critical role in the legume symbiosis (Gourion, Berrabah, Ratet, & Stacey, 2015; Zipfel & Oldroyd, 2017). PAMPs are recognized by pattern recognition receptors (PRRs) in the host plant and this initiates a defense response known as PAMP-triggered immunity (PTI). As recently noted, there are no reports of PAMP components from rhizobia shown to activate a defense reaction in its legume host (Berrabah et al., 2018). However it has been known for many years that symbiotic defects arise as a result of mutations that alter the bacterial surface lipopolysaccharides, cyclic glucans, or exopolysaccharides (Jones et al., 2007), and these findings gave rise to suggestions that rhizobial polysaccharides may act to suppress the host plant defense response (Jones et al., 2007; Mithöfer, 2002). Most recently, the role of EPS in legume-rhizobia recognition was dramatically strengthened by the identification of a membrane-spanning receptor-like protein (Epr3) from *Lotus* that interacts with rhizobial EPS. The Epr3 receptor contains three extracellular LysM domains and has significant similarity to Nod factor receptor proteins (Kawaharada et al., 2015); in the future, it will be interesting to determine if there are other receptors for bacterial polysaccharides and how these might down regulate plant defenses (Long, 2015).

In an experimental evolution approach, Masson-Boivin and colleagues are investigating the adaptations that occur in order for bacteria that receive symbiotic genes to express symbiotic traits (Marchetti et al., 2010, 2014). They transferred the *nod-nif* symbiotic plasmid from the β-rhizobia strain *C. taiwanensis* LMG19424 into the plant pathogen *Ralstonia solanacearum* GMI1000 that is also a member of the Burkholderiaceae family of the β-proteobacteria. The initial transconjugant strain was unable to form nodules on the *C. taiwanensis* LMG19424 legume host *M. pudica*, however spontaneous mutants able to nodulate and poorly infect nodules were identified by visually screening roots for nodules. These strains carried mutations that disrupted the virulence pathway of *R. solanacearum* (Marchetti et al., 2010). In further experiments, analysis of independent lineages following 16 serial passages through plants revealed strong and parallel evolution of symbiotic traits including, nodulation competitiveness, intracellular infection, and bacteroid persistence, and these evolved strains elicited a reduced plant defense reactions within nodules, suggesting that the expression of symbiotic competence requires a limited plant immunity response. The adaptive mutations were not observed in SNF genes on the Sym plasmid, rather they affected regulatory genes that control virulence (Marchetti et al., 2014).

Comparative genomics is also giving clues into the background genomic features that may enable bacteria to become a nitrogen-fixing plant symbiont following acquisition of the appropriate genes. A recent study utilized >1300 plant and soil-associated Rhizobiales isolates to demonstrate sweeping plant growth promotion phenotypes and enrichment in the plant and soil environments that was independent of the presence of *nod* or *nif* genes. Thus an ancestral state of adaptation to the root environment in the Rhizobiales predisposed these bacteria with a potential to shift from a commensal lifestyle in the plant rhizosphere (Garrido-Oter et al., 2018). Members of different lineages in the rhizobiales, already adapted as plant commensals, presumably transitioned to a sympiotic relationship upon acquisition of symbiotic genes by horizontal gene transfer. The authors were unable to identify genomic signatures of adaptation to a symbiotic lifestyle that differentiated nodulating from non-nodulating strains. What genomic adaptations that may have occurred to facilitate commensalism in the Rhizobiales, and whether specific background-dependent genes are required to allow a bacterium to become symbiotic remains to be discovered.

9. Rhizobia genome deconstruction

Research over the past sixty years on rhizobia appears to have identified most, if not all, of the major genetic components required for symbiotic N_2-fixation. Transcriptome, metabolome, and proteome studies have also generated system-level data and hence there is great knowledge as to what genes and proteins are expressed at different stages of the symbiosis. However, much remains to be uncovered, as exemplified by the recent discovery of the Epr3 receptor (Kawaharada et al., 2015), or the role of small RNAs derived from rhizobial tRNAs in regulating nodulation (Ren, Wang,

Duan, & Ma, 2019). Several genome level approaches that appear to hold much promise in uncovering novel aspects of the symbiosis include massively parallel transposon insertion sequencing (Tn-seq) (Perry, Akter, & Yost, 2016), and genome deconstruction approaches. Studies that employ Tn-seq or related methods on rhizobia have been recently reviewed (diCenzo et al., 2019), and will not be discussed further here.

Genome deconstruction-type approaches employ some classic genetic methodologies together with ongoing advances in synthetic biology, particularly in DNA synthesis and gene/genome assembly, that are still at early stages in their development. As a means of addressing the genomic features (genetic repertoire) that are required for SNF, in the future, we envisage the complete deconstruction of the genomes from several rhizobia and their synthetic reconstruction to build mosaic organisms with components (gene sets) that are mixed and matched with, for example, different polysaccharide biosynthetic gene clusters. Hitherto, this level of engineering was not possible at the level of the genome. New technology such as gene synthesis, yeast recombineering and specialized vectors, including those that utilize *repABC* modules to mimic rhizobial accessory plasmids (Brumwell et al., 2019; Döhlemann et al., 2017), will help facilitate synthetic reconstruction of rhizobial genomes. One possible step in such approaches is to minimize or remove what we loosely refer to as the accessory genome. In this regard, there are several reports in which much of the accessory genome of at least some rhizobial strains has been removed (Hynes & McGregor, 1990), and there is now considerable knowledge regarding the accessory genomes of species such as *S. meliloti* (Sugawara et al., 2013), or *R. etli* (González et al., 2010; Rogel et al., 2014). The 6530 kb *R. etli* CFN42 genome contains one chromosome (4382 kb) and six large plasmids (184−642 kb = 2148 kb) (González et al., 2006). Five of these plasmids and 90% of the p42e plasmid were removed without loss of cell viability; however, two genes of unknown function on the p42e plasmid were essential. Plasmid p42e also carries genes required for cytochrome *o* (*cyoABCDE*), septum formation (*minCDE*), and synthesis of cobalamin (vitamine B12), cardiolipin, NAD, and thiamine; however, cells lacking those genes could grow in complex media or minimal medium with appropriate supplements (Landeta et al., 2011). In the case of *S. meliloti* strain 1021, extensive deletion and bioinformatic analysis of the pSymA megaplasmid and pSymB chromid identified active toxin-antitoxin systems and a ~100 kb region of pSymB carrying two essential genes, tRNAarg and *engA* (G. diCenzo et al., 2012; Milunovic et al., 2014). This latter cluster of genes is present on the chromosome of other *Sinorhizobium* species and thus it translocated from the chromosome to pSymB in a *S. meliloti* ancestor strain (diCenzo et al., 2012). Manipulation of these loci and integration of the essential genes from pSymB into the chromosome in a strain that lacked pSymA (Oresnik, Liu, Yost, & Hynes, 2000) allowed the construction of derivatives in which both the pSymA megaplasmid and pSymB chromid (~40% of the genome) were removed (diCenzo, MacLean, Milunovic, Golding, & Finan, 2014). These strains (ΔpSymA, ΔpSymB, ΔpSymAB) are being used to identify the gene complement that is necessary and sufficient for the formation of N_2-fixing nodules (diCenzo, Zamani, Milunovic, & Finan, 2016). Once such minimal symbiotic genomes are defined, these will be used to uncover other genes that influence SNF, whether they are present on the chromosome or other replicons.

Acknowledgments

TMF is grateful for support from the Natural Sciences and Engineering Research Council of Canada. We apologize to all authors whose work was not cited because of space constraints.

References

Amadou, C., Pascal, G., Mangenot, S., Glew, M., Bontemps, C., Capela, D., et al. (2008). Genome sequence of the β-rhizobium *Cupriavidus taiwanensis* and comparative genomics of rhizobia. *Genome Research, 18*(9), 1472–1483.

Aoki, S., Ito, M., & Iwasaki, W. (2013). From β-to α-proteobacteria: The origin and evolution of rhizobial nodulation genes *nodIJ*. *Molecular Biology and Evolution, 30*(11), 2494–2508.

Ardourel, M., Demont, N., Debellé, F., Maillet, F., de Billy, F., Promé, J.-C., et al. (1994). *Rhizobium meliloti* lipooligosaccharide nodulation factors: Different structural requirements for bacterial entry into target root hair cells and induction of plant symbiotic developmental responses. *The Plant Cell, 6*(10), 1357–1374.

Baldani, J. I., Weaver, R. W., Hynes, M. F., & Eardly, B. D. (1992). Utilization of carbon substrates, electrophoretic enzyme patterns, and symbiotic performance of plasmid-cured clover rhizobia. *Applied and Environmental Microbiology, 58*(7), 2308–2314.

Barnett, M. J., Fisher, R. F., Jones, T., Komp, C., Abola, A. P., Barloy-Hubler, F., et al. (2001). Nucleotide sequence and predicted functions of the entire Sinorhizobium meliloti pSymA megaplasmid. *Proceedings of the National Academy of Sciences of the United States of America, 98*, 9883–9888.

Bartosik, D., Szymanik, M., & Wysocka, E. (2001). Identification of the partitioning site within the *repABC*-type replicon of the composite *Paracoccus versutus* plasmid pTAV1. *Journal of Bacteriology, 183*(21), 6234–6243.

Batut, J., Daveran-Mingot, M. L., David, M., Jacobs, J., Garnerone, A. M., & Kahn, D. (1989). *fixK*, a gene homologous with *fnr* and *crp* from *Escherichia coli*, regulates nitrogen fixation genes both positively and negatively in *Rhizobium meliloti*. *The EMBO Journal, 8*(4), 1279–1286.

Bauer, E., Kaspar, T., Fischer, H. M., & Hennecke, H. (1998). Expression of the *fixR-nifA* operon in *Bradyrhizobium japonicum* depends on a new response regulator, RegR. *Journal of Bacteriology, 180*(15), 3853–3863.

Beringer, J. E. (1974). R factor transfer in *Rhizobium leguminosarum. Microbiology, 84*(1), 188–198.

Berrabah, F., Balliau, T., Aït-Salem, E. H., George, J., Zivy, M., Ratet, P., et al. (2018). Control of the ethylene signaling pathway prevents plant defenses during intracellular accommodation of the rhizobia. *New Phytologist, 219*(1), 310–323.

Bobik, C., Meilhoc, E., & Batut, J. (2006). FixJ: A major regulator of the oxygen limitation response and late symbiotic functions of *Sinorhizobium meliloti. Journal of Bacteriology, 188*(13), 4890–4902.

Bournaud, C., de Faria, S. M., dos Santos, J. M. F., Tisseyre, P., Silva, M., Chaintreuil, C., et al. (2013). *Burkholderia* species are the most common and preferred nodulating symbionts of the *Piptadenia* group (tribe Mimoseae). *PLoS One, 8*(5), e63478.

Brewin, N. J., Beringer, J. E., & Johnston, A. W. B. (1980). Plasmid-mediated transfer of host-range specificity between two strains of *Rhizobium leguminosarum. Microbiology, 120*(2), 413–420.

Brumwell, S. L., MacLeod, M. R., Huang, T., Cochrane, R. R., Meaney, R. S., Zamani, M., et al. (2019). Designer *Sinorhizobium meliloti* strains and multi-functional vectors enable direct inter-kingdom DNA transfer. *PLoS One, 14*(6), e0206781 (h).

Burén, S., & Rubio, L. M. (2017). State of the art in eukaryotic nitrogenase engineering. *FEMS Microbiology Letters, 365*(2), fnx274.

Capela, D., Barloy-Hubler, F., Gouzy, J., Bothe, G., Ampe, F., Batut, J., et al. (2001). Analysis of the chromosome sequence of the legume symbiont *Sinorhizobium meliloti* strain 1021. *Proceedings of the National Academy of Sciences of the United States of America, 98*, 9877–9882.

diCenzo, G. C., Milunovic, B., Cheng, J., & Finan, T. M. (2012). tRNA^{arg} and *engA* are essential genes on the 1.7-Mb pSymB megaplasmid of Sinorhizobium meliloti and were translocated together from the chromosome in an ancestral strain. *Journal of Bacteriology, 195*(2), 202–212.

diCenzo, G. C., MacLean, A. M., Milunovic, B., Golding, G. B., & Finan, T. M. (2014). Examination of prokaryotic multipartite genome evolution through experimental genome reduction. *PLoS Genetics, 10*(10), e1004742.

diCenzo, G. C., Checcucci, A., Bazzicalupo, M., Mengoni, A., Viti, C., Dziewit, L., et al. (2016). Metabolic modelling reveals the specialization of secondary replicons for niche adaptation in *Sinorhizobium meliloti. Nature Communications,* 7(1), 12219.

diCenzo, G. C., Zamani, M., Milunovic, B., & Finan, T. M. (2016). Genomic resources for identification of the minimal N_2-fixing symbiotic genome. *Environmental Microbiology, 18*(8), 2534–2547.

diCenzo, G. C., & Finan, T. M. (2017). The divided bacterial genome: Structure, function, and evolution. *Microbiology and Molecular Biology Reviews, 81*(3). e00019-17.

diCenzo, G. C., Zamani, M., Checcucci, A., Fondi, M., Griffitts, J. S., Finan, T. M., et al. (2019). Multidisciplinary approaches for studying rhizobium–legume symbioses. *Canadian Journal of Microbiology, 65*(1), 1–33.

Cevallos, M. A., Cervantes-Rivera, R., & Gutiérrez-Ríos, R. M. (2008). The *repABC* plasmid family. *Plasmid, 60*(1), 19–37.

Chai, Y., & Winans, S. C. (2005). A small antisense RNA downregulates expression of an essential replicase protein of an *Agrobacterium tumefaciens* Ti plasmid. *Molecular Microbiology, 56*(6), 1574–1585.

Chen, I.-M. A., Chu, K., Palaniappan, K., Pillay, M., Ratner, A., Huang, J., et al. (2019). IMG/M v.5.0: An integrated data management and comparative analysis system for microbial genomes and microbiomes. *Nucleic Acids Research, 47*(D1), D666–D677. https://doi.org/10.1093/nar/gky901.

Chen, W.-M., James, E. K., Prescott, A. R., Kierans, M., & Sprent, J. I. (2003). Nodulation of *Mimosa* spp. by the β-proteobacterium *Ralstonia taiwanensis*. *Molecular Plant-Microbe Interactions, 16*(12), 1051–1061.
Cosseau, C., & Batut, J. (2004). Genomics of the *ccoNOQP*-encoded *cbb3* oxidase complex in bacteria. *Archives of Microbiology, 181*(2), 89–96.
Costas, A. M. G., Poudel, S., Miller, A.-F., Schut, G. J., Ledbetter, R. N., Fixen, K. R., et al. (2017). Defining electron bifurcation in the electron-transferring flavoprotein family. *Journal of Bacteriology, 199*(21). e00440-17.
Danino, V. E., Wilkinson, A., Edwards, A., & Downie, J. A. (2003). Recipient-induced transfer of the symbiotic plasmid pRL1JI in *Rhizobium leguminosarum* bv. *viciae* is regulated by a quorum-sensing relay. *Molecular Microbiology, 50*(2), 511–525.
De Meyer, S. E., Briscoe, L., Martínez-Hidalgo, P., Agapakis, C. M., de-los Santos, P. E., Seshadri, R., et al. (2016). Symbiotic *Burkholderia* species show diverse arrangements of *nif/fix* and *nod* genes and lack typical high-affinity cytochrome *cbb3* oxidase genes. *Molecular Plant-Microbe Interactions, 29*(8), 609–619.
Deakin, W. J., & Broughton, W. J. (2009). Symbiotic use of pathogenic strategies: Rhizobial protein secretion systems. *Nature Reviews Microbiology,* 7(4), 312.
Delaux, P.-M., Radhakrishnan, G. V., Jayaraman, D., Cheema, J., Malbreil, M., Volkening, J. D., et al. (2015). Algal ancestor of land plants was preadapted for symbiosis. *Proceedings of the National Academy of Sciences, 112*(43), 13390–13395.
Denarie, J., Debelle, F., & Prome, J.-C. (1996). Rhizobium lipo-chitooligosaccharide nodulation factors: Signaling molecules mediating recognition and morphogenesis. *Annual Review of Biochemistry, 65*(1), 503–535.
Ding, H., & Hynes, M. F. (2009). Plasmid transfer systems in the rhizobia. *Canadian Journal of Microbiology, 55*(8), 917–927.
Döhlemann, J., Wagner, M., Happel, C., Carrillo, M., Sobetzko, P., Erb, T. J., et al. (2017). A family of single copy *repABC* -type shuttle vectors stably maintained in the alpha-proteobacterium *Sinorhizobium meliloti*. *ACS Synthetic Biology, 6*(6), 968–984.
Dusha, I., Kovalenko, S., Banfalvi, Z., & Kondorosi, A. (1987). *Rhizobium meliloti* insertion element ISRm2 and its use for identification of the *fixX* gene. *Journal of Bacteriology, 169*(4), 1403–1409.
D'Haeze, W., Van Montagu, M., Promé, J.-C., & Holsters, M. (1999). Carbamoylation of azorhizobial Nod factors is mediated by NodU. *Molecular Plant-Microbe Interactions, 12*(1), 68–73.
Earl, C. D., Ronson, C. W., & Ausubel, F. M. (1987). Genetic and structural analysis of the *Rhizobium meliloti fixA, fixB, fixC,* and *fixX* genes. *Journal of Bacteriology, 169*(3), 1127–1136.
Edgren, T., & Nordlund, S. (2004). The *fixABCX* genes in *Rhodospirillum rubrum* encode a putative membrane complex participating in electron transfer to nitrogenase. *Journal of Bacteriology, 186*(7), 2052–2060.
Edgren, T., & Nordlund, S. (2005). Electron transport to nitrogenase in *Rhodospirillum rubrum*: Identification of a new *fdxN* gene encoding the primary electron donor to nitrogenase. *FEMS Microbiology Letters, 245*(2), 345–351.
Estrada-de los Santos, P., Palmer, M., Chávez-Ramírez, B., Beukes, C., Steenkamp, E., Briscoe, L., et al. (2018). Whole genome analyses suggests that *Burkholderia* sensu lato contains two additional novel genera (*Mycetohabitans* gen. Nov., and *Trinickia* gen. Nov.): Implications for the evolution of diazotrophy and nodulation in the *Burkholderiaceae*. *Genes, 9*(8), 389.
Fischer, H.-M. (1994). Genetic regulation of nitrogen fixation in rhizobia. *Microbiology and Molecular Biology Reviews, 58*(3), 352–386.
Foussard, M., Garnerone, A., Ni, F., Soupène, E., Boistard, P., & Batut, J. (1997). Negative autoregulation of the *Rhizobium meliloti fixK* gene is indirect and requires a newly

identified regulator, FixT. *Molecular Microbiology, 25*(1), 27–37. https://doi.org/10.1046/j.1365-2958.1997.4501814.x.

Gage, D. J. (2004). Infection and invasion of roots by symbiotic, nitrogen-fixing rhizobia during nodulation of temperate legumes. *Microbiology and Molecular Biology Reviews, 68*(2), 280–300.

Galardini, M., Pini, F., Bazzicalupo, M., Biondi, E. G., & Mengoni, A. (2013). Replicon-dependent bacterial genome evolution: The case of *Sinorhizobium meliloti*. *Genome Biology and Evolution, 5*(3), 542–558.

Galibert, F., Finan, T. M., Long, S. R., Pühler, A., Abola, P., Ampe, F., et al. (2001). The composite genome of the legume symbiont Sinorhizobium meliloti. *Science, 293*, 668–672.

Garnerone, A. M., Cabanes, D., Foussard, M., Boistard, P., & Batut, J. (1999). Inhibition of the FixL sensor kinase by the FixT protein in *Sinorhizobium meliloti*. *The Journal of Biological Chemistry, 274*(45), 32500–32506 (h).

Garrido-Oter, R., Nakano, R. T., Dombrowski, N., Ma, K.-W., Team, T. A., McHardy, A. C., et al. (2018). Modular traits of the Rhizobiales root microbiota and their evolutionary relationship with symbiotic rhizobia. *Cell Host & Microbe, 24*(1), 155–167.

Geddes, B. A., Ryu, M.-H., Mus, F., Costas, A. G., Peters, J. W., Voigt, C. A., et al. (2015). Use of plant colonizing bacteria as chassis for transfer of N_2-fixation to cereals. *Current Opinion in Biotechnology, 32*, 216–222.

Gibson, D. G., Glass, J. I., Lartigue, C., Noskov, V. N., Chuang, R.-Y., Algire, M. A., et al. (2010). Creation of a bacterial cell controlled by a chemically synthesized genome. *Science (New York, N.Y.), 329*(5987), 52–56.

Giraud, E., Moulin, L., Vallenet, D., Barbe, V., Cytryn, E., Avarre, J. C., et al. (2007). Legumes symbioses: Absence of *nod* genes in photosynthetic bradyrhizobia. *Science, 316*(5829), 1307–1312.

Gong, Z. Y., He, Z. S., Zhu, J. B., Yu, G. Q., & Zou, H. S. (2006). *Sinorhizobium meliloti nifA* mutant induces different gene expression profile from wild type in Alfalfa nodules. *Cell Research, 16*(10), 818.

González, V., Acosta, J. L., Santamaría, R. I., Bustos, P., Fernández, J. L., Hernández González, I. L., et al. (2010). Conserved symbiotic plasmid DNA Sequences in the multireplicon pangenomic structure of *Rhizobium etli*. *Applied and Environmental Microbiology, 76*(5), 1604–1614.

González, V., Santamaría, R. I., Bustos, P., Hernández-González, I., Medrano-Soto, A., Moreno-Hagelsieb, G., et al. (2006). The partitioned *Rhizobium etli* genome: Genetic and metabolic redundancy in seven interacting replicons. *Proceedings of the National Academy of Sciences, 103*(10), 3834–3839.

Gourion, B., Berrabah, F., Ratet, P., & Stacey, G. (2015). Rhizobium–legume symbioses: The crucial role of plant immunity. *Trends in Plant Science, 20*(3), 186–194.

Gully, D., Teulet, A., Busset, N., Nouwen, N., Fardoux, J., Rouy, Z., et al. (2017). Complete genome sequence of *Bradyrhizobium* sp. ORS285, a photosynthetic strain able to establish nod factor-dependent or nod factor-independent symbiosis with *Aeschynomene* legumes. *Genome Announcements, 5*(30). e00421-17.

Gyaneshwar, P., Hirsch, A. M., Moulin, L., Chen, W.-M., Elliott, G. N., Bontemps, C., et al. (2011). Legume-nodulating betaproteobacteria: Diversity, host range, and future prospects. *Molecular Plant-Microbe Interactions, 24*(11), 1276–1288.

Harrison, P. W., Lower, R. P. J., Kim, N. K. D., & Young, J. P. W. (2010). Introducing the bacterial 'chromid': Not a chromosome, not a plasmid. *Trends in Microbiology, 18*(4), 141–148.

Haskett, T. L., Ramsay, J. P., Bekuma, A. A., Sullivan, J. T., O'Hara, G. W., & Terpolilli, J. J. (2017). Evolutionary persistence of tripartite integrative and conjugative elements. *Plasmid, 92*, 30–36.

Haskett, T. L., Terpolilli, J. J., Bekuma, A., O'Hara, G. W., Sullivan, J. T., Wang, P., et al. (2016). Assembly and transfer of tripartite integrative and conjugative genetic elements. *Proceedings of the National Academy of Sciences, 113*(43), 12268–12273.

Haskett, T. L., Terpolilli, J. J., Ramachandran, V. K., Verdonk, C. J., Poole, P. S., O'Hara, G. W., et al. (2018). Sequential induction of three recombination directionality factors directs assembly of tripartite integrative and conjugative elements. *PLoS Genetics, 14*(3), e1007292.

Higashi, S. (1967). Transfer of clover infectivity of *Rhizobium trifolii* to *Rhizobium phaseoli* as mediated by an episomic factor. *The Journal of General and Applied Microbiology, 13*(4), 391–403.

Hooykaas, P. J. J., Snijdewint, F. G. M., & Schilperoort, R. A. (1982). Identification of the Sym plasmid of *Rhizobium leguminosarum* strain 1001 and its transfer to and expression in other rhizobia and *Agrobacterium tumefaciens*. *Plasmid, 8*(1), 73–82.

Hynes, M. F., & McGregor, N. F. (1990). Two plasmids other than the nodulation plasmid are necessary for formation of nitrogen fixing nodules by *Rhizobium leguminosarum*. *Molecular Microbiology, 4*, 567–574.

Hynes, M. F., Quandt, J., O'Connell, M. P., & Pühler, A. (1989). Direct selection for curing and deletion of Rhizobium plasmids using transposons carrying the *Bacillus subtilis sacB* gene. *Gene, 78*(1), 111–120.

Jabbouri, S., Fellay, R., Talmont, F., Kamalaprija, P., Burger, U., Relic, B., et al. (1995). Involvement of *nodS* in N-methylation and *nodU* in 6-O-carbamoylation of Rhizobium sp. NGR234 Nod factors. *Journal of Biological Chemistry, 270*(39), 22968–22973.

Jimenez-Guerrero, I., Acosta-Jurado, S., del Cerro, P., Navarro-Gómez, P., López-Baena, F., Ollero, F., et al. (2017). Transcriptomic studies of the effect of *nod* gene-inducing molecules in rhizobia: Different weapons, one purpose. *Genes, 9*(1), 1.

Jiménez-Guerrero, I., Pérez-Montaño, F., Medina, C., Ollero, F. J., & López-Baena, F. J. (2017). The *Sinorhizobium (Ensifer) fredii* HH103 nodulation outer protein NopI is a determinant for efficient nodulation of soybean and cowpea plants. *Applied and Environmental Microbiology, 83*(5). e02770-16.

Johnson, C. M., & Grossman, A. D. (2015). Integrative and conjugative elements (ICEs): What they do and how they work. *Annual Review of Genetics, 49*, 577–601.

Jones, K. M., Kobayashi, H., Davies, B. W., Taga, M. E., & Walker, G. C. (2007). How rhizobial symbionts invade plants: The *Sinorhizobium–Medicago* model. *Nature Reviews Microbiology, 5*(8), 619–633.

Jones, J. D. G., Vance, R. E., & Dangl, J. L. (2016). Intracellular innate immune surveillance devices in plants and animals. *Science, 354*(6316), aaf6395.

Kaneko, T., Nakamura, Y., Sato, S., Asamizu, E., Kato, T., Sasamoto, S., et al. (2000). Complete genome structure of the nitrogen-fixing symbiotic bacterium *Mesorhizobium loti*. *DNA Research, 7*, 331–338.

Kaneko, T., Nakamura, Y., Sato, S., Minamisawa, K., Uchiumi, T., Sasamoto, S., et al. (2002). Complete genomic sequence of nitrogen-fixing symbiotic bacterium *Bradyrhizobium japonicum* USDA110. *DNA Research, 9*(6), 189–197.

Kawaharada, Y., Kelly, S., Nielsen, M. W., Hjuler, C. T., Gysel, K., Muszynski, A., et al. (2015). Receptor-mediated exopolysaccharide perception controls bacterial infection. *Nature, 523*(7560), 308.

Kelly, S., Radutoiu, S., & Stougaard, J. (2017). Legume LysM receptors mediate symbiotic and pathogenic signalling. *Current Opinion in Plant Biology, 39*, 152–158.

Klipp, W., Reiländer, H., Schlüter, A., Krey, R., & Pühler, A. (1989). The *Rhizobium meliloti fdxN* gene encoding a ferredoxin-like protein is necessary for nitrogen fixation and is cotranscribed with *nifA* and *nifB*. *Molecular and General Genetics MGG, 216*(2–3), 293–302.

Klonowska, A., Melkonian, R., Miché, L., Tisseyre, P., & Moulin, L. (2018). Transcriptomic profiling of *Burkholderia phymatum* STM815, *Cupriavidus taiwanensis* LMG19424 and Rhizobium mesoamericanum STM3625 in response to *Mimosa pudica* root exudates illuminates the molecular basis of their nodulation competetiveness and symbiotic evolutionary history. *BMC Genomics, 19*(1), 105.

Kopat, V. V., Chirak, E. R., Kimeklis, A. K., Safronova, V. I., Belimov, A. A., Kabilov, M. R., et al. (2017). Evolution of *fixNOQP* genes encoding cytochrome oxidase with high affinity to oxygen in rhizobia and related bacteria. *Russian Journal of Genetics, 53*(7), 766–774.

Koper, P., Żebracki, K., Marczak, M., Skorupska, A., & Mazur, A. (2016). RepB proteins of the multipartite *Rhizobium leguminosarum* bv. *trifolii* genome discriminate between centromere-like *parS* sequences for plasmid segregational stability. *Molecular Microbiology, 102*(3), 446–466.

Krause, A., Doerfel, A., & Göttfert, M. (2002). Mutational and transcriptional analysis of the type III secretion system of *Bradyrhizobium japonicum*. *Molecular Plant-Microbe Interactions, 15*(12), 1228–1235.

Kulajta, C., Thumfart, J. O., Haid, S., Daldal, F., & Koch, H.-G. (2006). Multi-step assembly pathway of the *cbb3*-type cytochrome c oxidase complex. *Journal of Molecular Biology, 355*(5), 989–1004.

Kumar, S., Stecher, G., & Tamura, K. (2016). MEGA7: Molecular evolutionary genetics analysis version 7.0 for bigger datasets. *Molecular Biology and Evolution, 33*(7), 1870–1874. https://doi.org/10.1093/molbev/msw054.

Landeta, C., Dávalos, A., Cevallos, M.Á., Geiger, O., Brom, S., & Romero, D. (2011). Plasmids with a chromosome-like role in rhizobia. *Journal of Bacteriology, 193*(6), 1317–1326.

Lardi, M., Liu, Y., Purtschert, G., Bolzan de Campos, S., & Pessi, G. (2017). Transcriptome analysis of *Paraburkholderia phymatum* under nitrogen starvation and during symbiosis with *Phaseolus vulgaris*. *Genes, 8*(12), 389.

Layzell, D. B., Del Castillo, L. D., Hunt, S., Kuzma, M., Van Cauwenberghe, O., & Oresnik, I. (1993). The regulation of oxygen and its role in regulating nodule metabolism. In *New horizons in nitrogen fixation* (pp. 393–398). Springer.

Ledbetter, R. N., Garcia Costas, A. M., Lubner, C. E., Mulder, D. W., Tokmina-Lukaszewska, M., Artz, J. H., et al. (2017). The electron bifurcating FixABCX protein complex from *Azotobacter vinelandii*: Generation of low-potential reducing equivalents for nitrogenase catalysis. *Biochemistry, 56*(32), 4177–4190.

Lee, K.-B., De Backer, P., Aono, T., Liu, C.-T., Suzuki, S., Suzuki, T., et al. (2008). The genome of the versatile nitrogen fixer *Azorhizobium caulinodans* ORS571. *BMC Genomics, 9*(1), 271.

Lerouge, P., Roche, P., Faucher, C., Maillet, F., Truchet, G., Promé, J. C., et al. (1990). Symbiotic host-specificity of *Rhizobium meliloti* is determined by a sulphated and acylated glucosamine oligosaccharide signal. *Nature, 344*(6268), 781–784.

Ling, J., Wang, H., Wu, P., Li, T., Tang, Y., Naseer, N., et al. (2016). Plant nodulation inducers enhance horizontal gene transfer of *Azorhizobium caulinodans* symbiosis island. *Proceedings of the National Academy of Sciences, 113*(48), 13875–13880.

Lohar, D. P., Sharopova, N., Endre, G., Peñuela, S., Samac, D., Town, C., et al. (2006). Transcript analysis of early nodulation events in *Medicago truncatula*. *Plant Physiology, 140*(1), 221–234.

Long, S. R. (2015). Symbiosis: Receptive to infection. *Nature, 523*(7560), 298.

López-Lara, I. M., Blok-Tip, L., Quinto, C., Garcia, M. L., Stacey, G., Bloemberg, G. V., et al. (1996). NodZ of *Bradyrhizobium* extends the nodulation host range of *Rhizobium* by adding a fucosyl residue to nodulation signals. *Molecular Microbiology, 21*(2), 397–408.

MacLellan, S. R., Smallbone, L. A., Sibley, C. D., & Finan, T. M. (2005). The expression of a novel antisense gene mediates incompatibility within the large *repABC* family of α-proteobacterial plasmids. *Molecular Microbiology, 55*(2), 611–623.

MacLellan, S. R., Zaheer, R., Sartor, A. L., MacLean, A. M., & Finan, T. M. (2006). Identification of a megaplasmid centromere reveals genetic structural diversity within the *repABC* family of basic replicons. *Molecular Microbiology, 59*(5), 1559–1575.

Madsen, L. H., Tirichine, L., Jurkiewicz, A., Sullivan, J. T., Heckmann, A. B., Bek, A. S., et al. (2010). The molecular network governing nodule organogenesis and infection in the model legume *Lotus japonicus*. *Nature Communications, 1*, 10.

Marchetti, M., Capela, D., Glew, M., Cruveiller, S., Chane-Woon-Ming, B., Gris, C., et al. (2010). Experimental evolution of a plant pathogen into a legume symbiont. *PLoS Biology, 8*(1), e1000280.

Marchetti, M., Jauneau, A., Capela, D., Remigi, P., Gris, C., Batut, J., et al. (2014). Shaping bacterial symbiosis with legumes by experimental evolution. *Molecular Plant-Microbe Interactions, 27*(9), 956–964.

Marx, C. J., Bringel, F., Chistoserdova, L., Moulin, L., Haque, M. F. U., Fleischman, D. E., et al. (2012). Complete genome sequences of six strains of the genus *Methylobacterium*. *Journal of Bacteriology, 194*(17), 4746.

Masson-Boivin, C., Giraud, E., Perret, X., & Batut, J. (2009). Establishing nitrogen-fixing symbiosis with legumes: How many rhizobium recipes? *Trends in Microbiology, 17*(10), 458–466.

Mergaert, P., D'Haeze, W., Fernández-López, M., Geelen, D., Goethals, K., Promé, J., et al. (1996). Fucosylation and arabinosylation of Nod factors in *Azorhizobium caulinodans*: Involvement of *nolK nodZ* as well as *noeC* and/or downstream genes. *Molecular Microbiology, 21*(2), 409–419.

Mesa, S., Hauser, F., Friberg, M., Malaguti, E., Fischer, H.-M., & Hennecke, H. (2008). Comprehensive assessment of the regulons controlled by the FixLJ-FixK2-FixK1 cascade in *Bradyrhizobium japonicum*. *Journal of Bacteriology, 190*(20), 6568–6579.

Milunovic, B., Morton, R. A., & Finan, T. M. (2014). Cell growth inhibition upon deletion of four toxin-antitoxin loci from the megaplasmids of *Sinorhizobium meliloti*. *Journal of Bacteriology, 196*(4), 811–824.

Mithöfer, A. (2002). Suppression of plant defence in rhizobia–legume symbiosis. *Trends in Plant Science, 7*(10), 440–444.

Miwa, H., & Okazaki, S. (2017). How effectors promote beneficial interactions. *Current Opinion in Plant Biology, 38*, 148–154.

Moulin, L., Klonowska, A., Caroline, B., Booth, K., Vriezen, J. A. C., Melkonian, R., et al. (2014). Complete Genome sequence of *Burkholderia phymatum* STM815(T), a broad host range and efficient nitrogen-fixing symbiont of Mimosa species. *Standards in Genomic Sciences, 9*(3), 763–774.

Moulin, L., Munive, A., Dreyfus, B., & Boivin-Masson, C. (2001). Nodulation of legumes by members of the β-subclass of Proteobacteria. *Nature, 411*(6840), 948.

Mulligan, J. T., & Long, S. R. (1989). A family of activator genes regulates expression of *Rhizobium meliloti* nodulation genes. *Genetics, 122*(1), 7–18.

Mus, F., Crook, M. B., Garcia, K., Garcia Costas, A., Geddes, B. A., Kouri, E. D., et al. (2016). Symbiotic nitrogen fixation and the challenges to its extension to nonlegumes. *Applied and Environmental Microbiology, 82*(13), 3698–3710.

Nandasena, K. G., O'Hara, G. W., Tiwari, R. P., Sezmiş, E., & Howieson, J. G. (2007). *In situ* lateral transfer of symbiosis islands results in rapid evolution of diverse competitive strains of mesorhizobia suboptimal in symbiotic nitrogen fixation on the pasture legume *Biserrula pelecinus* L. *Environmental Microbiology, 9*(10), 2496–2511.

Nogales, J., Blanca-Ordóñez, H., Olivares, J., & Sanjuán, J. (2013). Conjugal transfer of the *Sinorhizobium meliloti* 1021 symbiotic plasmid is governed through the concerted action of one-and two-component signal transduction regulators. *Environmental Microbiology, 15*(3), 811–821.

Okazaki, S., Kaneko, T., Sato, S., & Saeki, K. (2013). Hijacking of leguminous nodulation signaling by the rhizobial type III secretion system. *Proceedings of the National Academy of Sciences, 110*(42), 17131–17136.

Okazaki, S., Nukui, N., Sugawara, M., & Minamisawa, K. (2004). Rhizobial strategies to enhance symbiotic interactions: Rhizobitoxine and 1-aminocyclopropane-1-carboxylate deaminase. *Microbes and Environments, 19*(2), 99–111.

Okazaki, S., Tittabutr, P., Teulet, A., Thouin, J., Fardoux, J., Chaintreuil, C., et al. (2016). Rhizobium–legume symbiosis in the absence of nod factors: Two possible scenarios with or without the T3SS. *The ISME Journal, 10*(1), 64.

Oresnik, I. J., Liu, S.-L., Yost, C. K., & Hynes, M. F. (2000). Megaplasmid pRme2011a of Sinorhizobium meliloti is not required for viability. *Journal of Bacteriology, 182*(12), 3582–3586.

Ormeño-Orrillo, E., Menna, P., Almeida, L. G. P., Ollero, F. J., Nicolás, M. F., Rodrigues, E. P., et al. (2012). Genomic basis of broad host range and environmental adaptability of *Rhizobium tropici* CIAT 899 and *Rhizobium* sp. PRF 81 which are used in inoculants for common bean (*Phaseolus vulgaris* L.). *BMC Genomics, 13*(1), 735.

Overmars, L., van Hijum, S. A. F. T., Siezen, R. J., & Francke, C. (2015). CiVi: Circular genome visualization with unique features to analyze sequence elements: Fig. 1. *Bioinformatics, 31*(17), 2867–2869.

Pérez-Mendoza, D., Sepúlveda, E., Pando, V., Munoz, S., Nogales, J., Olivares, J., et al. (2005). Identification of the *rctA* gene, which is required for repression of conjugative transfer of rhizobial symbiotic megaplasmids. *Journal of Bacteriology, 187*(21), 7341–7350.

Perret, X., Staehelin, C., & Broughton, W. J. (2000). Molecular basis of symbiotic promiscuity. *Microbiology and Molecular Biology Reviews, 64*(1), 180–201.

Perry, B. J., Akter, M. S., & Yost, C. K. (2016). The use of transposon insertion sequencing to interrogate the core functional genome of the legume symbiont *Rhizobium leguminosarum*. *Frontiers in Microbiology, 7*, 1873.

Petersen, J., Brinkmann, H., & Pradella, S. (2009). Diversity and evolution of *repABC* type plasmids in *Rhodobacterales*. *Environmental Microbiology, 11*(10), 2627–2638.

Petersen, J., Frank, O., Göker, M., & Pradella, S. (2013). Extrachromosomal, extraordinary and essential—the plasmids of the Roseobacter clade. *Applied Microbiology and Biotechnology, 97*(7), 2805–2815.

Pini, F., Galardini, M., Bazzicalupo, M., & Mengoni, A. (2011). Plant-bacteria association and symbiosis: Are there common genomic traits in alphaproteobacteria? *Genes, 2*(4), 1017–1032.

Pinto, U. M., Pappas, K. M., & Winans, S. C. (2012). The ABCs of plasmid replication and segregation. *Nature Reviews Microbiology, 10*(11), 755.

Poudel, S., Colman, D. R., Fixen, K. R., Ledbetter, R. N., Zheng, Y., Pence, N., et al. (2018). Electron transfer to nitrogenase in different genomic and metabolic backgrounds. *Journal of Bacteriology, 200*(10). e00757-17.

Preisig, O., Zufferey, R., & Hennecke, H. (1996). The Bradyrhizobium japonicum *fixGHIS* genes are required for the formation of the high-affinity *cbb3*-type cytochrome oxidase. *Archives of Microbiology, 165*(5), 297–305.

Preisig, O., Zufferey, R., Thöny-Meyer, L., Appleby, C. A., & Hennecke, H. (1996). A high-affinity *cbb3*-type cytochrome oxidase terminates the symbiosis-specific respiratory chain of *Bradyrhizobium japonicum*. *Journal of Bacteriology, 178*(6), 1532–1538.

Pueppke, S. G. (1996). The genetic and biochemical basis for nodulation of legumes by rhizobia. *Critical Reviews in Biotechnology, 16*(1), 1–51.

Ramsay, J. P., Major, A. S., Komarovsky, V. M., Sullivan, J. T., Dy, R. L., Hynes, M. F., et al. (2013). A widely conserved molecular switch controls quorum sensing and symbiosis island transfer in *Mesorhizobium loti* through expression of a novel antiactivator. *Molecular Microbiology, 87*(1), 1–13.

Ramsay, J. P., & Ronson, C. W. (2015). Silencing quorum sensing and ICE mobility through antiactivation and ribosomal frameshifting. *Mobile Genetic Elements, 5*(6), 103–108.

Reeve, W., Chain, P., O'Hara, G., Ardley, J., Nandesena, K., Bräu, L., et al. (2010). Complete genome sequence of the *Medicago* microsymbiont *Ensifer* (*Sinorhizobium*) *medicae* strain WSM419. *Standards in Genomic Sciences, 2*(1), 77–86.

Reeve, W., O'Hara, G., Chain, P., Ardley, J., Brau, L., Nandesena, K., et al. (2010). Complete genome sequence of *Rhizobium leguminosarum* bv. *trifolii* strain WSM1325, an effective microsymbiont of annual Mediterranean clovers. *Standards in Genomic Sciences, 2*(3), 347–356.

Reeve, W., Parker, M., Tian, R., Goodwin, L., Teshima, H., Tapia, R., et al. (2014). Genome sequence of *Microvirga lupini* strain LUT6 T , a novel *Lupinus* alphaproteobacterial microsymbiont from Texas. *Standards in Genomic Sciences, 9*(3), 1159.

Renalier, M., Batut, J., Ghai, J., Terzaghi, B., Gherardi, M., Garnerone, A., et al. (1987). A new symbiotic cluster on the pSym megaplasmid of *Rhizobium meliloti* 2011 carries a functional *fix* gene repeat and a *nod* locus. *Journal of Bacteriology, 169*, 2231–2238.

Ren, B., Wang, X., Duan, J., & Ma, J. (2019). Rhizobial tRNA-derived small RNAs are signal molecules regulating plant nodulation. *Science, 365*(6456), 919–922.

Reyes-González, A., Talbi, C., Rodríguez, S., Rivera, P., Zamorano-Sánchez, D., & Girard, L. (2016). Expanding the regulatory network that controls nitrogen fixation in *Sinorhizobium meliloti*: Elucidating the role of the two-component system hFixL-FxkR. *Microbiology, 162*(6), 979–988.

Reyrat, J. M., David, M., Blonski, C., Boistard, P., & Batut, J. (1993). Oxygen-regulated in vitro transcription of *Rhizobium meliloti nifA* and *fixK* genes. *Journal of Bacteriology, 175*(21), 6867–6872. https://doi.org/10.1128/jb.175.21.6867-6872.1993.

Roche, P., Debellé, F., Maillet, F., Lerouge, P., Faucher, C., Truchet, G., et al. (1991). Molecular basis of symbiotic host specificity in *Rhizobium meliloti: nodH* and *nodPQ* genes encode the sulfation of lipo-oligosaccharide signals. *Cell, 67*(6), 1131–1143.

Rogel, M. A., Bustos, P., Santamaría, R. I., González, V., Romero, D., Cevallos, M., et al. (2014). Genomic basis of symbiovar *mimosae* in *Rhizobium etli*. *BMC Genomics, 15*(1), 575.

Rosenberg, C., Boistard, P., Dénarié, J., & Casse-Delbart, F. (1981). Genes controlling early and late functions in symbiosis are located on a megaplasmid in Rhizobium meliloti. *Molecular and General Genetics, 184*(2), 326–333.

Rubio, L. M., & Ludden, P. W. (2008). Biosynthesis of the iron-molybdenum cofactor of nitrogenase. *Annual Review of Microbiology, 62*, 93–111.

Ruvkun, G. B., Sundaresan, V., & Ausubel, F. M. (1982). Directed transposon Tn*5* mutagenesis and complementation analysis of *Rhizobium meliloti* symbiotic nitrogen fixation genes. *Cell, 29*(2), 551–559.

Schlüter, A., Patschkowski, T., Quandt, J., Selinger, B., Weidner, S., Krämer, M., et al. (1997). Functional and regulatory analysis of the two copies of the *fixNOQP* operon of *Rhizobium leguminosarum* strain VF39. *Molecular Plant-Microbe Interactions, 10*, 605–616.

Schmeisser, C., Liesegang, H., Krysciak, D., Bakkou, N., Le Quéré, A., Wollherr, A., et al. (2009). *Rhizobium* sp. strain NGR234 possesses a remarkable number of secretion systems. *Applied and Environmental Microbiology, 75*(12), 4035–4045.

Seshadri, R., Reeve, W. G., Ardley, J. K., Tennessen, K., Woyke, T., Kyrpides, N. C., et al. (2015). Discovery of novel plant interaction determinants from the genomes of 163 root nodule bacteria. *Scientific Reports, 5*, 16825.

Slater, S. C., Goldman, B. S., Goodner, B., Setubal, J. C., Farrand, S. K., Nester, E. W., et al. (2009). Genome sequences of three agrobacterium biovars help elucidate the evolution of multichromosome genomes in bacteria. *Journal of Bacteriology, 191*(8), 2501–2511.

Soberón, N., Venkova-Canova, T., Ramırez-Romero, M. A., Téllez-Sosa, J., & Cevallos, M. A. (2004). Incompatibility and the partitioning site of the *repABC* basic replicon of the symbiotic plasmid from *Rhizobium etli*. *Plasmid, 51*(3), 203–216.

Spaink, H. P. (2000). Root nodulation and infection factors produced by rhizobial bacteria. *Annual Review of Microbiology, 54*, 257–288.

Stowers, M. D. (1985). Carbon metabolism in Rhizobium species. *Annual Review of Microbiology, 39*, 89–108.

Sugawara, M., Epstein, B., Badgley, B. D., Unno, T., Xu, L., Reese, J., et al. (2013). Comparative genomics of the core and accessory genomes of 48 *Sinorhizobium* strains comprising five genospecies. *Genome Biology, 14*(2).

Sullivan, J. T., Brown, S. D., & Ronson, C. W. (2013). The NifA-RpoN regulon of *Mesorhizobium loti* strain R7A and its symbiotic activation by a novel LacI/GalR-family regulator. *PLoS One, 8*(1), e53762.

Sullivan, J. T., Brown, S. D., Yocum, R. R., & Ronson, C. W. (2001). The *bio* operon on the acquired symbiosis island of *Mesorhizobium* sp. strain R7A includes a novel gene involved in pimeloyl-CoA synthesis. *Microbiology, 147*(5), 1315–1322.

Sullivan, J. T., & Ronson, C. W. (1998). Evolution of rhizobia by acquisition of a 500-kb symbiosis island that integrates into a *phe*-tRNA gene. *Proceedings of the National Academy of Sciences, 95*(9), 5145–5149.

Szeto, W. W., Zimmerman, J. L., Sundaresan, V., & Ausubel, F. M. (1984). A *Rhizobium meliloti* symbiotic regulatory gene. *Cell, 36*(4), 1035–1043.

Torres, M. J., Hidalgo-García, A., Bedmar, E. J., & Delgado, M. J. (2013). Functional analysis of the copy 1 of the *fixNOQP* operon of *Ensifer melilot* under free-living micro-oxic and symbiotic conditions. *Journal of Applied Microbiology, 114*(6), 1772–1781.

Truchet, G., Rosenberg, C., Vasse, J., Julliot, J.-S., Camut, S., & Denarie, J. (1984). Transfer of *Rhizobium meliloti* pSym genes into *Agrobacterium tumefaciens*: Host-specific nodulation by atypical infection. *Journal of Bacteriology, 157*(1), 134–142.

Venkova-Canova, T., Soberón, N. E., Ramírez-Romero, M. A., & Cevallos, M. A. (2004). Two discrete elements are required for the replication of a *repABCM* plasmid: An antisense RNA and a stem–loop structure. *Molecular Microbiology, 54*(5), 1431–1444.

Wang, L., Zhang, L., Liu, Z., Zhao, D., Liu, X., Zhang, B., et al. (2013). A minimal nitrogen fixation gene cluster from *Paenibacillus* sp. WLY78 enables expression of active nitrogenase in *Escherichia coli*. *PLoS Genetics, 9*(10), e1003865.

Wassem, R., Kobayashi, H., Kambara, K., Le Quéré, A., Walker, G. C., Broughton, W. J., et al. (2008). TtsI regulates symbiotic genes in *Rhizobium* species NGR234 by binding to *tts* boxes. *Molecular Microbiology, 68*(3), 736–748.

Wetzel, M. E., Olsen, G. J., Chakravartty, V., & Farrand, S. K. (2015). The *repABC* plasmids with quorum-regulated transfer systems in members of the *Rhizobiales* divide into two structurally and separately evolving groups. *Genome Biology and Evolution,* 7(12), 3337–3357.

Young, J. P. W., Crossman, L. C., Johnston, A. W., Thomson, N. R., Ghazoui, Z. F., Hull, K. H., et al. (2006). The genome of *Rhizobium leguminosarum* has recognizable core and accessory components. *Genome Biology and Evolution,* 7, R34.

Żebracki, K., Koper, P., Marczak, M., Skorupska, A., & Mazur, A. (2015). Plasmid-encoded *RepA* proteins specifically autorepress individual *repABC* operons in the multipartite *Rhizobium leguminosarum* bv. *trifolii* genome. *PLoS One, 10*(7), e0131907.

Zipfel, C., & Oldroyd, G. E. D. (2017). Plant signalling in symbiosis and immunity. *Nature, 543*(7645), 328.

Zurkowski, W. (1982). Molecular mechanism for loss of nodulation properties of *Rhizobium trifolii*. *Journal of Bacteriology, 150*(3), 999–1007.

CHAPTER NINE

Rhizobium diversity in the light of evolution

Mingxing Tang and Delphine Capela*
LIPM, Université de Toulouse, INRA, CNRS, Castanet-Tolosan, France
*Corresponding author: E-mail: delphine.capela@inra.fr

Contents

Advances in Botanical Research, Volume 94
ISSN 0065-2296
https://doi.org/10.1016/bs.abr.2019.09.006

Abstract

Rhizobia, nitrogen-fixing symbionts of legumes, are phylogenetically diverse soil bacteria. These bacteria do not share a unique genetic strategy to achieve symbiosis. Instead, they have evolved various mechanisms to induce and invade the root/stem organs where nitrogen fixation occurs. Most rhizobial lineages emerged from horizontal transfer of few essential symbiotic genes located on mobile genetic elements. Besides, rhizobia have extensively recruited their own native functions to optimize symbiosis and refine interaction with host plants. This has likely occurred following symbiotic gene transfer and via genomic adjustments under plant selection pressure. This two-step evolutionary scenario was recently validated by an evolution experiment that progressively turned a plant pathogen into legume symbionts. Acquisition and improvement of the first symbiotic stages, nodulation and infection, were most often gained via the rewiring of the virulence and/or metabolic regulatory network of the ancestral strain.

1. Introduction

The generic term rhizobium designates bacteria that share the capacity to establish a nitrogen-fixing symbiosis with legumes. Rhizobia induce, usually on the roots and occasionally on the stems of host plants, the formation of nodules they massively and intracellularly colonize. Several modes of entry of bacteria in plant tissues have been described depending on the symbiotic partner (Madsen et al., 2010). In temperate legumes such as *Medicago* or *Lotus*, rhizobia mainly invade the root at the tip of root hairs, which curl to entrap one or two bacteria and invaginate to form tubular structures called infection threads (ITs). Bacteria grow and divide inside ITs that elongate and ramify to cross the root cell layers toward cortical cells (Gage, 2002). Coordinately with ITs progression, root cortical cells at the basis of the infected root hair divide to form a nodule primordium. Then bacteria exit ITs to colonize the cytoplasm of nodule cells. In 25% of legumes, mainly tropical and subtropical legumes such as *Sesbania* or *Arachis*, bacteria enter the roots intercellularly by "crack entry" at the junctions of main and lateral roots or wounds. Intercellular bacteria trigger plant cell death and form infection pockets, from which bacteria spread either through the formation of ITs and release of bacteria in cortical cells or through a direct invasion of cortical cells (Capoen, Goormachtig, & Holsters, 2010; Madsen et al., 2010). Another simpler mode of infection was recently described in *Aeschynomene* species where bacteria enter intercellularly without ITs formation nor distant induction of nodule primordium (Bonaldi et al., 2011). In these cases, bacteria invade the roots by a mechanism that resembles the crack entry mechanism and then one bacterium is delivered into a single founder cortical cell, which subsequently divides to form a nodule primordium. In certain Aeschynomene species, in addition to the infection of the core nodule, superficial tissues are invaded by undifferentiated bacteria enclosed in tubular structures that may constitute a reservoir of viable bacteria able to infect new plants after the nodule senescence. Whatever the infection mode is used, all rhizobium-legume associations lead to the formation of symbiosomes. Symbiosomes are organelle-like structures in the cytosol of nodule cells where bacteria surrounded by a plant membrane differentiate into nitrogen-fixing bacteroids. Reduced dinitrogen into ammonia is directly exported and assimilated by the plant that in turn provides bacteria with carbon sources, leading to mutualistic interactions.

Rhizobia are very diverse bacteria that have been identified in distant phylogenetically branches, interspersed with non rhizobial bacterium species (Remigi, Zhu, Young, & Masson-Boivin, 2016). So far, up to 18 different rhizobium genera and hundreds of species have been described among α- and β-proteobacteria, thus now referred as α- and β-rhizobia (Table 1). The rhizobium genetic diversity is reflected by the diversity of molecular mechanisms these bacteria use during the symbiotic process. Comparative genomic studies of rhizobium symbiotic genes indeed do not support a unique shared genetic strategy for symbiosis in rhizobia (Amadou et al., 2008; Tian et al., 2012). Instead various functions, whose phylogenetic distribution ranges from ubiquitous to lineage- or even strain-specific, have been recruited for symbiosis.

The ability of rhizobia to induce a mutualistic symbiosis relies on their capacity to (i) trigger a plant developmental stage leading to nodule organogenesis and infection, (ii) suppress and/or avoid plant immunity to allow the massive and persistent invasion of bacteria and (iii) fix and export nitrogen in exchange to photosynthetates. In this review, we examined the different genetic strategies rhizobia have co-opted to achieve these steps. We explained this diversity by the evolutionary history of these bacteria, which mainly emerged from horizontal gene transfer (HGT) of key symbiotic genes followed by genome remodeling to activate and optimize the symbiotic potential. Finally, we reported a unique evolution experiment replaying the emergence of a new rhizobium under laboratory conditions and validating the proposed rhizobium evolutionary scenario.

Table 1 Bacterium genera containing rhizobia and examples of rhizobium species.

Class of proteobacteria	Genera[a]	Species
Alpha	*Allorhizobium*	*qilianshanense, taibaishanense, undicola*
	Aminobacter	*anthyllidis*
	Azorhizobium	*caulinodans, doebereinerae*
	Bradyrhizobium	*betae, canariense, denitrificans, elkanii, japonicum (diazoefficiens), liaoningense, retamae, yuanmingense*
	Devosia	*neptuniae*
	Mesorhizobium	*albiziae, amorphae, chacoense, ciceri, huakuii, loti, mediterraneum, plurifarium, septentrionale, temperatum, thiogangeticum, tianhanense, opportunistum*
	Methylobacterium	*nodulans*
	Microvirga	*lupini, lotononidis, ossetica, vignae, zambiensis*
	Neorhizobium	*galegae, huautlense*
	Ochrobactrum	*lupini, cytisi*
	Pararhizobium	*giardinii, herbae, helanshanense, sphaerophysae*
	Phyllobacterium	*salinisoli, sophorae, trifolii, zundukense*
	Rhizobium	*altiplani, cellulosilyticum, daejeonense, etli, freirei, gallicum, hainanense, huautlense, indigoferae, leguminosarum, leucaenae, loessense, lupinii, lusitanum, mesoamericanum, mongolense, sullae, tropici, yanlingense*
	Shinella	*kummerowiae*
	Sinorhizobium (Ensifer)	*adhaerens, americanum, alkalisoli, arboris, fredii, kostiense, kumerowiae, medicae, meliloti, mexicanus, morelense, saheli, sojae, terangae, xinjiangense*
Beta	*Cupriavidus*	*taiwanensis, necator, pinatubonensis*
	Paraburkholderia	*caribensis, diazotrophica, mimosarum, nodosa, phenoliruptrix, phymatum, piptadeniae, ribeironis, sabiae, tuberum*
	Trinickia	*symbiotica*

[a]Rhizobium taxonomy is based on (Estrada-de Los Santos et al., 2018; Mousavi, Willems, Nesme, de Lajudie, & Lindström, 2015; Ormeño-Orrillo et al., 2015).

2. Rhizobia achieve symbiosis via genetically diverse strategies

2.1 Nodulation: a widespread, yet not unique, nod factor strategy

The nodule developmental stage (organogenesis and root infection) is triggered via the activation of an ancestral plant symbiotic signaling pathway that was initially used in most land plants to establish ancient symbioses with mycorrhizal fungi. Notably, this pathway is also recruited in actinorhizal plants to form nitrogen-fixing nodules with *Frankia* actinobacteria (Gherbi et al., 2008; Hocher, Alloisio, Bogusz, & Normand, 2011; Svistoonoff et al., 2013). This common symbiosis signaling pathway (CSSP) is well characterized in the model legumes *Medicago truncatula* (Mt) and *Lotus japonicus* (Lj). It involves a first set of signal transduction proteins MtDMI2/LjSYMRK, and MtDMI1/LjCastor-LjPollux-LjNENA-LjNup85-LjNup133 that produces Ca^{2+} spiking signals following the perception of the microsymbiont, and a second complex composed of MtIPD3/LjCYCLOPS and MtDMI3/LjCCaMK that decodes the calcium signals. In turn, this second complex activates specific downstream transcription factors to induce either mycorrhizal symbiosis or root nodule symbiosis (Gough & Cullimore, 2011; Oldroyd, 2013).

Most rhizobia activate the CSSP pathway via the synthesis and secretion of Nod Factors (NFs) (D'Haeze & Holsters, 2002). Rhizobial NFs are the products of enzymes encoded by the nodulation (*nod*) genes (Fig. 1A). The expression of *nod* genes is induced by flavonoids, aromatic compounds exuded by host roots, and controlled by NodD transcription regulators that recognize conserved motifs (called *nod*-boxes) in their promoter region. Flavonoids attract bacteria near the roots and bind to the bacterial NodD proteins. In the case of compatible NodD-flavonoid interaction, this leads to DNA bending and further activation of the downstream genes. NodD regulators thus act both as sensors of the plant signal and as transcription regulators of *nod* loci (Schlaman, Okker, & Lugtenberg, 1992). The NF backbone consists of an oligomer of chitin acylated at the non-reducing terminal residue. This backbone is synthesized by the common NodABC proteins present in all NF-producing rhizobia. In addition, rhizobial strains possess specific pools of *nod*, *nol* or *noe* genes encoding enzymes involved in the attachment of various substituents (Table 2) such as methyl (*nodS*), carbamoyl (*nolO* and *nodU*), acetyl (*nodL, noeT*), sulfate (*nodH, nodPQ*), fucose (*nodZ*), O-methyl-fucose (*noeI*), O-sulfate-fucose (*noeE*), O-acetyl-

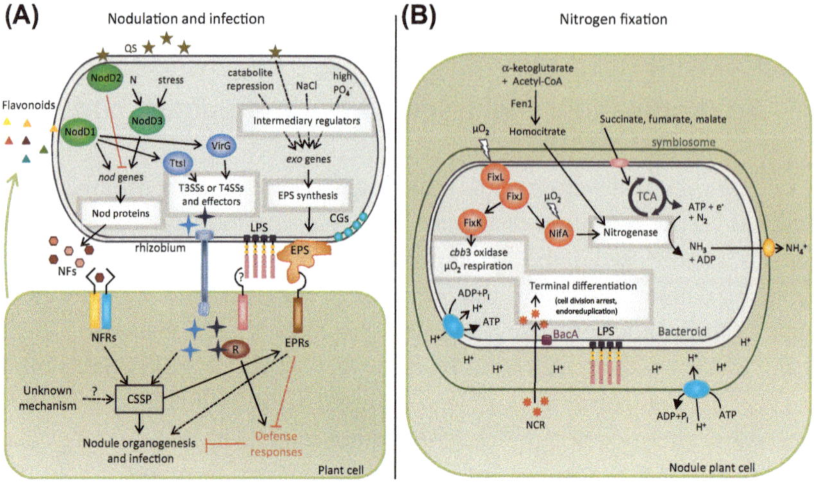

Fig. 1 ***Rhizobial genetic factors used in nodulation, infection and nitrogen fixation.*** **A.** In the rhizosphere, plants secrete flavonoids perceived by NodDs proteins which induce the expression of *nod* genes. *nod* gene expression leads to the production of Nod Factors specifically recognized by plant receptors and induces the symbiotic signaling pathway (CSSP). In the *nod*-independent rhizobia, the mechanism inducing the CSSP is not known. During the infection process, rhizobia also synthesize a panel of surface polysaccharides (EPS, LPS, CGs) to evade plant defense responses. In certain rhizobia, establishment of symbiosis is assisted by T3 or T4 secretion systems unless secreted effectors are recognized by plant resistance proteins. **B.** In nodule cells, released bacteria from infection threads remained surrounded by a peribacteroid membrane of plant origin thus forming symbiosomes in the plant cytosol. The peribacteroid space is probably an acidic compartment. LPS are required for the survival of bacteria following their release in nodule cells. In symbiosomes, bacteria differentiate into bacteroids able to fix atmospheric nitrogen into ammonia. Bacteroid terminal differentiation triggered by NCR peptides occurred in plants from IRLC and Dalbergioid clades and necessitates the bacterial BacA peptide transporter. Nitrogen fixation is possible in the microoxic environment of symbiosomes. A high oxygen affinity cytochrome *cbb*3 oxidase allowed the microaerobic respiration of bacteroids. Carbon sources are provided by the plant as C_4 dicarboxylic acids. The symbiotic determinants illustrated in panels A and B are from different bacteria. *NFs*, Nod Factors; *QS*, quorum sensing molecules; *T3SS*, type III secretion system; *EPS*, exopolysaccharides; *LPS*, lipopolysaccharides; *CGs*, cyclic β-glucans; *NFRs*, Nod factor receptors; *EPRs*, EPS receptors; *CSSP*, common symbiotic signaling pathway; *R*, resistance proteins; *TCA*, tricarboxylic acid cycle.

fucose (*nolL*) or arabinose (*noeC*) at different locations of the basal structure (Downie, 2010; Österman et al., 2014). NFs also vary in the length of the backbone and in the nature of the fatty acyl group (D'Haeze & Holsters, 2002). Usually rhizobia do not produce only one type of NFs but a set of

Table 2 Presence of some NF biosynthesis and export, EPS biosynthesis and nitrogen fixation genes in various rhizobia.

Functions	Genes [a]	Ac	Bd	Bs	Ml	Re	Rl	Sm	Sf	Ct	Pp
Nod factor synthesis and export genes	*nodA*	■	■		■	■	■	■	■	■	■
	nodB	■	■		■	■	■	■	■	■	■
	nodC	■	■		■	■	■	■	■	■	■
	nodE				■		■	■			
	nodF				■		■	■			
	nodG							■			
	nodH							■		■	■
	nodI	■	■		■	■	■	■	■	■	■
	nodJ	■	■		■	■	■	■	■	■	■
	nodL				■	■		■			
	nodM		■					■			
	nodN		■		■	■	■	■	■		
	nodO						■				
	nodP							■	■		
	nodQ							■	■	■	
	nodS	■	■		■	■			■	■	■
	nodT	■			■	■	■		■		
	nodU	■	■						■	■	■
	nodV		■								
	nodW	■	■			■			■		
	nodY		■								
	nodZ	■	■		■	■			■		
	noeC	■									
	noeE		■						■		
	noeI		■		■				■		
	noeK				■	■			■		
	noeL		■		■	■			■		
	noeM [b]									■	■
	nolA		■								
	nolE					■					
	nolF							■			
	nolG						■	■			
	nolK	■			■				■		
	nolL				■	■			■		
	nolO		■		■				■		
	nolT				■						
	nolU		■		■				■		
	nolV		■		■						
	nolW				■						
	nolX				■						
EPS biosynthesis genes	*exoA*			■	■			■	■		
	exoB	■	■	■	■	■	■	■	■		
	exoD	■				■	■	■	■		
	exoF				■	■	■	■	■		
	exoH								■		
	exoI			■	■	■	■	■	■		
	exoK, exoL				■			■	■		
	exoM, exoQ				■	■	■	■	■		
	exoN	■	■	■	■	■	■	■	■		
	exoO				■			■	■		
	exoP		■	■	■	■	■	■	■		
	exoR, exoS, chvI	■	■	■	■	■	■	■	■		
	exoT					■	■				
	exoU				■	■	■	■	■		
	exoV					■	■	■	■		
	exoX						■				
	exoY	■	■	■	■	■	■	■	■		
	exoZ				■	■	■	■	■		
	pssA	■	■	■	■	■	■	■	■	■	
	pssB	■	■	■	■	■	■	■	■		
	pssN	■	■	■	■	■	■	■	■		
	pssO					■	■				
	pssP		■	■	■	■	■	■	■		
	pssT		■	■		■	■				
	pssVSRMKLJIHGFCDE					■	■				
Nitrogen fixation genes	***fixL***	■	■	■	■	■	■	■			
	fixJ	■	■	■	■			■			
	fixK	■	■	■	■	■	■	■	■		
	ccoNOPQ	■	■	■	■	■	■	■	■	■	
	ccoG, ccoI	■	■	■	■	■	■	■	■	■	
	ccoH	■	■	■	■	■	■	■	■		
	ccoS	■	■	■	■	■		■		■	
	fixABCX	■	■	■	■	■	■	■	■	■	■
	fixM							■			
	fixR	■	■	■	■			■			
	fixU	■	■	■	■			■	■		■
	nifA	■	■	■	■	■	■	■	■	■	■
	nifHDKENB	■	■	■	■	■	■	■	■	■	■
	nifQ	■	■	■	■	■			■	■	■
	nifR	■	■	■	■	■	■	■	■		
	nifS	■	■	■	■	■	■	■	■		
	nifT [c]	■	■	■	■			■	■	■	■
	nifU	■	■	■	■	■	■		■		
	nifW, nifZ	■	■	■	■	■			■	■	■
	nifX	■	■	■	■	■		■	■	■	■

NFs whose diversity determine the host-range of the strain (D'Haeze & Holsters, 2002; Perret, Staehelin, & Broughton, 2000). NFs are exported via a type I secretion system encoded by the *nodIJ* genes, and recognized by a cognate host receptor complex, comprising MtNFP and MtLYK3 in *M. truncatula* (Mt) and LjNFR5 and LjNFR1 in *L. japonicus* (Lj). The perception of NFs initiates the phosphorylation of these receptors which induces the activation of the CSSP (Kouchi et al., 2010; Oldroyd, 2013). Then the CSSP activates the specific transcription factors NSP1, NSP2, NIN, ERN1 and ERN2 coordinating the nodule development and rhizobial infection in the different root tissues (Cerri et al., 2016, 2017; Vernié et al., 2015).

A NF-independent strategy was uncovered a decade ago in photosynthetic *Bradyrhizobium* lacking the *nodABC* genes and nodulating certain *Aeschynomene* species (Chaintreuil et al., 2018; Giraud et al., 2007; Okazaki et al., 2016). The bacterial factors involved in this alternative nodulation mechanism have not been elucidated yet. However two key proteins of the CSSP, SymRK and CCaMK, as well as the LHK1 cytokinin receptor required for *nod*-dependent nodulation process were shown to be also required for the *nod*-independent process (Fabre et al., 2015). Recently, a transcriptomic analysis of the *nod*-independent *Aeschynomene evenia - Bradyrhizobium* ORS278 symbiotic interaction identified transcripts of an NFP/NFR5 ortholog and orthologs of most of the genes belonging to the CSSP (MtDMI2/LjSYMRK, LjCastor-MtDMI1/LjPollux-LjNENA-LjNup85-LjNup133, MtIPD3/LjCYCLOPS and MtDMI3/LjCCaMK), supporting the involvement of the same plant signaling pathway in both *nod*-dependent and *nod*-independent symbioses (Gully et al., 2018). It is not clear for the moment whether this NF-independent strategy, so far restricted to some *Aeschynomene Bradyrhizobia*, is either a vestige of an ancient

Gray boxes mark the presence of the gene. *Ac, Azorhizobium caulinodans* ORS571. *Bd, Bradyrhizobium diazoefficiens* USDA110. *Bs, Bradyrhizobium* sp. ORS278. *Ml, Mesorhizobium loti* MAF303099. *Re, Rhizobium etli* CFN42. *Rl, Rhizobium leguminosarum* bv. viciae 3841. *Sm, Sinorhizobium meliloti* 1021. *Sf, Sinorhizobium fredii* NGR234. *Ct, Cupriavidus taiwanensis* LMG19424. *Pp, Paraburkholderia phymatum* STM815.

[a]Regulatory genes are in boldface.

[b](Daubech et al., 2019).

[c]*nifT* is also known as *fixU* or *fixT*.

Adapted and updated from Masson-Boivin, C., Giraud, E., Perret, X., and Batut, J. (2009). Establishing nitrogen-fixing symbiosis with legumes: How many rhizobium recipes? Trends in Microbiology, *17*, 458–466 and Black, M., Moolhuijzen, P., Chapman, B., Barrero, R., Howieson, J., Hungria, M., et al. (2012). The genetics of symbiotic nitrogen fixation: Comparative genomics of 14 rhizobia strains by resolution of protein clusters. Genes, 3, 138–166 and completed based on gene annotations from MaGe (http://www.genoscope.cns.fr/agc/microscope/home/index.php).

primitive mechanism or a more recently evolved mechanism (Okubo, Fukushima, & Minamisawa, 2012).

2.2 Infection: the requirement of surface polysaccharides

In addition to NFs, the progression of the symbiotic interaction requires the synthesis of surface polysaccharides that are essential for both early and late infection stages in all rhizobium-legume symbiotic interactions (Fig. 1A). Four main types of rhizobial surface polysaccharides have been shown to play a role in symbiosis: secreted exopolysaccharides (EPS), capsular polysaccharides (KPS) and lipopolysaccharides (LPS), both anchored in the outer membrane, and cyclic β-glucans mostly located in the periplasm (Downie, 2010; Fraysse, Couderc, & Poinsot, 2003; López-Baena, Ruiz-Sainz, Rodríguez-Carvajal, & Vinardell, 2016). The symbiotic role of these molecules often depends on the rhizobium species and plant host genotype.

Although their precise functions have not been resolved, it was proposed that surface polysaccharides protect rhizobia against a variety of biotic and abiotic stresses encountered under symbiotic conditions (osmotic stress, acidification, antimicrobial peptides, or reactive oxygen species) (Arnold, Penterman, Shabab, Chen, & Walker, 2018; D'Haeze & Holsters, 2004; Davies & Walker, 2007; Geddes, González, & Oresnik, 2014; Ingram-Smith & Miller, 1998; Lehman & Long, 2013; Miller-Williams, Loewen, & Oresnik, 2006). These molecules likely ensure the protection of rhizobia either by forming a physical diffusion barrier around bacteria or by actively inducing pathways that prevent plant defense responses, a prerequisite for nitrogen fixation (D'Antuono et al., 2008; D'Haeze & Holsters, 2004; Jaszek, Janczarek, Piersiak, & Grzywnowicz, 2014; Jones et al., 2008). The active signaling function of EPS was recently provided by the identification of an EPS receptor in *L. japonicus* (Kawaharada et al., 2015, 2017). Future studies should reveal whether other plant receptors not only to EPS but also to the other rhizobial polysaccharides, such as KPS and LPS, also contribute to the symbiosis signaling pathway.

2.2.1 Exopolysaccharides (EPS)

Rhizobial exopolysaccharides play a major role in the initiation and elongation of infection threads (ITs). In *Sinorhizobium meliloti*, two distinct EPSs have been reported, the EPS I or succinoglycan and EPS II or galactoglucan. The low molecular weight form of EPS I was shown to play a major role in the invasion of *Medicago* nodules while EPS II have a more moderate role (Battisti, Lara, & Leigh, 1992; Cheng & Walker, 1998). The *exo* genes required for the synthesis and export of EPS I are well conserved in *Sinorhizobium*, *Rhizobium*, and *Mesorhizobium* while most of them are absent in *Bradyrhizobium* and *Azorhizobium*, two more distant genera of rhizobia, as well as in β-rhizobia (Black et al., 2012) (Table 2). However EPS I are not always essential for infection. For example in *Sinorhizobium fredii*, mutations in *exo* genes do not alter symbiosis (Kim, Tully, & Keister, 1989). In this case, other surface polysaccharides, the KPS, are of main importance for infection. Also the *Vicia sativa* (vetch) - *Rhizobium leguminosarum* symbiotic interaction relies on another type of EPS synthesized by the products of *pss* genes specifically present in the *Rhizobium* genus (Table 2) (Black et al., 2012; van Workum, van Slageren, van Brussel, & Kijne, 1998).

A large diversity in EPS chemical structures can be found among rhizobia (van Workum et al., 1998; Wang et al., 2018). They consist in repeating unit of heteropolysaccharides which varies in the composition of sugars, their linkage within the unit, the degree of polymerization and the presence of non-carbohydrate substituents such as pyruvate, succinate or acetate (Skorupska, Janczarek, Marczak, Mazur, & Krol, 2006). A number of evidences indicated that the chemical structure of EPS influences infection efficiency in a host-dependent manner, suggesting an active role of EPS as signal molecules in the rhizobium-legume interactions. For example, an *exoU* mutant in *Mesorhizobium loti* R7A is unable to induce infected nodules on the Gifu ecotype of *L. japonicus,* while this mutant displayed only reduced infection efficiency on other ecotypes of *L. japonicus* such as MG-20 (Kelly et al., 2013). In the same line, complementation of EPS deficient strains by natural heterologous EPS produced by other rhizobia does not always lead to a successful infection. For examples, EPS from *Rhizobium tropici* or *Rhizobium etli* restore the infection ability of *Rhizobium leguminosarum* bv. *viciae pssD* mutants but EPS from *S. meliloti, S. fredii* NGR234 or *Bradyrhizobium diazoefficiens* cannot (van Workum et al., 1998). In the same way, EPS from *Rhizobium leguminosarum* bv. *trifolii* and *S. fredii* NGR234 cannot restore the defective infection phenotype of *S. meliloti exo* mutants on alfalfa (Battisti et al., 1992). The recent identification in *L. japonicus* of EPR3, a specific EPS receptor, that mediates the progression of ITs through the root cortex and nodule primordium confirmed the role of EPS as important signaling molecules (Kawaharada et al., 2015, 2017; Muszyński et al., 2016). This receptor is part of the symbiotic signal transduction pathway since its expression is activated downstream from the CSSP (Kawaharada et al., 2017). Interestingly, EPR3 orthologs were found in other legumes, such as *M. truncatula,* soybean as well as *Arachis hypogaea* whose infection is mediated by crack entry but not by ITs (Karmakar et al., 2019; Kawaharada et al., 2015).

2.2.2 Capsular polysaccharides (KPS)

Capsular polysaccharides are composed of dimeric repeating unit containing one hexose linked with a Kdo (3-deoxy-D-manno-oct-2-ulosonic acid). The number of *kps* gene clusters is variable among rhizobium genomes even within the same rhizobium species (Kiss et al., 2001; Sugawara et al., 2013). Their importance in symbiosis is also variable. For instance, in *S. fredii* HH103, KPS play a major role in nodule formation and growth (Parada et al., 2006) whereas in *S. meliloti* Rm1021, they have only a minor effect (Pellock, Cheng, & Walker, 2000).

2.2.3 Lipopolysaccharides (LPS)

LPS are major constituents of the outer membrane in gram-negative bacteria. They are composed of three parts, a core polysaccharide linked to an O-chain polysaccharide (or O-antigen) and a lipid A (Serrato, 2014). Mutants of rhizobia deficient in LPS biosynthesis are affected either in early infection stages with the formation of abortive ITs or in late infection stages with a rapid lysis of bacteria after their release in the cytoplasm of nodule cells both leading to the formation of non fixing nodules (Broughton et al., 2006; Campbell, Reuhs, & Walker, 2002; Noel, Forsberg, & Carlson, 2000). In gram-negative bacteria, the O-antigen is the main immunogenic feature and the most variable part of LPS. Mutants altering the composition of O-antigen in *R. etli* and *S. fredii* NGR234 are impaired in IT development and nitrogen fixation (Reuhs et al., 2005). Interestingly, in these two rhizobia, modifications in the O-chain of LPS occurred under symbiotic conditions (Broughton et al., 2006; Kobayashi, Graven, Broughton, & Perret, 2004; Noel, Box, & Bonne, 2004; Ojeda, Box, & Noel, 2010).

2.2.4 Cyclic β-glucans

All bacteria from the Rhizobiales order are able to synthesize cyclic β-glucans (CGs). CGs are neutral cyclic homopolymers of β-linked glucose residues often substituted by phosphoglycerol, phosphocholin or succinyl groups. The linkage between the glucose residues, either β(1,2) or β(1,3) or β(1,6), is variable among legume symbionts, however, there is no correlation between the nature of the linkage and their role on host plants (Breedveld & Miller, 1994). For instance, different soybean symbionts were shown to produce different CGs (Crespo-Rivas et al., 2009; Rolin et al., 1992). In α-rhizobia, CGs are essential for the invasion of nodules. Mutants in the synthesis or export in the periplasm of CGs elicit small, white pseudonodules containing abortive ITs and that fail to fix nitrogen (Bhagwat et al., 1992, 1996; Dylan, Nagpal, Helinski, & Ditta, 1990).

2.3 Nitrogen fixation: a similar set of *nif-fix* genes for all rhizobia

The reduction of dinitrogen into ammonia is achieved thanks to the nitrogenase enzymatic complex. All rhizobia possess the same type of nitrogenase, the molybdenum-dependent nitrogenase, encoded by the *nif* genes. The number of *nif* genes found in rhizobia is lower than that of free-living diazotrophs, such as *K. pneumoniae*, which possesses 18 structural *nif* genes. A minimal set of genes essential for nitrogen fixation *nifHDK* (nitrogenase enzyme complex) and *nifENB* (FeMoco biosynthesis) is present in all rhizobia. Some other *nif* genes, *nifTXQWZSUV*, involved in the FeMoco synthesis, the Fe—S cluster synthesis and electron transport have been found in rhizobium genomes but their presence is variable (Table 2). Missing *nif* genes compared to free-living diazotrophs might be replaced by housekeeping paralogs (*iscA*/*nifU* and *iscS*/*nifS*) or encoded by yet unidentified genes or, like *nifV*, can be complemented by the FEN1 homocitrate synthase gene of host plants (Hakoyama et al., 2009; Nouwen et al., 2017). Like in free-living diazotrophs, the expression of *nif* genes in rhizobia is directly controlled by the transcription regulator NifA. Both *nifA* transcription and NifA activity are only allowed under very low oxygen concentration (Bobik, Meilhoc, & Batut, 2006; Sciotti, Chanfon, Hennecke, & Fischer, 2003). This double control of NifA by oxygen may prevent the synthesis of nitrogenase in conditions where this enzyme would be inactive. Under symbiotic conditions, rhizobia have to cope with the conflicting situation where the nitrogenase is extremely sensitive to oxygen while bacteria are strictly aerobic. Moreover, the ATP-dependent and high energy-consuming process of nitrogen fixation necessitates rapid respiration rates of bacteroids. This problem is solved by the plant that provides bacteria with carbon sources and produces high amounts of leghemoglobins, plant proteins transporting and delivering oxygen to bacteroids while maintaining the concentration of free oxygen at very low levels. Under these microaerobic conditions, the *ccoNOQP* (formerly *fixNOQP*) and *ccoGHIS* (formerly *fixGHIS*) genes encoding a cytochrome *cbb*3 oxidase with high affinity for oxygen are essential for nitrogen fixation. In some rhizobia able to fix nitrogen under both free-living and symbiotic conditions, such as *Azorhizobium* *caulinodans* and the two beta-rhizobia *Paraburkholderia phymatum* and *Paraburkholderia tuberum*, the cytochrome *bd* terminal oxidase is also or alternatively used (De Meyer et al., 2016). Last, the *fixABCX* genes encoding a specific delivery system of electrons to nitrogenase (Ledbetter et al., 2017) are conserved in most rhizobia (Black et al., 2012).

2.4 Optimization of symbiosis through a diversity of molecular and regulatory mechanisms

2.4.1 Specific symbiotic functions

Besides the above-mentioned essential genetic determinants for symbiosis shared by most rhizobia, additional rhizobial factors allow or optimize symbiosis depending on the rhizobium strain and host genotype (Fig. 1).

Proteins secretion systems such as T3SS, T4SS and T6SS are examples of accessory functions that modulate symbiosis (Bladergroen, Badelt, & Spaink, 2003; Deakin & Broughton, 2009; Hubber, Vergunst, Sullivan, Hooykaas, & Ronson, 2004). Type III secretion systems (T3SS), found in many rhizobium strains belonging to *Sinorhizobium*, *Mesorhizobium*, *Bradyrhizobium*, and *Cupriavidus* genera (Amadou et al., 2008; Hubber et al., 2004; Krause, Doerfel, & Göttfert, 2002; Okazaki et al., 2009, 2016; Perret et al., 2000; Sánchez, Iannino, Deakin, Ugalde, & Lepek, 2009; Viprey, Del Greco, Golinowski, Broughton, & Perret, 1998), were shown to promote nodulation in several rhizobium-legume interactions. For example, inactivation of the T3SS in the wide host-range rhizobium *S. fredii* NGR234 decreases its nodulation efficiency on many host plants like *Tephrosia vogelii*, *Lablab purpureus* and *Flemingia congest* (Marie et al., 2003; Viprey et al., 1998). Likewise, mutations in the *M. loti* MAFF303099 T3SS affects its competitiveness on *Lotus* (Sánchez et al., 2009). Rhizobial T3SSs resemble those found in pathogenic bacteria and deliver effector proteins, called Nops (Nodule outer protein), directly into the host cells. The set of effector proteins largely vary between rhizobium species and strains. Few Nops have been characterized in details but some of them were clearly shown to interfere with plant defense responses. Expression of the NGR234 effector NopL in the non-host tobacco reduces the expression of pathogen-related proteins by interfering with the MAPK (mitogen-activated protein kinase) signaling while expression of NopM, an E3 ubiquitin ligase domain effector, reduces the production of reactive oxygen species in tobacco (Bartsev et al., 2004; Xin et al., 2012; Zhang, Chen, Lu, Xie, & Staehelin, 2011). However these secretion systems are not always essential for symbiosis and may even be incompatible with some hosts (Nelson & Sadowsky, 2015; Saad,

Crèvecoeur, Masson-Boivin, & Perret, 2012; Viprey et al., 1998; Zhao et al., 2018). In these latter cases, it is thought that plant R genes recognize bacterial effectors which induces plant defenses and prevents nodulation, like effector-triggered immune responses described in plant-pathogen interactions (Fan et al., 2017; Nelson & Sadowsky, 2015; Sugawara et al., 2018; Tang, Yang, & Zhu, 2016; Yang, Tang, Gao, Krishnan, & Zhu, 2010).

In addition to their role in the suppression of plant defense reactions, it was also recently proposed that Nop effectors may participate in the activation of the NF signaling pathway. This hypothesis came with the discovery that the T3SS of *Bradyrhizobium elkanii* USDA61 can bypass the NF-perception to induce nodule formation on soybean, a normally NF-dependent process (Okazaki, Kaneko, Sato, & Saeki, 2013). Still, the nodulation was poorly efficient in this case and bacterial invasion was achieved through an abnormal intercellular mode of infection of root tissues. Interestingly, the T3SS of some non-photosynthetic *Bradyrhizobium* sp. induces nodule formation on the NF-independent *Aeschynomene* legumes, supporting their putative role in the activation of the CSSP (Okazaki et al., 2016).

Another example of bacterial gene specifically required in particular rhizobium-host interactions is *bacA*/*bclA* (Guefrachi et al., 2015; Haag et al., 2011). These genes encode peptide transporters that are essential for symbioses with legumes producing antimicrobial peptides known as NCR (Nodule specific Cysteine-Rich) peptides. NCR-producing legumes were found in two distinct phylogenetic clades, the Inverted Repeat-Lacking Clade (IRLC) and *Aeschynomene* spp. from the Dalbergioid clade (Czernic et al., 2015; Mergaert et al., 2003). Rhizobia interacting with these two clades of legumes, *Sinorhizobium* and *Bradyrhizobium* species, respectively, undergo a terminal differentiation process in nodule cells which is induced by the action of hundreds of NCR peptides and requires the bacterial BacA/BclA peptide uptake system. Indeed mutants of these genes in *Sinorhizobium* and *Bradyrhizobium* do not survive in nodule cells (Van de Velde et al., 2010). Consistently, these genes are completely dispensable in rhizobia interacting with non NCR-producing legumes (Maruya & Saeki, 2010).

2.4.2 Fine-tuned regulation of symbiotic functions

A crucial level of optimization of symbiosis occurs through regulatory systems that adjust the expression of symbiotic functions to environmental conditions (Fig. 1). A strict and attuned regulation of Nod-factor production, exopolysaccharide biosynthesis and nitrogen fixation genes is required for successful symbiosis. Given the high plasticity of regulatory networks in bacteria, symbiotic regulators are plentiful and noticeably differ among rhizobia.

Host-mediated activation of nodulation is fulfilled by NodD proteins, LysR-type transcription regulators, whose activity relies on the binding with compatible compounds exuded by the plants, usually flavonoids or isoflavonoids. Rhizobia often possess multiple NodD proteins that integrate different signals, not only diverse plant inducers but also physiological signals like the nitrogen status of the cell (*e.g.* NodD3 in *S. meliloti*) or stress (e.g. NodD3 and NodD5 respond to salt stress in *R. tropici* CIAT899) (del Cerro et al., 2015; Dusha & Kondorosi, 1993; Kelly et al., 2018; Kobayashi et al., 2004). Some NodD proteins also act as repressors of *nod* genes. This is the case of NodD2 in *B. diazoefficiens*, which represses NF production in a quorum-sensing dependent way (Garcia, Dunlap, Loh, & Stacey, 1996; Loh et al., 2001; Loh & Stacey, 2003), or NodD2 in *S. fredii* NGR234 which down-regulates the *nod* gene activator NodD1 (Fellay et al., 1998). The co-modulation of *nod* gene expression by multiple NodDs regulators likely helps rhizobia to expand their host-range, nodulate under specific (harsh) environmental conditions and maintain a tight regulation of NF synthesis.

Biosynthesis of exopolysaccharides is a striking example of function subject to a complex regulation (Janczarek, 2011). In *S. meliloti,* in which EPS regulation was extensively studied, a dozen of regulatory proteins affecting either positively (MucR, SyrM, ExoD, SinI/SinR/ExpR for EPS I and WggR, SinI/SinR/ExpR, PhoB/PhoR for EPS II) or negatively (*e.g.* ExoX, ExsB, CbrA, NtrX, EmrR, ExoR/ExoS/ChvI, EmmABC for EPS I and MucR for EPS II) EPS production has been identified. All these regulators respond to a diversity of physiological conditions. EPS I synthesis is enhanced under ammonium or sulfur starvation, high phosphate concentration, hyperosmotic stress as well as high-cell density and repressed by succinate-mediated catabolite repression, while EPS-II production is promoted under phosphate starvation and high-cell density. Many rhizobial EPS regulatory systems are species-specific (Bonomi et al., 2012; Liu et al., 2018; Nishihata et al., 2018). The continued discovery of new regulators affecting the production of expolysaccharides in rhizobia accounts for the increasingly complex regulation of this function (Barnett & Long, 2018; Schäper et al., 2017). Most proteins controlling EPS synthesis encode global regulatory systems that not only control EPS synthesis but also affect multiple bacterial behaviors. Strikingly, expression of motility genes are frequently inversely correlated with EPS biosynthesis genes (Barnett & Long, 2018). The three component system ExoR/ExoS/ChvI (RSI) is one crucial regulatory system that switches bacteria from motile free-living cells to non-motile host-invading cells by up and down-regulating the expression of hundreds of genes, including EPS I and flagellum genes (Bélanger, Dimmick, Fleming, & Charles, 2009). ExoR is a protein secreted to the periplasm, whereas *exoS* and *chvI* encode a histidine kinase sensory protein and a response regulator, respectively (Cheng & Walker, 1998). Orthologs of the full set of RSI genes were uniquely found and conserved in Rhizobiales, an order containing numerous symbionts and pathogenic bacteria. The phylogenetic co-occurrence and conservation of the three components of the system strongly suggests that it is a key adaptive pathway in this order (Heavner, Qiu, & Cheng, 2015). Likewise, the other three-component system EmmABC, that plays an important role in adaptation in *S. meliloti* (Morris & González, 2009), was not found beyond the Rhizobiales (Barnett & Long, 2018).

Quorum sensing (QS) systems also affect EPS biosynthesis as well as a wide spectrum of other physiological traits that vary among rhizobial species, including biofilm formation, swarming motility, Type III secretion system, plasmid transfer, cell division, metabolism and transport (Calatrava-Morales, McIntosh, & Soto, 2018; Hoang, Becker, & González, 2004; Krysciak et al., 2014). Their impact on symbiosis can be either neutral (Krysciak et al., 2014), positive (Gurich & González, 2009; Yang et al., 2009; Zheng et al., 2006) or negative (Rodelas et al., 1999; Rosemeyer, Michiels, Verreth, & Vanderleyden, 1998). In *S. meliloti* and *R. etli*, mutants in autoinducer synthase genes, *sinI* and *cinI* respectively, were unable to invade nodules efficiently and were affected in symbiosome development and nitrogen fixation (Daniels et al., 2002; Gurich & González, 2009). Nodulation was also severely impaired in QS mutants of *M. loti* and *Mesorhizobium tianshanense* on their respective host plants *Lotus corniculatus* and *Glycyrrhiza uralensis* (Yang et al., 2009; Zheng et al., 2006). However there are also examples where mutations in QS systems have no impact on symbiosis. For examples, the TraI/R and NgrI/R QS systems in *S. fredii* NGR234 as well as the BraI/R QS system in the β-rhizobium *P. phymatum* do not affect nodulation on respective host plants, peas, common bean and *Mimosa pudica*, respectively, although these systems control exopolysaccharide production (Coutinho et al., 2013; Krysciak et al., 2014). Conversely, mutants in the autoinducer synthase genes *rhiI* and *raiI*, in *R. leguminosarum* and *R. etli* respectively, induce significantly more nodules than the wild-type strain on the host plants, peas and common bean (Rodelas et al., 1999; Rosemeyer et al., 1998). QS systems may thus be dispensable or not for symbiosis, probably depending on the genes they control and the complexity of the regulatory network to which they belong.

Another interesting example of regulatory process that modulates symbiosis in a rhizobium-host dependent manner is the ppGpp-mediated stringent response, which controls diverse cellular functions in response to environmental stresses, especially nutrient starvation (Irving & Corrigan, 2018). The level of intracellular ppGpp depends on the ppGpp synthase and hydrolase activities of enzymes of the RSH superfamily. Recently, a *relA* mutant in *S. meliloti*, that is unable to initiate the stringent response, fails to nodulate the host plant *M. sativa*, but successfully invades *M. truncatula* nodules with, however, a reduced nitrogen fixation capacity (Wippel & Long, 2018). In *R. etli* CE3, a *relA* mutant led to a strongly reduced capacity in both nodulation and nitrogen fixation (Calderón-Flores et al., 2005), while another strain of *R. etli,* CNPAF512, deficient in ppGpp production, is only impaired at later stage of symbiosis, showing abnormal bacteroid development and significant decrease in nitrogen fixation (Moris et al., 2005).

Finally, nitrogen fixation genes are under the control of the master regulator NifA in both α and β-rhizobia. Upstream from NifA, the regulatory cascade differs among rhizobia. At the top of this cascade, the FixLJ two component system, responding to microaerobic conditions, is conserved in α-rhizobia but absent in β-rhizobia (Amadou et al., 2008; Cosseau & Batut, 2004). In *S. meliloti*, it activates the two intermediate regulatory genes *fixK* and *nifA* (Bobik et al., 2006). FixK belongs to the Crp/Fnr family of transcriptional regulators and controls, among other genes, the *ccoNOQP* and *ccoGHIS* operons. In *A. caulinodans*, a rhizobium able to fix nitrogen under both free-living and symbiotic conditions, NifA is under the double control of oxygen and ammonia concentrations through the FixK and NtrBC regulators, respectively (Kaminski & Elmerich, 1998). In *B. diazoefficiens*, *nifA* transcription is largely independent from FixJ and oxygen concentration but instead is activated by the two-component system RegSR which responds to an unknown signal (Bauer, Kaspar, Fischer, & Hennecke, 1998). Interestingly, in this rhizobium, many genes not directly related to nitrogen fixation are controlled by NifA (Hauser et al., 2007).

3. Parallel evolution as a source of rhizobial diversity

3.1 The spread of mutualistic traits across diverse bacterial lineages occurred through repeated horizontal transfers of symbiotic genes

A large body of evidence indicates that NF-dependent rhizobia emerged once and have diversified from repeated and independent horizontal transfers of symbiosis genes across species and genera.

First, the sequencing of numerous rhizobium genomes revealed that essential symbiotic genes, in particular those involved in NFs synthesis (*nod* genes) and nitrogen fixation (*nif-fix* genes), very often, if not always for nodulation genes, reside in clusters on highly mobile genetic elements either genomic islands or plasmids (MacLean, Finan, & Sadowsky, 2007).

Moreover, the common rhizobial *nodABC* genes are monophyletic (Chen et al., 2003; Gyaneshwar et al., 2011; Laguerre et al., 2001), while rhizobia are polyphyletic, leading to incongruences between rhizobium core genes and symbiosis genes (Andrews et al., 2018). Indeed phylogenetic studies have highlighted the importance of horizontal transfers of symbiosis genes both within and between rhizobial genera.

Finally, transfers of symbiotic genomic islands from a rhizobium to non symbiotic soil bacteria were observed in nature. In the pioneer experiment of Sullivan et al., a stand of *L. corniculatus* was cultivated in a field devoid of natural rhizobia able to nodulate this legume and inoculated with one *M. loti* compatible symbiotic strain. After few years, the original inoculant could not be retrieved from *L. corniculatus* nodules. Instead nodules were occupied by local non symbiotic *Mesorhizobium* strains that had all acquired the symbiotic island of the original inoculant (Sullivan, Patrick, Lowther, Scott, & Ronson, 1995; Sullivan & Ronson, 1998). Similar observations were made with *Biserrula pelecinus* symbionts in Australia and soybean symbionts in Brazil (Barcellos, Menna, Batista, & Hungria, 2007; Nandasena, O'Hara, Tiwari, Sezmis, & Howieson, 2007). The sequencing of a rhizobium population isolated from the rhizosphere and nodules of beans cultivated in the same area provides another evidence that symbiosis plasmids and islands can spread rapidly within a soil population although these transfers occur preferentially between closely related species (Pérez Carrascal et al., 2016).

HGT have thus clearly played a major role in the dissemination of the rhizobial mutualistic traits among α- and β-proteobacteria. Although the origin of *nod* genes is not yet completely elucidated, it was recently proposed that the common canonical *nodABC* genes originate from certain *Frankia* symbionts (Ktari et al., 2017; Persson, 2015). In turn, accessory *nod* genes (involved in NF decorations) could have been recruited via duplications of resident genes encoding housekeeping functions during evolution. In *S. meliloti* strain 1021, the plasmidic *nodP1Q1* genes involved in the sulfation of NFs is thought to have arisen from a transposition of the ancestral genomic copy *nodP2Q2* (paralogs of the *cysDNC* genes) (Capela, Carrere, & Batut, 2005; Galibert et al., 2001). The NF transport system encoded by the highly conserved *nodIJ* genes is another example of genes that were first duplicated within β-rhizobia and then transferred to α-rhizobia (Aoki, Ito, & Iwasaki, 2013).

3.2 HGT is likely promoted in the plant environment

In nature, HGT can be mediated by different processes such as conjugation of plasmids, phage transduction, natural transformation, or integrative and conjugative elements (ICEs). Symbiosis genes cluster preferentially in ICEs (chromosomal islands) in rhizobial species of *Mesorhizobium, Bradyrhizobium* and *Azorhizobium* genera, while large symbiotic plasmids (>0.2 Mb) have been mainly found in rhizobial species of *Rhizobium*, *Sinorhizobium* (*Ensifer*) as well as the β-rhizobial *Paraburkholderia* and *Cupriavidus* genera.

Very low transfer frequencies of ICEs and symbiotic plasmids were observed under laboratory conditions, indicating that these transfers are tightly controlled (Bañuelos-Vazquez, Torres Tejerizo, & Brom, 2017; Delavat, Miyazaki, Carraro, Pradervand, & van der Meer, 2017; Ding & Hynes, 2009). Two main regulatory mechanisms of conjugative transfer were reported for rhizobial symbiotic plasmids in *Rhizobium* and *Sinorhizobium* genera. The first mechanism is mediated by the TraI/TraR/TraM quorum sensing system. In high cell density, quorum sensing molecules produced by the autoinducer synthase TraI binds to the regulator TraR which, in turn, activates the *tra/trb/virB* (encoding functions necessary for the DNA transfer including the conjugative T4SS) and *rep* (encoding genes necessary for segregation/partition and replication of the plasmid) genes as well as *traM,* a gene encoding an anti-activator of TraR. Additional levels of regulation acting either in parallel or upstream the TraI/TraR/TraM system were reported in several rhizobia. For examples, in *R. leguminosarum*, the BisR response regulator and the CinI/CinR quorum sensing system regulate the expression of *traR* (Danino, Wilkinson, Edwards, & Downie, 2003) while in *R. etli,* the LuxR-type regulator CinR positively regulate the expression of *traI* (Tun-Garrido, Bustos, González, & Brom, 2003). A second mechanism of regulation, independent of quorum sensing, described in *R. etli* CFN42, involves the RctA repressor of the *tra* and *virB* genes (Pérez-Mendoza et al., 2005) and another regulator RctB that antagonizes the repressive effect of RctA. Additionally, in *S. meliloti*, the expression of *rctB* is repressed by the regulator RctR and activated by the response regulator RctC in response to a yet unidentified signal (Nogales, Blanca-Ordóñez, Olivares, & Sanjuán, 2013). Homologs of RctR and RctC are present in *S. fredii* NGR234 but are absent in *R. etli*, showing the existence of different signaling cascades and regulation mechanisms of plasmid transfer in rhizobia.

The transfer of ICEs involves the excision and circularization of the DNA region which can replicate by a rolling circle mode. A single stranded DNA copy is then transferred to the recipient bacterium using a conjugative T4SS and integrated at specific conserved sites usually near tRNA genes. ICEs encode all the necessary functions for the excision, insertion and conjugation process. In *Mesorhizobium* species, symbiosis ICEs have been found either as a large genomic DNA region of *ca.* 500 kb or as a tripartite element located at three separate chromosomal regions, which reassemble into a single element before the transfer. The complex regulation of ICE transfer was deeply studied in *M. loti* R7A. Activation of excision and transfer of this ICE is mainly orchestrated by the TraI/TraR quorum sensing system which activates FseA, the master regulator of the *rdfS* excisionase gene. This activation is however prevented in most cells through three additional levels of regulation including i) an inhibition of the quorum sensor TraR by the regulator QseM, ii) a low frequency event of ribosomal frameshifting necessary for the translation of the complete FseA protein, and iii) an anti-activation of FseA by QseM preventing activation of *rdfS* (Ramsay et al., 2009, 2015). Thus the ICE of *M. loti* R7A transfer only occurs in a population of cells in which both *qseM* expression is repressed and FseA is translated.

These tight and complex regulations of DNA transfers contrast with the transfer frequencies observed in natural populations of legume symbionts. It suggests that either environmental factors in the plant rhizosphere promote the transfers of symbiosis genes or rare transfers are rapidly fixed in a population when it provides a strong adaptive advantage, such as the capacity to infect and multiply within legume plants. The role of plant factors in the stimulation of DNA transfer in the rhizosphere was recently provided by the study of Ling et al. demonstrating that the HGT of the symbiosis ICE of *A. caulinodans* is triggered by plant flavonoids sensed by a LysR-type transcription regulator, AhaR, controlling the expression of the *intC* integrase gene (Ling et al., 2016). Root exudates were also shown to enhance conjugal plasmid transfers among bacteria in the rhizosphere, indicating that HGTs can be favored in the rhizospheric environment (Mølbak, Molin, & Kroer, 2007).

3.3 Optimization of the symbiosis potential via post-transfer adaptations

Horizontal acquisition of symbiosis genes is not always sufficient to immediately convert a recipient bacterium into an efficient legume symbiont and multiple genomic adjustments may be required to achieve symbiosis (Fig. 2A). In line with this, HGT observed in fields sometimes results in the evolution of novel rhizobia that are suboptimal for nitrogen fixation. For instance, introduction of the herbaceous legume *B. pelecinus* in Western Australian together with its compatible symbiont *Mesorhizobium ciceri* bv. *biserrulae* led to the rapid evolution of novel *Mesorhizobium* strains that have acquired the symbiosis genomic island from *M. ciceri* inoculant but were either ineffective or fix nitrogen poorly. This suggests that the symbiotic potential newly acquired need to be optimized in these strains to produce efficient rhizobia (Nandasena et al., 2007). Also experimental transfers of symbiotic genes to distant non rhizobial bacteria in the lab were usually unsuccessful leading either to non-nodulating strains or to strains able to induce ineffective pseudonodules (Hirsch et al., 1984; Marchetti et al., 2010).

One optimization step may rely on the integration of symbiosis genes into the endogenous regulatory circuit. This phenomenon is illustrated by the diversity of regulatory systems controlling nitrogen fixation genes in rhizobia (Dixon & Kahn, 2004). Yet the incorporation of laterally acquired genes into the existing regulatory network was shown to be a long process taking many million years (Lercher & Pál, 2008).

In addition to the transferred symbiotic genes, rhizobia largely recruit resident genes for symbiosis. Missing genes for nodulation or nitrogen fixation can be recruited from housekeeping genes either directly (*glmS* can replace *nodM, iscA* and *iscS* can replace *nifU* and *nifS,* respectively) or following a duplication and integration into the symbiotic region (*glmS*/*nodM* or *cysDNC*/*nodPQ*) (Galibert et al., 2001; Schwedock & Long, 1992). As mentioned in the previous section, protein secretion systems, although not essential for symbiosis, were frequently recruited in rhizobia to assist NFs to activate the CSSP and/or reduce plant defense responses. Integration of these systems in the nodulation regulatory circuit to

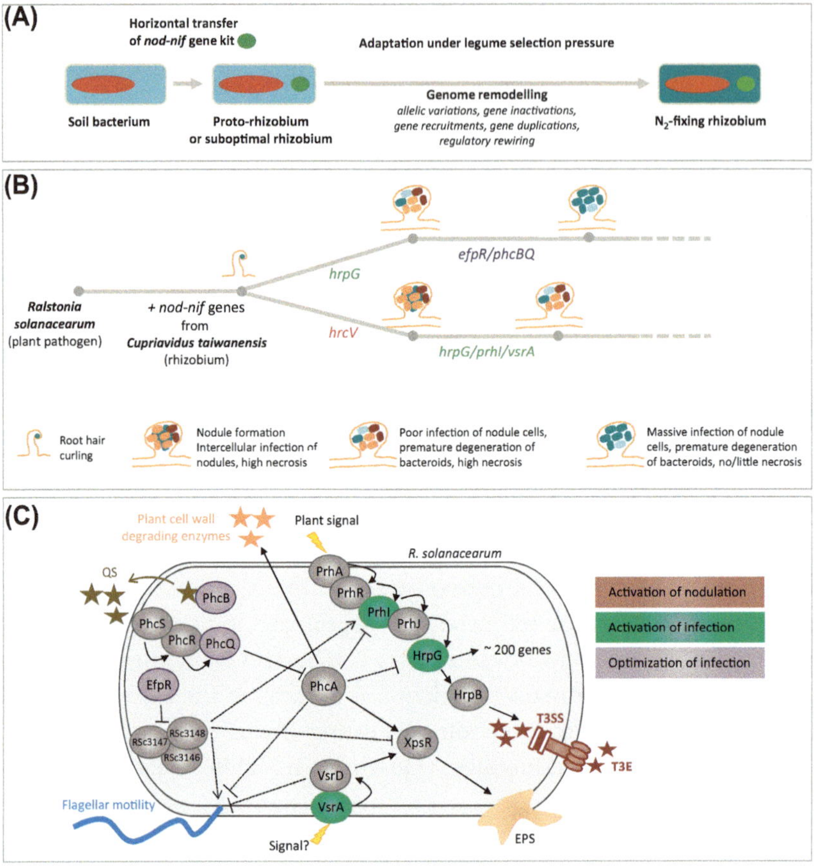

Fig. 2 ***Experimental evolution of legume symbionts.*** **(A) Proposed evolutionary model for the emergence of NF-dependent rhizobia.** Symbiotic plasmids or islands carrying essential symbiotic genes are horizontally transferred in the soil from rhizobia to non-rhizobium strains. These transfers may lead to a proto-rhizobium that needs additional genome remodeling under the plant selection pressure to turn into an effective nitrogen fixing rhizobium. **(B) Experimental evolution of the plant pathogen R. solanacearum into legume symbionts** (Masson-Boivin & Sachs, 2018). The *nod-nif* symbiotic genes from *Cupriavidus taiwanensis*, the symbiont of *Mimosa pudica*, were introduced into the plant pathogen *Ralstonia solanacearum*. The resulting chimeric strain was able to produce Nod Factors and induce root hair curling but unable to form nodules on *M. pudica*. The ability to perform the two first steps of symbiosis, nodulation and intracellular infection, was progressively acquired in parallel lineages evolved by repeated inoculation to *M. pudica* (Marchetti et al., 2010, 2017). Mutations allowing the main symbiotic shifts were identified (Capela et al., 2017; Guan et al., 2013; Marchetti et al., 2010). Nodulation was acquired through inactivation of the T3SS either via a *hrcV* or a *hrpG* stop mutation. A first level of intracellular infection was acquired via a *hrpG* stop mutation or a combination of a *hrcV* mutation and inactivation of other virulence regulators such as *hrpG*, *prhI* or *vsrA*. Optimization of intracellular infection was observed following mutations affecting global virulence regulatory pathways such as

coordinate their expression with *nod* genes confirms their important role in the symbiotic adaptation (Hubber, Sullivan, & Ronson, 2007; Kobayashi et al., 2004; Nelson, Chun, & Sadowsky, 2017). In addition, several examples of ubiquitous functions required for nitrogen fixation, like heme biosynthesis or central metabolism functions, are encoded by native genes that are used to fulfill symbiosis. The phylogenetic analysis of *ca.* 500 genes known to be involved in symbiosis collected from 41 rhizobial strains revealed extensive recruitment of lineage-specific genes (Tian et al., 2012). Striking examples of utilization of local genes for symbiosis are the *exo*, *exs*, *lps*, or *pss* genes, essential for infection and whose phylogeny is consistent with the phylogeny of species (Black et al., 2012; Wang et al., 2018). Some rhizobia have also exploited their intrinsic metabolic properties, such as methylotrophy in *Methylobacterium nodulans* or photosynthesis in *Bradyrhizobium* sp. ORS278, to serve symbiosis (Giraud, Hannibal, Fardoux, Vermeglio, & Dreyfus, 2000; Jourand et al., 2005).

Other post-transfer adaptive evolution may occur through diverse processes such as allelic variation, gene inactivation or deletion, or additional horizontal gene acquisition. The extent of genomic modifications required to adapt to legume symbiosis is not known and probably depends on the recipient genome and the host genotype. We can hypothesize that symbiosis gene transfers between close and pre-adapted genomes could be immediately efficient while transfers between distant genera would necessitate further adaptation steps to enable symbiosis. Long-term co-evolution of rhizobia and their hosts should also increase the number of genome adjustments to refine the mutualistic interactions.

---

efpR or the *phc* quorum-sensing pathway. **(C) Rewiring of the *R. solanacearum* virulence regulatory network leading to symbiotic adaptation.** The virulence master regulator HrpG not only controls the T3SS via HrpB but also *ca.* 200 other genes some of them interfere with intracellular infection of *M. pudica* nodules. The Phc quorum sensing system controls the activity of PhcA, which in turn regulates the expression of hundreds of genes including components of the HrpG regulatory cascade, EPS production, motility and cell wall degrading enzymes. VsrA and EfpR are two other virulence global regulators controlling similar functions as PhcA. Genes mutated in the evolution experiment are colored according to their associated symbiotic phenotypes. *QS*, quorum sensing molecules; *EPS*, exopolysaccharides; *T3SS*, type 3 secretion system; *T3E*, type 3 effectors. Dashed and straight lines indicate indirect and direct controls, respectively. *(A) Adapted from Masson-Boivin, C., & Sachs, J.L. (2018). Symbiotic nitrogen fixation by rhizobia-the roots of a success story. Current Opinion in Plant Biology, 44, 7–15.*

4. Insights into the emergence of rhizobia through experimental evolution

4.1 An experimental design reproducing the natural evolutionary scenario of rhizobia

Some years ago, an evolution experiment was designed to mimic the two-step natural evolutionary scenario of rhizobia and replay the emergence of a new rhizobial genus under laboratory conditions. This experiment consisted in the transfer of essential symbiotic genes from a legume symbiont into a non rhizobium strain followed by further evolution of the resulting strain under the plant selection pressure. The rhizobium *Cupriavidus taiwanensis* and the plant pathogen *Ralstonia solanacearum*, two phylogenetically relatively distant bacteria, were chosen as donor and recipient of symbiosis genes, respectively. The symbiotic plasmid (pRalta) of *C. taiwanensis*, was introduced into *R. solanacearum*, resulting in a chimeric strain unable to nodulate *M. pudica,* the *C. taiwanensis* host. Subsequent repeated inoculation trials on hundreds of *M. pudica* plants finally produced 3 rare nodules, from which three distinct nodulating clones have been isolated (Marchetti et al., 2010). The nodulation efficiency of these clones was however much lower than that of the natural *M. pudica* symbiont, *C. taiwanensis.* These three first nodulating strains were then used as ancestors to evolve several parallel lineages of bacteria through serial cycles of nodulation. At each cycle, nodules were collected, surface sterilized and crushed to release nodule bacteria, which were directly re-inoculated onto new plants. These evolution cycles roughly approximated what happens in nature, *i.e.* when nodules degenerate, bacteria are released in the soil where they can survive as a free-living state or infect new plants. In a first experiment, 16 cycles of nodulation on *M. pudica* were performed in 18 parallel lineages of evolved populations along which the evolution of symbiotic properties was analyzed.

4.2 Infection, a symbiotic trait drastically improved along lab-evolution

Beside the acquisition of nodulation, a second crucial symbiotic property, intracellular infection of nodules, was rapidly acquired. A first rudimentary level of intracellular infection was indeed observed in nodules induced by two of the three nodulating ancestors as well as by evolved bacteria derived from the third ancestor after few evolution cycles (Guan et al., 2013; Marchetti et al., 2010). These nodules however always contained necrotic zones, indicative of plant defense reactions. Strikingly, this property drastically improved within few additional cycles in a majority of parallel lineages. Nodules formed by most final evolved clones were nicely invaded and devoid of necrosis (Marchetti et al., 2014, 2017). Thus three distinct levels of infection were obtained during the course of this experiment, extracellular, partially intracellular and nicely intracellular.

4.3 Regulatory rewiring as a main adaptive change

Resequencing the genomes of evolved bacteria and reconstructing the detected mutations allowed the identification of adaptive mutations responsible for main symbiotic phenotypic shifts. The capacity to nodulate was activated through the inactivation of the T3SS of *R. solanacearum* either via a stop mutation in *hrcV*, an essential structural gene of the T3SS, or via stop mutations in the master regulator of virulence *hrpG*, that not only controls the T3SS and effectors but also hundreds of other genes (Valls, Genin, & Boucher, 2006). Rudimentary intracellular infection ability was gained through *hrpG* inactivation or via a combination of the *hrcV* mutation and inactivation of various virulence regulators either *vsrA*, or *prhI* or *hrpG* (Fig. 2B) (Guan et al., 2013; Marchetti et al., 2010). Tremendous improvement of intracellular infection together with drastic reduction of nodule necrosis then occurred via a rewiring of the virulence regulatory network of *R. solanacearum* in several independent lineages. Mutations down-regulating efpR, a virulence regulator and metabolic repressor, appeared five times independently to promote infection. Repression of EfpR led to a constitutive induction of three intermediary genes encoding membrane components (Rsc3146-3148) that in turn, likely indirectly, deregulate the expression of hundreds of genes and allows the utilization of tens of new carbon, nitrogen and phosphorous compounds (Fig. 2BCE) (Capela et al., 2017). In other lines, mutations that only slightly modulate the responsiveness of *R. solanacearum* to cell density (*phcBQ* mutations) were sufficient to allow nice intracellular infection (M. Tang, unpublished). Interestingly, both repression of EfpR and the delayed QS-response modify the expression of hundreds of genes that many were in common and lead to a similar metabolic profile (Fig. 2C).

Thus, adaptation to symbiosis occurred via diverse genetic changes, either inactivation of structural genes (*hrcV*), or inactivation of regulators (*hrpG*, *vsrA*, *prhI*, *efpR*) and recruitment of constitutively expressed target genes (Rsc3146-3148), or modulation of regulatory paths to environmental signals (*phcBQ*). The rewiring of the virulence regulatory network of *R. solanacearum* appeared as the main adaptive mechanism to convert this pathogenic bacterium into legume symbionts, improving its symbiotic infection ability while decreasing its pathogenicity. These results illustrated how modifications in the regulatory circuitry of a bacterium can rapidly operate drastic changes in its lifestyle and contribute to ecological transitions.

4.4 Parallels between experimental and natural evolution of *Mimosa pudica* symbionts

To evaluate to what extent such artificial evolution experiment can tell us about the natural evolutionary history of legume symbionts, Clerissi et al. recently compared the emergence of *M. pudica* symbionts in natural populations of *C. taiwanensis* with the experimental conversion of *R. solanacearum* into *Mimosa* symbionts (Clerissi et al., 2018). In both cases the symbiotic transition was driven by the transfer of a symbiotic plasmid. However the two processes displayed many fundamental differences in terms of genetic backgrounds, *Ralstonia* and *Cupriavidus*, of time frame, more than 10 MY in one case and few years in the case of lab-evolution, of environmental conditions, complex and changing in the first case and simplified and constant in the second case, and finally of symbiosis achievement, mutualism was not achieved in the evolution experiment while *C. taiwanensis* are effective and well adapted symbionts. Yet a significant number of parallels were pointed out between the two evolutionary processes. First symbiotic adaptation occurred on the recipient genomes and not on the symbiotic plasmid. Both processes are also characterized by a dominance of purifying selection (*i.e.* a purge of non synonymous mutations). This phenomenon may be due to an *imuABC* cassette encoding stress-responsive error-prone DNA polymerases present in most extant strains of *C. taiwanensis* and generally present on the symbiotic plasmid. In the evolution experiment, this cassette was shown to increase the mutation rate of the recipient genome in the rhizosphere thus triggering a transient hypermutagenesis that accelerated the adaptive evolution of the chimeric strain (Remigi et al., 2014). The co-transfer of this cassette together with symbiosis genes was proposed to have enhanced the dissemination of symbiotic traits not only in the *C. taiwanensis* branch but more broadly among α- and β-proteobacteria by facilitating the post-HGT adaptations (Remigi et al., 2014). Finally, another interesting parallel was the co-option of the Phc quorum-sensing system for symbiosis in both natural and experimental evolution. All together these parallels demonstrated the relevance of experimental evolution approaches to potentially provide important insights into natural evolutionary processes.

5. Concluding remarks

Rhizobia arose independently in distant bacterial lineages mostly following HGT of mutualistic traits. Their parallel evolution from different ancestors accounts for the current diversity of the genetic strategies these bacteria use to establish symbiosis with legumes. During evolution, extensive recruitment, inactivation or modulation of indigenous functions together with deep regulatory rewiring has achieved the ecological transition toward mutualism initiated by HGT. The success of this 'molecular tinkering' essentially depends on the recipient genetic background as well as the chance a beneficial mutation occurs. It also strongly relies on the strict selection process exerted by host plant that only allows the entry of compatible variants into root tissues. The diversity of symbiotic molecular mechanisms identified in rhizobia nicely illustrates the plasticity of bacterial genomes and the range of possibilities to achieve a complex biological process. This also suggests that it would be possible to evolve nitrogen-fixing symbiotic partners of non-legume crops engineered to nodulate.

References

Amadou, C., Pascal, G., Mangenot, S., Glew, M., Bontemps, C., Capela, D., et al. (2008). Genome sequence of the beta-rhizobium Cupriavidus taiwanensis and comparative genomics of rhizobia. *Genome Research, 18*, 1472–1483.

Andrews, M., De Meyer, S., James, E. K., Stępkowski, T., Hodge, S., Simon, M. F., et al. (2018). Horizontal transfer of symbiosis genes within and between rhizobial genera: Occurrence and importance. *Genes, 9*, 321.

Aoki, S., Ito, M., & Iwasaki, W. (2013). From beta- to alpha-proteobacteria: The origin and evolution of rhizobial nodulation genes nodIJ. *Molecular Biology and Evolution, 30*, 2494–2508.

Arnold, M. F. F., Penterman, J., Shabab, M., Chen, E. J., & Walker, G. C. (2018). Important late-stage symbiotic role of the *Sinorhizobium meliloti* exopolysaccharide succinoglycan. *Journal of Bacteriology, 200*. e00665–17.

Bañuelos-Vazquez, L. A., Torres Tejerizo, G., & Brom, S. (2017). Regulation of conjugative transfer of plasmids and integrative conjugative elements. *Plasmid, 91*, 82–89.

Barcellos, F. G., Menna, P., Batista, J. S. D., & Hungria, M. (2007). Evidence of horizontal transfer of symbiotic genes from a *Bradyrhizobium japonicum* inoculant strain to indigenous diazotrophs *Sinorhizobium (Ensifer) fredii* and *Bradyrhizobium elkanii* in a Brazilian Savannah soil. *Applied and Environmental Microbiology, 73*, 2635–2643.

Barnett, M. J., & Long, S. R. (2018). Novel genes and regulators that influence production of cell surface exopolysaccharides in. *Sinorhizobium meliloti*. *J Bacteriol, 200*, e00501–e00517.

Bartsev, A. V., Deakin, W. J., Boukli, N. M., McAlvin, C. B., Stacey, G., Malnoe, P., et al. (2004). NopL, an effector protein of *Rhizobium* sp. NGR234, thwarts activation of plant defense reactions. *Plant Physiology, 134*, 871–879.

Battisti, L., Lara, J. C., & Leigh, J. A. (1992). Specific oligosaccharide form of the *Rhizobium meliloti* exopolysaccharide promotes nodule invasion in alfalfa. *Proceedings of the National Academy of Sciences of the United States of America, 89*, 5625–5629.

Bauer, E., Kaspar, T., Fischer, H. M., & Hennecke, H. (1998). Expression of the *fixR-nifA* operon in *Bradyrhizobium japonicum* depends on a new response regulator, RegR. *Journal of Bacteriology, 180*, 3853–3863.

Bélanger, L., Dimmick, K. A., Fleming, J. S., & Charles, T. C. (2009). Null mutations in *Sinorhizobium meliloti exoS* and *chvI* demonstrate the importance of this two-component regulatory system for symbiosis. *Molecular Microbiology, 74*, 1223–1237.

Bhagwat, A. A., Gross, K. C., Tully, R. E., & Keister, D. L. (1996). Beta-glucan synthesis in *Bradyrhizobium japonicum*: Characterization of a new locus (*ndvC*) influencing beta-(1->6) linkages. *Journal of Bacteriology, 178*, 4635–4642.

Bhagwat, A. A., Tully, R. E., & Keister, D. L. (1992). Isolation and characterization of an *ndvB* locus from *Rhizobium fredii*. *Mol Microbiol, 6*, 2159–2165.

Black, M., Moolhuijzen, P., Chapman, B., Barrero, R., Howieson, J., Hungria, M., et al. (2012). The genetics of symbiotic nitrogen fixation: Comparative genomics of 14 rhizobia strains by resolution of protein clusters. *Genes, 3*, 138–166.

Bladergroen, M. R., Badelt, K., & Spaink, H. P. (2003). Infection-blocking genes of a symbiotic *Rhizobium leguminosarum* strain that are involved in temperature-dependent protein secretion. *Molecular Plant-Microbe Interactions, 16*, 53–64.

Bobik, C., Meilhoc, E., & Batut, J. (2006). FixJ: A major regulator of the oxygen limitation response and late symbiotic functions of Sinorhizobium meliloti. *Journal of Bacteriology, 188*, 4890–4902.

Bonaldi, K., Gargani, D., Prin, Y., Fardoux, J., Gully, D., Nouwen, N., et al. (2011). Nodulation of *Aeschynomene afraspera* and *A. indica* by photosynthetic *Bradyrhizobium* sp. strain ORS285: The Nod-dependent versus the Nod-independent symbiotic interaction. *Molecular Plant-Microbe Interactions, 24*, 1359–1371.

Bonomi, H. R., Posadas, D. M., Paris, G., Carrica, M.e. C., Frederickson, M., Pietrasanta, L. I., et al. (2012). Light regulates attachment, exopolysaccharide production, and nodulation in *Rhizobium leguminosarum* through a LOV-histidine kinase photoreceptor. *Proceedings of the National Academy of Sciences of the United States of America, 109*, 12135–12140.

Breedveld, M. W., & Miller, K. J. (1994). Cyclic beta-glucans of members of the family Rhizobiaceae. *Microbiological Reviews, 58*, 145–161.

Broughton, W. J., Hanin, M., Relic, B., Kopcińska, J., Golinowski, W., Simsek, S., et al. (2006). Flavonoid-inducible modifications to rhamnan O antigens are necessary for *Rhizobium* sp. strain NGR234-legume symbioses. *Journal of Bacteriology, 188*, 3654–3663.

Calatrava-Morales, N., McIntosh, M., & Soto, M. J. (2018). Regulation mediated by N-acyl homoserine lactone quorum sensing signals in the *Rhizobium*-Legume symbiosis. *Genes, 9*, 263.

Calderón-Flores, A., Du Pont, G., Huerta-Saquero, A., Merchant-Larios, H., Servín-González, L., & Durán, S. (2005). The stringent response is required for amino acid and nitrate utilization, nod factor regulation, nodulation, and nitrogen fixation in *Rhizobium etli*. *J Bacteriol, 187*, 5075–5083.

Campbell, G. R. O., Reuhs, B. L., & Walker, G. C. (2002). Chronic intracellular infection of alfalfa nodules by *Sinorhizobium meliloti* requires correct lipopolysaccharide core. *Proceedings of the National Academy of Sciences of the United States of America, 99*, 3938–3943.

Capela, D., Carrere, S., & Batut, J. (2005). Transcriptome-based identification of the *Sinorhizobium meliloti* NodD1 regulon. *Applied and Environmental Microbiology, 71*, 4910–4913.

Capela, D., Marchetti, M., Clerissi, C., Perrier, A., Guetta, D., Gris, C., et al. (2017). Recruitment of a lineage-specific virulence regulatory pathway promotes intracellular infection by a plant pathogen experimentally evolved into a legume symbiont. *Molecular Biology and Evolution, 34*, 2503–2521.

Capoen, W., Goormachtig, S., & Holsters, M. (2010). Water-tolerant legume nodulation. *Journal of Experimental Botany, 61*, 1251–1255.

Cerri, M. R., Frances, L., Kelner, A., Fournier, J., Middleton, P. H., Auriac, M. C., et al. (2016). The symbiosis-related ERN transcription factors act in concert to coordinate rhizobial host root infection. *Plant Physiology, 171*, 1037–1054.

Cerri, M. R., Wang, Q., Stolz, P., Folgmann, J., Frances, L., Katzer, K., et al. (2017). The ERN1 transcription factor gene is a target of the CCaMK/CYCLOPS complex and controls rhizobial infection in. *Lotus japonicus. New Phytol, 215*, 323–337.

del Cerro, P., Rolla-Santos, A. A., Gomes, D. F., Marks, B. B., del Rosario Espuny, M., Rodríguez-Carvajal, M., et al. (2015). Opening the "black box" of *nodD3, nodD4* and *nodD5* genes of *Rhizobium tropici* strain CIAT 899. *BMC Genomics, 16*, 864.

Chaintreuil, C., Perrier, X., Martin, G., Fardoux, J., Lewis, G. P., Brottier, L., et al. (2018). Naturally occurring variations in the *nod*-independent model legume *Aeschynomene evenia* and relatives: A resource for nodulation genetics. *BMC Plant Biology, 18*, 54.

Cheng, H. P., & Walker, G. C. (1998). Succinoglycan is required for initiation and elongation of infection threads during nodulation of alfalfa by *Rhizobium meliloti*. *J Bacteriol, 180*, 5183–5191.

Chen, W. M., Moulin, L., Bontemps, C., Vandamme, P., Bena, G., & Boivin-Masson, C. (2003). Legume symbiotic nitrogen fixation by beta-proteobacteria is widespread in nature. *Journal of Bacteriology, 185*, 7266–7272.

Clerissi, C., Touchon, M., Capela, D., Tang, M., Cruveiller, S., Parker, M. A., et al. (2018). Parallels between experimental and natural evolution of legume symbionts. *Nature Communications, 9*, 2264.

Cosseau, C., & Batut, J. (2004). Genomics of the *ccoNOQP*-encoded *cbb*(3) oxidase complex in bacteria. *Archives of Microbiology, 181*, 89–96.

Coutinho, B. G., Mitter, B., Talbi, C., Sessitsch, A., Bedmar, E. J., Halliday, N., et al. (2013). Regulon studies and in planta role of the BraI/R quorum-sensing system in the plant-beneficial *Burkholderia* cluster. *Applied and Environmental Microbiology, 79*, 4421–4432.

Crespo-Rivas, J. C., Margaret, I., Hidalgo, A., Buendía-Clavería, A. M., Ollero, F. J., López-Baena, F. J., et al. (2009). *Sinorhizobium fredii* HH103 *cgs* mutants are unable to nodulate determinate- and indeterminate nodule-forming legumes and overproduce an altered EPS. *Molecular Plant-Microbe Interactions, 22*, 575–588.

Czernic, P., Gully, D., Cartieaux, F., Moulin, L., Guefrachi, I., Patrel, D., et al. (2015). Convergent evolution of endosymbiont differentiation in Dalbergioid and inverted repeat-lacking clade legumes mediated by nodule-specific cysteine-rich peptides. *Plant Physiology, 169*, 1254–1265.

D'Antuono, A. L., Ott, T., Krusell, L., Voroshilova, V., Ugalde, R. A., Udvardi, M., et al. (2008). Defects in rhizobial cyclic glucan and lipopolysaccharide synthesis alter legume gene expression during nodule development. *Molecular Plant-Microbe Interactions, 21*, 50–60.

D'Haeze, W., & Holsters, M. (2002). Nod factor structures, responses, and perception during initiation of nodule development. *Glycobiology, 12*, 79R–105R.

D'Haeze, W., & Holsters, M. (2004). Surface polysaccharides enable bacteria to evade plant immunity. *Trends in Microbiology, 12*, 555–561.

Daniels, R., De Vos, D. E., Desair, J., Raedschelders, G., Luyten, E., Rosemeyer, V., et al. (2002). The *cin* quorum sensing locus of *Rhizobium etli* CNPAF512 affects growth and symbiotic nitrogen fixation. *Journal of Biological Chemistry, 277*, 462–468.

Danino, V. E., Wilkinson, A., Edwards, A., & Downie, J. A. (2003). Recipient-induced transfer of the symbiotic plasmid pRL1JI in *Rhizobium leguminosarum* bv. *viciae* is regulated by a quorum-sensing relay. *Molecular Microbiology, 50*, 511–525.

Daubech, B., Poinsot, V., Klonowska, A., Capela, D., Chaintreuil, C., Moulin, L., Marchetti, M., & Masson-Boivin, C. (2019). noeM, a new nodulation gene involved in the biosynthesis of nod factors with an open-chain oxidized terminal residue and in the symbiosis with Mimosa pudica. *Molecular Plant Microbe Interactions, 32*, 1635–1648.

Davies, B. W., & Walker, G. C. (2007). Identification of novel *Sinorhizobium meliloti* mutants compromised for oxidative stress protection and symbiosis. *Journal of Bacteriology, 189*, 2110–2113.

De Meyer, S. E., Briscoe, L., Martínez-Hidalgo, P., Agapakis, C. M., de-Los Santos, P. E., Seshadri, R., et al. (2016). Symbiotic *Burkholderia* species show diverse arrangements of *nif/fix* and *nod* genes and lack typical high-affinity cytochrome *cbb*3 oxidase genes. *Molecular Plant-Microbe Interactions, 29*, 609–619.

Deakin, W. J., & Broughton, W. J. (2009). Symbiotic use of pathogenic strategies: Rhizobial protein secretion systems. *Nature Reviews Microbiology, 7*, 312–320.

Delavat, F., Miyazaki, R., Carraro, N., Pradervand, N., & van der Meer, J. R. (2017). The hidden life of integrative and conjugative elements. *FEMS Microbiology Reviews, 41*, 512–537.

Ding, H., & Hynes, M. F. (2009). Plasmid transfer systems in the rhizobia. *Canadian Journal of Microbiology, 55*, 917–927.

Dixon, R., & Kahn, D. (2004). Genetic regulation of biological nitrogen fixation. *Nature Reviews Microbiology, 2*, 621–631.

Downie, J. A. (2010). The roles of extracellular proteins, polysaccharides and signals in the interactions of rhizobia with legume roots. *FEMS Microbiology Reviews, 34*, 150–170.

Dusha, I., & Kondorosi, A. (1993). Genes at different regulatory levels are required for the ammonia control of nodulation in. *Rhizobium meliloti. Mol Gen Genet, 240*, 435–444.

Dylan, T., Nagpal, P., Helinski, D. R., & Ditta, G. S. (1990). Symbiotic pseudorevertants of *Rhizobium meliloti ndv* mutants. *Journal of Bacteriology, 172*, 1409–1417.

Estrada-de Los Santos, P., Palmer, M., Chávez-Ramírez, B., Beukes, C., Steenkamp, E. T., Briscoe, L., et al. (2018). Whole genome analyses suggests that *Burkholderia sensu lato* contains two additional novel genera (*mycetohabitans* gen. Nov., and *trinickia* gen. Nov.): Implications for the evolution of diazotrophy and nodulation in the burkholderiaceae. *Genes, 9*, 389.

Fabre, S., Gully, D., Poitout, A., Patrel, D., Arrighi, J. F., Giraud, E., et al. (2015). Nod Factor-independent nodulation in *Aeschynomene evenia* required the common plant-microbe symbiotic toolkit. *Plant Physiology, 169*, 2654–2664.

Fan, Y., Liu, J., Lyu, S., Wang, Q., Yang, S., & Zhu, H. (2017). The soybean *rfg1* gene restricts nodulation by *Sinorhizobium fredii* USDA193. *Frontiers of Plant Science, 8*, 1548.

Fellay, R., Hanin, M., Montorzi, G., Frey, J., Freiberg, C., Golinowski, W., et al. (1998). *nodD2* of *Rhizobium* sp. NGR234 is involved in the repression of the *nodABC* operon. *Molecular Microbiology, 27*, 1039–1050.

Fraysse, N., Couderc, F., & Poinsot, V. (2003). Surface polysaccharide involvement in establishing the rhizobium-legume symbiosis. *European Journal of Biochemistry, 270*, 1365–1380.

Gage, D. J. (2002). Analysis of infection thread development using Gfp- and DsRed-expressing *Sinorhizobium meliloti*. *J Bacteriol, 184*, 7042–7046.

Galibert, F., Finan, T. M., Long, S. R., Puhler, A., Abola, P., Ampe, F., et al. (2001). The composite genome of the legume symbiont *Sinorhizobium meliloti*. *Science, 293*, 668–672.

Garcia, M., Dunlap, J., Loh, J., & Stacey, G. (1996). Phenotypic characterization and regulation of the *nolA* gene of *Bradyrhizobium japonicum*. *Mol Plant Microbe Interact, 9*, 625–636.

Geddes, B. A., González, J. E., & Oresnik, I. J. (2014). Exopolysaccharide production in response to medium acidification is correlated with an increase in competition for nodule occupancy. *Molecular Plant-Microbe Interactions, 27*, 1307–1317.

Gherbi, H., Markmann, K., Svistoonoff, S., Estevan, J., Autran, D., Giczey, G., et al. (2008). SymRK defines a common genetic basis for plant root endosymbioses with arbuscular mycorrhiza fungi, rhizobia, and Frankia bacteria. *Proceedings of the National Academy of Sciences of the United States of America, 105*, 4928–4932.
Giraud, E., Hannibal, L., Fardoux, L., Vermeglio, A., & Dreyfus, B. (2000). Effect of *Bradyrhizobium* photosynthesis on stem nodulation of Aeschynomene sensitiva. *Proceedings of the National Academy of Sciences of the United States of America, 97*, 14795–14800.
Giraud, E., Moulin, L., Vallenet, D., Barbe, V., Cytryn, E., Avarre, J. C., et al. (2007). Legumes symbioses: Absence of *nod* genes in photosynthetic bradyrhizobia. *Science, 316*, 1307–1312.
Gough, C., & Cullimore, J. (2011). Lipo-chitooligosaccharide signaling in endosymbiotic plant-microbe interactions. *Molecular Plant-Microbe Interactions, 24*, 867–878.
Guan, S. H., Gris, C., Cruveiller, S., Pouzet, C., Tasse, L., Leru, A., et al. (2013). Experimental evolution of nodule intracellular infection in legume symbionts. *The ISME Journal, 7*, 1367–1377.
Guefrachi, I., Pierre, O., Timchenko, T., Alunni, B., Barrière, Q., Czernic, P., et al. (2015). *Bradyrhizobium* BclA Is a peptide transporter required for bacterial differentiation in symbiosis with *Aeschynomene* legumes. *Molecular Plant-Microbe Interactions, 28*, 1155–1166.
Gully, D., Czernic, P., Cruveiller, S., Mahé, F., Longin, C., Vallenet, D., et al. (2018). Transcriptome profiles of Nod Factor-independent symbiosis in the tropical legume *Aeschynomene evenia*. *Science Report, 8*, 10934.
Gurich, N., & González, J. E. (2009). Role of quorum sensing in *Sinorhizobium meliloti*-Alfalfa symbiosis. *Journal of Bacteriology, 191*, 4372–4382.
Gyaneshwar, P., Hirsch, A. M., Moulin, L., Chen, W. M., Elliott, G. N., Bontemps, C., et al. (2011). Legume-nodulating betaproteobacteria: Diversity, host range, and future prospects. *Molecular Plant-Microbe Interactions, 24*, 1276–1288.
Haag, A. F., Baloban, M., Sani, M., Kerscher, B., Pierre, O., Farkas, et al. (2011). Protection of *Sinorhizobium* against host cysteine-rich antimicrobial peptides is critical for symbiosis. *PLoS Biology, 9*, e1001169.
Hakoyama, T., Niimi, K., Watanabe, H., Tabata, R., Matsubara, J., Sato, S., et al. (2009). Host plant genome overcomes the lack of a bacterial gene for symbiotic nitrogen fixation. *Nature, 462*, 514–517.
Hauser, F., Pessi, G., Friberg, M., Weber, C., Rusca, N., Lindemann, A., et al. (2007). Dissection of the *Bradyrhizobium japonicum* NifA+sigma54 regulon, and identification of a ferredoxin gene (*fdxN*) for symbiotic nitrogen fixation. *Molecular Genetics and Genomics, 278*, 255–271.
Heavner, M. E., Qiu, W.-G., & Cheng, H.-P. (2015). Phylogenetic co-occurrence of ExoR, ExoS, and ChvI, components of the RSI bacterial invasion switch, suggests a key adaptive mechanism regulating the transition between free-living and host-invading phases in Rhizobiales. *PLoS One, 10*, e0135655.
Hirsch, A. M., Wilson, K. J., Jones, J. D. G., Bang, M., Walker, V. V., & Ausubel, F. M. (1984). *Rhizobium meliloti* nodulation genes allow *Agrobacterium tumefaciens* and *Escherichia coli* to form pseudonodules on alfalfa. *Journal of Bacteriology, 158*, 1133–1143.
Hoang, H. H., Becker, A., & González, J. E. (2004). The LuxR homolog ExpR, in combination with the Sin quorum sensing system, plays a central role in *Sinorhizobium meliloti* gene expression. *Journal of Bacteriology, 186*, 5460–5472.
Hocher, V., Alloisio, N., Bogusz, D., & Normand, P. (2011). Early signaling in actinorhizal symbioses. *Plant Signaling & Behavior, 6*, 1377–1379.
Hubber, A. M., Sullivan, J. T., & Ronson, C. W. (2007). Symbiosis-induced cascade regulation of the *Mesorhizobium loti* R7A VirB/D4 type IV secretion system. *Molecular Plant-Microbe Interactions, 20*, 255–261.

Hubber, A., Vergunst, A. C., Sullivan, J. T., Hooykaas, P. J. J., & Ronson, C. W. (2004). Symbiotic phenotypes and translocated effector proteins of the *Mesorhizobium loti* strain R7A VirB/D4 type IV secretion system. *Molecular Microbiology, 54*, 561–574.

Ingram-Smith, C., & Miller, K. J. (1998). Effects of ionic and osmotic strength on the glucosyltransferase of *Rhizobium meliloti* responsible for cyclic beta-(1,2)-glucan biosynthesis. *Applied and Environmental Microbiology, 64*, 1290–1297.

Irving, S. E., & Corrigan, R. M. (2018). Triggering the stringent response: Signals responsible for activating (p)ppGpp synthesis in bacteria. *Microbiology, 164*, 268–276.

Janczarek, M. (2011). Environmental signals and regulatory pathways that influence exopolysaccharide production in rhizobia. *International Journal of Molecular Sciences, 12*, 7898–7933.

Jaszek, M., Janczarek, M., Piersiak, T., & Grzywnowicz, K. (2014). The response of the *Rhizobium leguminosarum* bv. *trifolii* wild-type and exopolysaccharide-deficient mutants to oxidative stress. *Plant and Soil, 376*, 75–94.

Jones, K. M., Sharopova, N., Lohar, D. P., Zhang, J. Q., VandenBosch, K. A., & Walker, G. C. (2008). Differential response of the plant *Medicago truncatula* to its symbiont *Sinorhizobium meliloti* or an exopolysaccharide-deficient mutant. *Proceedings of the National Academy of Sciences of the United States of America, 105*, 704–709.

Jourand, P., Renier, A., Rapior, S., de Faria, S. M., Prin, Y., Galiana, A., et al. (2005). Role of methylotrophy during symbiosis between *Methylobacterium nodulans* and *Crotalaria podocarpa*. *Mol Plant-Microbe Interact, 18*, 1061–1068.

Kaminski, P. A., & Elmerich, C. (1998). The control of *Azorhizobium caulinodans nifA* expression by oxygen, ammonia and by the HF-I-like protein, NrfA. *Molecular Microbiology, 28*, 603–613.

Karmakar, K., Kundu, A., Rizvi, A. Z., Dubois, E., Severac, D., Czernic, P., et al. (2019). Transcriptomic analysis with the progress of symbiosis in 'crack-entry' legume *Arachis hypogaea* highlights its contrast with 'infection thread' adapted legumes. *Molecular Plant-Microbe Interactions, 32*, 271–285. MPMI06180174R.

Kawaharada, Y., Kelly, S., Nielsen, M. W., Hjuler, C. T., Gysel, K., Muszyński, A., et al. (2015). Receptor-mediated exopolysaccharide perception controls bacterial infection. *Nature, 523*, 308–312.

Kawaharada, Y., Nielsen, M. W., Kelly, S., James, E. K., Andersen, K. R., Rasmussen, S. R., et al. (2017). Differential regulation of the Epr3 receptor coordinates membrane-restricted rhizobial colonization of root nodule primordia. *Nature Communications, 8*, 14534.

Kelly, S. J., Muszyński, A., Kawaharada, Y., Hubber, A. M., Sullivan, J. T., Sandal, N., et al. (2013). Conditional requirement for exopolysaccharide in the *Mesorhizobium-Lotus* symbiosis. *Molecular Plant-Microbe Interactions, 26*, 319–329.

Kelly, S., Sullivan, J. T., Kawaharada, Y., Radutoiu, S., Ronson, C. W., & Stougaard, J. (2018). Regulation of Nod factor biosynthesis by alternative NodD proteins at distinct stages of symbiosis provides additional compatibility scrutiny. *Environmental Microbiology, 20*, 97–110.

Kim, C. H., Tully, R. E., & Keister, D. L. (1989). Exopolysaccharide-deficient mutants of *Rhizobium fredii* HH303 which are symbiotically effective. *Applied and Environmental Microbiology, 55*, 1852–1854.

Kiss, E., Kereszt, A., Barta, F., Stephens, S., Reuhs, B. L., Kondorosi, A., et al. (2001). The *rkp*-3 gene region of *Sinorhizobium meliloti* Rm41 contains strain-specific genes that determine K antigen structure. *Molecular Plant-Microbe Interactions, 14*, 1395–1403.

Kobayashi, H., Graven, Y. N., Broughton, W. J., & Perret, X. (2004). Flavonoids induce temporal shifts in gene-expression of *nod*-box controlled loci in *Rhizobium* sp. NGR234. *Molecular Microbiology, 51*, 335–347.

Kouchi, H., Imaizumi-Anraku, H., Hayashi, M., Hakoyama, T., Nakagawa, T., Umehara, Y., et al. (2010). How many peas in a pod? Legume genes responsible for mutualistic symbioses underground. *Plant and Cell Physiology, 51*, 1381–1397.

Krause, A., Doerfel, A., & Göttfert, M. (2002). Mutational and transcriptional analysis of the type III secretion system of *Bradyrhizobium japonicum. Mol Plant Microbe Interact, 15*, 1228–1235.

Krysciak, D., Grote, J., Rodriguez Orbegoso, M., Utpatel, C., Förstner, K. U., Li, L., et al. (2014). RNA sequencing analysis of the broad-host-range strain *Sinorhizobium fredii* NGR234 identifies a large set of genes linked to quorum sensing-dependent regulation in the background of a *traI* and *ngrI* deletion mutant. *Applied and Environmental Microbiology, 80*, 5655–5671.

Ktari, A., Nouioui, I., Furnholm, T., Swanson, E., Ghodhbane-Gtari, F., Tisa, L. S., et al. (2017). Permanent draft genome sequence of *Frankia* sp. NRRL B-16219 reveals the presence of canonical *nod* genes, which are highly homologous to those detected in Candidatus *Frankia* Dg1 genome. *Stand Genomic Sci, 12*, 51.

Laguerre, G., Nour, S. M., Macheret, V., Sanjuan, J., Drouin, P., & Amarger, N. (2001). Classification of rhizobia based on *nodC* and *nifH* gene analysis reveals a close phylogenetic relationship among *Phaseolus vulgaris* symbionts. *Microbiology, 147*, 981–993.

Ledbetter, R. N., Garcia Costas, A. M., Lubner, C. E., Mulder, D. W., Tokmina-Lukaszewska, M., Artz, J. H., et al. (2017). The electron bifurcating FixABCX protein complex from *Azotobacter vinelandii*: Generation of low-potential reducing equivalents for nitrogenase catalysis. *Biochemistry, 56*, 4177–4190.

Lehman, A. P., & Long, S. R. (2013). Exopolysaccharides from *Sinorhizobium meliloti* can protect against H_2O_2-dependent damage. *Journal of Bacteriology, 195*, 5362–5369.

Lercher, M. J., & Pál, C. (2008). Integration of horizontally transferred genes into regulatory interaction networks takes many million years. *Molecular Biology and Evolution, 25*, 559–567.

Ling, J., Wang, H., Wu, P., Li, T., Tang, Y., Naseer, N., et al. (2016). Plant nodulation inducers enhance horizontal gene transfer of *Azorhizobium caulinodans* symbiosis island. *Proceedings of the National Academy of Sciences of the United States of America, 113*, 13875–13880.

Liu, W., Sun, Y., Shen, R., Dang, X., Liu, X., Sui, F., et al. (2018). A chemotaxis-like pathway of *Azorhizobium caulinodans* controls flagella-driven motility, which regulates biofilm formation, exopolysaccharide biosynthesis, and competitive nodulation. *Molecular Plant-Microbe Interactions, 31*, 737–749.

Loh, J., & Stacey, G. (2003). Nodulation gene regulation in *Bradyrhizobium japonicum*: A unique integration of global regulatory circuits. *Applied and Environmental Microbiology, 69*, 10–17.

Loh, J. T., Yuen-Tsai, J. P., Stacey, M. G., Lohar, D., Welborn, A., & Stacey, G. (2001). Population density-dependent regulation of the *Bradyrhizobium japonicum* nodulation genes. *Molecular Microbiology, 42*, 37–46.

López-Baena, F. J., Ruiz-Sainz, J. E., Rodríguez-Carvajal, M. A., & Vinardell, J. M. (2016). Bacterial molecular signals in the *Sinorhizobium fredii*-Soybean symbiosis. *International Journal of Molecular Sciences, 17*.

MacLean, A. M., Finan, T. M., & Sadowsky, M. J. (2007). Genomes of the symbiotic nitrogen-fixing bacteria of legumes. *Plant Physiology, 144*, 615–622.

Madsen, L. H., Tirichine, L., Jurkiewicz, A., Sullivan, J. T., Heckmann, A. B., Bek, A. S., et al. (2010). The molecular network governing nodule organogenesis and infection in the model legume *Lotus japonicus. Nat Commun, 1*, 10.

Marchetti, M., Capela, D., Glew, M., Cruveiller, S., Chane-Woon-Ming, B., Gris, C., et al. (2010). Experimental evolution of a plant pathogen into a legume symbiont. *PLoS Biology, 8*, e1000280.

Marchetti, M., Clerissi, C., Yousfi, Y., Gris, C., Bouchez, O., Rocha, E., et al. (2017). Experimental evolution of rhizobia may lead to either extra- or intracellular symbiotic adaptation depending on the selection regime. *Molecular Ecology, 26*, 1818–1831.

Marchetti, M., Jauneau, A., Capela, D., Remigi, P., Gris, C., Batut, J., et al. (2014). Shaping bacterial symbiosis with legumes by experimental evolution. *Molecular Plant-Microbe Interactions, 27*, 956–964.

Marie, C., Deakin, W. J., Viprey, V., Kopciñska, J., Golinowski, W., Krishnan, H. B., et al. (2003). Characterization of Nops, nodulation outer proteins, secreted via the type III secretion system of NGR234. *Molecular Plant-Microbe Interactions, 16*, 743–751.

Maruya, J., & Saeki, K. (2010). The bacA gene homolog, mlr7400, in *Mesorhizobium loti* MAFF303099 is dispensable for symbiosis with *Lotus japonicus* but partially capable of supporting the symbiotic function of *bacA* in. *Sinorhizobium meliloti. Plant Cell Physiol, 51*, 1443–1452.

Masson-Boivin, C., Giraud, E., Perret, X., & Batut, J. (2009). Establishing nitrogen-fixing symbiosis with legumes: How many rhizobium recipes? *Trends in Microbiology, 17*, 458–466.

Masson-Boivin, C., & Sachs, J. L. (2018). Symbiotic nitrogen fixation by rhizobia-the roots of a success story. *Current Opinion in Plant Biology, 44*, 7–15.

Mergaert, P., Nikovics, K., Kelemen, Z., Maunoury, N., Vaubert, D., Kondorosi, A., et al. (2003). A novel family in *Medicago truncatula* consisting of more than 300 nodule-specific genes coding for small, secreted polypeptides with conserved cysteine motifs. *Plant Physiology, 132*, 161–173.

Miller-Williams, M., Loewen, P. C., & Oresnik, I. J. (2006). Isolation of salt-sensitive mutants of *Sinorhizobium meliloti* strain Rm1021. *Microbiology, 152*, 2049–2059.

Mølbak, L., Molin, S., & Kroer, N. (2007). Root growth and exudate production define the frequency of horizontal plasmid transfer in the rhizosphere. *FEMS Microbiology Ecology, 59*, 167–176.

Moris, M., Braeken, K., Schoeters, E., Verreth, C., Beullens, S., Vanderleyden, J., et al. (2005). Effective symbiosis between *Rhizobium etli* and *Phaseolus vulgaris* requires the alarmone ppGpp. *Journal of Bacteriology, 187*, 5460–5469.

Morris, J., & González, J. E. (2009). The novel genes *emmABC* are associated with exopolysaccharide production, motility, stress adaptation, and symbiosis in *Sinorhizobium meliloti. J Bacteriol, 191*, 5890–5900.

Mousavi, S. A., Willems, A., Nesme, X., de Lajudie, P., & Lindström, K. (2015). Revised phylogeny of rhizobiaceae: Proposal of the delineation of *pararhizobium* gen. Nov., and 13 new species combinations. *Systematic & Applied Microbiology, 38*, 84–90.

Muszyński, A., Heiss, C., Hjuler, C. T., Sullivan, J. T., Kelly, S. J., Thygesen, M. B., et al. (2016). Structures of exopolysaccharides involved in receptor-mediated perception of *Mesorhizobium loti* by *Lotus japonicus. J Biol Chem, 291*, 20946–20961.

Nandasena, K. G., O'Hara, G. W., Tiwari, R. P., Sezmis, E., & Howieson, J. G. (2007). In situ lateral transfer of symbiosis islands results in rapid evolution of diverse competitive strains of mesorhizobia suboptimal in symbiotic nitrogen fixation on the pasture legume *Biserrula pelecinus* L. *Environmental Microbiology, 9*, 2496–2511.

Nelson, M. S., Chun, C. L., & Sadowsky, M. J. (2017). Type IV effector proteins involved in the *Medicago-Sinorhizobium* Symbiosis. *Molecular Plant-Microbe Interactions, 30*, 28–34.

Nelson, M. S., & Sadowsky, M. J. (2015). Secretion systems and signal exchange between nitrogen-fixing rhizobia and legumes. *Frontiers of Plant Science, 6*, 491.

Nishihata, S., Kondo, T., Tanaka, K., Ishikawa, S., Takenaka, S., Kang, C. M., et al. (2018). *Bradyrhizobium diazoefficiens* USDA110 PhaR functions for pleiotropic regulation of cellular processes besides PHB accumulation. *BMC Microbiology, 18*, 156.

Noel, K. D., Box, J. M., & Bonne, V. J. (2004). 2-O-methylation of fucosyl residues of a rhizobial lipopolysaccharide is increased in response to host exudate and is eliminated in a symbiotically defective mutant. *Applied and Environmental Microbiology, 70*, 1537–1544.

Noel, K. D., Forsberg, L. S., & Carlson, R. W. (2000). Varying the abundance of O antigen in *Rhizobium etli* and its effect on symbiosis with *Phaseolus vulgaris*. *J Bacteriol, 182*, 5317–5324.

Nogales, J., Blanca-Ordóñez, H., Olivares, J., & Sanjuán, J. (2013). Conjugal transfer of the *Sinorhizobium meliloti* 1021 symbiotic plasmid is governed through the concerted action of one- and two-component signal transduction regulators. *Environmental Microbiology, 15*, 811–821.

Nouwen, N., Arrighi, J. F., Cartieaux, F., Chaintreuil, C., Gully, D., Klopp, C., et al. (2017). The role of rhizobial (NifV) and plant (FEN1) homocitrate synthases in *Aeschynomene*/photosynthetic *Bradyrhizobium* symbiosis. *Scientific Reports, 7*, 448.

Ojeda, K. J., Box, J. M., & Noel, K. D. (2010). Genetic basis for Rhizobium etli CE3 O-antigen O-methylated residues that vary according to growth conditions. *Journal of Bacteriology, 192*, 679–690.

Okazaki, S., Kaneko, T., Sato, S., & Saeki, K. (2013). Hijacking of leguminous nodulation signaling by the rhizobial type III secretion system. *Proceedings of the National Academy of Sciences of the United States of America, 110*, 17131–17136.

Okazaki, S., Tittabutr, P., Teulet, A., Thouin, J., Fardoux, J., Chaintreuil, C., et al. (2016). Rhizobium-legume symbiosis in the absence of nod factors: Two possible scenarios with or without the T3SS. *The ISME Journal, 10*, 64–74.

Okazaki, S., Zehner, S., Hempel, J., Lang, K., & Göttfert, M. (2009). Genetic organization and functional analysis of the type III secretion system of *Bradyrhizobium elkanii*. *FEMS Microbiol Lett, 295*, 88–95.

Okubo, T., Fukushima, S., & Minamisawa, K. (2012). Evolution of *bradyrhizobium-aeschynomene* mutualism: Living testimony of the ancient world or highly evolved state? *Plant and Cell Physiology, 53*, 2000–2007.

Oldroyd, G. E. (2013). Speak, friend, and enter: Signalling systems that promote beneficial symbiotic associations in plants. *Nature Reviews Microbiology, 11*, 252–263.

Ormeño-Orrillo, E., Servín-Garcidueñas, L. E., Rogel, M. A., González, V., Peralta, H., Mora, J., et al. (2015). Taxonomy of rhizobia and agrobacteria from the Rhizobiaceae family in light of genomics. *Systematic & Applied Microbiology, 38*, 287–291.

Österman, J., Marsh, J., Laine, P. K., Zeng, Z., Alatalo, E., Sullivan, J. T., et al. (2014). Genome sequencing of two *Neorhizobium galegae* strains reveals a *noeT* gene responsible for the unusual acetylation of the nodulation factors. *BMC Genomics, 15*, 500.

Parada, M., Vinardell, J. M., Ollero, F. J., Hidalgo, A., Gutiérrez, R., Buendía-Clavería, A. M., et al. (2006). *Sinorhizobium fredii* HH103 mutants affected in capsular polysaccharide (KPS) are impaired for nodulation with soybean and Cajanus cajan. *Molecular Plant-Microbe Interactions, 19*, 43–52.

Pellock, B. J., Cheng, H. P., & Walker, G. C. (2000). Alfalfa root nodule invasion efficiency is dependent on *Sinorhizobium meliloti* polysaccharides. *Journal of Bacteriology, 182*, 4310–4318.

Pérez Carrascal, O. M., VanInsberghe, D., Juárez, S., Polz, M. F., Vinuesa, P., & González, V. (2016). Population genomics of the symbiotic plasmids of sympatric nitrogen-fixing *Rhizobium* species associated with. *Phaseolus vulgaris. Environ Microbiol, 18*, 2660–2676.

Pérez-Mendoza, D., Sepúlveda, E., Pando, V., Muñoz, S., Nogales, J., Olivares, J., et al. (2005). Identification of the *rctA* gene, which is required for repression of conjugative transfer of rhizobial symbiotic megaplasmids. *Journal of Bacteriology, 187*, 7341–7350.

Perret, X., Staehelin, C., & Broughton, W. J. (2000). Molecular basis of symbiotic promiscuity. *Microbiology and Molecular Biology Reviews, 64*, 180–201.

Persson, T., Battenberg, K., Demina, I. V., Vigil-Stenman, T., Vanden Heuvel, B., Pujic, P., et al. (2015). Candidatus *Frankia Datiscae* Dg1, the actinobacterial microsymbiont of *Datisca glomerata*, expresses the canonical *nod* Genes *nodABC* in symbiosis with its host plant. *PLoS One, 10*, e0127630.

Ramsay, J. P., Sullivan, J. T., Jambari, N., Ortori, C. A., Heeb, S., Williams, P., et al. (2009). A LuxRI-family regulatory system controls excision and transfer of the *Mesorhizobium loti* strain R7A symbiosis island by activating expression of two conserved hypothetical genes. *Molecular Microbiology, 73*, 1141–1155.

Ramsay, J. P., Tester, L. G. L., Major, A. S., Sullivan, J. T., Edgar, C. D., Kleffmann, T., et al. (2015). Ribosomal frameshifting and dual-target antiactivation restrict quorum-sensing-activated transfer of a mobile genetic element. *Proceedings of the National Academy of Sciences of the United States of America, 112*, 4104–4109.

Remigi, P., Capela, D., Clerissi, C., Tasse, L., Torchet, R., Bouchez, O., et al. (2014). Transient hypermutagenesis accelerates the evolution of legume endosymbionts following horizontal gene transfer. *PLoS Biology, 12*, e1001942.

Remigi, P., Zhu, J., Young, J. P. W., & Masson-Boivin, C. (2016). Symbiosis within symbiosis: Evolving nitrogen-fixing legume symbionts. *Trends in Microbiology, 24*, 63–75.

Reuhs, B. L., Relić, B., Forsberg, L. S., Marie, C., Ojanen-Reuhs, T., Stephens, S. B., et al. (2005). Structural characterization of a flavonoid-inducible *Pseudomonas aeruginosa* A-band-like O antigen of *Rhizobium* sp. strain NGR234, required for the formation of nitrogen-fixing nodules. *Journal of Bacteriology, 187*, 6479–6487.

Rodelas, B., Lithgow, J. K., Wisniewski-Dye, F., Hardman, A., Wilkinson, A., Economou, A., et al. (1999). Analysis of quorum-sensing-dependent control of rhizosphere-expressed (*rhi*) genes in *Rhizobium leguminosarum* bv. viciae. *Journal of Bacteriology, 181*, 3816–3823.

Rolin, D. B., Pfeffer, P. E., Osman, S. F., Szwergold, B. S., Kappler, F., & Benesi, A. J. (1992). Structural studies of a phosphocholine substituted beta-(1,3);(1,6) macrocyclic glucan from *Bradyrhizobium japonicum* USDA 110. *Biochimica et Biophysica Acta, 1116*, 215–225.

Rosemeyer, V., Michiels, J., Verreth, C., & Vanderleyden, J. (1998). *luxI*- and *luxR*-homologous genes of *Rhizobium etli* CNPAF512 contribute to synthesis of autoinducer molecules and nodulation of Phaseolus vulgaris. *Journal of Bacteriology, 180*, 815–821.

Saad, M. M., Crèvecoeur, M., Masson-Boivin, C., & Perret, X. (2012). The T3SS of *Cupriavidus taiwanensis* strain LMG19424 compromizes symbiosis with *Leucaena leucocephala*. *Applied and Environmental Microbiology, 78*, 7476–7479.

Sánchez, C., Iannino, F., Deakin, W. J., Ugalde, R. A., & Lepek, V. C. (2009). Characterization of the *Mesorhizobium loti* MAFF303099 type-three protein secretion system. *Molecular Plant-Microbe Interactions, 22*, 519–528.

Schäper, S., Steinchen, W., Krol, E., Altegoer, F., Skotnicka, D., Søgaard-Andersen, L., et al. (2017). AraC-like transcriptional activator CuxR binds c-di-GMP by a PilZ-like mechanism to regulate extracellular polysaccharide production. *Proceedings of the National Academy of Sciences of the United States of America, 114*, E4822–E4831.

Schlaman, H. R., Okker, R. J., & Lugtenberg, B. J. (1992). Regulation of nodulation gene expression by NodD in rhizobia. *Journal of Bacteriology, 174*, 5177–5182.

Schwedock, J. S., & Long, S. R. (1992). *Rhizobium meliloti* genes involved in sulfate activation: The two copies of *nodPQ* and a new locus. *Saa. Genetics, 132*, 899–909.

Sciotti, M. A., Chanfon, A., Hennecke, H., & Fischer, H. M. (2003). Disparate oxygen responsiveness of two regulatory cascades that control expression of symbiotic genes in *Bradyrhizobium japonicum*. *J Bacteriol, 185*, 5639–5642.

Serrato, R. V. (2014). Lipopolysaccharides in diazotrophic bacteria. *Front Cell Infect Microbiol, 4*, 119.

Skorupska, A., Janczarek, M., Marczak, M., Mazur, A., & Krol, J. (2006). Rhizobial exopolysaccharides: Genetic control and symbiotic functions. *Microbial Cell Factories, 5*, 7.

Sugawara, M., Epstein, B., Badgley, B. D., Unno, T., Xu, L., Reese, J., et al. (2013). Comparative genomics of the core and accessory genomes of 48 *Sinorhizobium* strains comprising five genospecies. *Genome Biology, 14*, R17.

Sugawara, M., Takahashi, S., Umehara, Y., Iwano, H., Tsurumaru, H., Odake, H., et al. (2018). Variation in bradyrhizobial NopP effector determines symbiotic incompatibility with Rj2-soybeans via effector-triggered immunity. *Nature Communications, 9*, 3139.

Sullivan, J. T., Patrick, H. N., Lowther, W. L., Scott, D. B., & Ronson, C. W. (1995). Nodulating strains of *Rhizobium loti* arise through chromosomal symbiotic gene transfer in the environment. *Proceedings of the National Academy of Sciences of the United States of America, 92*, 8985–8989.

Sullivan, J., & Ronson, C. (1998). Evolution of rhizobia by acquisition of a 500-kb symbiosis island that integrates into a phe-tRNA gene. *Proceedings of the National Academy of Sciences of the United States of America, 95*, 5145–5149.

Svistoonoff, S., Benabdoun, F. M., Nambiar-Veetil, M., Imanishi, L., Vaissayre, V., Cesari, S., et al. (2013). The independent acquisition of plant root nitrogen-fixing symbiosis in Fabids recruited the same genetic pathway for nodule organogenesis. *PLoS One, 8*, e64515.

Tang, F., Yang, S., & Zhu, H. (2016). Functional analysis of alternative transcripts of the soybean Rj2 gene that restricts nodulation with specific rhizobial strains. *Plant Biology, 18*, 537–541.

Tian, C. F., Zhou, Y. J., Zhang, Y. M., Li, Q. Q., Zhang, Y. Z., Li, D. F., et al. (2012). Comparative genomics of rhizobia nodulating soybean suggests extensive recruitment of lineage-specific genes in adaptations. *Proceedings of the National Academy of Sciences of the United States of America, 109*, 8629–8634.

Tun-Garrido, C., Bustos, P., González, V., & Brom, S. (2003). Conjugative transfer of p42a from *Rhizobium etli* CFN42, which is required for mobilization of the symbiotic plasmid, is regulated by quorum sensing. *Journal of Bacteriology, 185*, 1681–1692.

Valls, M., Genin, S., & Boucher, C. (2006). Integrated regulation of the type III secretion system and other virulence determinants in *Ralstonia solanacearum*. *PLoS Pathog, 2*, 798–807.

Van de Velde, W., Zehirov, G., Szatmari, A., Debreczeny, M., Ishihara, H., Kevei, Z., et al. (2010). Plant peptides govern terminal differentiation of bacteria in symbiosis. *Science, 327*, 1122–1126.

Vernié, T., Kim, J., Frances, L., Ding, Y., Sun, J., Guan, D., et al. (2015). The NIN transcription factor coordinates diverse nodulation programs in different tissues of the *Medicago truncatula* root. *Plant Cell, 27*, 3410–3424.

Viprey, V., Del Greco, A., Golinowski, W., Broughton, W. J., & Perret, X. (1998). Symbiotic implications of type III protein secretion machinery in *Rhizobium*. *Mol Microbiol, 28*, 1381–1389.

Wang, D., Couderc, F., Tian, C. F., Gu, W., Liu, L. X., & Poinsot, V. (2018). Conserved composition of Nod factors and exopolysaccharides produced by different phylogenetic lineage. *Frontiers in Microbiology, 9*, 2852.

Wippel, K., & Long, S. R. (2018). Symbiotic performance of *Sinorhizobium meliloti* lacking ppGpp depends on the *Medicago* host species. *Molecular Plant-Microbe Interactions*. https://doi.org/10.1094/MPMI-11-18-0306-R.

van Workum, W. A. T., van Slageren, S., van Brussel, A. A. N., & Kijne, J. W. (1998). Role of exopolysaccharides of *Rhizobium leguminosarum* bv. *viciae* as host plant-specific molecules required for infection thread formation during nodulation of Vicia sativa. *Molecular Plant-Microbe Interactions, 11*, 1233–1241.

Xin, D. W., Liao, S., Xie, Z. P., Hann, D. R., Steinle, L., Boller, T., et al. (2012). Functional analysis of NopM, a novel E3 ubiquitin ligase (NEL) domain effector of *Rhizobium* sp. strain NGR234. *PLoS Pathogens, 8*, e1002707.

Yang, M., Sun, K., Zhou, L., Yang, R., Zhong, Z., & Zhu, J. (2009). Functional analysis of three AHL autoinducer synthase genes in *Mesorhizobium loti* reveals the important role of quorum sensing in symbiotic nodulation. *Canadian Journal of Microbiology, 55*, 210–214.

Yang, S. M., Tang, F., Gao, M. Q., Krishnan, H. B., & Zhu, H. Y. (2010). R gene-controlled host specificity in the legume-rhizobia symbiosis. *Proceedings of the National Academy of Sciences of the United States of America, 107*, 18735–18740.

Zhang, L., Chen, X. J., Lu, H. B., Xie, Z. P., & Staehelin, C. (2011). Functional analysis of the type 3 effector nodulation outer protein L (NopL) from *rhizobium* sp. NGR234: Symbiotic effects, phosphorylation, and interference with mitogen-activated protein kinase signaling. *Journal of Biological Chemistry, 286*, 32178–32187.

Zhao, R., Liu, L. X., Zhang, Y. Z., Jiao, J., Cui, W. J., Zhang, B., et al. (2018). Adaptive evolution of rhizobial symbiotic compatibility mediated by co-evolved insertion sequences. *The ISME Journal, 12*, 101–111.

Zheng, H., Zhong, Z., Lai, X., Chen, W. X., Li, S., & Zhu, J. (2006). A LuxR/LuxI-type quorum-sensing system in a plant bacterium, *Mesorhizobium tianshanense*, controls symbiotic nodulation. *Journal of Bacteriology, 188*, 1943–1949.

CHAPTER TEN

Gene regulation by extracytoplasmic function (ECF) σ factors in alpha-rhizobia

Doreen Meier, Delia Casas-Pastor, Georg Fritz and Anke Becker*
Center for Synthetic Microbiology (SYNMIKRO) and Faculty of Biology, Philipps-Universität Marburg, Marburg, Germany
*Corresponding author: E-mail: anke.becker@synmikro.uni-marburg.de

Contents

Advances in Botanical Research, Volume 94
ISSN 0065-2296
https://doi.org/10.1016/bs.abr.2019.09.012

Abstract

Bacteria have evolved efficient mechanisms to sense environmental changes to which they respond with high sensitivity and specificity. Reprogramming gene transcription by directing RNA polymerase to specific promoters through alternative σ subunits is a wide-spread bacterial strategy of rapid adaptation to environmental challenges. Extracytoplasmic function (ECF) σ factors represent the most minimalistic and diverse group of the σ^{70} protein family. These σ factors are typically stress-inducible and non-essential. Most bacterial species possess multiple ECF σ factors. Here, we review the current knowledge on promoter specificity, regulatory control and biological function of ECF σ factors in members of the α-proteobacterial Rhizobiales, which enter a root nodule symbiosis with leguminous host plants.

1. Introduction

The ability to sense and to respond to changes in the environment is essential for the survival of bacteria in their ecological niches. Transcriptional regulation plays a key role for coping with these challenges. σ factors are dissociable subunits of the core RNA polymerase and important for promoter recognition and transcription initiation (Helmann & Chamberlin, 1988; Hook-Barnard & Hinton, 2007). In addition to the housekeeping σ factor which is responsible for the expression of most genes under favorable conditions, bacteria can draw on a variable number of alternative σ factors. These recruit the RNA polymerase (RNAP) to distinct promoters, thereby mediating expression of a subset of genes, whose products support adaptation to changes in growth conditions, such as stress factors, or promote morphological development (Gruber & Gross, 2003).

1.1 Alternative σs within the σ^{70} family

The σ^{70} family of σ factors encompasses the housekeeping σs as well as structurally related alternative σ factors. They can be divided into four different groups based on phylogenetic relatedness and protein domain architecture (Gruber & Gross, 2003; Helmann, 2002; Lonetto, Gribskov, & Gross, 1992; Lonetto, Brown, Rudd, & Buttner, 1994). Essential housekeeping σs are group 1 members. They contain four distinct helical structured domains (designated σ_1 to σ_4) and, in some cases, a non-conserved region (NCR). Upon binding to RNAP, σ_2, σ_3 and σ_4 recognize specific promoter elements (Fig. 1) (Gruber & Gross, 2003; Paget, 2015). Group 2 σs include non-essential paralogues of the housekeeping sigma factors. The most famous representative is probably σ^S - the master regulator of the general stress response in *Escherichia coli* and related γ-proteobacteria (Battesti, Majdalani, & Gottesman, 2011). Group 3 σs are known to be involved in heat shock response, flagellar synthesis and sporulation. Compared to members of group 1 and 2, they lack the σ_1-region and the NCR (Gruber & Gross, 2003). σ factors of group 4, called ECF (extracytoplasmic function) σs, represent the most diverse and minimalistic group of

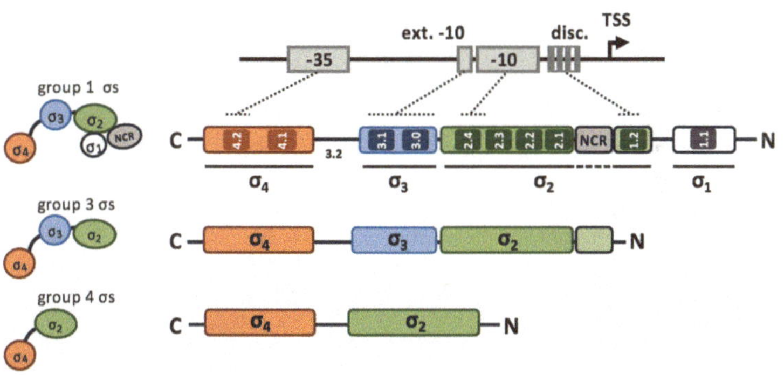

Fig. 1 Domain architecture and promoter recognition of σ^{70} family members. Domains σ_1-σ_4 and the NCR (non-conserved region) are labeled. A schematic representation of an *E. coli* promoter is shown above. Typical DNA motifs are shown: -10, -35 (-10 and -35 motif), ext. -10 (extended -10 motif), disc. (discriminator motif), TSS (transcription start site). Dashed lines between promoter elements and σ-regions denote DNA binding determinants that are involved in promoter recognition (Gruber & Gross, 2003; Paget, 2015). *Modified after Bervoets I., Van Brempt M., Van Nerom K., Van Hove B., Maertens J., De Mey M., et al. (2018). A sigma factor toolbox for orthogonal gene expression in* Escherichia coli. Nucleic Acids Research, *46(4), 2133–2144; Paget M. S. (2015). Bacterial sigma factors and anti-sigma factors: Structure, function and distribution.* Biomolecules, *5(3), 1245–1265.*

the σ^{70} family. They only possess the domains σ_2 and σ_4, which bind the promoter -10 and -35 elements and are sufficient for interaction with RNAP (Fig. 1) (Gruber & Gross, 2003; Staron et al., 2009). Lonetto et al. (1994) initially proposed the term ECF, as early findings pointed toward their role in cell surface or transport related processes. However, some ECF σs sense stimuli intracellularly (Staron et al., 2009).

1.2 ECF-based signal transduction

Previously, diverse mechanisms mediating the transduction of extracytoplasmic or cytoplasmic stimuli to ECF σs have been analyzed (Brooks & Buchanan, 2008; Mascher, 2013). These studies showed a pivotal role of anti-σ factors - negative regulators of ECF activity - in these processes. Anti-σ factors are often membrane anchored proteins. The ECF σ together with its cognate anti-σ factor is usually encoded by an operon, which is frequently positively autoregulated by the ECF σ itself. In absence of a stimulus, the ECF σ is kept in an inactive state by the bound cognate anti-σ factor. Signal sensing at the cell surface or intracellularly leads to inactivation of the corresponding anti-σ factor (Fig. 2), e.g. by proteolysis or through conformational changes, and consequently to ECF σ release. Finally, binding of the released ECF σ to the RNA polymerase core mediates increased

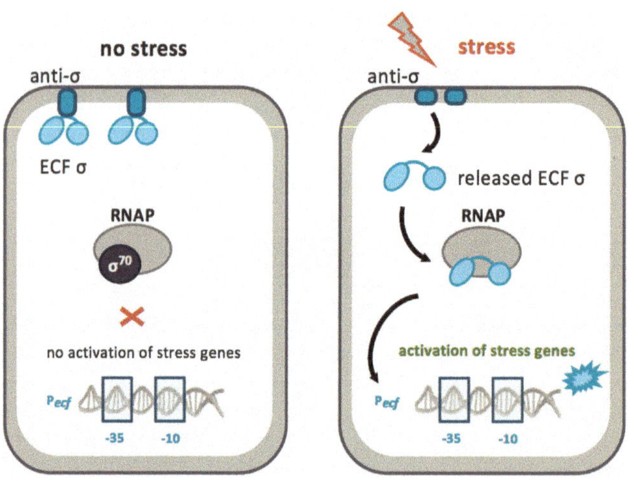

Fig. 2 ECF-dependent signal transduction. Schematic representations of the ECF σ factor, the anti-σ factor, the RNAP (RNA polymerase) and the primary σ^{70} are labeled. The RNAP is either associated with the primary σ^{70} under non-stress conditions or with the ECF σ factor under stress conditions. An ECF σ factor-dependent promoter with core promoter motifs (-35 and -10) is shown. See text for details.

expression of the *ecf* operon and other stress related genes (Brooks & Buchanan, 2008; Mascher, 2013; Staron et al., 2009). Genes that are activated by an ECF σ define a regulon and their number can vary enormously (Lang et al., 2018).

1.3 ECFs in nitrogen-fixing α-rhizobia

Several members of the Rhizobiales, an order of Gram-negative α-proteobacteria, are known for their ability to establish a nitrogen-fixing symbiosis with legumes. These α-rhizobia can undergo a drastic change in their lifestyle. They either live as saprophytes in the soil or as symbionts in root nodules where they differentiate into bacteroids which fix molecular nitrogen to the benefit of the host plant. On their part, α-rhizobia take advantage of carbohydrates provided by the host (Poole, Ramachandran, & Terpolilli, 2018). A comprehensive list of plant-nodulating bacteria and their hosts has recently been published (Andrews & Andrews, 2017). Among the Rhizobiales, representatives can most often be found in genera such as *Bradyrhizobium*, *Sinorhizobium* (*Ensifer*), *Rhizobium* or *Mesorhizobium*. Under free living conditions, rhizobia must survive the rather rigorous conditions in the soil and the rhizosphere where they co-exist with competing microbes and face different stresses, like desiccation, oxidative stress and plant defense responses. Besides, they undergo different physiological and morphological adaptations as they establish a symbiosis with their host plants (Damiani, Pauly, Puppo, Brouquisse, & Boscari, 2016; Gibson, Kobayashi, & Walker, 2008; Poole et al., 2018). ECF σ factor-mediated transcription contributes to the ability of rhizobia to cope with those challenges (Gourion et al., 2009; Ledermann, Bartsch, Muller, Wulser, & Fischer, 2018; Martinez-Salazar, Salazar, Encarnacion, Ramirez-Romero, & Rivera, 2009a; Sauviac, Philippe, Phok, & Bruand, 2007).

Across the bacterial world previous computational analyses identified more than 90 phylogenetically different ECF groups (Huang, Pinto, Fritz, & Mascher, 2015; Jogler et al., 2012; Staron et al., 2009). For many groups a conserved genomic context and consensus promoter motifs were predicted and some of them were verified (Rhodius et al., 2013; Staron et al., 2009). However, so far, ECF σs have been studied in detail only in a few model nitrogen-fixing α-rhizobia, namely *Rhizobium leguminosarum, Sinorhizobium meliloti, Bradyrhizobium japonicum* and *Rhizobium etli.* These bacteria are characterized by large genomes that contain several ECF σ-encoding genes (Galibert et al., 2001; Gonzalez et al., 2006; Kaneko et al., 2002; Young et al., 2006).

An overview of ECF σs, their relatedness and their classification into major ECF groups is shown for *R. leguminosarum, S. meliloti, B. japonicum* and *R. etli* (Fig. 3). More than half of these ECF groups (groups 15, 16, 26, 41, 42) are represented in all four species, which often possess two or even more members from one ECF group. ECF group 15 includes σ factors of the general stress response (GSR) in α-proteobacteria (Staron et al., 2009). These are the best-characterized ECF σs within the α-rhizobia (see sections below). More recent experimental data provide further insights into the function of σs from ECF groups 16, 33, 26 and for some unclassified ones (see sections below).

Here, we review the current knowledge on promoter specificity, regulatory control and biological function of ECF σs in members of the α-proteobacterial Rhizobiales, able to enter a root nodule symbiosis with their host plants.

2. Group ECF15 σs

ECF15 σs — also known as EcfG — are restricted to α-proteobacteria where they mediate the GSR enabling bacteria to cope with a number of different stresses (Francez-Charlot, Kaczmarczyk, Fischer, & Vorholt, 2015; Staron et al., 2009). In this taxonomic class, these ECF15 σs fulfill a function that is usually covered by other σ^{70} family members. In *E. coli* and related γ-proteobacteria σ^{S}, a group 2 σ factor, which is conserved in β- and δ-proteobacteria as well, is the master regulator of GSR (Battesti et al., 2011; Hengge, 2011). In Gram-positive bacteria such as *Bacillus subtilis* and other firmicutes, the group 3 σ factor σ^{B} is the main actor in this response (Hecker, Pane-Farre, & Volker, 2007; Price, 2011).

2.1 Core regulators of GSR in α-proteobacteria

2.1.1 EcfG σ factors

About half of the α-proteobacteria possess more than one *ecf* from the ECF15 group (see also Fig. 3). The primary *ecf* copies usually share a conserved genomic context reflecting their regulation via a unique signal transduction cascade. This genetic context includes genes coding for a NepR-like anti-σ factor, a PhyR-like two component response regulator and, frequently, a sensor histidine kinase, albeit the role of the latter is less well investigated (Francez-Charlot et al., 2015; Staron & Mascher, 2010)

(A)

	ECF11	ECF15	ECF16	ECF18	ECF20	ECF26	ECF29	ECF33	ECF41	ECF42	UG
S. meliloti		2	1			4	1		1	1	1
R. leguminosarum	1	1	1	1	1	4			1	2	3
R. etli	1	2	1		1	3	1		1	3	4
B. japonicum		1	2	1	1	3	1	2	1	3	2

(B)

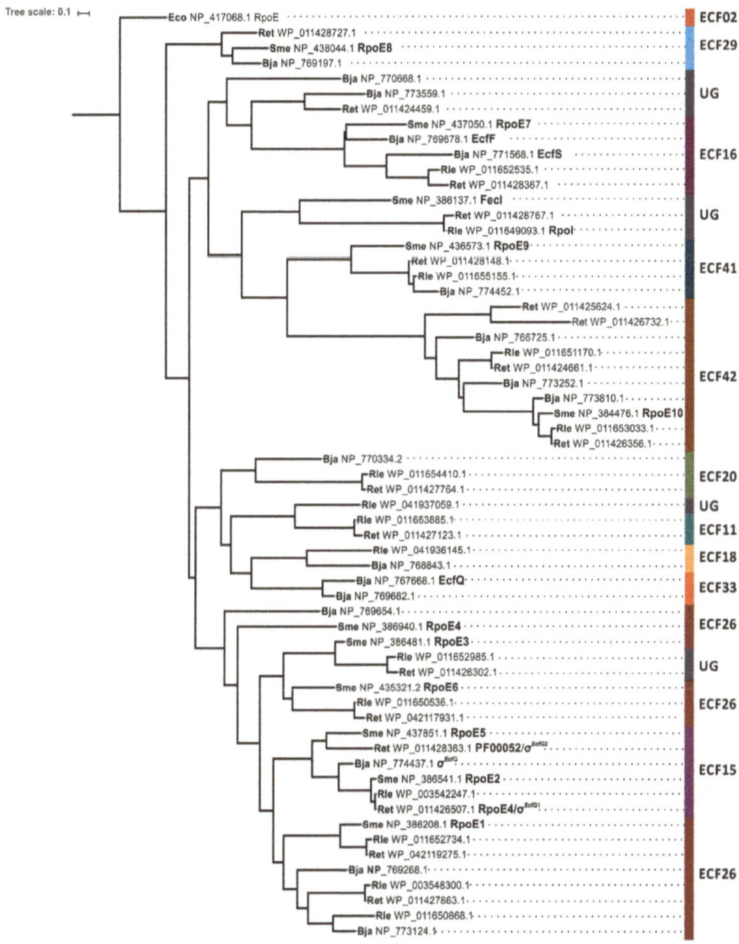

Fig. 3 (A) Assignment of ECF σ factors encoded by representative α-rhizobia to major ECF groups, UG: unclassified. (B) Similarity of ECF σ factors from selected species (see A). The classification into major ECF groups is shown on the right. The phylogenetic tree was built from the concatenated σ_2 and σ_4 domains of the corresponding ECFs. Data were generated by IQ-tree with 1000 ultrafast bootstraps (Nguyen, Schmidt, von Haeseler, & Minh, 2015) and visualized in iTOL (Letunic & Bork, 2019). RpoE from *E. coli* is a distantly related ECF used as outlier for rooting the phylogenetic tree. Eco: *E. coli*, Ret: *R. etli*, Sme: *S. meliloti*, Bja: *Bradyrhizobium japonicum*.

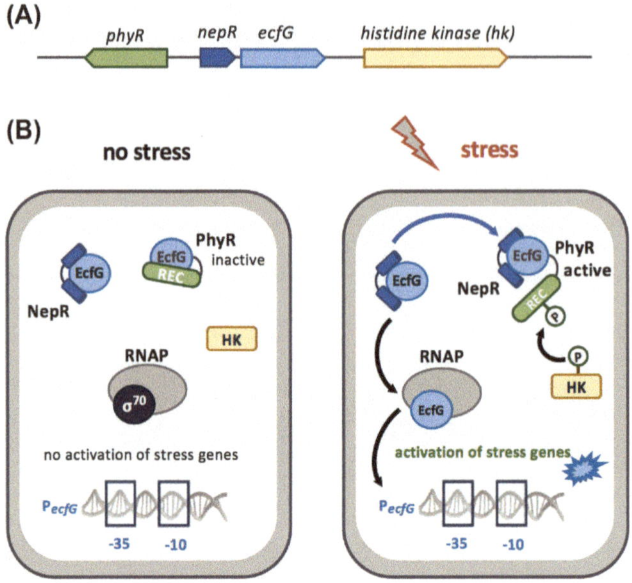

Fig. 4 (A) Schematic representation of a typical *phyR* locus. The *phyR* gene is divergently transcribed to the *nepR-ecfG* operon. Genes encoding histidine kinases (hk) are less conserved but can often be found in the vicinity. (B) Schematic representation of the basic GSR transduction cascade in α-proteobacteria. Under stress conditions, the PhyR receiver domain (REC) becomes phosphorylated by histidine kinases (HK) and sequesters NepR via its EcfG-like output domain. Released EcfG σ factor binds to the RNA polymerase (RNAP) instead of the primary σ^{70} and recruits the RNAP to GSR-related genes to activate their expression. See text for details.

(Fig. 4). Activation of EcfG σs enables the expression of hundreds of genes, some of which can be assigned to certain stress tolerances (Gourion et al., 2009; Jans et al., 2013; Lang et al., 2018). Also, ECF15 σ-dependent promoters were identified upstream of direct target genes in several α-proteobacterial species (Francez-Charlot, Frunzke, Zingg, Kaczmarczyk, & Vorholt, 2016; Gourion et al., 2009; Jans et al., 2013; Lang et al., 2018). The ECF15 group specific promoter model differs from those of other ECF groups or the primary σ^{70} (Rhodius et al., 2013; Staron et al., 2009).

2.1.2 NepR like anti-σ factors

The small cytoplasmic NepR like anti-σ factors are a unique feature of the ECF15 group and do not exhibit any similarities to canonical anti-σ factors (Francez-Charlot et al., 2009; Sauviac et al., 2007; Staron et al., 2009). Instead of a four helical anti-σ domain (ASD) directly connected to a signal sensory domain (Campbell et al., 2007), NepR like anti-σ factors feature a helix-loop-helix structure that extends over the surface of the σ_2 and σ_4 domains of EcfG (Campagne et al., 2012). Initial studies, especially in plant associated α-proteobacteria, have proven that theses minimalistic anti-σ factors sequester cognate ECF σs efficiently and prevent them from binding to the RNAP under non-stress conditions (Bastiat, Sauviac, & Bruand, 2010; Francez-Charlot et al., 2009; Gourion et al., 2009).

2.1.3 PhyR like response regulators

A PhyR response regulator, initially described in *Methylobacterium extorquens*, plays a crucial role for the GSR in α-proteobacteria. Besides its deficiency in plant colonization, a *phyR* deletion strain is more sensitive to various stresses, as induction of stress-related genes is impaired (Gourion et al., 2006, 2008). PhyR-like response regulators are conserved among α-proteobacteria. They contain a C-terminal receiver domain, which is suspected to be phosphorylated by cognate sensor histidine kinases, and a N-terminal output domain with sequence similarities to ECF σs (Galperin, 2006; Gourion et al., 2006, 2008). A function of PhyR as DNA-binding transcriptional regulator seemed to be unlikely, since the σ factor-like domain lacks promoter binding determinants in region 2.4, and region 4.2 is degenerated (Staron et al., 2009). In a pivotal study, PhyR was shown to bind the NepR anti-σ factor upon phosphorylation of a conserved aspartate residue in its receiver domain, thereby acting as anti-anti-σ factor in a partner switch that controls EcfG σ activity (Francez-Charlot et al., 2009).

2.2 The general transduction mechanism of GSR in α-proteobacteria

Based on genetic and biochemical experiments, a general transduction mechanism, in which NepR tightly binds to EcfG under non-stress conditions, was proposed. In response to stress, PhyR becomes phosphorylated and its ECF σ factor-like output domain interacts with NepR, thereby releasing EcfG. This in turn mediates transcription of the GSR associated regulon (Bastiat et al., 2010; Francez-Charlot et al., 2009; Gourion et al., 2009) (Fig. 4).

A high-resolution crystal structure of unphosphorylated PhyR from *Caulobacter crescentus* (Herrou, Foreman, Fiebig, & Crosson, 2010) and an NMR structure of NepR from *Sphingomonas* sp Fr.1 in complex with the constitutively active EcfG-like domain of PhyR (Campagne et al., 2012) provided further evidence for this model. The receiver domain of PhyR in its unphosphorylated form was shown to contact the σ factor-like output domain, thereby stabilizing a closed conformation that prevents binding of NepR. Conformational changes upon phosphorylation of the PhyR receiver domain were supposed to destabilize these intramolecular interactions, thereby allowing an association of the PhyR output domain with the anti-σ factor NepR (Herrou et al., 2010). A requirement for a functional partner switch under stress conditions would be a higher affinity of NepR for the EcfG-like output domain of PhyR than for EcfG itself. Indeed, homology modeling indicated that NepR binds EcfG with the same binding determinants but likely with lower affinity due the absence of specific polar contacts and a lower density of hydrophobic interactions (Campagne et al., 2012).

Compared to the conserved core regulators, little is known about histidine kinases that mediate phosphorylation of PhyR-like response regulators upon stress. *Cis*-encoded kinases, frequently encoded by genes in the vicinity of *phyR* loci, were suspected to fulfill such a function (Gourion et al., 2008; Staron & Mascher, 2010). These cytoplasmic or periplasmic sensing kinases lack a common domain of dimerization and histidine transfer but instead harbor a Pfam: HWE_HK or a Pfam: His_KA_2 domain (Staron & Mascher, 2010). As suggested by Sauviac & Bruand, (2014), we will refer to these kinases as Hw- and Hk-type kinases. Recent studies confirmed their function, either as kinase or phosphatase or as a bifunctional enzyme, in mediating the phosphorylation state of PhyR – similar to the role of response regulators (Kaczmarczyk et al., 2011; Kim, Caswell, Foreman, Roop, & Crosson, 2013; Lourenco, Kohler, & Gomes, 2011; Sauviac & Bruand, 2014). But how exactly they perceive activating signals from the environment still remains enigmatic.

2.3 Variations of the GSR transduction mechanism in nitrogen-fixing α-rhizobia

Even though the general GSR activating mechanism is conserved among α-proteobacteria, there are species-specific variations regarding the regulatory pathway in terms of environmental triggers and the number and contribution of participating core regulators. This likely reflects the different needs under which the response is crucial for a certain organism and may provide advantages for its survival in a given ecological niche. Even among closely related nitrogen-fixing plant symbionts like *S. meliloti*, *R. etli* and *B. japonicum*, striking differences in the GSR transduction cascade exist (Fig. 5).

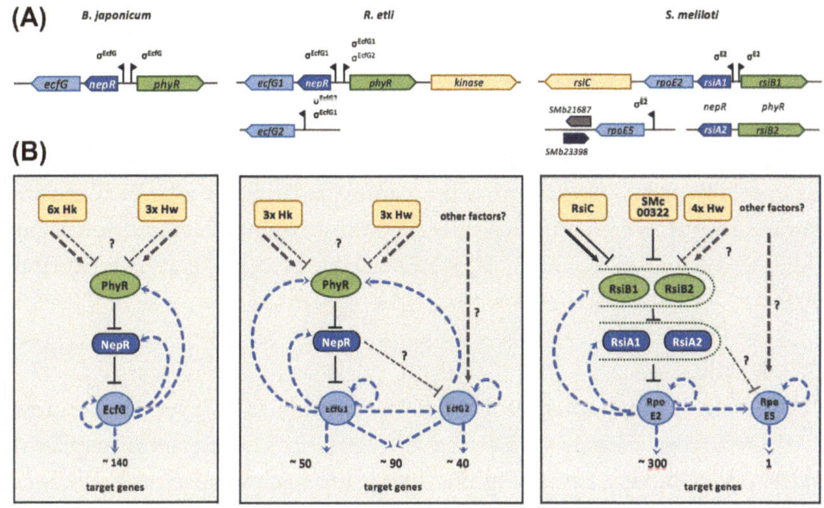

Fig. 5 Organization of GSR loci and regulatory control of GSR in *B. japonicum*, *R. etli* and *S. meliloti*. (A) Genes encoding *phyR*-like response regulators, *nepR*-like anti-σ factors, *ecfG* σ factors and *phyR*-associated kinases are shown. The *S. meliloti SMb23398* gene product has similarities to *RsiA1/RsiA2* and was therefore included. (B) schematic representation of the regulatory cascade underlying the GSR. Black arrows/T-lines represent experimentally confirmed activating/inhibiting relationships. Dotted arrows/T-lines associated with a question mark are hypothetical. Fine dotted arrows indicate transcriptional activation. See text for further details.

2.3.1 B. japonicum

At first glance, *B. japonicum* has the most minimalistic set of GSR regulators, since *ecfG*, *phyR* and *nepR* genes occur only once in the genome. The *ecfG* locus is organized as in most α-proteobacteria, where the *phyR* gene is divergently transcribed to the *nepR-ecfG* operon in an EcfG-dependent way (Gourion et al., 2009). *B. japonicum* can draw on a large set of sensor histidine kinases of the Hk-type (6) and Hw-type (3), even though the encoding genes are not located in the vicinity of *phyR* (Gourion et al., 2009; Staron & Mascher, 2010). So far, the role of these histidine kinases has not been experimentally addressed. It is therefore unclear, how PhyR becomes phosphorylated under stress conditions. Apart from this, the regulatory mechanism underlying the GSR strongly resembles the general path as described in the above section.

Under carbon starvation, 116 and 138 genes were significantly downregulated in the *phyR* and the *ecfG* mutant strains, respectively (Gourion et al., 2009). Not surprisingly, the PhyR and the EcfG regulons show a high degree of overlap, including 112 genes. More than 75 % of these encode hypothetical or conserved hypothetical proteins of unknown function. Among the annotated genes are at least ten that encode putative regulatory proteins (Gourion et al., 2009). Thus, EcfG is likely on top of a regulatory cascade, meaning that a proportion of the GSR regulon in *B. japonicum* is indirectly controlled by this ECF. This is a common theme in the GSR also in *R. etli* and *S. meliloti,* providing an explanation of why EcfG related promoter motifs were identified only for a subset of the regulated genes (Jans et al., 2013; Lang et al., 2018; Martinez-Salazar et al., 2009a; Sauviac et al., 2007).

2.3.2 R. etli

In contrast to *B. japonicum, R. etli* harbors two genes assigned to the group ECF15 σ factors, namely *rpoE4* (also known as *ecfG1*) and the megaplasmid (p42f)-localized copy *ecfG2* (Fig. 5). The primary chromosomal GSR locus comprises *rpoE4* and *phyR*- and *nepR*-like genes showing the typical organization (Jans et al. 2013; Martinez-Salazar et al., 2009a). Downstream of *phyR,* a Hw-type histidine kinase gene was identified (Staron & Mascher, 2010). In total, *R. etli* possesses three Hw- and three Hk-type histidine kinases, but so far, no functional roles in the GSR were assigned to them (Staron & Mascher, 2010).

The two group ECF15 σs seem to act rather in parallel than in series (Jans et al., 2013). PhyR, NepR and EcfG1 are at the core of the GSR while EcfG2 has more accessory functions. These conclusions were drawn based on several observations. First, EcfG1 strongly activates the expression of *phyR,* while EcfG2 contributes only modestly. Second, EcfG1-mediated promoter activation is strongly reduced in absence of *phyR*, an effect which is less severe for EcfG2-mediated regulation. Third, EcfG1 and EcfG2 are both autoregulatory, however, only EcfG1 is able to cross-activate the non-cognate *ecf15*-operon. Since significant *ecfG2* expression is detectable in the absence of *phyR* and *ecfG1,* EcfG2 is likely involved in GSR-independent stress resistance pathways as well (Jans et al., 2013). Together, EcfG1 and EcfG2 control more than 100 genes. In accordance with the findings detailed above, their regulons comprise both, an unique and an overlapping set of genes encoding for example cell envelope proteins, proteins involved in stress responses, as well as small regulatory RNAs with unassigned function (Jans et al., 2013; Martinez-Salazar et al., 2009a).

In *R. etli,* the regulation of the GSR response is further influenced by another factor. In the stationary growth phase, *ecfG1/G2* expression was shown to be positively controlled by the alarmone (p)ppGpp. A similar dependency was observed for many genes of the EcfG1 regulon, including those involved in cell envelope biosynthesis (Vercruysse et al., 2011). This is consistent with findings in *E. coli*, where the alarmone promotes the transcription from promoters that depend on alternative σ factors, including RpoS (Magnusson, Farewell, & Nystrom, 2005). However, a comparable role of (p)ppGpp in the GSR of *S. meliloti* could not be confirmed (Krol & Becker, 2011).

2.3.3 S. meliloti

Apart from a canonical *GSR* locus, including *rpoE2,* the *nepR*-like gene *rsiA1* and the *phyR*-like gene *rsiB1*, the *S. meliloti* genome contains another *ecfG* gene on megaplasmid pSymB, namely *rpoE5*, as well as paralogues of *rsiA1* and *rsiB1*, known as *rsiA2* and *rsiB2*, respectively (Fig. 5). Thus, *S. meliloti* reveals the highest complexity regarding the encoded GSR core components. Furthermore, the organism can draw on a set of six Hw-like histidine kinases (Staron & Mascher, 2010).

RpoE2 is the primary EcfG σ factor in *S. meliloti* and controls a large regulon of up to 300 genes. Some of these mediate tolerances to distinct stresses (see next section), other genes encode regulatory proteins or small RNAs and there is a proportion of genes with unassigned function (Lang et al., 2018; Sauviac et al., 2007; Schlüter et al., 2013). The presence of two EcfG σ factors and the fact that *rpoE5* is part of the RpoE2 regulon strongly resemble the situation in *R. etli* at first glance (Sauviac et al., 2007). However, RpoE5 is neither autoregulatory, nor does it control a considerable regulon, since only one gene with unassigned function was activated upon *rpoE5* overexpression (Lang et al., 2018). Nevertheless, one may speculate that RpoE5 significantly contributes to the GSR under yet unknown conditions. Remarkably, Lang et al. (2018) noted an ORF downstream of *rpoE5* on the opposite strand whose product shares some similarities with RsiA1/RsiA2.

The presence of additional *phyR* and *nepR* copies in the genome of α-proteobacteria is rather unusual (Staron & Mascher, 2010). In *S. meliloti*, both PhyR and NepR paralogous were proven to participate in the partner-switching mechanisms (Bastiat et al., 2010). Since *rsiA2* and *rsiB2* are not subject to positive autoregulation by RpoE2 one may suggest an unequal contribution to the GSR in *S. meliloti* (Lang et al., 2018; Sauviac et al., 2007). Accordantly, while deletion of the *nepR*-like *rsiA1* gene is lethal, likely because of uncontrolled overactivation of RpoE2 and its targets, cells can survive without an *rsiA2* copy (Bastiat et al., 2010). Nevertheless, redundant PhyR and NepR paralogues may allow *S. meliloti* to modulate the GSR response depending on the integration of different stimuli.

In contrast to *R. etli* and *B. japonicum*, the role of histidine kinases in the GSR has been experimentally addressed in *S. meliloti* (Sauviac & Bruand, 2014). The cis-encoded kinase RsiC was shown to act preferentially as negative regulator of RpoE2 activity by decreasing RsiB1/RsiB2 phosphorylation *in vivo*. However, acting in concert with the Hw-like kinase SMc0322, RsiC positively affected RpoE2 activity under heat stress, likely by phosphorylation of RsiB1/RsiB2 (Sauviac & Bruand, 2014). Since *S. meliloti* can draw on four additional HW-type kinases, one may speculate that these trigger the GSR cascade under yet unknown conditions (Staron & Mascher, 2010).

Additionally, Hfq, known to act as an RNA chaperone that facilitates binding between trans-encoded sRNAs and their mRNA-targets, was shown to positively affect the expression of *rpoE2* and some target genes in the stationary growth phase (Barra-Bily et al., 2010). Thus, it is tempting to speculate that sRNAs may promote the accumulation of RpoE2 in an Hfq-dependent manner - as was shown for RpoS in *E. coli* (Repoila, Majdalani, & Gottesman, 2003).

2.4 Environmental triggers of GSR and biological functions

The GSR is a global but rather unspecific response to unfavorable conditions. In the laboratory, the GSR has been analyzed after cell exposure to various stress conditions including, heat, pH changes, salt, osmotic or oxidative stress, desiccation, starvation and stationary phase growth. These are common stresses nitrogen-fixing bacteria encounter either in the soil or upon entry of a root nodule symbiosis with their hosts. Relevant triggers of the GSR in *R. etli, S. meliloti* and *B. japonicum* and its significance - as deduced from mutant phenotypes - will be outlined in this section.

2.4.1 B. japonicum

Carbon starvation is an efficient trigger of the GSR in *B. japonicum* and was applied to identify PhyR and EcfG regulons. An increased sensitivity of *phyR* and *ecfG* mutants toward heat and desiccation has only been detected after carbon starvation but not for exponentially growing cells (Gourion et al., 2009). The observation that cross-resistances are obtained upon starvation or stationary phase growth can be traced back to GSR activation. Accordantly, stationary phase bacteria of the plant symbiont *R. leguminosarum* are more resistant to heat, salt or acidic stress than exponentially growing cells (Thorne & Williams, 1997).

Compared to *S. meliloti* and *R. etli*, the EcfG regulon of *B. japonicum* comprises the highest proportion of genes encoding hypothetical proteins or proteins of unknown functions. It is therefore difficult to make predictions regarding their contribution to certain stress resistances. However, *B. japonicum* is the only nitrogen-fixing α-rhizobium studied so far whose GSR mutants displayed symbiotic defects, even though the EcfG homologs RpoE2 and EcfG1 were suspected to be active in *S. meliloti* and *R. etli* during symbiosis as well (Martinez-Salazar et al., 2009a; Sauviac et al., 2007). Compared to the wild-type, *B. japonicum phyR* and *ecfG* mutants induced fewer nodules with aberrant morphologies on *Glycine max* (soybean) and *Vigna radiata* (mungbeans), and showed delayed nitrogen fixation (Gourion et al., 2009). In a recent study, EcfG activity was detected early during symbiotic interactions, especially when bacteria get entrapped in curled root hairs and invade roots via infection threads (Ledermann et al., 2018). Consistently, the formation of infection threads by *ecfG* mutants was impaired (Ledermann et al., 2018). How exactly the GSR is triggered during early stages of symbiosis remains still enigmatic.

2.4.2 R. etli

In *R. etli*, various stimuli were confirmed to trigger the GSR response, including oxidative, saline, heat and osmotic stresses as well as stationary phase growth (Martinez-Salazar et al., 2009a; Vercruysse et al., 2011). The *ecfG1* and *ecfG2* mutants display different phenotypes upon exposure to certain stresses. Both mutants were more sensitive toward oxidative stress than the wild-type strain, though, this was more severe upon deletion of *ecfG2* (Vercruysse et al., 2011). This likely correlates with the EcfG2-dependent expression of *katG*, a gene known to be important for oxidative stress resistance in *R. etli* (Garcia-de Los Santos et al., 2008; Jans et al., 2013; Vargas Mdel et al., 2003). An increased sensitivity of the *ecfG1* but not to the *ecfG2* mutant strain toward osmotic stress and elevated temperatures was reported (Vercruysse et al., 2011). Accordantly, EcfG1 promotes transcription of the heat shock σ factor gene *rpoH2*, which is involved in osmotic tolerance and contributes to heat resistance (Jans et al., 2013; Martinez-Salazar et al., 2009b).

2.4.3 S. meliloti

In a first comprehensive study on the GSR in *S. meliloti*, RpoE2 was found to be activated upon salt and osmotic stress, heat treatment or entry into the stationary growth phase following carbon or nitrogen starvation (Sauviac et al., 2007). As in *R. etli*, genes encoding a heat shock σ factor (*rpoH2*) and a catalase (*katC*) were initially identified to be part of the GSR regulon (Jans et al., 2013; Martinez-Salazar et al., 2009a; Sauviac et al., 2007). Accordingly, *rpoE2* and *katC* mutant strains were found to be more sensitive toward H_2O_2 exposure in the stationary growth phase (Barra-Bily et al., 2010; Flechard, Fontenelle, Trautwetter, Ermel, & Blanco, 2009). In *S. meliloti*, RpoH2 activity seems to be less important for heat-shock tolerance but rather contributes to stationary phase growth and osmotic tolerance. This may account for the observed growth defect of an *rpoE2* mutant in hyperosmotic medium (Barnett, Bittner, Toman, Oke, & Long, 2012; Flechard et al., 2010; Oke, Rushing, Fisher, Moghadam-Tabrizi, & Long, 2001). However, RpoE2-dependent expression of *otsA*, a gene involved in synthesis of the osmoprotective compound trehalose, likely plays a major role in osmotic adaptation (Flechard et al., 2010). Reduced trehalose levels in an *rpoE2* mutant strain may also explain its increased sensitivity to desiccation (Flechard et al., 2010; Humann, Ziemkiewicz, Yurgel, & Kahn, 2009), as it has been reported for an *R. leguminosarum* mutant strain lacking several trehalose synthesis genes (McIntyre et al., 2007).

3. Group ECF26 σs

ECF26 σs form a large group and appear almost exclusively in proteobacteria (Staron et al., 2009). After the ECF15 σs, they are the second most common σ factors in the members of the Rhizobiales, and co-occurrence of several different ECF26 σs in one species is widespread (D. Pastor and G. Fritz, unpublished data). This is also true for *S. meliloti*, *R. leguminosarum*, *B. japonicum* and *R. etli* (Fig. 3). But so far, ECF26 σs have not been characterized in great detail.

3.1 Regulons and promoter specificity of ECF26 σs in *S. meliloti*

Studies of ECF26 σs in Rhizobiales are restricted to the four representatives in *S. meliloti* (RpoE1, RpoE3, RpoE4 and RpoE6). Recently, target genes were mapped for each of them. While RpoE1, RpoE3 and RpoE4 control a rather small number of 10 or fewer genes, RpoE6 mediates expression of about 40 genes (Lang et al., 2018). The regulons of RpoE1, RpoE3, RpoE4 and RpoE6 partially overlap and RpoE6 activates a substantial number of RpoE2 targets, albeit to much lesser extent than does RpoE2 (Lang et al., 2018). This might be explained by the relatedness of ECF15 and

ECF26 σs and a significant overlap of group specific consensus promoter motifs (Rhodius et al., 2013; Staron et al., 2009).

A closer look at the genes most strongly activated by the ECF26 σs in *S. meliloti* revealed sub-specificities, suggesting only little functional overlap in spite of their overall similarity (Fig. 6A). Even though some cross-activations occur, RpoE1, RpoE4 and RpoE6 revealed a strong preference for promoters of their own operon and/or for those of neighboring genes (Fig. 6B). RpoE3, however, is an exception since *rpoE3* transcription is neither autoregulated nor are any of the identified target genes in close proximity to the *ecf* locus. Accordantly, a promoter that is rather dependent on RpoD was found upstream of the *rpoE3* operon (Lang et al., 2018). Inspection of the most prominent target promoters suggests that only a few differences in their sequences may account for the preferred activation by one of the ECF26 σ factors (Fig. 6C). For example, the nature of a sole flipped out base in the -10 region is critical for promoter recognition and transcription activation by ECF σs (Campagne, Marsh, Capitani, Vorholt, & Allain, 2014). Furthermore, ECF σs are known to tolerate a rather narrow spacer length between the -35 and -10 promoter motifs (Rhodius et al., 2013). Consistently, the strongest RpoE6 target promoters are characterized by a shortened spacer. This exemplifies that transcriptional insulation of signaling pathways can easily be achieved even between closely related ECF σs of one phylogenetic group.

3.2 Activation and biological function of ECF26 σs in *S. meliloti*

All ECF26 σ factors in *S. meliloti* are associated with a canonical anti-σ factor encompassing an N-terminal ASD domain, a transmembrane domain, and a C-terminal sensory cog5662 domain, which has some similarity to that of RsiW in *B. subtilis* (Marchler-Bauer et al., 2017; Staron et al., 2009). However, this does not necessarily point toward sensing of the same stimuli. RpoE1, RpoE3 and RpoE4 were suggested to be active during stationary growth phase, since an association with the RNAP was proven under these conditions (Bastiat, Sauviac, Picheraux, Rossignol, & Bruand, 2012). Furthermore, Hfq promotes expression of *rpoE1, rpoE3, rpoE4* but not *rpoE6* (Barra-Bily et al., 2010). Conditions under which RpoE6 is active could not be defined so far.

Activation of RpoE4 and RpoE1 during the stationary growth phase may be traced back to the accumulation of sulfite (Bastiat et al., 2012). However, it is unknown how this is exactly perceived and transduced by the

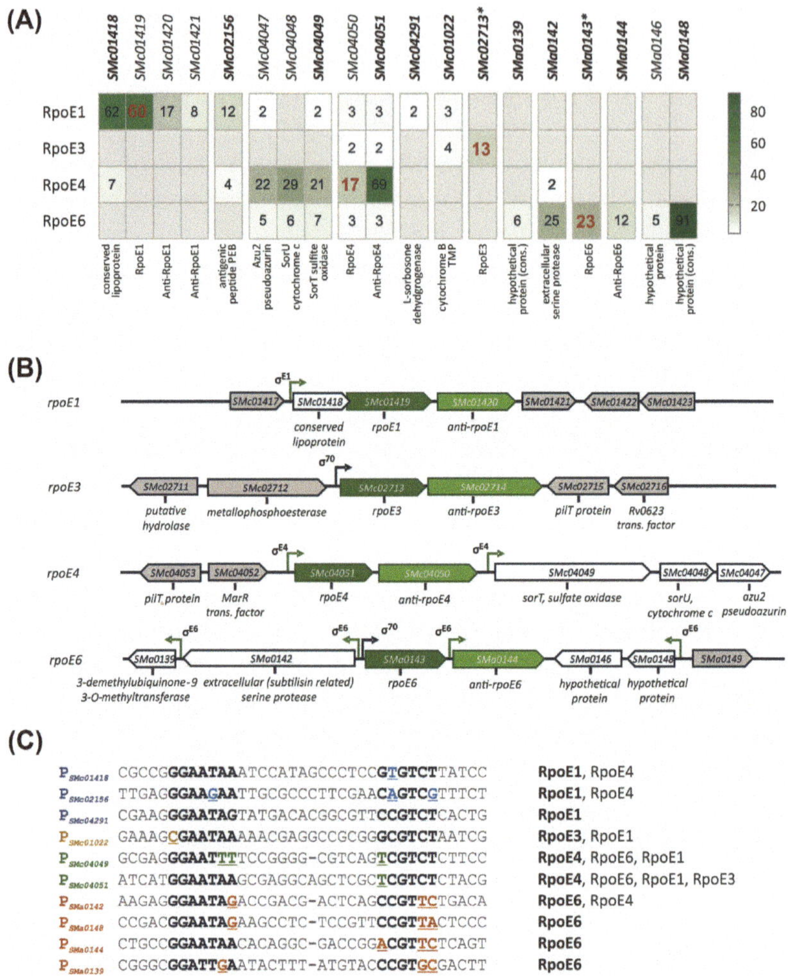

Fig. 6 ECF26 σs in *S. meliloti* – promoter activation and genomic context. (A) Overview of the highest expressed target genes of ECF26 σs in *S. meliloti*. The analysis is based on Affymetrix transcriptome data that were compiled from Lang et al. (2018). Numbers represent fold induction of gene expression. Operons are separated from each other by a gap. Gene names in bold letters refer to the presence of an upstream ECF26-dependent promoter motif (Rhodius et al., 2013). In case of *Smc02713* (*rpoE3*) and *SMa0143* (*rpoE6*) (labeled with a *) an RpoD promoter motif was assigned. Enlarged numbers within the heat map indicate ectopic overexpression of *ecf* genes (Lang et al., 2018). (B) Genomic context of *rpoE1*, *rpoE3*, *rpoE4* and *rpoE6* in *S. meliloti*. ECF σ factor target genes apart from the *ecf* operon are shown in white. Arrows represent location and orientation of putative promoters (Lang et al., 2018). (C) Sequence of mapped ECF26-dependent promoters and their preferential activation by RpoE1, RpoE3, RpoE4, RpoE5. Underlined residues within the -35/-10 promoter motifs (bold letters) correlate with a preferred activation by a certain ECF26 σ factor.

cognate anti-σ factors. RpoE4 and RpoE1 and in turn their main target genes (*sorT, rpoE4, SMc04050* and *SMc01418, rpoE1, SMc02156,* respectively) become activated in the presence of taurine as sole carbon source (Bastiat et al., 2012). Taurine is an organic sulfur compound and many soil bacteria are able to utilize it as carbon, nitrogen or sulfur source (Bruggemann, Denger, Cook, & Ruff, 2004; Cook & Denger, 2006; Kertesz, 2000), which is likely advantageous for survival. Upon taurine catabolism, sulfite, a highly reactive and therefore cytotoxic compound, accumulates (Cook & Denger, 2006; Feng, Tollin, & Enemark, 2007; Kappler, 2011; Kappler, Friedrich, Truper, & Dahl, 2001; Kertesz, 2000). Expression of *sorT*, encoding a periplasmic sulfite-oxidizing enzyme responsible for the conversion of sulfite, mainly depends on RpoE4 (Bastiat et al., 2012; Lang et al., 2018; Wilson & Kappler, 2009). Beside a detoxification function it was hypothesized that SorT contributes to energy production during the stationary growth phase (Bastiat et al., 2012). Indeed, *sorT* is co-transcribed with two genes encoding a cytochrome *c* (Smc04048, SorU) and a pseudoazurin (Smc04047, Azu2), which may function as electron acceptors (Wilson & Kappler, 2009). A recent study showed that SorT-mediated sulfite oxidation is coupled to its cognate cytochrome *c*, which likely transfers electrons to the respiratory chain (Low, Ryan Kilmartin, Paul, & Ulrike, 2011; McGrath et al., 2015).

Considering that the *sorT* operon is the strongest RpoE4 target, it is not surprising that growth of *rpoE4* and *sorT* mutant strains was severely affected when taurine was provided as sole carbon source. Even though independent studies revealed a contribution of RpoE1 to the expression of the *sorT* operon, this was not sufficient to complement the observed *rpoE4* mutant phenotype (Bastiat et al., 2012; Lang et al., 2018).

A functional role for RpoE3 and RpoE6 has so far not been defined.

3.3 Genomic context conservation of ECF26 σ factors

The broad co-occurrence of ECF26 σ factors in α-rhizobia motivated us to further inspect the genomic context of ECF26 σ factor encoding genes for conservation based on the four studied examples in *S. meliloti.* A renewed classification of ECF σs including a higher number of ECF sequences (D. Pastor and G. Fritz, unpublished data), allows a more detailed evaluation.

The *rpoE1* locus matches a common genomic arrangement characterized by a lipoprotein encoding gene upstream of the *ecf* gene (Staron et al., 2009) (Fig. 6B). According to the renewed classification, this syntenic organization was found in 20% of bacteria harboring ECF26 members and is most common within the Rhizobiacea. A similar locus organization of one *ecf26* gene was also found in *R. leguminosarum* (RL_RS04130/WP_011650536.1) and in *R. etli* (RHE_RS14220/WP_042119275.1). However, the functional relevance of the encoded lipoprotein remains elusive. Only a small set of *ecf26* genes (6%) including *rpoE3*, feature an upstream-encoded metallophosphoesterase gene. A similar arrangement occurs at one *ecf26* locus in *B. japonicum* (bll3014/NE_769,654.1) and in case of two ungrouped *ecf26*-like *genes* in *R. leguminosarum* (RL_RS18070/WP_011652985.1) and *R. etli* (RHE_RS03775/WP_011426302.1).

Since RpoE4 and to some extent RpoE1 contribute to sulfite detoxification and respiration by activating the *sorT-sorU-azu2* operon downstream of *rpoE4* (Bastiat et al., 2012; Low et al., 2011; McGrath et al., 2015; Wilson & Kappler, 2009), we carefully analyzed the distribution of this genomic context. However, this whole syntenic region was not conserved beyond the *Sinorhizobium* genus. It should be noted that in *Starkeya novella,* the taurine inducible *sorAB* genes, which encode a soluble sulfite:cytochrome *c* oxireductase representing another sulfite oxidizing system, are in close vicinity to the *ecf26* gene *sigE* (Kappler et al., 2001). The presence of a typical ECF26-type promoter upstream of *sorA* suggests its SigE-dependent expression. Therefore, one may hypothesize that ECF-mediated sulfite sensing is not a rare event among ECF26 σs.

The *rpoE6* operon is preceded by a protease encoding gene of the subtilase family. This genomic arrangement was found in 20% of bacteria harboring ECF26 members, but was not observed in *R. etli, R. leguminosarum* or *B. japonicum.*

Overall, *ecf26* σ factor genes show a high variability regarding their genomic context (Pinto & Mascher, 2016, D. Pastor and G. Fritz, unpublished data; Staron et al., 2009). More efforts are necessary to experimentally address their physiological role in α-rhizobia.

4. Group ECF16 and group ECF33 σ factors

Both ECF groups are restricted to proteobacteria and only a few members have been characterized to date (Staron et al., 2009). Among α-rhizobia, a functional role of ECF16 and ECF33 σs was only defined in *B. japonicum.* Its genome contains two *ecf16* and two *ecf33* genes (Fig. 3). However, only one representative from each group, namely *ecfF* (ECF16) end *ecfQ* (ECF33), was induced in response to reactive oxygen species

(Masloboeva et al., 2012). The second *ecf16* gene (*ecfS*) was further shown to be required for establishing a functional symbiosis with soybean (Stockwell, Reutimann, & Guerinot, 2012). In *S. meliloti*, the regulon of the ECF16 σ factor RpoE7 was determined, but a functional role could not be inferred (Lang et al., 2018).

4.1 Regulatory control of ECF16 and ECF33 σs and activation of target genes

4.1.1 ECF33 σ factors

In *B. japonicum*, the EcfQ regulon comprises ~35 genes, among them 27 positively controlled ones. Most of these genes encode hypothetical or functionally unknown proteins. Despite the relatively large number of regulated genes, it was not possible to determine a common promoter motif that might be recognized by EcfQ (Masloboeva et al., 2012), suggesting that many of these genes are indirectly controlled by EcfQ. Furthermore, *ecfQ* has some unusual features. It is neither autoregulated nor is it co-transcribed with a cognate *anti-σ* gene (Masloboeva et al., 2012). In contrast, the second *ecf33* copy (*blr3042 or ecf33_375*, see Fig. 3) has a canonical *anti-σ* gene in its proximity and the encoded *σ* factor likely controls its own transcription, as it was demonstrated in the heterologous host *E. coli* (Rhodius et al., 2013). Harboring two ECF33 σ factor encoding genes, one with associated anti-σ factor gene and one without such gene, was found to be the predominant situation in species using this ECF σ factor group (Masloboeva et al., 2012; Staron et al., 2009). However, based on the findings of Masloboeva et al. (2012), it is rather unlikely that the two ECF33 σ factors act in concert. Expression of *ecfQ* and related *ecf33* genes lacking a cognate anti-σ factor gene was supposed to be dependent on a yet undefined transcription factor, since conserved sequence motifs are present in the promoter regions (Masloboeva et al., 2012).

4.1.2 ECF16 σ factors

ECF16 σ factors feature an unusual type of anti-σ factor characterized by a conserved huge membrane anchored domain of unknown function (DUF1109) (Staron et al., 2009). Biochemical and genetic experiments in *C. crescentus* and *B. japonicum* confirmed a negative regulatory role for the anti-σ factor, likely through direct interaction with the cognate ECF σ factor. Two conserved cysteine-residues are important for the anti-σ factor activity, presumably for sensing and transducing defined stress signals (Kohler, Lourenco, Avelar, & Gomes, 2012; Masloboeva et al., 2012).

In *B. japonicum*, EcfF and EcfS target gene candidates have been identified in a microarray-based transcriptome study of strains that either lack or overexpress the cognate anti-σ factor encoding genes *osrA* and *tmrS*. Masloboeva et al. (2012) found a set of nine overlapping genes that were downregulated in an *ecfF-osrA* mutant strain under stress conditions and upregulated in an *osrA* mutant strain presumably displaying elevated levels of EcfF activity. Except for one, all genes were also activated in the wild-type strain upon H_2O_2 exposure and three of them encode methionine sulfoxide reductases (Masloboeva et al., 2012). The putative EcfS regulon was determined in a similar way. The lists of down- and upregulated genes in an *ecfS* mutant and a *tmrS* overexpression strain shared 40 genes. Most of them have no functions assigned or encode transport proteins (Stockwell et al., 2012). In *S. meliloti*, the RpoE7 regulon is rather small and comprises six target genes. The four most strongly RpoE7-regulated genes are situated in vicinity of the *rpoE7* operon and encode conserved hypothetical proteins (Lang et al., 2018).

While RpoE7 and EcfF activate their own promoters, this was not proven for EcfS. Consensus promoter motifs were identified upstream of *rpoE7* and *ecfF*, but not preceding *ecfS*, and these motifs share striking similarities with the ECF16 group specific motif (Masloboeva et al., 2012; Rhodius et al., 2013; Staron et al., 2009; Stockwell et al., 2012).

4.2 Biological function of ECF16 and ECF33 σ factors in α-rhizobia

In *B. japonicum*, group 16 EcfF and group 33 EcfQ contribute to oxidative stress tolerance as confirmed by the observation that mutants lacking either of the encoding genes were more sensitive to singlet oxygen in their free-living lifestyle (Masloboeva et al., 2012). Rhizobia encounter reactive oxygen species not only intrinsically in aerobic metabolism but also during symbiosis (Pauly et al., 2006). However, *ecfF* and *ecfQ* are not crucial for a functional symbiosis with soybean. Notably, the EcfF and EcfQ regulons show little overlap, suggesting redundant but independent protective systems against oxidative stress in *B. japonicum* (Masloboeva et al., 2012).

EcfS represents the second ECF16 member in *B. japonicum* and is necessary for establishing a proficient symbiosis. Soybean plants inoculated with the *ecfS* mutant strain formed small nodules unable to fix nitrogen, likely because the bacteria are unable to invade nodules (Stockwell et al., 2012). But how exactly and at which stage during symbiosis EcfS activity is triggered remains elusive.

To date, responsiveness of ECF16 σ factors to oxidative stress has been confirmed in *B. japonicum* (EcfF) and in *C. crescentus* (σ^F) (Alvarez-Martinez, Baldini, & Gomes, 2006; Masloboeva et al., 2012). Nevertheless, members of this ECF group likely have broader functions. In *C. crescentus*, σ^F is further involved in the response to heavy metal stress (Kohler et al., 2012). And even though a functional role for RpoE7 in *S. meliloti* was not defined, *rpoE7* seems to be dispensable for protection against heavy metals and reactive oxygen species as well as for symbiosis (Lang et al., 2018). This suggests another role in stress response.

5. ECF σs involved in iron uptake

FecI like proteins were assigned to six different ECF groups (5–10) and most of them are found in proteobacteria where they regulate iron acquisition (Staron et al., 2009). The best studied member is probably FecI from *E. coli*. It directs transcription of a large operon encoding proteins of a transport machinery that mediates the uptake of iron as siderophore complex (Braun & Mahren, 2005; Brooks & Buchanan, 2008). Regarding α-rhizobia, experimental data are available for two ECF σs that fulfill a role in the uptake of iron: RpoI from *R. leguminosarum* and FecI from *S. meliloti* (Lang et al., 2018; Staron et al., 2009; Yeoman et al., 1999, 2003). Due to a lack of sequence similarity these ECF σs remained ungrouped (Staron et al., 2009). However, they show the highest sequence similarities to ECF09 σs (Lang et al., 2018; Staron et al., 2009; Visca, Leoni, Wilson, & Lamont, 2002).

PvdS from *Pseudomonas aeruginosa* is a well-studied ECF09 group σ factor. This protein is involved in siderophore biosynthesis but not directly in the uptake of iron (Visca et al., 2002). *R. leguminosarum* produces the siderophore vicibactin. Genes of four different operons are necessary for its biosynthesis (Carter et al., 2002). RpoI drives the expression of two of these operons, *vbsADL* and *vbsGSO*, which are localized in vicinity of the *rpoI* gene (Yeoman et al., 1999, 2003). Thus, this ECF σ factor fulfills a similar function as PvdS in *P. aeruginosa*. The *rpoI* gene is neither autoregulatory, a common theme of genes coding for FecI like ECF σs, nor is it co-transcribed with an anti-σ factor gene similar to other *ecf09* σ factor genes (Staron et al., 2009; Yeoman et al., 2003). However, *rpoI* expression is induced in iron-depleted medium and depends on the transcriptional regulator RirA (Yeoman, Curson, Todd, Sawers, & Johnston, 2004). RirA acts as repressor that binds to IRO (iron-responsive operator) sequences in the presence of high iron concentrations, while under low iron conditions it has a much lower affinity to DNA and likely dissociates (Pellicer Martinez et al., 2017; Yeoman et al., 2004). RirA is conserved among rhizobia and fulfills a function that is usually accomplished by FurA, a widely distributed iron-response regulator in Gram-negative bacteria (Rudolph, Hennecke, & Fischer, 2006).

Despite the high sequence similarity, FecI from *S. meliloti* differs from ECF09 σs in that a gene encoding a putative FecR-like anti-σ factor is located downstream of the *fecI* gene (Staron et al., 2009). Recently, *SMc4206*, encoding a putative hemolysin-type calcium binding protein, was identified as one of two FecI regulated genes whose expression is induced under iron-limiting conditions in an RirA-dependent manner (Chao, Buhrmester, Hansmeier, Puhler, & Weidner, 2005; Lang et al., 2018). This might favor a putative role of FecI in the iron metabolism of *S. meliloti* – however, *fecI* expression itself is neither RirA-dependent nor is it activated in iron-depleted media (Chao et al., 2005).

5.1 Other ECF σ groups present in species of selected nitrogen-fixing rhizobia

The selected α-rhizobia highlighted in this chapter also feature a range of ECFs from phylogenetic groups (ECF groups 11, 18, 20, 29, 41, 42 see Fig. 3) other than the ones described above. So far, these ECF σs have only barely been studied in nitrogen-fixing bacteria. Here, we briefly describe main characteristics of the remaining ECF groups and refer to experimentally studied examples.

5.1.1 Group ECF11 σ factors

This ECF group is restricted to proteobacteria and its members have been identified in several plant-nodulating bacteria, among them *Mesorhizobium-loti, R. etli* and *R. leguminosarum*. ECF11 σs are associated with cytoplasmic anti-σ factors, which implies sensing of intracellular stimuli (Staron et al., 2009). Consistent with this assumption, the best known representative, RpoE from *Rhodobacter sphaeroides*, orchestrates the response to reactive oxygen species (Dufour, Landick, & Donohue, 2008).

5.1.2 Group ECF18 σ factors

ECF18 σs predominantly occur in proteobacterial species, among them the nitrogen-fixing bacteria *B. japonicum* and *R. leguminosarum*

(*Staron* et al., 2009). RpoT from *Pseudomonas putida* was shown to mediate tolerance to toluene and other organic solvents (Duque et al., 2007).

5.1.3 Group ECF20 σ factors

Members of group ECF20 σs are mainly found in proteobacteria with a strong bias toward α-proteobacteria. ECF20 σ encoding genes occur in α-rhizobia able to engage in symbioses with leguminous host plants, e.g. *B. japonicum, R. etli, R. leguminosarum* and *M. loti* (Staron et al., 2009). Two members of this ECF group have been studied so far. CnrH from the β-proteobacterium *Cupriavidus metallidurans* is essential for nickel resistance (Grosse, Friedrich, & Nies, 2007), whereas MibX from the actinobacterium *Microbispora caralina* is involved in the production of the peptide antibiotic microbisporicin (Foulston & Bibb, 2011).

5.1.4 Group ECF29 σ factors

So far, no biological function has been assigned to ECF29 σs. These lack a cognate anti-σ factor and are mainly found in proteobacteria. Among plant-nodulating bacteria, they were identified in *M. loti, S. meliloti, R. etli* and *B. japonicum*. In these bacteria, the ECF29 σ factor encoding genes share a conserved genomic context including genes coding for an outer membrane and a blue copper like protein (Staron et al., 2009). In a *S. meliloti rpoE8* overexpression strain, a search for RpoE8 target genes did not reveal a single candidate (Lang et al., 2018).

5.1.5 Group ECF41 σ factors

This ECF group has about 400 members, is widely distributed in 10 different phyla, and many plant-nodulating α-rhizobia harbor a representative (Staron et al., 2009; Wecke et al., 2012). ECF41 σs lack an anti-σ factor, but are characterized by a C-terminal extension that modulates σ factor activity (Wecke et al., 2012; Wu et al., 2019). ECF41 σs in *Bacillus licheniformis, R. sphaeroides* and *S. meliloti* (RpoE9) where found to activate expression of a carboxymuconolactone decarboxylase encoding gene, whose genomic localization in vicinity of the ECF41 σ factor encoding gene is conserved (Lang et al., 2018; Staron et al., 2009; Wecke et al., 2012). The carboxymuconolactone decarboxylase was proposed to be involved in oxidative stress response, however, a functional link between the ECF41 homologs in *B. licheniformis, R. sphaeroides* and *S. meliloti* and the tolerance to reactive oxygen species could not be confirmed (Lang et al., 2018; Wecke et al., 2012).

5.1.6 Group ECF42 σs

Similar to ECF41 σs, members of group ECF42 lack an anti-σ factor but share a long C-terminal extension that was shown to be required for σ factor activity (Liu, Pinto, & Mascher, 2018; Staron et al., 2009; Wu et al., 2019). ECF42 σs have a broad taxonomic distribution. They occur in 11 different phyla, but predominantly in Actinobacteria and Proteobacteria (Liu et al., 2018). Usually, more than one representative member of this ECF group is present in plant-nodulating bacteria, such as *B. japonicum* (3), *M. loti* (3), *R. etli* (4) *R. leguminosarum* (2), *S. melioti* (1). In these bacteria, the ECF42 σ factor encoding genes are co-transcribed with one or more genes coding for so-far uncharacterized DGPF proteins, which are named according to a conserved amino acid motif (Liu et al., 2018; Staron et al., 2009). In the vicinity of these genes, in several cases genes coding for glyoxalase/bleomycin resistance proteins are located, which is also the case for *rpoE10* in *S. meliloti* (Liu et al., 2018). However, while RpoE10 failed to activate expression of its own gene and that of the DGPF protein encoding gene, the latter (either linked to an *ecf42* locus or not) are the main targets of ECF42 σs in *Streptomyces venezuelae* (Lang et al., 2018; Liu et al., 2018). Remarkably, two out of six identified RpoE10-dependent target genes encode glyoxalase/bleomycin resistance proteins, one in the vicinity of the chromosomal *rpoE10* locus and one on the megaplasmid pSymB (Lang et al., 2018). The identification of ECF42 regulons did not help to unravel their biological function in *S. venezuelae* (Liu et al., 2018) or in *S. meliloti* (Lang et al., 2018). However, another characterized ECF42 σ factor from *P. putida* mediates antibiotic stress resistance and is involved in biofilm formation (Tettmann, Dotsch, Armant, Fjell, & Overhage, 2014).

6. Conclusions and future perspectives

Plant-symbiotic α-rhizobia, such as *S. meliloti, R. etli, R. leguminosarum, M. loti* or *B. japonicum*, are characterized by large genomes known for their plasticity and redundancy (MacLean, Finan, & Sadowsky, 2007). They can draw on a high number (>10) of ECF σs and display a significant overlap in the presence of major phylogenetic ECF groups. Apart from ECF15 σs, the regulators of general stress response, only a few members from other ECF groups have been experimentally addressed in α-rhizobia. Compared to the regulons of EcfG σ factors from group ECF15, they usually control a smaller set of genes as it was reported e.g. for *S. meliloti* (Lang et al., 2018). Accordantly, these alternative σ factors control rather distinct functions: they mediate the resistance to certain metal ions or reactive oxygen species, are involved in the uptake of iron or have a particular role in oxidation processes. Although for some of these σ factors a physiological role has been inferred, only little is known about the perception of stimuli and the exact mechanisms of σ factor activation, which will probably be subject of future studies. Notably, it is not always possible to draw general conclusions on the function of ECF σs from one phylogenetic group despite their overall similarity. This likely reflects a functional diversification during evolution and may point toward the existence of sub-groups with particular roles, as it can be seen for members of groups ECF16, ECF26 or ECF33.

ECF σs usually having narrow functions and redundancy in protective systems may explain why a functional role has only been assigned to a few of these proteins. Notably, a *S. meliloti* mutant strain deleted for all *ecf* genes did neither show any severe growth defects under various tested conditions nor was it symbiotically impaired (Lang et al., 2018), suggesting ECF σs to have only accessory functions. The importance of such functions for survival or competitive capacity of α-rhizobia in their ecological niches is corroborated by the observation that these bacteria retained multiple *ecf* genes.

Many independent studies clarified that ECF15 σs are the regulators of general stress response in α-proteobacteria. Although the main signal transduction cascade leading to EcfG activation seems to be conserved, there are significant differences between model plant-associated α-rhizobia regarding the environmental triggers of GSR, the involvement of (redundant) core regulators and the target genes. Future studies are necessary to clarify how environmental triggers finally lead to the activation of kinases and how other regulatory systems, represented by the alarmone ppGpp or by small RNAs are integrated and modulate the rather unspecific GSR. Besides these highly relevant studies of the regulatory interplay with other signaling systems, more efforts are necessary to experimentally address the physiological role of the regulated targets, given the high number of activated genes with yet unassigned functions. Such studies will be key to provide further explanations for the different phenotypes known to be associated with *ecfG* mutants in plant-nodulating bacteria.

Acknowledgment

Part of the work presented here, carried out at the Philipps-Universität Marburg, Germany, has been funded by the Federal Ministry of Education and Research (BMBF) (grant 031L0010B).

References

Alvarez-Martinez, C. E., Baldini, R. L., & Gomes, S. L. (2006). A *Caulobacter crescentus* extracytoplasmic function sigma factor mediating the response to oxidative stress in stationary phase. *Journal of Bacteriology, 188*(5), 1835–1846.

Andrews, M., & Andrews, M. E. (2017). Specificity in legume-rhizobia symbioses. *International Journal of Molecular Sciences, 18*(4).

Barnett, M. J., Bittner, A. N., Toman, C. J., Oke, V., & Long, S. R. (2012). Dual RpoH sigma factors and transcriptional plasticity in a symbiotic bacterium. *Journal of Bacteriology, 194*(18), 4983–4994.

Barra-Bily, L., Fontenelle, C., Jan, G., Flechard, M., Trautwetter, A., Pandey, S. P., et al. (2010). Proteomic alterations explain phenotypic changes in *Sinorhizobium meliloti* lacking the RNA chaperone Hfq. *Journal of Bacteriology, 192*(6), 1719–1729.

Bastiat, B., Sauviac, L., & Bruand, C. (2010). Dual control of *Sinorhizobium meliloti* RpoE2 sigma factor activity by two PhyR-type two-component response regulators. *Journal of Bacteriology, 192*(8), 2255–2265.

Bastiat, B., Sauviac, L., Picheraux, C., Rossignol, M., & Bruand, C. (2012). *Sinorhizobium meliloti* sigma factors RpoE1 and RpoE4 are activated in stationary phase in response to sulfite. *PLoS One,* 7(11). e50768.

Battesti, A., Majdalani, N., & Gottesman, S. (2011). The RpoS-mediated general stress response in *Escherichia coli*. *Annual Review of Microbiology, 65*, 189–213.

Bervoets, I., Van Brempt, M., Van Nerom, K., Van Hove, B., Maertens, J., De Mey, M., et al. (2018). A sigma factor toolbox for orthogonal gene expression in *Escherichia coli*. *Nucleic Acids Research, 46*(4), 2133–2144.

Braun, V., & Mahren, S. (2005). Transmembrane transcriptional control (surface signalling) of the *Escherichia coli* Fec type. *FEMS Microbiology Reviews, 29*(4), 673–684.

Brooks, B. E., & Buchanan, S. K. (2008). Signaling mechanisms for activation of extracytoplasmic function (ECF) sigma factors. *Biochimica et Biophysica Acta, 1778*(9), 1930–1945.

Bruggemann, C., Denger, K., Cook, A. M., & Ruff, J. (2004). Enzymes and genes of taurine and isethionate dissimilation in *Paracoccus denitrificans*. *Microbiology, 150*(Pt 4), 805–816.

Campagne, S., Damberger, F. F., Kaczmarczyk, A., Francez-Charlot, A., Allain, F. H., & Vorholt, J. A. (2012). Structural basis for sigma factor mimicry in the general stress response of Alphaproteobacteria. *Proceedings of the National Academy of Sciences of the United States of America, 109*(21), E1405–E1414.

Campagne, S., Marsh, M. E., Capitani, G., Vorholt, J. A., & Allain, F. H. (2014). Structural basis for -10 promoter element melting by environmentally induced sigma factors. *Nature Structural & Molecular Biology, 21*(3), 269–276.

Campbell, E. A., Greenwell, R., Anthony, J. R., Wang, S., Lim, L., Das, K., et al. (2007). A conserved structural module regulates transcriptional responses to diverse stress signals in bacteria. *Molecular Cell, 27*(5), 793–805.

Carter, R. A., Worsley, P. S., Sawers, G., Challis, G. L., Dilworth, M. J., Carson, K. C., et al. (2002). The *vbs* genes that direct synthesis of the siderophore vicibactin in *Rhizobium leguminosarum*: Their expression in other genera requires ECF sigma factor RpoI. *Molecular Microbiology, 44*(5), 1153–1166.

Chao, T. C., Buhrmester, J., Hansmeier, N., Puhler, A., & Weidner, S. (2005). Role of the regulatory gene *rirA* in the transcriptional response of *Sinorhizobium meliloti* to iron limitation. *Applied and Environmental Microbiology, 71*(10), 5969–5982.

Cook, A. M., & Denger, K. (2006). Metabolism of taurine in microorganisms: A primer in molecular biodiversity? *Advances in Experimental Medicine & Biology, 583*, 3–13.

Damiani, I., Pauly, N., Puppo, A., Brouquisse, R., & Boscari, A. (2016). Reactive oxygen species and nitric oxide control early steps of the legume - rhizobium symbiotic interaction. *Frontiers of Plant Science,* 7, 454.

Dufour, Y. S., Landick, R., & Donohue, T. J. (2008). Organization and evolution of the biological response to singlet oxygen stress. *Journal of Molecular Biology, 383*(3), 713–730.

Duque, E., Rodriguez-Herva, J. J., de la Torre, J., Dominguez-Cuevas, P., Munoz-Rojas, J., & Ramos, J. L. (2007). The RpoT regulon of *Pseudomonas putida* DOT-T1E and its role in stress endurance against solvents. *Journal of Bacteriology, 189*(1), 207–219.

Feng, C., Tollin, G., & Enemark, J. H. (2007). Sulfite oxidizing enzymes. *Biochimica et Biophysica Acta, 1774*(5), 527–539.

Flechard, M., Fontenelle, C., Blanco, C., Goude, R., Ermel, G., & Trautwetter, A. (2010). RpoE2 of *Sinorhizobium meliloti* is necessary for trehalose synthesis and growth in hyperosmotic media. *Microbiology, 156*(Pt 6), 1708–1718.

Flechard, M., Fontenelle, C., Trautwetter, A., Ermel, G., & Blanco, C. (2009). *Sinorhizobium meliloti rpoE2* is necessary for H_2O_2 stress resistance during the stationary growth phase. *FEMS Microbiology Letters, 290*(1), 25–31.

Foulston, L., & Bibb, M. (2011). Feed-forward regulation of microbisporicin biosynthesis in *Microbispora corallina. Journal of Bacteriology, 193*(12), 3064–3071.

Francez-Charlot, A., Frunzke, J., Reichen, C., Ebneter, J. Z., Gourion, B., & Vorholt, J. A. (2009). Sigma factor mimicry involved in regulation of general stress response. *Proceedings of the National Academy of Sciences of the United States of America, 106*(9), 3467–3472.

Francez-Charlot, A., Frunzke, J., Zingg, J., Kaczmarczyk, A., & Vorholt, J. A. (2016). Multiple sigma EcfG and NepR proteins are involved in the general stress response in *Methylobacterium extorquens. PLoS One, 11*(3). e0152519.

Francez-Charlot, A., Kaczmarczyk, A., Fischer, H. M., & Vorholt, J. A. (2015). The general stress response in Alphaproteobacteria. *Trends in Microbiology, 23*(3), 164–171.

Galibert, F., Finan, T. M., Long, S. R., Puhler, A., Abola, P., Ampe, F., et al. (2001). The composite genome of the legume symbiont *Sinorhizobium meliloti. Science, 293*(5530), 668–672.

Galperin, M. Y. (2006). Structural classification of bacterial response regulators: Diversity of output domains and domain combinations. *Journal of Bacteriology, 188*(12), 4169–4182.

Garcia-de Los Santos, A., Lopez, E., Cubillas, C. A., Noel, K. D., Brom, S., & Romero, D. (2008). Requirement of a plasmid-encoded catalase for survival of *Rhizobium etli* CFN42 in a polyphenol-rich environment. *Applied and Environmental Microbiology, 74*(8), 2398–2403.

Gibson, K. E., Kobayashi, H., & Walker, G. C. (2008). Molecular determinants of a symbiotic chronic infection. *Annual Review of Genetics, 42*, 413–441.

Gonzalez, V., Santamaria, R. I., Bustos, P., Hernandez-Gonzalez, I., Medrano-Soto, A., Moreno-Hagelsieb, G., et al. (2006). The partitioned *Rhizobium etli* genome: Genetic and metabolic redundancy in seven interacting replicons. *Proceedings of the National Academy of Sciences of the United States of America, 103*(10), 3834–3839.

Gourion, B., Francez-Charlot, A., & Vorholt, J. A. (2008). PhyR is involved in the general stress response of *Methylobacterium extorquens* AM1. *Journal of Bacteriology, 190*(3), 1027–1035.

Gourion, B., Rossignol, M., & Vorholt, J. A. (2006). A proteomic study of *Methylobacterium extorquens* reveals a response regulator essential for epiphytic growth. *Proceedings of the National Academy of Sciences of the United States of America, 103*(35), 13186–13191.

Gourion, B., Sulser, S., Frunzke, J., Francez-Charlot, A., Stiefel, P., Pessi, G., et al. (2009). The PhyR-sigma(EcfG) signalling cascade is involved in stress response and symbiotic efficiency in *Bradyrhizobium japonicum. Molecular Microbiology, 73*(2), 291–305.

Grosse, C., Friedrich, S., & Nies, D. H. (2007). Contribution of extracytoplasmic function sigma factors to transition metal homeostasis in *Cupriavidus metallidurans* strain CH34. *Journal of Molecular Microbiology and Biotechnology, 12*(3–4), 227–240.

Gruber, T. M., & Gross, C. A. (2003). Multiple sigma subunits and the partitioning of bacterial transcription space. *Annual Review of Microbiology, 57*, 441–466.

Hecker, M., Pane-Farre, J., & Volker, U. (2007). SigB-dependent general stress response in *Bacillus subtilis* and related gram-positive bacteria. *Annual Review of Microbiology, 61*, 215–236.

Helmann, J. D. (2002). The extracytoplasmic function (ECF) sigma factors. *Advances in Microbial Physiology, 46*, 47–110.

Helmann, J. D., & Chamberlin, M. J. (1988). Structure and function of bacterial sigma factors. *Annual Review of Biochemistry, 57*, 839–872.

Hengge, R. (2011). The general stress response in Gram-negative bacteria. In H. R. Storz G (Ed.), *Bacterial stress responses* (2nd ed., pp. 251–289). Washington D.C.: ASM Press.

Herrou, J., Foreman, R., Fiebig, A., & Crosson, S. (2010). A structural model of anti-anti-sigma inhibition by a two-component receiver domain: The PhyR stress response regulator. *Molecular Microbiology, 78*(2), 290–304.

Hook-Barnard, I. G., & Hinton, D. M. (2007). Transcription initiation by mix and match elements: Flexibility for polymerase binding to bacterial promoters. *Gene Regulation and Systems Biology, 1*, 275–293.

Huang, X., Pinto, D., Fritz, G., & Mascher, T. (2015). Environmental sensing in Actinobacteria: A comprehensive survey on the signaling capacity of this phylum. *Journal of Bacteriology, 197*(15), 2517–2535.

Humann, J. L., Ziemkiewicz, H. T., Yurgel, S. N., & Kahn, M. L. (2009). Regulatory and DNA repair genes contribute to the desiccation resistance of *Sinorhizobium meliloti* Rm1021. *Applied and Environmental Microbiology, 75*(2), 446–453.

Jans, A., Vercruysse, M., Gao, S., Engelen, K., Lambrichts, I., Fauvart, M., et al. (2013). Canonical and non-canonical EcfG sigma factors control the general stress response in *Rhizobium etli*. *Microbiologyopen, 2*(6), 976–987.

Jogler, C., Waldmann, J., Huang, X., Jogler, M., Glockner, F. O., Mascher, T., et al. (2012). Identification of proteins likely to be involved in morphogenesis, cell division, and signal transduction in Planctomycetes by comparative genomics. *Journal of Bacteriology, 194*(23), 6419–6430.

Kaczmarczyk, A., Campagne, S., Danza, F., Metzger, L. C., Vorholt, J. A., & Francez-Charlot, A. (2011). Role of *Sphingomonas sp.* strain Fr1 PhyR-NepR-sigmaEcfG cascade in general stress response and identification of a negative regulator of PhyR. *Journal of Bacteriology, 193*(23), 6629–6638.

Kaneko, T., Nakamura, Y., Sato, S., Minamisawa, K., Uchiumi, T., Sasamoto, S., et al. (2002). Complete genomic sequence of nitrogen-fixing symbiotic bacterium *Bradyrhizobium japonicum* USDA110 (supplement). *DNA Research, 9*(6), 225–256.

Kappler, U. (2011). Bacterial sulfite-oxidizing enzymes. *Biochimica et Biophysica Acta, 1807*(1), 1–10.

Kappler, U., Friedrich, C. G., Truper, H. G., & Dahl, C. (2001). Evidence for two pathways of thiosulfate oxidation in *Starkeya novella* (formerly *Thiobacillus novellus*). *Archives of Microbiology, 175*(2), 102–111.

Kertesz, M. A. (2000). Riding the sulfur cycle - metabolism of sulfonates and sulfate esters in Gram-negative bacteria. *FEMS Microbiology Reviews, 24*(2), 135–175.

Kim, H. S., Caswell, C. C., Foreman, R., Roop, R. M., 2nd, & Crosson, S. (2013). The *Brucella abortus* general stress response system regulates chronic mammalian infection and is controlled by phosphorylation and proteolysis. *Journal of Biological Chemistry, 288*(19), 13906–13916.

Kohler, C., Lourenco, R. F., Avelar, G. M., & Gomes, S. L. (2012). Extracytoplasmic function (ECF) sigma factor σ^F is involved in *Caulobacter crescentus* response to heavy metal stress. *BMC Microbiology, 12*, 210.

Krol, E., & Becker, A. (2011). ppGpp in *Sinorhizobium meliloti*: biosynthesis in response to sudden nutritional downshifts and modulation of the transcriptome. *Molecular Microbiology, 81*(5), 1233–1254.

Lang, C., Barnett, M. J., Fisher, R. F., Smith, L. S., Diodati, M. E., & Long, S. R. (2018). Most *Sinorhizobium meliloti* extracytoplasmic function sigma factors control accessory functions. *mSphere, 3*(5).

Ledermann, R., Bartsch, I., Muller, B., Wulser, J., & Fischer, H. M. (2018). A functional general stress response of *Bradyrhizobium diazoefficiens* is required for early stages of host plant infection. *Molecular Plant-Microbe Interactions, 31*(5), 537–547.

Letunic, I., & Bork, P. (2019). Interactive tree of life (iTOL) v4: Recent updates and new developments. *Nucleic Acids Research, 47*(W1), W256–W259.

Liu, Q., Pinto, D., & Mascher, T. (2018). Characterization of the widely distributed novel ECF42 group of extracytoplasmic function sigma factors in *Streptomyces venezuelae*. *Journal of Bacteriology, 200*(21).

Lonetto, M., Gribskov, M., & Gross, C. A. (1992). The sigma 70 family: Sequence conservation and evolutionary relationships. *Journal of Bacteriology, 174*(12), 3843–3849.

Lonetto, M. A., Brown, K. L., Rudd, K. E., & Buttner, M. J. (1994). Analysis of the *Streptomyces coelicolor sigE* gene reveals the existence of a subfamily of eubacterial RNA polymerase sigma factors involved in the regulation of extracytoplasmic functions. *Proceedings of the National Academy of Sciences of the United States of America, 91*(16), 7573–7577.

Lourenco, R. F., Kohler, C., & Gomes, S. L. (2011). A two-component system, an anti-sigma factor and two paralogous ECF sigma factors are involved in the control of general stress response in *Caulobacter crescentus*. *Molecular Microbiology, 80*(6), 1598–1612.

Low, L., Ryan Kilmartin, J., Paul, V. B., & Ulrike, K. (2011). How are "atypical" sulfite dehydrogenases linked to cell metabolism? Interactions between the SorT sulfite dehydrogenase and small redox proteins. *Frontiers in Microbiology, 2*, 58.

MacLean, A. M., Finan, T. M., & Sadowsky, M. J. (2007). Genomes of the symbiotic nitrogen-fixing bacteria of legumes. *Plant Physiology, 144*(2), 615–622.

Magnusson, L. U., Farewell, A., & Nystrom, T. (2005). ppGpp: a global regulator in *Escherichia coli*. *Trends in Microbiology, 13*(5), 236–242.

Marchler-Bauer, A., Bo, Y., Han, L., He, J., Lanczycki, C. J., Lu, S., et al. (2017). CDD/SPARCLE: Functional classification of proteins via subfamily domain architectures. *Nucleic Acids Research, 45*(D1), D200–D203.

Martinez-Salazar, J. M., Salazar, E., Encarnacion, S., Ramirez-Romero, M. A., & Rivera, J. (2009a). Role of the extracytoplasmic function sigma factor RpoE4 in oxidative and osmotic stress responses in *Rhizobium etli*. *Journal of Bacteriology, 191*(13), 4122–4132.

Martinez-Salazar, J. M., Sandoval-Calderon, M., Guo, X., Castillo-Ramirez, S., Reyes, A., Loza, M. G., et al. (2009b). The *Rhizobium etli* RpoH1 and RpoH2 sigma factors are involved in different stress responses. *Microbiology, 155*(Pt 2), 386–397.

Mascher, T. (2013). Signaling diversity and evolution of extracytoplasmic function (ECF) sigma factors. *Current Opinion in Microbiology, 16*(2), 148–155.

Masloboeva, N., Reutimann, L., Stiefel, P., Follador, R., Leimer, N., Hennecke, H., et al. (2012). Reactive oxygen species-inducible ECF sigma factors of *Bradyrhizobium japonicum*. *PLoS One, 7*(8). e43421.

McGrath, A. P., Laming, E. L., Casas Garcia, G. P., Kvansakul, M., Guss, J. M., Trewhella, J., et al. (2015). Structural basis of interprotein electron transfer in bacterial sulfite oxidation. *Elife, 4*. e09066.

McIntyre, H. J., Davies, H., Hore, T. A., Miller, S. H., Dufour, J. P., & Ronson, C. W. (2007). Trehalose biosynthesis in *Rhizobium leguminosarum* bv. *trifolii* and its role in desiccation tolerance. *Applied and Environmental Microbiology, 73*(12), 3984–3992.

Nguyen, L. T., Schmidt, H. A., von Haeseler, A., & Minh, B. Q. (2015). IQ-tree: A fast and effective stochastic algorithm for estimating maximum-likelihood phylogenies. *Molecular Biology and Evolution, 32*(1), 268–274.

Oke, V., Rushing, B. G., Fisher, E. J., Moghadam-Tabrizi, M., & Long, S. R. (2001). Identification of the heat-shock sigma factor RpoH and a second RpoH-like protein in *Sinorhizobium meliloti*. *Microbiology, 147*(Pt 9), 2399–2408.

Paget, M. S. (2015). Bacterial sigma factors and anti-sigma factors: Structure, function and distribution. *Biomolecules, 5*(3), 1245–1265.

Pauly, N., Pucciariello, C., Mandon, K., Innocenti, G., Jamet, A., Baudouin, E., et al. (2006). Reactive oxygen and nitrogen species and glutathione: Key players in the legume-rhizobium symbiosis. *Journal of Experimental Botany, 57*(8), 1769–1776.

Pellicer Martinez, M. T., Martinez, A. B., Crack, J. C., Holmes, J. D., Svistunenko, D. A., Johnston, A. W. B., et al. (2017). Sensing iron availability via the fragile [4Fe-4S] cluster of the bacterial transcriptional repressor RirA. *Chemical Science, 8*(12), 8451–8463.

Pinto, D., & Mascher, T. (2016). The ECF classification: A phylogenetic reflection of the regulatory diversity in the extracytoplasmic function σ factor protein family. In F. J. de Bruijn (Ed.), *Stress and environmental regulation of gene expression and adaptation in bacteria* (pp. 64–96). Hoboken, NJ: Wiley-Blackwell.

Poole, P., Ramachandran, V., & Terpolilli, J. (2018). Rhizobia: From saprophytes to endosymbionts. *Nature Reviews Microbiology, 16*(5), 291–303.

Price, C. W. (2011). General stress response in *Bacillus subtilis* and related gram positive bacteria. In H. R. Storz G (Ed.), *Bacterial stress responses* (2nd ed., pp. 301–318). Washington D.C.: ASM Press.

Repoila, F., Majdalani, N., & Gottesman, S. (2003). Small non-coding RNAs, co-ordinators of adaptation processes in *Escherichia coli*: The RpoS paradigm. *Molecular Microbiology, 48*(4), 855–861.

Rhodius, V. A., Segall-Shapiro, T. H., Sharon, B. D., Ghodasara, A., Orlova, E., Tabakh, H., et al. (2013). Design of orthogonal genetic switches based on a crosstalk map of sigmas, anti-sigmas, and promoters. *Molecular Systems Biology, 9*, 702.

Rudolph, G., Hennecke, H., & Fischer, H. M. (2006). Beyond the Fur paradigm: Iron-controlled gene expression in rhizobia. *FEMS Microbiology Reviews, 30*(4), 631–648.

Sauviac, L., & Bruand, C. (2014). A putative bifunctional histidine kinase/phosphatase of the HWE family exerts positive and negative control on the *Sinorhizobium meliloti* general stress response. *Journal of Bacteriology, 196*(14), 2526–2535.

Sauviac, L., Philippe, H., Phok, K., & Bruand, C. (2007). An extracytoplasmic function sigma factor acts as a general stress response regulator in *Sinorhizobium meliloti*. *Journal of Bacteriology, 189*(11), 4204–4216.

Schlüter, J. P., Reinkensmeier, J., Barnett, M. J., Lang, C., Krol, E., Giegerich, R., et al. (2013). Global mapping of transcription start sites and promoter motifs in the symbiotic alpha-proteobacterium *Sinorhizobium meliloti* 1021. *BMC Genomics, 14*, 156.

Staron, A., & Mascher, T. (2010). General stress response in alpha-proteobacteria: PhyR and beyond. *Molecular Microbiology, 78*(2), 271–277.

Staron, A., Sofia, H. J., Dietrich, S., Ulrich, L. E., Liesegang, H., & Mascher, T. (2009). The third pillar of bacterial signal transduction: Classification of the extracytoplasmic function (ECF) sigma factor protein family. *Molecular Microbiology, 74*(3), 557–581.

Stockwell, S. B., Reutimann, L., & Guerinot, M. L. (2012). A role for *Bradyrhizobium japonicum* ECF16 sigma factor EcfS in the formation of a functional symbiosis with soybean. *Molecular Plant-Microbe Interactions, 25*(1), 119–128.

Tettmann, B., Dotsch, A., Armant, O., Fjell, C. D., & Overhage, J. (2014). Knockout of extracytoplasmic function sigma factor ECF-10 affects stress resistance and biofilm formation in *Pseudomonas putida* KT2440. *Applied and Environmental Microbiology, 80*(16), 4911–4919.

Thorne, S. H., & Williams, H. D. (1997). Adaptation to nutrient starvation in *Rhizobium leguminosarum* bv. *phaseoli*: Analysis of survival, stress resistance, and changes in macromolecular synthesis during entry to and exit from stationary phase. *Journal of Bacteriology, 179*(22), 6894–6901.

Vargas Mdel, C., Encarnacion, S., Davalos, A., Reyes-Perez, A., Mora, Y., Garcia-de los Santos, A., et al. (2003). Only one catalase, *katG*, is detectable in *Rhizobium etli*, and is encoded along with the regulator OxyR on a plasmid replicon. *Microbiology, 149*(Pt 5), 1165–1176.

Vercruysse, M., Fauvart, M., Jans, A., Beullens, S., Braeken, K., Cloots, L., et al. (2011). Stress response regulators identified through genome-wide transcriptome analysis of the (p)ppGpp-dependent response in *Rhizobium etli*. *Genome Biology, 12*(2), R17.

Visca, P., Leoni, L., Wilson, M. J., & Lamont, I. L. (2002). Iron transport and regulation, cell signalling and genomics: Lessons from *Escherichia coli* and *Pseudomonas*. *Molecular Microbiology, 45*(5), 1177–1190.

Wecke, T., Halang, P., Staron, A., Dufour, Y. S., Donohue, T. J., & Mascher, T. (2012). Extracytoplasmic function sigma factors of the widely distributed group ECF41 contain a fused regulatory domain. *Microbiologyopen, 1*(2), 194–213.

Wilson, J. J., & Kappler, U. (2009). Sulfite oxidation in *Sinorhizobium meliloti*. *Biochimica et Biophysica Acta, 1787*(12), 1516–1525.

Wu, H., Liu, Q., Casas-Pastor, D., Durr, F., Mascher, T., & Fritz, G. (2019). The role of C-terminal extensions in controlling ECF sigma factor activity in the widely conserved groups ECF41 and ECF42. *Molecular Microbiology*.

Yeoman, K. H., Curson, A. R., Todd, J. D., Sawers, G., & Johnston, A. W. (2004). Evidence that the Rhizobium regulatory protein RirA binds to cis acting iron responsive operators (IROs) at promoters of some Fe-regulated genes. *Microbiology, 150*(Pt 12), 4065–4074.

Yeoman, K. H., May, A. G., deLuca, N. G., Stuckey, D. B., & Johnston, A. W. (1999). A putative ECF sigma factor gene, *rpol*, regulates siderophore production in *Rhizobium leguminosarum*. *Molecular Plant-Microbe Interactions, 12*(11), 994–999.

Yeoman, K. H., Mitelheiser, S., Sawers, G., & Johnston, A. W. (2003). The ECF sigma factor RpoI of *R. leguminosarum* initiates transcription of the *vbsGSO* and *vbsADL* siderophore biosynthetic genes *in vitro*. *FEMS Microbiology Letters, 223*(2), 239–244.

Young, J. P., Crossman, L. C., Johnston, A. W., Thomson, N. R., Ghazoui, Z. F., Hull, K. H., et al. (2006). The genome of *Rhizobium leguminosarum* has recognizable core and accessory components. *Genome Biology, 7*(4), R34.

CHAPTER ELEVEN

Partner preference in the legume-rhizobia symbiosis and impact on legume inoculation strategies

Stephane Boivin and Marc Lepetit*
Laboratoire des Symbioses Tropicales et Méditerranéennes, INRA, Montpellier, France
*Corresponding author: E-mail: marc.lepetit@inra.fr

Contents

Abstract

Legumes interact with diverse soil bacteria of the genus *Rhizobium* to form dinitrogen (N_2)-fixing symbiotic root nodules. This symbiotic interaction occurs under plant nitrogen (N)-deficit and only between compatible partners. When young nodules are formed, bacteroids begin to fix atmospheric N_2 and provide ammonium to the plant which in return provides phytosynthates to the bacteria. Ability and Competitiveness to Form root Nodules (AFN and CFN) within natural populations have a strong impact on early partner choice. In soil, there are frequently multiple compatible partners and legume roots are often colonized by multiple rhizobia displaying different Symbiotic N_2 Fixation (SNF) efficiencies. In mature nodules, Post-Infection Partner Preference (PIPP)

Advances in Botanical Research, Volume 94
ISSN 0065-2296
https://doi.org/10.1016/bs.abr.2019.09.016

shape partner choice by favoring effective symbiotic N_2fixing structures to the expense of inefficient ones. Little is known about mechanisms involved in legume-rhizobium partner choices despite the relevance of this trait to design new strategies to improve inoculation of legume crops.

1. Introduction

Legumes have the unique ability to associate with rhizobia within specific symbiotic root organs, called nodules. The legume-rhizobium symbiosis has the emerging property to use dinitrogen (N_2) present in the air as an unlimited nitrogen (N) source and therefore to circumvent a soil mineral N limitation. Atmospheric N_2 fixation takes place within bacteroids where it is transformed into ammonium and translocated to the legume host roots. In return, the plant provides photosynthates to the bacteria (Oldroyd, Murray, Poole, & Downie, 2011). Legume-rhizobium interactions are however not ubiquitous as these symbioses happen only between compatible partners.

In soil and rhizosphere, bacteria form complex populations which composition vary depending on environmental conditions (Fierer & Jackson, 2006; Kaiser et al., 2016; Rousk et al., 2010). Rhizobia represent a little part of soil bacteria (<1%) but high local variations in population size and symbiotic performances are often observed (Wakelin et al., 2018). Although different protocols were designed to study rhizobia populations from soil samples, their identification and direct quantification remain difficult, mainly because of their low density in soils combined with quality issues of the DNA extracted from soil samples (Laguerre, Bardin, & Amarger, 1993; Sullivan, Eardly, van Berkum, & Ronson, 1996; Zézé, Mutch, & Young, 2001). Taking advantage of their capacity to associate with plants within nodules, rhizobia have been mostly collected from root nodules, which were considered to be generally occupied by a dominant rhizobium genotype (Bourion et al., 2018; Laguerre et al., 2012). This so-called "trapping" strategy coupled with molecular methods allowed drawing indirect pictures of the rhizobia diversity in soils (Kumar et al., 2015; Laguerre, Allard, Revoy, & Amarger, 1994; Young & Wexler, 1988). Indeed, legume-rhizobium partner choice mechanisms biased the diversity assessment. Although a large genetic diversity is frequently found at a local scale in soils (Bailly et al., 2011; Kumar et al., 2015), some rhizobia are occasionally absent, and therefore some specific legume hosts cannot be nodulated in these soils. For example, soybean is native from East Asia where its symbionts are naturally found in soils, but the compatible bacteria *Bradyrhizobium japonicum* are not indigenous in western Europe and inoculation is then required (Revellin, Pinochet, Beauclair, & Catroux, 1996).

The partner choice influences the symbiotic interaction from early steps, i.e. in relation with the infection of the plant by symbiotic bacteria, as well as later steps, i.e. in N_2 fixing nodules. The intrinsic Ability to Form root Nodules (AFN) is generally estimated as the capacity to form nodules with a legume in mono-inoculation experiments. As the host spectrum of some rhizobia is larger than for others, AFN specificities are then observed at various phylogenetic levels, from specific rhizobium genotypes to different rhizobium complex species. However, rhizobia in a given soil potentially share an equivalent AFN, but display a wide range of Competitiveness to Form root Nodules (CFN) with their compatible host plants. This CFN is estimated as the capacity of a specific genotype to form nodules when multiple different compatible partners co-exist. Therefore, not only AFN but also CFN determine the early partner choice (Amarger, 1981; Triplett & Sadowsky, 1992). Once young nodules are formed after the selection of both compatible and competitive rhizobium genotypes, the Symbiotic N_2 Fixation (SNF) begins progressively. The SNF efficiency of individual bacteria strongly varies within rhizobial populations, as well as in function of the legume host and of environmental conditions. Because CFN and SNF efficiency are genetically independent, young nodules with different levels of SNF efficiency are formed on the same host root system (Bourion et al., 2018). Consequently, a mosaic of symbiotic nodule organs, displaying various levels of efficiency, is frequently observed in natural conditions on legume root systems (Laguerre, Louvrier, Allard, & Amarger, 2003; Mutch & Young, 2004). At this stage, Post-Infection Partner Preference (PIPP) mechanisms influence both the dynamic of the interaction and nodule development, in order to (1) satisfy the plant N demand and (2) prevent dissipation of carbon (C) metabolites toward inefficient symbiotic structures. Understanding genetic and molecular bases of AFN, CFN and PIPP is therefore highly relevant to select the best symbiotic partners, design inoculation strategies, and predict their benefits for agro-ecological systems.

This chapter reviews our current knowledge on symbiotic partner choices in conditions where multiple symbiotic partners can interact with their host plants. The first part focuses on the early partner choice and describes mechanisms modulating the initiation of the legume-rhizobium symbiotic interaction (i.e. AFN and CFN). The second part focuses on PIPP that occurs later in nodules when legume hosts are associated with multiple partners. Mechanisms related to SNF efficiency and plant N signaling, potentially involved in PIPP, are additionally presented. To conclude, consequences of partner choice for the improvement of inoculation strategies in legume crops is discussed.

2. Early partner choice in the legume-rhizobium symbiosis

2.1 Legume and rhizobial symbiotic mechanisms controlling the AFN

2.1.1 Symbiosis initiation by the perception of legume plant flavonoid spectra

Under N deficit, legumes associate with compatible rhizobia, and the Ability to Form root Nodules (AFN) mainly depends on two conditions: the capacity of rhizobia to detect plant signals (i.e. flavonoids) and in return, the capacity of legume plants to recognize Nod Factors (NFs) and bacterial surface polysaccharides (SPs; Fig. 1; Oldroyd et al., 2011). Legume roots release a wide range of flavonoids in the rhizosphere. The amount and the chemical structure of secreted flavonoids impact bacterial symbiotic responses, as some chemical species are more effective than others to trigger responses on specific rhizobia (Jiménez-Guerrero et al., 2017; Maj, Wielbo, Marek-Kozaczuk, & Skorupska, 2010). Synergy between the different types of flavonoids exsuded can modulate the symbiotic initiation (Begum, Leibovitch, Migner, & Zhang, 2001; Firmin, Wilson, Rossen, & Johnston, 1986). Flavonoids interact with the LyR transcriptional activator nodulation D (NodD) protein which binds to specific Nod Boxes (NBs) DNA motifs located upstream of a number of bacterial nodulation genes (*nod* genes) involved in the production, modification or transport of NFs (Hong, Burn, & Johnston, 1987; Peck, Fisher, & Long, 2006). The broad host-range *Rhizobium sp.* NGR234 strain possess two different *nodD* alleles, called *NodD1* and *NodD2*, which potentially interact with 19 NBs (Fellay et al., 1998; Kobayashi, Naciri-Graven, Broughton, & Perret, 2004), whereas the *Rhizobium leguminosarum* symbiovar *viciae* (*Rlv*) strain 3841, which forms nodule only with plants of the Fabeae tribe, has only one NodD protein that potentially interact with 6 NBs (4 validated and 2 putatives; Young et al., 2006). Therefore, plasticity and diversity of flavonoids produced by the plant and their perception by rhizobia influence the AFN. Moreover, the regulation of *nod* gene expression and nodulation via NodD can be also affected by non-flavonoid compounds secreted by legumes (i.e. jasmonic acid,

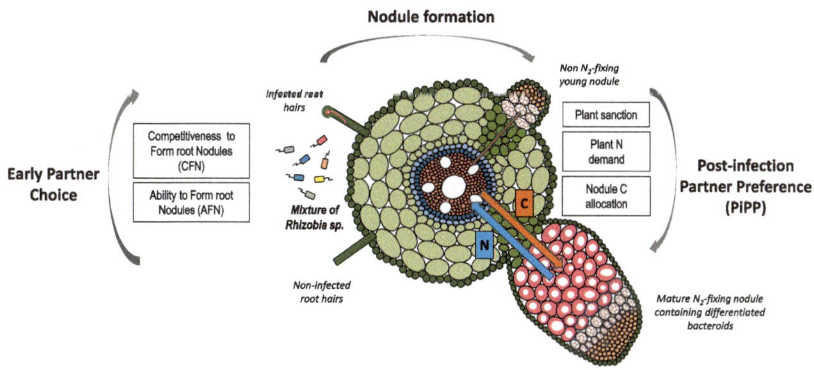

Fig. 1 ***Early partner choice and post-infection partner preference in the legume-rhizobium symbiosis in a context of multiple partners.*** The schematized transversal root section represents the main developmental steps of the legume-rhizobium symbiosis, from the root hairs infection to the N_2 fixing nodule. Rhizobia in soils are diverse and display variable Abilities and Competitiveness to Form root Nodules (AFN and CFN), as well as variable capacities to drive efficient Symbiotic N_2 Fixation (SNF). Nod factors, surface polysaccharides, and secretion systems from rhizobia, as well as flavonoids and LysM-RLKs from legume hosts, are the main factors modulating the early partner choice between symbiotic partners. Symbiosis initiation and nodule formation result in young nodules that are non N_2 fixing and differentiate into expanded mature nodules with a large central zone containing N_2 fixing bacteroids. In these nodules, large fluxes of nitrogen (N) from the nodules to the plant roots and carbon (C) phytosynthates from the plant to the nodules are exchanged. Evidences suggest that the local N_2 fixation capacity, the whole plant N demand, and C allocation may pilot the proliferation of efficient N_2-fixing nodules to the expenses of the inefficient ones, and are therefore potential drivers of Post-Infection Partner Preference (PIPP). Dark green: root hairs and epidermis; pale green: cortex; middle green: division of cortical cells; light blue: endodermis; dark blue: pericycle; pink, stele tissues including vascular bundles; orange: nodule meristem; pink points: nodule infection zone; pink circles: nodule N_2 fixation zone; blue arrow: N flux; orange arrow: C flux. Different rhizobium genotypes are represented with different colors.

stachydrine, trigonelline, tetronic and erythronic acids) and non-legume plants (i.e. vanillin and isovanillin), suggesting that a large set of plant-secreted molecules can influence the bacterial NodD function (Gagnon & Ibrahim, 1998; Le Strange, Bender, Djordjevic, Rolfe, & Redmond, 1990; Sun et al., 2006).

2.1.2 Rhizobial AFN is generally determined by the range of secreted NFs

The best characterized genes regulated by flavonoids and NodD are *nod* genes, grouped together in clusters generally located on a symbiotic plasmid or in specific mobile islands of the chromosome (Kaneko et al., 2002; Kondorosi, Banfalvi, & Kondorosi, 1984; Young et al., 2006). The typical set of *nod* genes include conserved core genes and additional genes specific to some rhizobia (Mergaert, Montagu, & Holsters, 1997; Triplett & Sadowsky, 1992). In *Rlv*, the *nod* gene cluster contains the *nodD, nodO* genes, and the *nodABCIJ(X), nodFEL, nodMN(T)* operons. These transcriptional units are under the control of 4 NBs (Spaink, Okker, Wijffelman, Pees, & Lugtenberg, 1987). The *nodABC, nodFEL* and *nodMN* operons are responsible of the biosynthesis of lipo-chitooligosaccharidic NF molecules that are composed of four to five N-acetyl glucosamines (forming the NF sugar backbone) and a terminal N-acylated sugar carrying a fatty acid chain of 16−18 C residues (Mergaert et al., 1997). Different species of rhizobia produce different NFs characterized by various chemical substitutions (i.e. acetyl, carbamoyl, sulfate, fucosyl, arabinosyl and methyl residues) on the sugar backbone, and by different unsaturation levels of the acyl chain (Cooper, 2004). Specific *nod* genes are involved in these modifications and hence, depending on strains and species, this allows modulating the rhizobial AFN with compatible legumes. For example, some *Rlv* strains are able to specifically interact with the *Pisum sativum* cultivar 'Afghanistan' because they produce a specific NF resulting of the *nodX* gene function (Firmin, Wilson, Carlson, Davies, & Downie, 1993). As NFs are important factors in the legume-rhizobium symbiotic initiation, their secretion in the extracellular medium is a key step for legume NF perception and is mediated by the transmembrane proteins NodI and NodJ (Spaink et al., 1995). Some genes are present in the *nod* gene cluster of certain rhizobia, such as *nodT* and *nodO* in *R. leguminosarum* (Rivilla, Sutton, & Downie, 1995; Surin, Watson, Hamilton, Economou, & Downie, 1990). These two genes are involved in the secretion of NFs as well as in the modulation of the host-specificity and AFN (Djordjevic, Schofield, & Rolfe, 1985; Lewis-Henderson & Djordjevic, 1991).

The AFN with compatible legumes (generally defined as "symbiovars") is generally well-correlated with the *nod* gene phylogenetic diversity, but very poorly with the whole chromosome genetic variation (Kumar et al., 2015). Consequently, closely-related chromosome backgrounds can be found in different "symbiovars". Symbiovars are well discriminated by their *nod* genes present on plasmids or on mobile islands. Because *nod* genes are generally located on symbiotic plasmids, horizontal gene transfer (HGT) is one of the main drivers of AFN genetic evolution in populations. One example illustrating the impact of HGTs on AFN has been evidenced in

New Zealand where *Rhizobium loti* is absent in most of uncultivated soils (Patrick & Lowther, 1992). A strain of *R. loti*, which is able to nodulate *Lotus corniculatus*, was introduced in a field and root nodule trapping was performed seven years later at the same place (Sullivan, Patrick, Lowther, Scott, & Ronson, 1995). Molecular genotyping revealed the transfer of symbiotic genes from the introduced strain to native non-symbiotic rhizobia already present in the soil. HGTs have been well studied in pure cultures and bioinformatic tools are now available to investigate them across various microbe genomes (Li et al., 2018; Soucy, Huang, & Gogarten, 2015). However, HGTs can only happen between compatible bacteria, and mechanisms controlling gene exchanges within complex bacterial populations are not well understood (Klümper et al., 2015). The Quorum Sensing (QS) is a gene expression regulatory system acting at the level of bacterial populations (Papenfort & Bassler, 2016). In *R. leguminosarum*, the transfer of the symbiotic plasmid carrying *nod* genes is mediated by LuxR-type QS regulators and therefore, the QS system is predicted to influence the diversity and the abundance of rhizobial genotypes in soil populations (McAnulla, Edwards, Sanchez-Contreras, Sawers, & Downie, 2007).

2.1.3 The genetic diversity of legume LysM-RLK receptors impacts rhizobial AFN

Perception of NFs is mediated by plant receptors that trigger the first steps of rhizobial infection (Fig. 1). They belong to the large and diverse family of LysM receptor-like kinases (LysM-RLKs) (Gough, Cottret, Lefebvre, & Bono, 2018), which possess several extracellular LysM domains that are able to bind NFs, a transmembrane domain, and an intracellular kinase domain which transduces the signal. In *Medicago truncatula* and *Lotus japonicus* model legumes, two different LysM-RLK co-receptors, likely forming dimers, have been identified (MtLYK3 and MtNFP, and LjNFR1 and LjNFR5 respectively Dénarié, Debellé, & Promé, 1996; Madsen et al., 2011; Radutoiu et al., 2007). The MtLYK3/LjNFR1 receptors belong to the "LysM domain-containing RLK" (LYK) family of proteins that have functional kinase domains, whereas MtNFP/LjNFR5 belong to the "LYK-related" (LYR) family of proteins that have lost an activation loop motif, leading to unfunctional kinase domains. Although a role in specific NF perception is yet demonstrated only for few LysM-RLKs, it is likely that polymorphism of these protein families has an important role for the variation and diversity of AFN. The ectopic expression of *L. japonicus* LjNFR1 and LjNFR5 receptors in transgenic *M. truncatula* roots indeed allowed the formation of nodules by *Mesorhizobium loti* that is not a natural host of *M. truncatula* (Radutoiu et al., 2007). Evidences for a positive selection pressure on residues of the LysM domain of LYK3 have been shown, suggesting a role of LYK3 in controlling host specificity (Sulima et al., 2017). In the case of the specific interaction between the *P. sativum* cultivar 'Afghanistan' and the *Rlv* strain TOM, the pea *sym2* allele, encoding a LysM-RLK, prevented the nodulation of European *Rlv* strains lacking the *nodX* gene (Davis et al., 1988; Firmin et al., 1993; Lie, 1984; Sulima et al., 2017). However, all symbiotic functions of LysM-RLKs may be not exclusively related to NFs and some SPs directly interact with LysM domains (Kawaharada et al., 2015). SPs are composed of extracellular polysaccharides (EPS), capsular polysaccharides (KPS) and lipopolysaccharides (LPS), and are implicated at various stages of the legume-rhizobium interaction, from root attachment to nodule colonization (for a review see Janczarek, Rachwał, Marzec, Grządziel, & Palusińska-Szysz, 2015; Kelly, Radutoiu, & Stougaard, 2017). The chemical structure of SPs, mainly composed of a mono or polysaccharide core unit decorated with various substitutions, can be variable and therefore may modulate the partner choice.

2.1.4 AFN can depend on the secretion of bacterial effectors

Bacterial effectors are released by secretion systems in the extracellular medium or directly inside plant root cells, and some of these are proposed to modulate the legume-rhizobium symbiosis (Costa et al., 2015; Miwa & Okazaki, 2017). Until today, type I, III and VI secretion systems were involved in legume-rhizobium symbioses. The type I secretion system (T1SS), composed of transmembrane proteins such as PrsD and PrsE, can directly translocate proteins into the extracellular environment. This is the case for the Ca^{2+}-binding protein NodO, as well as for glycanases and adhesins (Economou, Hamilton, Johnston, & Downie, 1990; Russo et al., 2006). The type III secretion system (T3SS) is a complex of about twenty different proteins which form a syringe embedded in both bacterial and plant root host cell membranes (Miwa & Okazaki, 2017; Staehelin & Krishnan, 2015). The T3SS has been initially implicated in the secretion of bacterial effectors during plant pathogenic interactions (Miwa & Okazaki, 2017; Staehelin & Krishnan, 2015). T3SS effectors were identified in several rhizobial species like *B. japonicum*, *B. elkani*, *S. fredii*, but not in *R. leguminosarum* and *Sinorhizobium meliloti*. Proteins secreted via the T3SS in host root cells, which are called Nops for Nodulation outer proteins, influence the symbiosis establishment (Miwa & Okazaki, 2017). NodD regulates the *type III secretion system 1* (*tts1*) bacterial transcription factor that drives effector gene expression (Okazaki, Kaneko, Sato, & Saeki, 2013). Interestingly, in *B. japonicum*, the two effectors NopE1 and NopE2 stimulate nodulation in soybean but repress it in mungbean (Krause, Doerfel, & Göttfert, 2002), illustrating that T3SS can modulate the AFN. Some effectors are released through a type VI secretion system (T6SS), composed of membrane and tail complexes that translocate bacterial effectors into host cells, and which was initially linked to pathogenic bacteria effector secretion (for review see Costa et al., 2015). Despite the T6SS was suspected to inhibit the *R. leguminosarum* – pea symbiosis (Bernal, Llamas, & Filloux, 2018) the specific contribution of such secreted effectors in Fabeae-rhizobium early partner choice remains to be demonstrated.

Some effectors released in legume host cells modulate plant immunity, illustrating that a crosstalk between immunity and early partner choice likely exist during nodulation (Gourion, Berrabah, Ratet, & Stacey, 2015; Miwa & Okazaki, 2017). A soybean R gene (*rj2*) was implicated in the decrease of AFN, suggesting that establishment of the root nodule symbiosis requires the evasion of plant immune responses triggered by rhizobial effectors (Yang, Tang, Gao, Krishnan, & Zhu, 2010). Several evidences confirmed the role of secretion systems in the regulation of plant immune responses. The NopL and NopM T3SS-secreted proteins of the rhizobium NGR234 strain were indeed shown to suppress defense responses in *N. benthamiana* (Bartsev et al., 2004; Xin et al., 2012). In addition, T3SS-secreted effector proteins of the *B. elkanii* USDA61 bacterial strain were shown to highjack the NF-dependent pathway by suppressing plant defense responses in the soybean cultivar Clark-rj-1 (Okazaki et al., 2013).

2.2 Competitiveness to form root nodules (CFN) is a determinant of early partner choice in rhizobial populations

The capacity of a rhizobium to form nodules with a legume in mono-inoculation experiments does not necessarily predict the competitive success of the interaction in a context of multiple compatible partners (Fig. 1). Different rhizobia can share similar AFN but display a wide range of CFN in mixture, depending on the legume host (Bourion et al., 2018). Although *Rlv* are generally able to form nodules with pea or fababean when mono-inoculated, these two plant species exposed to complex soil *Rlv* populations select preferentially different *Rlv* genotypes (Laguerre et al., 2003; Mutch &

Young, 2004). Preferential nodulation within compatible partners populations was also evidenced in the soybean-Bradyrhizobium symbiosis (Cregan, Keyser, & Sadowsky, 1989; Keyser & Cregan, 1987). More recently, a co-inoculation of *Rlv* genotypes in mixture with a panel of pea accessions resulted in preferential symbiotic associations depending on pea genotypes (Bourion et al., 2018). Although the general concept of CFN has been described since several decades, very little is known about mechanisms controlling this process (Amarger, 1981; Triplett & Sadowsky, 1992). The hypothesis of a differential proliferation of competitive bacterial genotypes in the host rhizosphere cannot be ruled out, but received little experimental support (Moawad, Ellis, & Schmidt, 1984). In most rhizobia, a preferential association between the partner choice and the NF gene diversity supports the hypothesis that signaling pathways associated to early symbiont recognition are the major drivers of CFN (Kumar et al., 2015; Laguerre et al., 2003; Triplett & Sadowsky, 1992). As described above, mechanisms controlling the symbiotic establishment can drastically differ among both host legumes and rhizobia genotypes (i.e. flavonoids, NFs, LysM-RLKs, SPs, TSSs) and may explain some variations in rhizobial CFN in natural populations. For instance, when the *Rlv* strain TOM is mixed with strains that do not express NodX and inoculated with the *P. sativum* cultivar 'Afghanistan' (nodulating only with the strain TOM), the NFs produced by the non-nodulating strain compete with NFs produce by TOM and drastically reduce its nodulation capacity (Dowling, Stanley, & Broughton, 1989; Hogg, Davies, Wilson, Bisseling, & Downie, 2002; Winarno & Lie, 1979). In soil bacterial populations, NFs may be provided in *trans* between individuals, allowing bacteria unable to produce NFs by themselves to form nodules (Borthakur, Barker, Latchford, Rossen, & Johnston, 1988). Other mechanisms such as antibiosis (antibiotics released in the extracellular medium by rhizobia and non-rhizobia) may also modulate the CFN (Triplett & Barta, 1987; Robleto, Scupham, & Triplett, 1997; for a review see; Naamala, Jaiswal, & Dakora, 2016).

3. Post-infection partner preference in the legume-rhizobium symbiosis

3.1 Integration of legume-rhizobium interactions at the whole plant level

Symbiotic N_2 Fixation (SNF) efficiency of the legume-rhizobium associations varies among bacteria able to infect legume roots and form nodules (Laguerre et al., 2012; Terpolilli, O'Hara, Tiwari, Dilworth, & Howieson, 2008; Triplett & Sadowsky, 1992). Moreover, a single strain may result in different levels of N acquisition when mono-inoculated on different compatible host legumes (Bourion et al., 2018; Kazmierczak et al., 2017). These different capacities depend on symbiotic partner couples as, for example, the *S. meliloti* strain RCR2011 was characterized as a poor N_2 fixer in *M. truncatula* but as an efficient fixer in *M. sativa* (Terpolilli et al., 2008). As bacterial populations in the soil contain multiple competitive rhizobia, it is rather frequent that bacteria with different levels of N_2 fixation efficiency form nodules on roots (Laguerre et al., 2003; Mutch & Young, 2004). Therefore, symbiotic legume root systems in soils are expected to be a mosaic of nodule organs with various level of SNF efficiency. However, the plant influences the dynamic of the interaction and activates Post-Infection Partner Preference (PIPP) mechanisms that favor the most efficient symbiotic organs to the expenses of the inefficient ones, in order to satisfy the plant N demand (Fig. 1).

Although plant N deficit is required to enable rhizobium-legume symbiosis (Streeter & Wong, 1988), there is little evidence suggesting its role in early choices between various compatible symbiotic partners (Grillo, Stinchcombe, & Heath, 2016). Indeed, the capacity of the legume-rhizobium association to fix N_2 has little impact on partner choice until differentiation of the mature N_2 fixing symbiotic organ occurs, and variations in early partner choice depend mainly on the capacities of bacteria to infect and colonize root cells rather than on whole plant C—N trade-offs. Indeed, studies on *M. truncatula*/*Sinorhizobium* (Amarger, 1981; Laguerre et al., 2012) or pea/*R. leguminosarum* (Westhoek et al., 2017; Bourion et al., 2018) showed that inefficient N fixing bacteria genotypes, either corresponding to artificial mutants or natural isolates, can be as compatible and competitive as efficient N_2 fixing bacteria, indicating that CFN is likely independent of SNF efficiency. Consequently, N_2 fixing inefficient bacteria, so-called "cheaters", are frequently present in a fraction of the nodules of an host plant (Fujita, Aoki, & Kawaguchi, 2014). In addition, mixtures of *M. truncatula* compatible rhizobia displaying different levels of SNF efficiency were inoculated on plants under contrasted N regimes (Laguerre et al., 2012), revealing an effect of the N status of the plant on nodule numbers but not on CFN. The plant is therefore most probably unable to discriminate between inefficient and efficient bacteria as a function of its N demand before N_2 fixation is indeed operating. However, as the mature symbiotic organs become efficient for N_2 fixation, fluxes of N and C metabolites between the plant and the mature symbiotic organs impact significantly the whole plant growth and consequently its nutritional demand (Fig. 1). This then allows triggering plant N signaling mechanisms that modulate the symbiotic interaction and the partner preference. When a mixture of mature N_2 fixing efficient and inefficient nodules is present within the same root system, the plant develop mechanisms stimulating the development of efficient structures to the expense of inefficient ones, in order to satisfy its N demand (Laguerre et al., 2012). It has been shown that regulation of plant nitrate acquisition is highly integrated at the whole plant level (for reviews see Bouguyon, Gojon, & Nacry, 2012; Ruffel, Gojon, & Lejay, 2014). The presence of nitrate induces both local and whole plant responses: nitrate stimulates locally root proliferation, nitrate uptake and nitrate assimilation whereas accumulation of downstream N metabolites by the whole plant triggers a systemic feedback repression of nitrate acquisition and root development when the N demand is satisfied. C metabolites accumulation also contributes to the regulation of nitrate acquisition: the level of photosynthates, translocated from shoots to roots, stimulates the root capacity to acquire nitrate (Lejay et al., 2003). It is well known from early studies that the addition of mineral N or of N metabolites strongly inhibits nodulation (Parsons, Stanforth, Raven, & Sprent, 1993; Streeter & Wong, 1988) but mechanisms adjusting symbiotic capacities to the nutritional demand of the legume plant are not well documented. However, several evidences argue that reminiscent mechanisms also exist in the case of SNF.

3.2 Control of the symbiosis by local N availability

Evidences for a local regulation of mature symbiotic organs by the efficiency of SNF have been reported. Suppressing N_2 fixation by Ar/O_2 treatments result locally in a rapid inhibition of nodule growth (Kiers, Rousseau, West, & Denison, 2003; Jeudy et al., 2010). It was proposed that the plant activates nodule autonomous mechanisms to locally restrict the development of nodules formed with ineffective bacteria (Kiers et al., 2003; Oono, Anderson, & Denison, 2011). Long term Ar/O_2 treatments in determinate legumes resulted in a decrease of bacteroids fitness in nodules, associated with an early senescence (Kiers et al., 2003; Oono, Denison, & Kiers, 2009). This result was interpreted as a "host sanction" toward the less beneficial partners. Indeed, in an evolutionary point of view, such sanction would tend to limit ineffective rhizobia multiplication and dispersion when they are released in the soil (Fig. 1). The concept has been extended to indeterminate nodules although bacteroids are, in this case, terminally differentiated and that only undifferentiated bacteria are able to multiply (Oono & Denison, 2010; Oono et al., 2011). The general small size and the early developmental arrest of nodules formed by fix$^-$ mutant bacteria is globally in agreement with a concept that an absence of N_2 fixation is associated to an inhibition of nodule development. Although reports in soybean/*Bradyrhizobium* and *Medicago*/*Sinorhizobium* symbioses indicate that mutations suppressing N_2 fixation do not necessarily impact the reproductive fitness of rhizobia present in nodules (Laguerre et al., 2012; Marco et al., 2009), evidences for a long-term stimulation of the reproductive fitness of fix$^+$ *Cupriavidus taiwanensis* bacteria versus fix$^-$ bacteria in co-infected root systems of *Mimosa pudica* have also been reported (Daubech et al., 2017). Overall, these data demonstrate a control of symbiotic organ development by SNF efficiency. Mechanisms related to nodule oxygen permeability or pH have been additionally proposed to be instrumental in the local control of legume-rhizobium symbiotic organs by N signaling (Hunt & Layzell, 1993; Kiers et al., 2003) but, up to date, mechanisms involved remain to be precisely elucidated.

3.3 Control of symbiosis by the whole plant N demand

Symbiotic organ development is also under the control of systemic signals integrated at the whole plant level. Nodule formation is inhibited by pre-existing nodules through the autoregulation of nodulation (AON) pathway (review by Ferguson et al., 2019). A role of the N status of the plant in AON was suggested by the ability of several AON mutants to nodulate in the presence of high concentrations of nitrate that normally systemically inhibits nodulation in wild type plants. Conversely, a plant N deficit releases this repressive mechanism resulting in the stimulation of nodule formation (Jeudy et al., 2010). This contributes to the adjustment of nodule number to the N demand of the plant, and may explain the stimulation of nodulation frequently observed in plant mono-inoculated with poorly N_2 fixing rhizobia (Bourion et al., 2018). Systemic regulations of mature symbiotic organs function and development by the whole plant N demand have been evidenced in *M. truncatula* using split root systems (Jeudy et al., 2010; Laguerre et al., 2012; Ruffel et al., 2008). The whole plant N satiety systemic signaling results in a rapid repression of the N_2 fixation activity as well as in the initiation of nodule senescence. Whether this repression is the consequence of senescence or a direct modulation of the N_2 fixation activity is not known. A change in permeability to oxygen within the nodule (so called the "oxygen barrier") has been also proposed (Hunt & Layzell,

1993), but whether this change is the consequence or the cause of the reduction of N_2 fixation remains once more a matter of debate. A whole plant N deficit obtained either by Ar/O_2 treatments or by inoculation with fix^- mutant bacteria (fix^- roots) in a split-root experimental system triggers a compensatory response on the remaining fix^+ roots (Jeudy et al., 2010; Laguerre et al., 2012). It is firstly observed a systemic stimulation of nodule expansion, and secondly the formation of additional symbiotic organs (Jeudy et al., 2010; Laguerre et al., 2012). Whether this developmental response is associated to the upregulation of the specific N_2 fixation activity (per gram of nodule) remains controversial (Laguerre et al., 2012; Ruffel et al., 2008; Sulieman & Schulze, 2010). Systemic mechanisms adjusting the N_2 fixation capacity to the symbiotic plant N demand are instrumental for the root nutrient "foraging", which is relevant to optimize plant growth in soil conditions that are intrinsically highly variable in time and space. These systemic regulations enable the plant to compensate a lower N_2 fixation in the inefficient part of its symbiotic root system by stimulating the symbiotic development of the remaining beneficial roots. In contrast to the local regulation described above ("sanction"), this mechanism integrating the N_2 fixation capacity to the plant N demand is likely to operate at the scale of the root bundle rather than at the scale of the individual nodule (Laguerre et al., 2012). A plant facing a uniform reduction of its symbiotic root capacity (e.g. having both fix+ and fix-nodules uniformly distributed on its root system) is unable to triggers this systemic response probably because it is not possible to discriminate between efficient and inefficient root bundles and therefore to allocate resources preferentially to the efficient symbiotic roots (Laguerre et al., 2012).

Very little is known about the nature of systemic N signals (Parsons et al., 1993). As addition of amino acids in plant nutrient media generally rapidly represses N_2 fixation, a pioneer hypothesis was that products of N assimilation, namely amino acid translocated through the phloem from shoots to roots, may act as mobile signals (Bacanamwo & Harper, 1997; Sulieman, Fischinger, Gresshoff, & Schulze, 2010; Sulieman & Schulze, 2010). However, this hypothesis was never supported by a clear demonstration, as amino acids are also N sources rapidly assimilated and are able to trigger a repression of N_2 fixation without inducing a direct signaling effect. Mechanisms behind the systemic regulation of symbiosis by N therefore remain mostly not known. AON is suggested to be implicated in the regulation of nodule formation by the N demand (Fig. 1; reviewed by Ferguson et al., 2019), as AON hypernodulating mutants are generally able to form nodules independently of the presence of a high level of mineral N, as previously mentioned (Krusell et al., 2002; Olsson, Nakao, Bohlool, & Gresshoff, 1989). However (1) N is not the only driver of AON, as the AON pathway is activated far before the nodule becomes N_2 fixing (Kosslak & Bohlool, 1984; Kosslak, Bohlool, Dowdle, & Sadowsky, 1983); and (2), the nodule expansion response to plant N deficit remains active in hypernodulating mutants, suggesting that additional AON-independent mechanisms are implicated (Jeudy et al., 2010; Kassaw, Bridges, & Frugoli, 2015). Interestingly, another systemic mechanism acting in legumes in parallel of the classical AON pathway to positively regulate nodule formation was discovered recently (Huault et al., 2014; Laffont et al., 2019), but its function in symbiotic N signaling remain to be fully understood.

3.4 Differential allocation of C metabolites by the plant to the nodules

Symbiotic organs require large fluxes of C metabolites (namely sucrose and organic acids) translocated from photosynthetic shoots to nodules through the phloem. The main part of these C fluxes is required to fuel N_2 fixation both as energy, as reducing power sources, and as C acceptors for ammonium assimilation (Oldroyd et al., 2011). C metabolites are also major resources required for root and nodule development (Fig. 1). Amino acids are products of C and N metabolisms that are tightly adjusted (Voisin, Salon, Jeudy, & Warembourg, 2003). A high CO_2 level increases photosynthesis and C metabolites production in the shoot, therefore enhancing N_2 fixation in roots and the nodule biomass (Rogers et al., 2006; Rogers, Ainsworth, & Leakey, 2009). As an elevation of CO_2 is not generally associated to an increase of nitrate uptake from the soil, it was suggested that symbiotic N_2 fixation will become more ecologically competitive compared to nitrate acquisition in a context of climate change associated to a strong increase of atmospheric CO_2 concentrations (Rogers et al., 2009). It is generally assumed that the C provision to the nodule limits N_2 fixation efficiency (Rogers et al., 2006). Variations of N_2 fixation capacity of symbiotic associations may be associated to differences in the C use efficiency of bacteroids. For example, the activity of so-called the gamma-aminobutyric acid "shunt" of the bacteroid tricarboxylic acid cycle (Sulieman & Schulze, 2010), or plant/bacteroid amino acid exchanges (Lodwig et al., 2003; Prell et al., 2009) have been implicated in these variations. The hypothesis that differential allocations of C metabolites may be a driver to pilot symbiotic organ development and function has been frequently suggested (Voisin et al., 2003 and references therein), but only few studies have directly investigated this hypothesis (Jeudy et al., 2010). Indeed, as C metabolites fuel nodule functioning and development, C allocation and N_2 fixation are always correlated and therefore, specific signaling associated to C allocation is difficult to discriminate from N signaling (Fig. 1). Nevertheless, a ^{13}C labeling experiment has shown that the systemic stimulation of nodule development in response to a local N deficit is correlated to very rapid changes of C metabolite allocation toward symbiotic organs before any measurable response on nodule development can be detected (Jeudy et al., 2010). This is in agreement with the hypothesis that plant C metabolites allocation to the bacteroid mediates at least a part of the plant N demand systemic response. According to this model, C allocation to nodules may be a driver allowing the plant to pilot the whole plant systemic preference toward N_2 fixing efficient structures (i.e. with efficient partners) to the expense of the inefficient ones (i.e. with inefficient partners). However, how the plant perceive individual nodule N_2 fixation efficiency and how the plant controls the differential C metabolite allocation toward N_2 fixing efficient versus inefficient nodules is unknown and will deserve future investigations.

4. Conclusions

Symbiotic legumes are strategical for sustainable agriculture as well as a source of proteins for both human and animal food. Genetic selection during domestication improved their performances to agriculture needs (i.e. increased grain yields and protein content; Duc et al., 2015; Zhou et al., 2015), however the symbiotic interaction with rhizobia was rarely considered as a selection trait. Consequently, plant selection was even occasionally associated to a reduction of symbiotic capacities (Pérez-Jaramillo, Mendes, & Raaijmakers, 2016; Weese, Heath, Dentinger, & Lau, 2015). Inoculation strategies were then designed to bring in field more efficient N-fixing rhizobia for different legume crops (Triplett & Sadowsky, 1992). Nevertheless, inoculation strategies frequently failed because highly efficient selected strains were outcompeted by indigenous bacteria, which were highly competitive but poorly N_2 fixing efficient with host plants (Triplett & Sadowsky, 1992). Identifying rhizobium genetic determinants controlling AFN and CFN is therefore a key to allow selecting bacteria well adapted to inoculation strategies, i.e. both competitive and efficient. As the effect of biotic and abiotic soil environmental variations on AFN and CFN are almost unknown, a new frontier is to decipher and predict their impacts on early partner choice, in order to improve and stabilize inoculation strategies.

Legume genetic selection seems to have sometime reduced their wild capacities to interact with the rhizobial genetic diversity existing in natural soil populations (Bourion et al., 2018; Kim et al., 2014; Mutch & Young, 2004), and in laboratory conditions legume productivity can be decreased when associated with a mixture of rhizobium genotypes compared to mono-inoculations (Barrett, Bever, Bissett, & Thrall, 2015). Some domesticated soybean cultivars have an improved SNF when mono-inoculated with elite bacteria than wild soybeans (Muñoz et al., 2016). However, the gain is lower for the domesticated genotypes than for wild soybeans when a mixture of effective and ineffective rhizobial strains is used as inoculant (Kiers, Hutton, & Denison, 2007). As legume genetic selection programs using single-strain inoculation tend to select specific traits associated to a specific bacterial strain, selection of new cultivars able to maintain high symbiotic performances upon multi-inoculation may be a new strategy to improve symbiotic capacities of legumes. The establishment of rhizobial core collections covering geographical, genomic, and symbiotic (i.e. *nod* genes ...) diversities, could be interesting tools to measure legume agronomic traits when challenged with multiple different partners.

Rhizobia are not obligate symbionts and can then be found as free-living bacteria in soils, meaning that their interaction with legumes is therefore only optional for their maintenance, survival and multiplication. Consequently, the dynamic of rhizobial populations cannot be explained only on the basis of their symbiotic interaction traits, and other selection pressures drive genetic population structures in soils. Notably, certain genotypes of rhizobia are able to colonize the endorhizosphere of non-legumes and to act potentially as Plant-Growth Promoting Rhizobacteria (PGPR; Tan, Radziah, Halimi, Khairuddin, & Shamsuddin, 2015; Adnan et al., 2016; Santoyo, Moreno-Hagelsieb, del Carmen Orozco-Mosqueda, & Glick, 2016; Vargas et al., 2017). In this case, the plant endorhizosphere likely plays the role of an ecological niche that favors bacterial proliferation, similarly as legume nodules. Therefore, soils contain a large set of rhizobia having heterogeneous behaviors with various plants, including non-legumes, and their soil symbiotic capacities are then very difficult to predict. New sequencing technologies, such as metagenomics and metabarcoding, have allowed to better describe dynamics of soil bacterial communities (Daniel, 2005; Taberlet, Coissac, Pompanon, Brochmann, & Willerslev, 2012; Tringe et al., 2005), but these methods cannot predict symbiotic functions yet.

However, the use of bacterial genetic markers, associated with symbiotic traits such as AFN, CFN and PIPP, should be a way for improving the prediction of soil symbiotic capacities.

References

Adnan, M., Shah, Z., Arif, M., Khan, M. J., Mian, I. A., Sharif, M., et al. (2016). Impact of rhizobial inoculum and inorganic fertilizers on nutrients (NPK) availability and uptake in wheat crop. *Canadian Journal of Soil Science, 96*(2), 169–176. https://doi.org/10.1139/cjss-2016-0012.

Amarger, N. (1981). Competition for nodule formation between effective and ineffective strains of Rhizobium meliloti. *Soil Biology and Biochemistry, 13*(6), 475–480. https://doi.org/10.1016/0038-0717(81)90037-7.

Bacanamwo, M., & Harper, J. E. (1997). The feedback mechanism of nitrate inhibition of nitrogenase activity in soybean may involve asparagine and/or products of its metabolism. *Physiologia Plantarum, 100*(2), 371–377. https://doi.org/10.1111/j.1399-3054.1997.tb04795.x.

Bailly, X., Giuntini, E., Sexton, M. C., Lower, R. P., Harrison, P. W., Kumar, N., et al. (2011). Population genomics of Sinorhizobium medicae based on low-coverage sequencing of sympatric isolates. *The ISME Journal, 5*(11), 1722–1734. https://doi.org/10.1038/ismej.2011.55.

Barrett, L. G., Bever, J. D., Bissett, A., & Thrall, P. H. (2015). Partner diversity and identity impacts on plant productivity in Acacia–rhizobial interactions. *Journal of Ecology, 103*(1), 130–142. https://doi.org/10.1111/1365-2745.12336.

Bartsev, A. V., Deakin, W. J., Boukli, N. M., McAlvin, C. B., Stacey, G., Malnoë, P., et al. (2004). NopL, an effector protein of Rhizobium sp. NGR234, thwarts activation of plant defense reactions. *Plant Physiology, 134*(2), 871–879. https://doi.org/10.1104/pp.103.031740.

Begum, A. A., Leibovitch, S., Migner, P., & Zhang, F. (2001). Specific flavonoids induced nod gene expression and pre-activated nod genes of Rhizobium leguminosarum increased pea (Pisum sativum L.) and lentil (Lens culinaris L.) nodulation in controlled growth chamber environments. *Journal of Experimental Botany, 52*(360), 1537–1543. https://doi.org/10.1093/jexbot/52.360.1537.

Bernal, P., Llamas, M. A., & Filloux, A. (2018). Type VI secretion systems in plant-associated bacteria. *Environmental Microbiology, 20*(1), 1–15. https://doi.org/10.1111/1462-2920.13956.

Borthakur, D., Barker, R. F., Latchford, J. W., Rossen, L., & Johnston, A. W. B. (1988). Analysis of pss genes of Rhizobium leguminosarum required for exopolysaccharide synthesis and nodulation of peas : Their primary structure and their interaction with psi and other nodulation genes. *Molecular and General Genetics MGG, 213*(1), 155–162. https://doi.org/10.1007/BF00333413.

Bouguyon, E., Gojon, A., & Nacry, P. (2012). Nitrate sensing and signaling in plants. *Seminars in Cell & Developmental Biology, 23*(6), 648–654. https://doi.org/10.1016/j.semcdb.2012.01.004.

Bourion, V., Heulin-Gotty, K., Aubert, V., Tisseyre, P., Chabert-Martinello, M., Pervent, M., et al. (2018). Co-inoculation of a pea core-collection with diverse rhizobial strains shows competitiveness for nodulation and efficiency of nitrogen fixation are distinct traits in the interaction. *Frontiers of Plant Science, 8*. https://doi.org/10.3389/fpls.2017.02249.

Cooper, J. E. (2004). Multiple responses of rhizobia to flavonoids during legume root infection. In *Incorporating advances in plant pathology: Vol. 41. Advances in botanical research* (pp. 1–62). https://doi.org/10.1016/S0065-2296(04)41001-5.

Costa, T. R. D., Felisberto-Rodrigues, C., Meir, A., Prevost, M. S., Redzej, A., Trokter, M., et al. (2015). Secretion systems in Gram-negative bacteria : Structural and mechanistic insights. *Nature Reviews Microbiology, 13*(6), 343–359. https://doi.org/10.1038/nrmicro3456.

Cregan, P. B., Keyser, H. H., & Sadowsky, M. J. (1989). Soybean genotype restricting nodulation of a previously unrestricted serocluster 123 bradyrhizobia. *Crop Science, 29*(2), 307–312. https://doi.org/10.2135/cropsci1989.0011183X002900020015x.

Daniel, R. (2005). The metagenomics of soil. *Nature Reviews Microbiology, 3*(6), 470–478. https://doi.org/10.1038/nrmicro1160.

Daubech, B., Remigi, P., Doin de Moura, G., Marchetti, M., Pouzet, C., Auriac, M.-C., et al. (2017). Spatio-temporal control of mutualism in legumes helps spread symbiotic nitrogen fixation. *ELife, 6*. https://doi.org/10.7554/eLife.28683.

Davis, E. O., Evans, I. J., & Johnston, A. W. (1988). Identification of nodX, a gene that allows Rhizobium leguminosarum biovar viciae strain TOM to nodulate Afghanistan peas. *Molecular & General Genetics: MGG, 212*(3), 531–535.

Dénarié, J., Debellé, F., & Promé, J. C. (1996). Rhizobium lipo-chitooligosaccharide nodulation factors : Signaling molecules mediating recognition and morphogenesis. *Annual Review of Biochemistry, 65*, 503–535. https://doi.org/10.1146/annurev.bi.65.070196.002443.

Djordjevic, M. A., Schofield, P. R., & Rolfe, B. G. (1985). Tn5 mutagenesis of Rhizobium trifolii host-specific nodulation genes result in mutants with altered host-range ability. *Molecular and General Genetics MGG, 200*(3), 463–471. https://doi.org/10.1007/BF00425732.

Dowling, D. N., Stanley, J., & Broughton, W. J. (1989). Competitive nodulation blocking of Afghanistan pea is determined by nodDABC and nodFE alleles in Rhizobium leguminosarum. *Molecular and General Genetics MGG, 216*(1), 170–174. https://doi.org/10.1007/BF00332247.

Duc, G., Agrama, H., Bao, S., Berger, J., Bourion, V., Ron, A. M. D., et al. (2015). Breeding annual grain legumes for sustainable Agriculture : New methods to approach complex traits and target new cultivar ideotypes. *Critical Reviews in Plant Sciences, 34*(1–3), 381–411. https://doi.org/10.1080/07352689.2014.898469.

Economou, A., Hamilton, W. D., Johnston, A. W., & Downie, J. A. (1990). The Rhizobium nodulation gene nodO encodes a Ca2(+)-binding protein that is exported without N-terminal cleavage and is homologous to haemolysin and related proteins. *The EMBO Journal, 9*(2), 349–354. https://doi.org/10.1002/j.1460-2075.1990.tb08117.x.

Fellay, R., Hanin, M., Montorzi, G., Frey, J., Freiberg, C., Golinowski, W., et al. (1998). nodD2 of Rhizobium sp. NGR234 is involved in the repression of the nodABC operon. *Molecular Microbiology, 27*(5), 1039–1050.

Ferguson, B. J., Mens, C., Hastwell, A. H., Zhang, M., Su, H., Jones, C. H., et al. (2019). Legume nodulation : The host controls the party. *Plant, Cell and Environment, 42*(1), 41–51. https://doi.org/10.1111/pce.13348.

Fierer, N., & Jackson, R. B. (2006). The diversity and biogeography of soil bacterial communities. *Proceedings of the National Academy of Sciences of the United States of America, 103*(3), 626–631. https://doi.org/10.1073/pnas.0507535103.

Firmin, J. L., Wilson, K. E., Carlson, R. W., Davies, A. E., & Downie, J. A. (1993). Resistance to nodulation of cv. Afghanistan peas is overcome by nodX, which mediates an O-acetylation of the Rhizobium leguminosarum lipo-oligosaccharide nodulation factor. *Molecular Microbiology, 10*(2), 351–360. https://doi.org/10.1111/j.1365-2958.1993.tb01961.x.

Firmin, J. L., Wilson, K. E., Rossen, L., & Johnston, A. W. B. (1986). Flavonoid activation of nodulation genes in Rhizobium reversed by other compounds present in plants. *Nature, 324*(6092), 90. https://doi.org/10.1038/324090a0.

Fujita, H., Aoki, S., & Kawaguchi, M. (2014). Evolutionary dynamics of nitrogen fixation in the legume—rhizobia symbiosis. *PLoS One, 9*(4), e93670. https://doi.org/10.1371/journal.pone.0093670.

Gagnon, H., & Ibrahim, R. K. (1998). Aldonic Acids : A novel family of nod gene inducers of Mesorhizobium loti, Rhizobium lupini, and Sinorhizobium meliloti. *Molecular Plant-Microbe Interactions, 11*(10), 988—998. https://doi.org/10.1094/MPMI.1998.11.10.988.

Gough, C., Cottret, L., Lefebvre, B., & Bono, J.-J. (2018). Evolutionary history of plant LysM receptor proteins related to root endosymbiosis. *Frontiers of Plant Science, 9*. https://doi.org/10.3389/fpls.2018.00923.

Gourion, B., Berrabah, F., Ratet, P., & Stacey, G. (2015). Rhizobium—legume symbioses : The crucial role of plant immunity. *Trends in Plant Science, 20*(3), 186—194. https://doi.org/10.1016/j.tplants.2014.11.008.

Grillo, M. A., Stinchcombe, J. R., & Heath, K. D. (2016). Nitrogen addition does not influence pre-infection partner choice in the legume-rhizobium symbiosis. *American Journal of Botany, 103*(10), 1763—1770. https://doi.org/10.3732/ajb.1600090.

Hogg, B., Davies, A. E., Wilson, K. E., Bisseling, T., & Downie, J. A. (2002). Competitive nodulation blocking of cv. Afghanistan pea is related to high levels of nodulation factors made by some strains of *Rhizobium leguminosarum* bv. Viciae. *Molecular Plant-Microbe Interactions, 15*(1), 60—68. https://doi.org/10.1094/MPMI.2002.15.1.60.

Hong, G. F., Burn, J. E., & Johnston, A. W. (1987). Evidence that DNA involved in the expression of nodulation (nod) genes in Rhizobium binds to the product of the regulatory gene nodD. *Nucleic Acids Research, 15*(23), 9677—9690.

Huault, E., Laffont, C., Wen, J., Mysore, K. S., Ratet, P., Duc, G., et al. (2014). Local and systemic regulation of plant root system Architecture and symbiotic nodulation by a receptor-like kinase. *PLoS Genetics, 10*(12), e1004891. https://doi.org/10.1371/journal.pgen.1004891.

Hunt, S., & Layzell, D. B. (1993). Gas exchange of legume nodules and the regulation of nitrogenase activity. *Annual Review of Plant Physiology and Plant Molecular Biology, 44*(1), 483—511. https://doi.org/10.1146/annurev.pp.44.060193.002411.

Janczarek, M., Rachwał, K., Marzec, A., Grządziel, J., & Palusińska-Szysz, M. (2015). Signal molecules and cell-surface components involved in early stages of the legume-rhizobium interactions. *Applied Soil Ecology, 85*, 94—113.

Jeudy, C., Ruffel, S., Freixes, S., Tillard, P., Santoni, A. L., Morel, S., et al. (2010). Adaptation of Medicago truncatula to nitrogen limitation is modulated via local and systemic nodule developmental responses. *New Phytologist, 185*(3), 817—828. https://doi.org/10.1111/j.1469-8137.2009.03103.x.

Jiménez-Guerrero, I., Acosta-Jurado, S., Del Cerro, P., Navarro-Gómez, P., López-Baena, F. J., Ollero, F. J., et al. (2017). Transcriptomic studies of the effect of nod gene-inducing molecules in Rhizobia : Different weapons, one purpose. *Genes, 9*(1). https://doi.org/10.3390/genes9010001.

Kaiser, K., Wemheuer, B., Korolkow, V., Wemheuer, F., Nacke, H., Schöning, I., et al. (2016). Driving forces of soil bacterial community structure, diversity, and function in temperate grasslands and forests. *Scientific Reports, 6*, 33696. https://doi.org/10.1038/srep33696.

Kaneko, T., Nakamura, Y., Sato, S., Minamisawa, K., Uchiumi, T., Sasamoto, S., et al. (2002). Complete genomic sequence of nitrogen-fixing symbiotic Bacterium Bradyrhizobium japonicum USDA110. *DNA Research, 9*(6), 189—197. https://doi.org/10.1093/dnares/9.6.189.

Kassaw, T., Bridges, W., & Frugoli, J. (2015). Multiple autoregulation of nodulation (AON) signals identified through split root analysis of Medicago truncatula sunn and rdn1 mutants. *Plants, 4*(2), 209–224. https://doi.org/10.3390/plants4020209.

Kawaharada, Y., Kelly, S., Nielsen, M. W., Hjuler, C. T., Gysel, K., Muszyński, A., et al. (2015). Receptor-mediated exopolysaccharide perception controls bacterial infection. *Nature, 523*(7560), 308–312. https://doi.org/10.1038/nature14611.

Kazmierczak, T., Nagymihaly, M., Lamouche, F., BARRIERE, Q., Guefrachi, I., Alunni, B., et al. (2017). Specific host-responsive associations between Medicago truncatula accessions and Sinorhizobium strains. *Molecular Plant-Microbe Interactions, 30*(5), 399–409. https://doi.org/10.1094/MPMI-01-17-0009-R.

Kelly, S., Radutoiu, S., & Stougaard, J. (2017). Legume LysM receptors mediate symbiotic and pathogenic signalling. *Current Opinion in Plant Biology, 39*, 152–158. https://doi.org/10.1016/j.pbi.2017.06.013.

Keyser, H. H., & Cregan, P. B. (1987). Nodulation and competition for nodulation of selected soybean genotypes among Bradyrhizobium japonicum serogroup 123 isolates. *Applied and Environmental Microbiology, 53*(11), 2631–2635.

Kiers, E. T., Hutton, M. G., & Denison, R. F. (2007). Human selection and the relaxation of legume defences against ineffective rhizobia. *Proceedings of the Royal Society B: Biological Sciences, 274*(1629), 3119–3126. https://doi.org/10.1098/rspb.2007.1187.

Kiers, E. T., Rousseau, R. A., West, S. A., & Denison, R. F. (2003). Host sanctions and the legume-rhizobium mutualism. *Nature, 425*(6953), 78–81. https://doi.org/10.1038/nature01931.

Kim, D. H., Kaashyap, M., Rathore, A., Das, R. R., Parupalli, S., Upadhyaya, H. D., et al. (2014). Phylogenetic diversity of Mesorhizobium in chickpea. *Journal of Biosciences, 39*(3), 513–517.

Klümper, U., Riber, L., Dechesne, A., Sannazzarro, A., Hansen, L. H., Sørensen, S. J., et al. (2015). Broad host range plasmids can invade an unexpectedly diverse fraction of a soil bacterial community. *The ISME Journal, 9*(4), 934–945. https://doi.org/10.1038/ismej.2014.191.

Kobayashi, H., Naciri-Graven, Y., Broughton, W. J., & Perret, X. (2004). Flavonoids induce temporal shifts in gene-expression of nod-box controlled loci in Rhizobium sp. NGR234. *Molecular Microbiology, 51*(2), 335–347.

Kondorosi, E., Banfalvi, Z., & Kondorosi, A. (1984). Physical and genetic analysis of a symbiotic region of Rhizobium meliloti : Identification of nodulation genes. *Molecular and General Genetics MGG, 193*(3), 445–452. https://doi.org/10.1007/BF00382082.

Kosslak, R. M., & Bohlool, B. B. (1984). Suppression of nodule development of one side of a split-root system of soybeans caused by prior inoculation of the other side. *Plant Physiology, 75*(1), 125–130.

Kosslak, R. M., Bohlool, B. B., Dowdle, S., & Sadowsky, M. J. (1983). Competition of Rhizobium japonicum strains in early stages of soybean nodulation. *Applied and Environmental Microbiology, 46*(4), 870–873.

Krause, A., Doerfel, A., & Göttfert, M. (2002). Mutational and transcriptional analysis of the type III secretion system of Bradyrhizobium japonicum. *Molecular Plant-Microbe Interactions: MPMI, 15*(12), 1228–1235.

Krusell, L., Madsen, L. H., Sato, S., Aubert, G., Genua, A., Szczyglowski, K., et al. (2002). Shoot control of root development and nodulation is mediated by a receptor-like kinase. *Nature, 420*(6914), 422–426. https://doi.org/10.1038/nature01207.

Kumar, N., Lad, G., Giuntini, E., Kaye, M. E., Udomwong, P., Shamsani, N. J., et al. (2015). Bacterial genospecies that are not ecologically coherent : Population genomics of Rhizobium leguminosarum. *Open Biology, 5*(1), 140133. https://doi.org/10.1098/rsob.140133.

Laffont, C., Huault, E., Gautrat, P., Endre, G., Kalo, P., Bourion, V., et al. (2019). Independent regulation of symbiotic nodulation by the SUNN negative and CRA2 positive systemic pathways. *Plant Physiology, 180*(1), 559–570. https://doi.org/10.1104/pp.18.01588.

Laguerre, G., Allard, M.-R., Revoy, F., & Amarger, N. (1994). Rapid identification of rhizobia by restriction fragment length polymorphism analysis of PCR-amplified 16S rRNA genes. *Applied and Environmental Microbiology, 60*(1), 56–63.

Laguerre, G., Bardin, M., & Amarger, N. (1993). Isolation from soil of symbiotic and nonsymbiotic Rhizobium leguminosarum by DNA hybridization. *Canadian Journal of Microbiology, 39*(12), 1142–1149. https://doi.org/10.1139/m93-172.

Laguerre, G., Heulin-Gotty, K., Brunel, B., Klonowska, A., Le Quéré, A., Tillard, P., et al. (2012). Local and systemic N signaling are involved in Medicago truncatula preference for the most efficient Sinorhizobium symbiotic partners. *New Phytologist, 195*(2), 437–449. https://doi.org/10.1111/j.1469-8137.2012.04159.x.

Laguerre, G., Louvrier, P., Allard, M.-R., & Amarger, N. (2003). Compatibility of rhizobial genotypes within natural populations of Rhizobium leguminosarum biovar viciae for nodulation of host legumes. *Applied and Environmental Microbiology, 69*(4), 2276–2283.

Le Strange, K. K., Bender, G. L., Djordjevic, M. A., Rolfe, B. G., & Redmond, J. W. (1990). The Rhizobium strain NGR234 nodD1 gene product responds to activation by the simple phenolic compounds vanillin and isovanillin present in wheat seedling extracts. *Molecular Plant-Microbe Interactions, 3*(4), 214–220.

Lejay, L., Gansel, X., Cerezo, M., Tillard, P., Müller, C., Krapp, A., et al. (2003). Regulation of root ion transporters by photosynthesis: Functional importance and relation with hexokinase. *The Plant Cell, 15*(9), 2218–2232. https://doi.org/10.1105/tpc.013516.

Lewis-Henderson, W. R., & Djordjevic, M. A. (1991). NodT, a positively-acting cultivar specificity determinant controlling nodulation of Trifolium subterraneum by Rhizobium leguminosarum biovar trifolii. *Plant Molecular Biology, 16*(4), 515–526.

Li, X., Tong, W., Wang, L., Rahman, S. U., Wei, G., & Tao, S. (2018). A novel strategy for detecting recent horizontal gene transfer and its application to Rhizobium strains. *Frontiers in Microbiology, 9*, 973. https://doi.org/10.3389/fmicb.2018.00973.

Lie, T. A. (1984). Host genes in Pisum sativum L. conferring resistance to European Rhizobium leguminosarum strains. *Plant and Soil, 82*(3), 415–425. https://doi.org/10.1007/BF02184279.

Lodwig, E. M., Hosie, A. H. F., Bourdès, A., Findlay, K., Allaway, D., Karunakaran, R., et al. (2003). Amino-acid cycling drives nitrogen fixation in the legume-rhizobium symbiosis. *Nature, 422*(6933), 722–726. https://doi.org/10.1038/nature01527.

Madsen, E. B., Antolín-Llovera, M., Grossmann, C., Ye, J., Vieweg, S., Broghammer, A., et al. (2011). Autophosphorylation is essential for the in vivo function of the Lotus japonicus Nod factor receptor 1 and receptor-mediated signalling in cooperation with Nod factor receptor 5. *The Plant Journal, 65*(3), 404–417. https://doi.org/10.1111/j.1365-313X.2010.04431.x.

Maj, D., Wielbo, J., Marek-Kozaczuk, M., & Skorupska, A. (2010). Response to flavonoids as a factor influencing competitiveness and symbiotic activity of Rhizobium leguminosarum. *Microbiological Research, 165*(1), 50–60. https://doi.org/10.1016/j.micres.2008.06.002.

Marco, D. E., Carbajal, J. P., Cannas, S., Pérez-Arnedo, R., Hidalgo-Perea, A., Olivares, J., et al. (2009). An experimental and modelling exploration of the host-sanction hypothesis in legume-rhizobia mutualism. *Journal of Theoretical Biology, 259*(3), 423–433. https://doi.org/10.1016/j.jtbi.2009.03.033.

McAnulla, C., Edwards, A., Sanchez-Contreras, M., Sawers, R. G., & Downie, J. A. (2007). Quorum-sensing-regulated transcriptional initiation of plasmid transfer and replication

genes in Rhizobium leguminosarum biovar viciae. *Microbiology (Reading, England), 153*(Pt 7), 2074–2082. https://doi.org/10.1099/mic.0.2007/007153-0.
Mergaert, P., Montagu, M. V., & Holsters, M. (1997). Molecular mechanisms of Nod factor diversity. *Molecular Microbiology, 25*(5), 811–817. https://doi.org/10.1111/j.1365-2958.1997.mmi526.x.
Miwa, H., & Okazaki, S. (2017). How effectors promote beneficial interactions. *Current Opinion in Plant Biology, 38*, 148–154. https://doi.org/10.1016/j.pbi.2017.05.011.
Moawad, H. A., Ellis, W. R., & Schmidt, E. L. (1984). Rhizosphere response as a factor in competition among three serogroups of indigenous Rhizobium japonicum for nodulation of field-grown soybeans. *Applied and Environmental Microbiology, 47*(4), 607–612.
Muñoz, N., Qi, X., Li, M.-W., Xie, M., Gao, Y., Cheung, M.-Y., et al. (2016). Data from : Improvement in nitrogen fixation capacity could be part of the domestication process in soybean. *Heredity*. https://doi.org/10.5061/dryad.vp6fs.
Mutch, L. A., & Young, J. P. W. (2004). Diversity and specificity of Rhizobium leguminosarum biovar viciae on wild and cultivated legumes. *Molecular Ecology, 13*(8), 2435–2444. https://doi.org/10.1111/j.1365-294X.2004.02259.x.
Naamala, J., Jaiswal, S. K., & Dakora, F. D. (2016). Antibiotics resistance in Rhizobium : Type, process, mechanism and benefit for agriculture. *Current Microbiology, 72*(6), 804–816. https://doi.org/10.1007/s00284-016-1005-0.
Okazaki, S., Kaneko, T., Sato, S., & Saeki, K. (2013). Hijacking of leguminous nodulation signaling by the rhizobial type III secretion system. *Proceedings of the National Academy of Sciences of the United States of America, 110*(42), 17131–17136. https://doi.org/10.1073/pnas.1302360110.
Oldroyd, G. E. D., Murray, J. D., Poole, P. S., & Downie, J. A. (2011). The rules of engagement in the legume-rhizobial symbiosis. *Annual Review of Genetics, 45*, 119–144. https://doi.org/10.1146/annurev-genet-110410-132549.
Olsson, J. E., Nakao, P., Bohlool, B. B., & Gresshoff, P. M. (1989). Lack of systemic suppression of nodulation in split root systems of supernodulating soybean (Glycine max [L.] Merr.) mutants. *Plant Physiology, 90*(4), 1347–1352. https://doi.org/10.1104/pp.90.4.1347.
Oono, R., Anderson, C. G., & Denison, R. F. (2011). Failure to fix nitrogen by non-reproductive symbiotic rhizobia triggers host sanctions that reduce fitness of their reproductive clonemates. *Proceedings. Biological Sciences, 278*(1718), 2698–2703. https://doi.org/10.1098/rspb.2010.2193.
Oono, R., & Denison, R. F. (2010). Comparing symbiotic efficiency between swollen versus nonswollen rhizobial Bacteroids1[C][W][OA]. *Plant Physiology, 154*(3), 1541–1548. https://doi.org/10.1104/pp.110.163436.
Oono, R., Denison, R. F., & Kiers, E. T. (2009). Controlling the reproductive fate of rhizobia : How universal are legume sanctions? *The New Phytologist, 183*(4), 967–979. https://doi.org/10.1111/j.1469-8137.2009.02941.x.
Papenfort, K., & Bassler, B. L. (2016). Quorum sensing signal–response systems in Gram-negative bacteria. *Nature Reviews Microbiology, 14*(9), 576–588. https://doi.org/10.1038/nrmicro.2016.89.
Parsons, R., Stanforth, A., Raven, J. A., & Sprent, J. I. (1993). Nodule growth and activity may be regulated by a feedback mechanism involving phloem nitrogen*. *Plant, Cell and Environment, 16*(2), 125–136. https://doi.org/10.1111/j.1365-3040.1993.tb00854.x.
Patrick, H. N., & Lowther, W. L. (1992). Response of *Lotus corniculatus* to inoculation and pelleting on a range of Otago tussock grassland environments. *Proceedings of the New Zealand Grassland Association*, (54), 105–109.
Peck, M. C., Fisher, R. F., & Long, S. R. (2006). Diverse flavonoids stimulate NodD1 binding to nod gene promoters in Sinorhizobium meliloti. *Journal of Bacteriology, 188*(15), 5417–5427. https://doi.org/10.1128/JB.00376-06.

Pérez-Jaramillo, J. E., Mendes, R., & Raaijmakers, J. M. (2016). Impact of plant domestication on rhizosphere microbiome assembly and functions. *Plant Molecular Biology, 90*, 635–644. https://doi.org/10.1007/s11103-015-0337-7.

Prell, J., White, J. P., Bourdes, A., Bunnewell, S., Bongaerts, R. J., & Poole, P. S. (2009). Legumes regulate Rhizobium bacteroid development and persistence by the supply of branched-chain amino acids. *Proceedings of the National Academy of Sciences of the United States of America, 106*(30), 12477–12482. https://doi.org/10.1073/pnas.0903653106.

Radutoiu, S., Madsen, L. H., Madsen, E. B., Jurkiewicz, A., Fukai, E., Quistgaard, E. M., et al. (2007). LysM domains mediate lipochitin–oligosaccharide recognition and Nfr genes extend the symbiotic host range. *The EMBO Journal, 26*(17), 3923–3935. https://doi.org/10.1038/sj.emboj.7601826.

Revellin, C., Pinochet, X., Beauclair, P., & Catroux, G. (1996). Influence of soil properties and soya bean cropping history on the Bradyrhizobium japonicum population in some French soils. *European Journal of Soil Science, 47*(4), 505–510. https://doi.org/10.1111/j.1365-2389.1996.tb01850.x.

Rivilla, R., Sutton, J. M., & Downie, J. A. (1995). Rhizobium leguminosarum NodT is related to a family of outer-membrane transport proteins that includes TolC, PrtF, CyaE and AprF. *Gene, 161*(1), 27–31.

Robleto, E. A., Scupham, A. J., & Triplett, E. W. (1997). Trifolitoxin production in Rhizobium etli strain CE3 increases competitiveness for rhizosphere colonization and root nodulation of Phaseolus vulgaris in soil. *Molecular Plant-Microbe Interactions, 10*(2), 228–233. https://doi.org/10.1094/MPMI.1997.10.2.228.

Rogers, A., Ainsworth, E. A., & Leakey, A. D. B. (2009). Will elevated carbon dioxide concentration amplify the benefits of nitrogen fixation in legumes? *Plant Physiology, 151*(3), 1009–1016. https://doi.org/10.1104/pp.109.144113.

Rogers, A., Gibon, Y., Stitt, M., Morgan, P. B., Bernacchi, C. J., Ort, D. R., et al. (2006). Increased C availability at elevated carbon dioxide concentration improves N assimilation in a legume. *Plant, Cell and Environment, 29*(8), 1651–1658.

Rousk, J., Bååth, E., Brookes, P. C., Lauber, C. L., Lozupone, C., Caporaso, J. G., et al. (2010). Soil bacterial and fungal communities across a pH gradient in an arable soil. *The ISME Journal, 4*(10), 1340–1351. https://doi.org/10.1038/ismej.2010.58.

Ruffel, S., Freixes, S., Balzergue, S., Tillard, P., Jeudy, C., Martin-Magniette, M. L., et al. (2008). Systemic signaling of the plant nitrogen status triggers specific transcriptome responses depending on the nitrogen source in Medicago truncatula. *Plant Physiology, 146*(4), 2020–2035. https://doi.org/10.1104/pp.107.115667.

Ruffel, S., Gojon, A., & Lejay, L. (2014). Signal interactions in the regulation of root nitrate uptake. *Journal of Experimental Botany, 65*(19), 5509–5517. https://doi.org/10.1093/jxb/eru321.

Russo, D. M., Williams, A., Edwards, A., Posadas, D. M., Finnie, C., Dankert, M., et al. (2006). Proteins exported via the PrsD-PrsE type I secretion system and the acidic exopolysaccharide are involved in biofilm formation by Rhizobium leguminosarum. *Journal of Bacteriology, 188*(12), 4474–4486. https://doi.org/10.1128/JB.00246-06.

Santoyo, G., Moreno-Hagelsieb, G., del Carmen Orozco-Mosqueda, M., & Glick, B. R. (2016). Plant growth-promoting bacterial endophytes. *Microbiological Research, 183*, 92–99. https://doi.org/10.1016/j.micres.2015.11.008.

Soucy, S. M., Huang, J., & Gogarten, J. P. (2015). Horizontal gene transfer : Building the web of life. *Nature Reviews Genetics, 16*(8), 472–482. https://doi.org/10.1038/nrg3962.

Spaink, H. P., Bloemberg, G. V., Van Brussel, A. A. N., Lugtenberg, B. J. J., van der Drift, K. M. G. M., Haverkamp, J., & Thomas-Oates, J. E. (1995). *Molecular Plant-Microbe Interactions, 8*, 155–164.

Spaink, H. P., Okker, R. J. H., Wijffelman, C. A., Pees, E., & Lugtenberg, B. J. J. (1987). Promoters in the nodulation region of the Rhizobium leguminosarum Sym plasmid pRL1JI. *Plant Molecular Biology, 9*(1), 27–39. https://doi.org/10.1007/BF00017984.

Staehelin, C., & Krishnan, H. B. (2015). Nodulation outer proteins : Double-edged swords of symbiotic rhizobia. *Biochemical Journal, 470*(3), 263–274. https://doi.org/10.1042/BJ20150518.

Streeter, J., & Wong, P. P. (1988). Inhibition of legume nodule formation and N2 fixation by nitrate. *Critical Reviews in Plant Sciences,* 7(1), 1–23. https://doi.org/10.1080/07352688809382257.

Sulieman, S., Fischinger, S. A., Gresshoff, P. M., & Schulze, J. (2010). Asparagine as a major factor in the N-feedback regulation of N2 fixation in Medicago truncatula. *Physiologia Plantarum, 140*(1), 21–31. https://doi.org/10.1111/j.1399-3054.2010.01380.x.

Sulieman, S., & Schulze, J. (2010). Phloem-derived γ-aminobutyric acid (GABA) is involved in upregulating nodule N2 fixation efficiency in the model legume Medicago truncatula. *Plant, Cell and Environment, 33*(12), 2162–2172. https://doi.org/10.1111/j.1365-3040.2010.02214.x.

Sulima, A. S., Zhukov, V. A., Afonin, A. A., Zhernakov, A. I., Tikhonovich, I. A., & Lutova, L. A. (2017). Selection signatures in the first exon of paralogous receptor kinase genes from the Sym2 region of the Pisum sativum L. Genome. *Frontiers of Plant Science, 8.* https://doi.org/10.3389/fpls.2017.01957.

Sullivan, J. T., Eardly, B. D., van Berkum, P., & Ronson, C. W. (1996). Four unnamed species of nonsymbiotic rhizobia isolated from the rhizosphere of *Lotus corniculatus. Applied and Environmental Microbiology, 62*(8), 2818–2825.

Sullivan, J. T., Patrick, H. N., Lowther, W. L., Scott, D. B., & Ronson, C. W. (1995). Nodulating strains of Rhizobium loti arise through chromosomal symbiotic gene transfer in the environment. *Proceedings of the National Academy of Sciences of the United States of America, 92*(19), 8985–8989.

Sun, J., Cardoza, V., Mitchell, D. M., Bright, L., Oldroyd, G., & Harris, J. M. (2006). Crosstalk between jasmonic acid, ethylene and Nod factor signaling allows integration of diverse inputs for regulation of nodulation. *The Plant Journal, 46*(6), 961–970. https://doi.org/10.1111/j.1365-313X.2006.02751.x.

Surin, B. P., Watson, J. M., Hamilton, W. D. O., Economou, A., & Downie, J. A. (1990). Molecular characterization of the nodulation gene, nodT, from two biovars of Rhizobium leguminosarum. *Molecular Microbiology, 4*(2), 245–252. https://doi.org/10.1111/j.1365-2958.1990.tb00591.x.

Taberlet, P., Coissac, E., Pompanon, F., Brochmann, C., & Willerslev, E. (2012). Towards next-generation biodiversity assessment using DNA metabarcoding. *Molecular Ecology, 21*(8), 2045–2050. https://doi.org/10.1111/j.1365-294X.2012.05470.x.

Tan, K. Z., Radziah, O., Halimi, M. S., Khairuddin, A. R., & Shamsuddin, Z. H. (2015). Assessment of plant growth-promoting rhizobacteria (PGPR) and rhizobia as multi-strain biofertilizer on growth and N2 fixation of rice plant. *Australian Journal of Crop Science, 9*(12), 1257.

Terpolilli, J. J., O'Hara, G. W., Tiwari, R. P., Dilworth, M. J., & Howieson, J. G. (2008). The model legume Medicago truncatula A17 is poorly matched for N2 fixation with the sequenced microsymbiont Sinorhizobium meliloti 1021. *New Phytologist, 179*(1), 62–66. https://doi.org/10.1111/j.1469-8137.2008.02464.x.

Tringe, S. G., Mering, C. von, Kobayashi, A., Salamov, A. A., Chen, K., Chang, H. W., et al. (2005). Comparative metagenomics of Microbial communities. *Science, 308*(5721), 554–557. https://doi.org/10.1126/science.1107851.

Triplett, E. W., & Barta, T. M. (1987). Trifolitoxin production and nodulation are necessary for the expression of superior nodulation competitiveness by Rhizobium leguminosarum bv. Trifolii strain T24 on clover 1. *Plant Physiology, 85*(2), 335–342.

Triplett, E. W., & Sadowsky, M. J. (1992). Genetics of competition for nodulation of legumes. *Annual Review of Microbiology, 46*, 399–428. https://doi.org/10.1146/annurev.mi.46.100192.002151.

Vargas, L. K., Volpiano, C. G., Lisboa, B. B., Giongo, A., Beneduzi, A., & Passaglia, L. M. P. (2017). Potential of rhizobia as plant growth-promoting rhizobacteria. In A. Zaidi, M. S. Khan, & J. Musarrat (Eds.), *Microbes for legume improvement* (pp. 153–174). https://doi.org/10.1007/978-3-319-59174-2_7.

Voisin, A. S., Salon, C., Jeudy, C., & Warembourg, F. R. (2003). Symbiotic N2 fixation activity in relation to C economy of Pisum sativum L. as a function of plant phenology. *Journal of Experimental Botany, 54*(393), 2733–2744. https://doi.org/10.1093/jxb/erg290.

Wakelin, S., Tillard, G., van Ham, R., Ballard, R., Farquharson, E., Gerard, E., et al. (2018). High spatial variation in population size and symbiotic performance of Rhizobium leguminosarum bv. Trifolii with white clover in New Zealand pasture soils. *PLoS One, 13*(2), e0192607. https://doi.org/10.1371/journal.pone.0192607.

Weese, D. J., Heath, K. D., Dentinger, B. T. M., & Lau, J. A. (2015). Long-term nitrogen addition causes the evolution of less-cooperative mutualists. *Evolution, 69*(3), 631–642. https://doi.org/10.1111/evo.12594.

Westhoek, A., Field, E., Rehling, F., Mulley, G., Webb, I., Poole, P. S., et al. (2017). Policing the legume-rhizobium symbiosis : A critical test of partner choice. *Scientific Reports,* 7(1), 1–10. https://doi.org/10.1038/s41598-017-01634-2.

Winarno, R., & Lie, T. A. (1979). Competition between Rhizobium strains in nodule formation : Interaction between nodulating and non-nodulating strains. *Plant and Soil, 51*(1), 135–142. https://doi.org/10.1007/BF02205933.

Xin, D.-W., Liao, S., Xie, Z.-P., Hann, D. R., Steinle, L., Boller, T., et al. (2012). Functional analysis of NopM, a novel E3 ubiquitin ligase (NEL) domain effector of Rhizobium sp. Strain NGR234. *PLoS Pathogens, 8*(5), e1002707. https://doi.org/10.1371/journal.ppat.1002707.

Yang, S., Tang, F., Gao, M., Krishnan, H. B., & Zhu, H. (2010). R gene-controlled host specificity in the legume-rhizobia symbiosis. *Proceedings of the National Academy of Sciences of the United States of America, 107*(43), 18735–18740. https://doi.org/10.1073/pnas.1011957107.

Young, J. P. W., Crossman, L. C., Johnston, A. W., Thomson, N. R., Ghazoui, Z. F., Hull, K. H., et al. (2006). The genome of Rhizobium leguminosarum has recognizable core and accessory components. *Genome Biology,* 7(4), R34. https://doi.org/10.1186/gb-2006-7-4-r34.

Young, J. P. W., & Wexler, M. (1988). Sym plasmid and chromosomal genotypes are correlated in field populations of Rhizobium leguminosarum. *Microbiology, 134*(10), 2731–2739. https://doi.org/10.1099/00221287-134-10-2731.

Zézé, A., Mutch, L. A., & Young, J. P. (2001). Direct amplification of nodD from community DNA reveals the genetic diversity of Rhizobium leguminosarum in soil. *Environmental Microbiology, 3*(6), 363–370.

Zhou, Z., Jiang, Y., Wang, Z., Gou, Z., Lyu, J., Li, W., et al. (2015). Resequencing 302 wild and cultivated accessions identifies genes related to domestication and improvement in soybean. *Nature Biotechnology, 33*(4), 408–414. https://doi.org/10.1038/nbt.3096.